Computerized Tomography

Computerized Tomography

Proceedings of the Fourth
International Symposium
Novosibirsk, Russia

Editor-in-Chief: M.M. Lavrent'ev

///VSP///
Utrecht, The Netherlands, 1995

VSP BV
P.O. Box 346
3700 AH Zeist
The Netherlands

First published in 1995

ISBN 90-6764-187-1

CIP-DATA KONINKLIJKE BIBLIOTHEEK, DEN HAAG

Computerized

Computerized tomography : proceedings of the fourth
international symposium held in Novosibirsk (Russia) on
10-14 August, 1993 / ed.-in-chief M.M. Lavrent'ev. -
Utrecht : VSP
ISBN 906764-187-1 bound
NUGI 812
Subject headings: computerized tomography.

Printed in The Netherlands by Koninklijke Wöhrmann BV, Zutphen.

CONTENTS

Preface

This volume contains the Proceedings of the Fourth International Symposium on Computerized Tomography (CT-93) which was held in Novosibirsk (Russia) on 10–14 August, 1993. The CT-93 was managed and sponsored by

- Siberian Branch of the Russian Academy of Sciences;
- Institute of Mathematics, Novosibirsk;
- Institute of Theoretical and Applied Mechanics, Novosibirsk;
- Institute of Automatics and Electrometry, Novosibirsk;
- Computing Center, Novosibirsk;
- Institute of Applied Physics, Novosibirsk;
- Novosibirsk State University.

The Ct-93 was held under protection and in cooperation with the following associations:

- Russian Society "Tomography Corporation", Moscow;
- Tomography Society, Novosibirsk;
- Russian Chapter of the International Society for Optical Engineering (SPIE/RUSS), Moscow.

A great interest to theoretical and numerical analysis of tomography problems and their applications leads to sufficiently large amount of participants from Australia, Belorussia, Canada, Estonia, France, Germany, Israel, Italy, Japan, Kazakhstan, Kirgizstan, Russia, The Netherlands, Ukraina, USA, Uzbekistan. The symposium had three plenary sessions concerned with

- mathematical problems of computerized tomography;
- algorithms of computerized tomography;
- tomography applications in physics, geophysics, industry, and medicine.

Some lectures and talks presented to the CT-93 are included to this book. Regretfully not all plenary lecturers prepared their lectures for publication and some limitations did not allow all authors of communications to take part in this volume.

The Editorial Board refrained from changing the authors' terminology due to the wide range of computerized tomography applications.

Acknowledgement

We wish to acknowledge with thanks all contributors and participants, who made the CT-93 a success.

It is our pleasure to express our gratitude to the group at Institute of Mathematics that has prepared these proceedings for publication. Of those people whose contribution to this work was of particular importance we would like to mention T. V. Bugueva, I. G. Kabanikhina, S. V. Martakov, D. V. Nechaev.

Professor M. M. Lavrent'ev,
Editor-in-Chief

Professor S. I. Kabanikhin,
Editor.

Image Reconstruction of Celestial Objects from a Few Strip-Integrated Projections by Lunar Occultations

M. I. AGAFONOV and K. S. STANKEVICH

Radiophysical Research Institute
B. Pecherskaya st. 25/14, Nizhny Novgorod 603600, Russia

ABSTRACT

In this paper we consider the technique of image reconstruction based on iterative algorithms and perform a numerical analysis of the solution convergence. We represent the results of numerical experiments.

The brightness images of celestial objects are very important for studying physical processes in astrophysics. In certain cases, usually for low frequencies, the angular resolution of up-to-date radiotelescopes is not enough to get sufficiently accurate images. Nevertheless, observations of the objects in their Lunar occultations allow us to obtain integral brightness profiles with a high angular resolution. One-dimensional $(1 - D)$ projections of several occultations may be used for the $2 - D$ image reconstruction. The Lunar limb is approximated by a plane screen moving at different position angles.

A similar problem of image reconstruction arises in observing radiosources with the knife diagram. The stretched beam of the plane reflector of RATAN–600 radiotelescope can change its position in a limited range of position angles. The irregular distribution of position angles such that a sector of about 100 degrees is not filled is typical both for the case of Lunar occultations of radiosources and for the observations by the knife–beam radiotelescopes.

We present here:

1. the technique of image reconstruction from a limited number of projections using iterative algorithms with nonlinear constrains (Agafonov and Podvoiskaya, 1989, 1990);

2. the results of numerical experiments using the known iterative algorithms such as the standard CLEAN (Hogbom, 1974) and the Trim Contour CLEAN (TC–CLEAN) (Steer et al., 1984); the study of the solution convergence by minimizing the error of initial and control profiles with variations of the loop gain (g) and the trim contour level (TC);

3. the experimental results of the Crab Nebula images reconstruction on frequencies of 178 and 750 MHz with a high angular resolution from four one-dimensional projections of Lunar occultations which were obtained using the 70-meter radiotelescope (RT-70) (Agafonov et al., 1984).

The Deconvolution Problem

The problem is reduced to the classical deconvolution equation $G = H * F$(+noise), where $F(x, y)$ is the brightness distribution of the observed object on the celestial sphere, $H(x, y)$ is the fan beam (the instrument point response function or dirty beam), $G(x, y)$ is the dirty map or summary image. The functions $G(x, y)$ and $H(x, y)$ are calculated by summing the corresponding values of one-dimensional profiles. The classical case by Bracewell (Bracewell and Riddle, 1967) with the function $H(x, y)$ filling completely the region of spatial frequencies of UV-plane up to the radius with the boundary frequency f needs the number of projections N to be not less than $3,14\ Df$, where D is the diameter of the considered region. To achieve the desired angular resolution $1/f$ from a limited number of projections, it is required to extrapolate the solution and to use nonlinear processing methods.

Iterative algorithms with nonlinear constraints (see, for example, Vasilenko and Taratorin, 1986) are the most effective in the image reconstruction from a few strip-integrated projections.

The paper by Agafonov and Podvoiskaya (1989) is devoted to solving the reconstruction problems by the iterative algorithms used in radioastronomy. The study of reconstruction possibilities was accomplished by the numerical experiments based on the following scheme of calculations. One-dimensional profiles and the dirty map were obtained from a $2-D$ object model. To clean the dirty map, we applied independently both of the iterative algorithms CLEAN (Hogbom, 1974) and TC–CLEAN (Steer et al., 1984), with the use of a fan beam. After the scanning of the obtained images, the error of control and initial projections (ERROR) was calculated. The solution convergence was investigated by the ERROR minimization over the variable parameters g and TC.

The stand CLEAN algorithm (Hogbom, 1974) is the realization iterative algorithm which has been used extensively in radioastronomy since 1974. It needs only one component of the image net at every iterative step. The low computational rate and probable distortions - stripes and ridges which may appear in the case of extended areas - are the shortages of the method. The algorithm does not provide with a clear criterion for the choice of the parameter g (loop gain), but this choice was found to influence the result in a complicated manner. The Trim Contour CLEAN algorithm (Steer et al., 1984) differs essentially from the former. The Trim contour method is used here for selecting components at each iteration. Its advantage is particularly seen when processing extended areas. A great number of components exceeding the fixed TC level are considered simultaneously at each iteration thus preventing from the appearance of defects. The Trim Contour CLEAN inspired a hope for improving the image quality on extended areas, but it needs a study in different practical cases (Cornwell, 1988).

We studied the solution convergence by the ERROR minimization over the variable parameters g (loop gain) and TC (trim contour level). The known maps of radioastronomical objects were taken as initial models. The numerical experiments were carried out on the net of 64×64 pixels. The BESM–6 computer was used .

It was shown (Agafonov and Podvoiskaya, 1989) that the search for the solution $F_{\rm opt}(x, y)$ using the above algorithms is to be made by the optimization of the parameters of iterative processes: the loop gain g (in both of the algorithms) and the TC level (in the TC–CLEAN case only). The parameters are chosen in such a way to minimize the error of initial and control profiles. This corresponds to a discrepancy solution. The process was illustrated in details in the paper by Agafonov and Podvoiskaya (1989). It was shown also that the TC–CLEAN is more effective than the former algorithm and, moreover, it is stable and prevents from image defects in the case of extended areas. The TC–CLEAN gains an advantage in computational time. The reconstruction of a complicated object is performed by the TC–CLEAN in 13–18 iterations but needs 400–600 iterations using the standard CLEAN. Moreover, the parameters are chosen more easily in the TC–CLEAN algorithm.

The minimum error of initial and control one-dimensional profiles for stable solutions (in the CLEAN case it needs sometimes additional efforts (Agafonov and Podvoiskaya, 1989)) may be approximately the same for both of the algorithms. Then we prefer the result reflecting the physical peculiarities of the object to a great extend or assume the existence of the class of probable solutions ranged from "obtuse" (TC–CLEAN) to "sharp" (CLEAN) variants. As is known, the CLEAN summarizes peaks in the solution, which explains the result trend towards the sharpest variant (as far as it is possible for the fixed constraints). The TC–CLEAN, on the contrary, accumulates the result by involving the whole areas into the domain of the given constraints and beginning, as a rule, with the most extended components. It is natural that in this case a more smoothed solution should be expected.

Thus, the two methods are proposed to study the solution existence in complicated cases and, in particular, for insufficient preliminary information. For simple objects, the domain of the solution uncertainty narrows. The results obtained by both of the algorithms coincide.

In the paper by Agafonov and Podvoiskaya (1990), the examples of typical different structures are considered. The trivial object consisting of peaks may be successfully restored by the standard CLEAN. The results obtained by the two methods practically coincide for the simple object consisting of independent components of finite dimensions. But the TC–CLEAN is more effective. A difference of the results is observed when considering a more complicated object having strong components on an extended area. The image reconstructed by the standard CLEAN has a grooved structure on extended areas. For smoothed $1 - D$ profiles with small hillocks it was possible to get a solution both from isolated, contrasting components (the CLEAN under stability conditions) and from more smoothed components (the TC–CLEAN). Since the errors of initial and control profiles had the same value, the preference was not given to one of these variants. The standard CLEAN increased the contrast of small components, but the quality of the extended background deteriorated considerably because of grooves. Thus, the pair of solutions obtained in the complicated case by the two methods allows us to estimate the range of possible change of the structure. As a result, the conclusions about the components may be drawn. At the same time, the availability of preliminary information on the object may indicate the solution in accordance to the physical peculiarities of the object.

The questions as how to improve the reconstruction quality from a limited number of projections are of great importance both in radioastronomy and tomography (Vishnyakov et al., 1985). We hope that the presented technique based on the two above algorithms will be useful for researchers.

Images Reconstruction

A detailed study of the Crab Nebula structure on the frequencies less than 1 GHz is of great interest since good synthesis telescope maps are absent. The brightness profiles of the Lunar occultations were obtained by means of the radiotelescope RT-70 (West

Crimea). Two images with a high angular resolution in this range were obtained on the basis of these observations.

Maloney and Gottesman (1979) succeeded in restoring the Crab Nebula image on a frequency of 114 MHz from four projections by the standard algorithm CLEAN. The reconstruction of the structure on a frequency of 750 MHz with the angular resolution of 20×35 arcsec was also made using the standard CLEAN in Agafonov et al. (1986). It should be noted that the Nebula diameter is equal to about 6 arcmin. The images presented in Maloney and Gottesman (1979) and Agafonov et al. (1986) have some defects on extended areas which are typical for the standard CLEAN.

The results of numerical experiments (Agafonov and Podvoiskaya, 1989, 1990) which have been briefly described above clearly demonstrate the effectiveness of the TC–CLEAN (Steer et al., 1984) as compared with the CLEAN. Using the TC–CLEAN yields high-quality images of extended areas which are typical for the Nebula. Images on frequencies of 178 and 750 MHz were reconstructed with the angular resolutions of 20×35 and 45×65 arcsec, respectively (Agafonov et al., 1984). Four profiles obtained in the Lunar occultations of 1982 and 1983 years were used on every frequency. The numerical experiments using the well-known image of the Nebula on a frequency of 1,4 GHz (Agafonov and Podvoiskaya, 1989) allowed us to estimate beforehand possibilities of the object structure restoration by a low UV-plane filling. The earlier restoration (Agafonov et al., 1986) by the standard CLEAN sharpened the contrast of small components. This was useful for establishing the correlation with the position of filaments in optics. However, the numerical experiments (Agafonov and Podvoiskaya, 1989, 1990) showed that the image obtained by the TC–CLEAN reflect the physical features of the Nebula to a large extend.

The described technique may be used for the image reconstruction from the Lunar occultations profiles and may be adapted for observations by the radiotelescope RATAN-600 and for solution of tomographic problems.

REFERENCES

Agafonov M. I., Aslanyan A. M., Gulyan A. G., Ivanov V. P., et al. (1986), *Pis'ma v Astronom. zhurnal,* vol. 12, 275.

Agafonov M. I., Ivanov V. P., Podvoiskaya O. A. (1984), *Astronomicheskii zhurnal,* vol. 137, 159.

Agafonov M. I., Podvoiskaya O. A. (1989), *Izvestiya Vyssh. Uchebn. Zaved.,* Radiofizika, vol. 32, 742.

Agafonov M. I., Podvoiskaya O. A. (1990), *Izvestiya Vyssh. Uchebn. Zaved.,* Radiofizika, vol. 33, 1185.

Bracewell R. N., Riddle A. C. (1967), *Astrophys. J.,* vol. 150, 427.

Cornwell T. J. (1988), *Astron. Astrophys.,* vol. 202, 316.

Hogbom J. A. (1964), *Astron. Astrophys. Suppl. Ser.,* vol. 15, 417.

Maloney F. P., Gottesman S. T. (1979), *Astrophys. J.,* vol. 234, 485.

Steer D. G, Dewdney P. E, Ito M. R. (1984), *Astron. Astrophys.,* vol. 137, 159.

Vasilenko G. I., Taratorin A. M. (1986), Image Restoration, *"Radio i svyaz",* Moscow. (in Russian)

Vishnyakov G. N., Gil'man G. A., Levin G. G. (1985), *Opt. Spectroscopy,* vol. 58, 406.

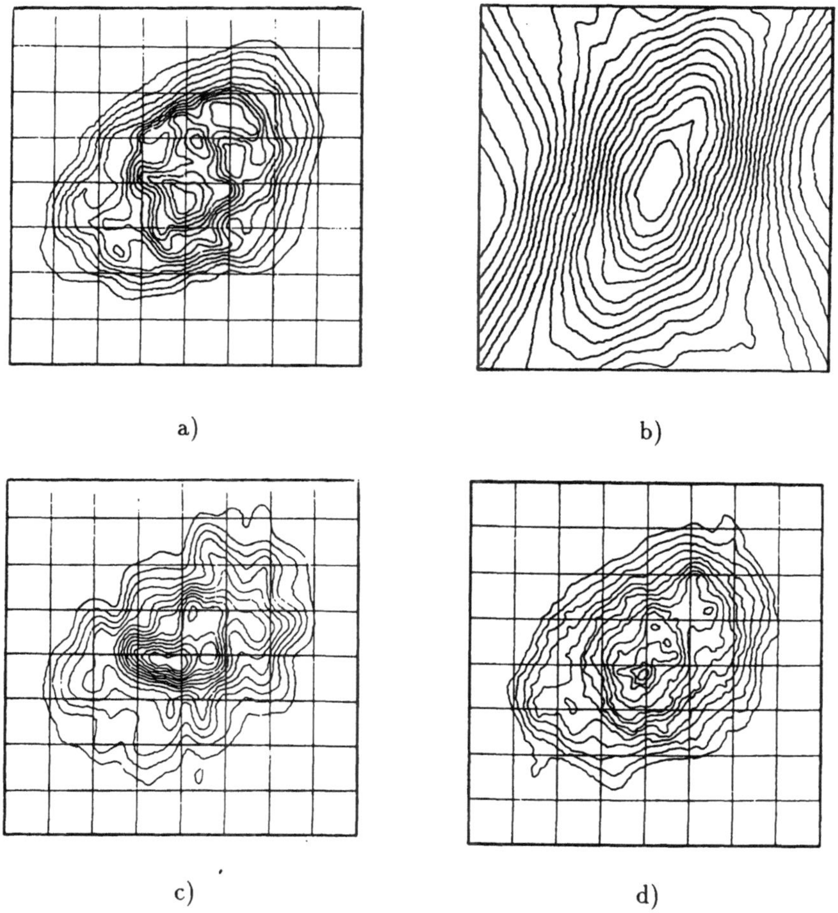

FIGURE 1. The example of numerical modeling: a) TEST OBJECT (Crab Nebula map 1.4 GHz, VLA); b) DIRTY MAP; c) standard CLEAN reconstruction; d) *TC*-CLEAN reconstruction (*TC*-0.79, g-0.26, 16 iterations).

a)

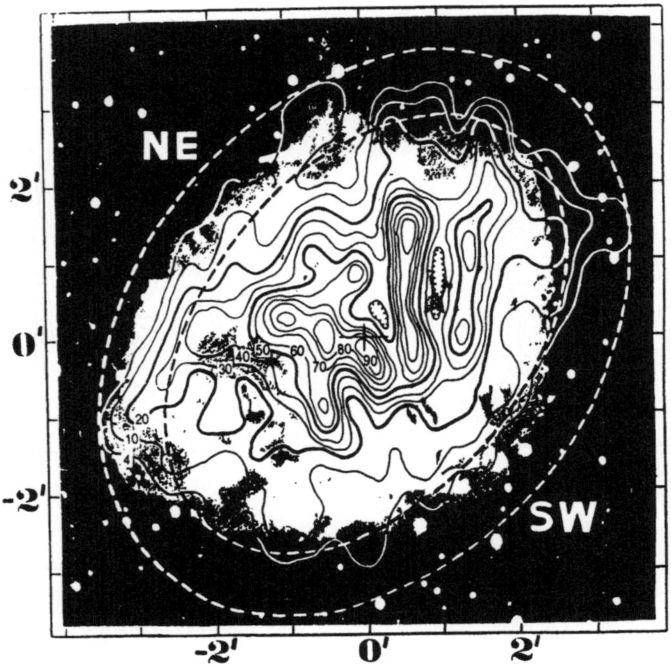

b)

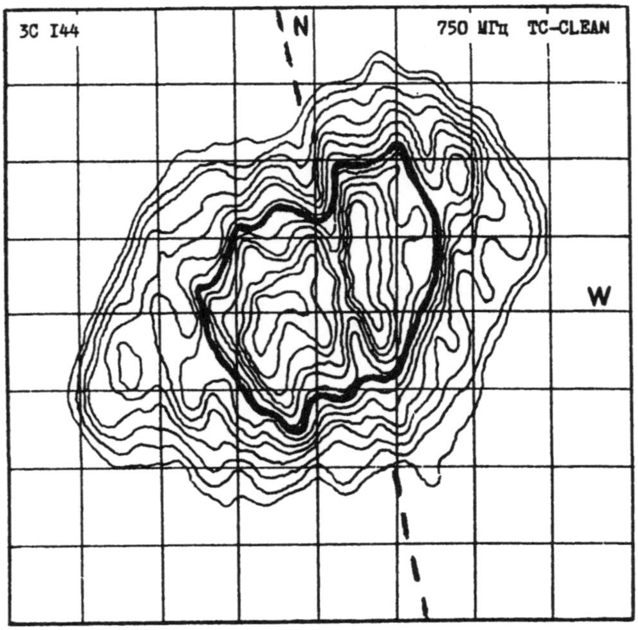

FIGURE 2. The obtained results of the Crab Nebula reconstruction on a frequency of 750 MHz from four profiles of Lunar occultations: a) CLEAN and optic image; b) *TC*-CLEAN.

Numerical Study of a Nonlinear Sound Radiating Inverse Problem of Underwater Acoustics

G. V. ALEKSEEV

Institute of Applied Mathematics, Far Eastern Branch of Russian Academy of Sciences
690041, Vladivostok, Russia

ABSTRACT

In this paper we consider a nonlinear inverse extremal problem of sound radiating. This problem is to minimize the power of the sound field,which is radiated by a source into the far zone of a regular acoustic waveguide, by locating a number of secondary point sources. The near field technique is used to calculate the power output for an arbitrary number of primary and secondary point sources. This power output is a function nonconvex in coordinates of the secondary point sources. For given complex strengths of sources this function reaches its (nonunique) minimum for an optimal location of the secondary sources. Results are presented for the case when the sound field is emitted by a single primary point source and by a number of secondary point sources located in an optimal way. In particular, it is shown that a substantial reduction in power output can be achieved even for a few secondary sources.

Let $D = D_H = \{\mathbf{x} = (x,z) : 0 < z < H, \ -\infty < x < \infty\}$ be a regular plane acoustic waveguide of the depth H with parameters

$$\omega > 0, \quad \rho(z), c(z) \in L^\infty(0,H), \quad \rho \geq \rho_0 > 0, \quad c(z) \geq c_0 > 0. \tag{1}$$

Introduce a $2 \times N$ - matrix (array)

$$\mathbf{Z} = \begin{pmatrix} x_1 & x_2 & \dots & x_N \\ z_1 & z_2 & \dots & z_N \end{pmatrix}$$

and denote by $(\mathbf{Z}, \mathbf{q})$, $\mathbf{q} = (q_1, q_2, \dots, q_N) \in C^N$, a discrete antenna which consists of N monopoles with a density

$$f(x,z) = \sum_{j=1}^{N} q_j \delta(x - x_j, z - z_j).$$

Let $D^+ = \{\mathbf{x} = (x,z) \in D : x > x^+ > x_j, \ j = 1,2,\dots,N\}$,

$$\Gamma^+ = \{\mathbf{x} \in D : x = x^+\}, \quad a_{nj} = \xi_n^{-1/2} q_j \phi_n(z_j) \exp(-\mathrm{i}\,\xi_n x_j). \tag{2}$$

It is well-known (see Alekseev, 1988) that the field p generated by a pair $(\mathbf{Z}, \mathbf{q})$ in D^+ has a form

$$p(x,z) = \frac{\mathrm{i}}{2} \sum_{n=1}^{\infty} \sum_{j=1}^{N} \xi_n^{-1/2} a_{nj} \phi_n(z) \exp(\mathrm{i}\,\xi_n x), \tag{3}$$

and its power which is transferred through Γ^+ into D^+ equals (Alekseev and Komarov, 1991)

$$\mathcal{N}(p) = \frac{1}{8\omega}|[A\mathbf{q}]|^2 = \frac{1}{8\omega}\sum_{n=1}^{M}|(A\mathbf{q})_n|^2. \tag{4}$$

Here ξ_n^2, ϕ_n are eigenvalues and eigenfunctions of the spectral problem

$$\rho(z)\frac{\mathrm{d}}{\mathrm{d}\,z}\left[\frac{1}{\rho(z)}\frac{\mathrm{d}\,\phi}{\mathrm{d}\,z}\right] + [k^2(z) - \xi^2]\phi = 0, \quad z \in (0, H), \quad k^2 = \frac{\omega^2}{c^2}, \tag{5}$$

$$\phi(0) = 0, \quad \phi'(H) = 0, \tag{6}$$

corresponding to the waveguide D_H, $M > 0$ is an integer such that

$$\max_{z\in[0,H]} k^2(z) > \xi_1^2 > \xi_2^2 > \ldots > \xi_M^2 > 0 > \xi_{M+1}^2 > \ldots \to -\infty, \tag{7}$$

$A = A(\mathbf{Z}) : C^N \to C^M$ is a $M \times N$-matrix with the components a_{nj}. If the waveguide D^+ has not only the field (3) but also a field p_b generated by a noising object,

$$p_b(x, z) = -\frac{\mathrm{i}}{2}\sum_{n=1}^{\infty}\xi_n^{-1/2}\beta_n\phi_n(z)\exp(\mathrm{i}\,\xi_n x) \tag{8}$$

with given coefficients β_n, then the power $\mathcal{N}_b$ of the field p_b and the power $\mathcal{N}(p + p_b)$ of the total field $p + p_b$ are defined in D^+ as

$$\mathcal{N}_b = \frac{1}{8\omega}|[A\mathbf{q}]|^2, \quad \mathcal{N}(p + p_b)) = \frac{1}{8\omega}|[A\mathbf{q} - \mathbf{b}]|^2, \quad \mathbf{b} = (\beta_1, \beta_2, \ldots, \beta_M). \tag{9}$$

The problem of active suppression of the sound field (8) by using an additional discrete antenna $(\mathbf{Z}, \mathbf{q})$ is of great practical importance. The case when the coordinates (x_j, z_j) are fixed and it is required to determine the complex strengths q_j so as to minimize (8) is studied in details in Alekseev and Komarov (1991). Below we consider an opposite (in some sense) case, when the strengths q_j are given and it is required to determine unknown coordinates (x_j, z_j) of the point sources, i.e., the array $\mathbf{Z}$. Set $\Omega_{\alpha\delta} = [-\alpha, \alpha] \times [\delta, H - \delta]$ and assume that all the point sources are situated in the parallelepiped $\Omega_{\alpha\delta}$, that is, $\mathbf{Z} \in \Omega_{\alpha\delta}$.

Problem 1. Given a vector $\mathbf{q}$ in C^N, find an array $\mathbf{Z}$ such that

$$J(\mathbf{Z}) \equiv |[A(\mathbf{Z})\mathbf{q} - \mathbf{b}]|^2 \to \inf. \tag{10}$$

It is well-known that under conditions (1) the eigenfunctions $\phi_n(z)$ belong to $C^1[0, H]$. Therefore, from (10) it follows that J is a continuously differentiable function of the variables z_j, $j = 1, 2, \ldots, N$, and infinitely differentiable function of the variables x_j, $j = 1, 2, \ldots, N$. Hence, Problem 1 is solvable. However, as the functional J is not convex, this problem is a multiextremal one, which makes a search for its solution essentially more difficult.

For numerical solution of Problem 1 an algorithm was developed which is based on the multistart method of the global optimization and a modified Newton method of finding a local extremum. Numerical experiments showed that considerable suppression of the primary (parasitic) field p_b can be achieved even for a small number N of secondary sources due to the optimal choice of their coordinates. This fact, together with the results by Alekseev and Komarov (1991), confirms that it is the location of the secondary sources (but not the number N) that plays the leading role when suppressing the primary field (Alekseev and Komarov, 1991).

In this paper we refrain from outlining this algorithm and analyzing in details the results of numerical experiments. We only discuss a test example for the case of plane ideal acoustic waveguide D with parameters

$$H = 200\,m, \quad \rho = 1\,g \cdot cm^{-3}, \quad c = 1450\,m\,s^{-1}, \; \omega = 20\pi, (k = 0.043\,m^{-1}).$$

For such a waveguide the eigenvalues ξ_n^2 and the eigenfunctions $\phi_n(z)$ are defined by

$$\xi_n^2 = k^2 - \mu_n^2, \quad \mu_n = \left(n - \frac{1}{2}\right)\frac{\pi}{H}, \quad \phi_n(z) = \frac{2}{H}\sin \mu_n z.$$

As a primary sound field we chose the field

$$p_b(x,z) = -\frac{\mathrm{i}}{2}\sum_{n=1}^{\infty} \xi_n^{-1}\phi_n(z_0)\phi_n(z)\exp(\mathrm{i}\,\xi_n x) \tag{11}$$

which is generated by a unit point source $q_0((x_0,z_0), \delta(x-x_0)\delta(z-z_0))$ located at the point (x_0, z_0), where $x_0 = -25\,m$ and $z_0 = -100\,m$. The number N of sources was changed from 1 to 10, and their strengths q_j were assumed as constant and satisfying the condition

$$\sum_{j=1}^{N} q_j \equiv Nq = \lambda,$$

where a parameter $\lambda > 0$ was set.

The results of experiments are presented in Figures 1–6.

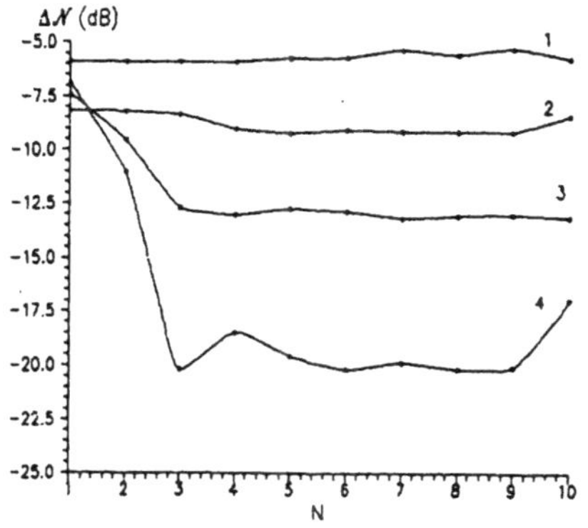

Figure 1

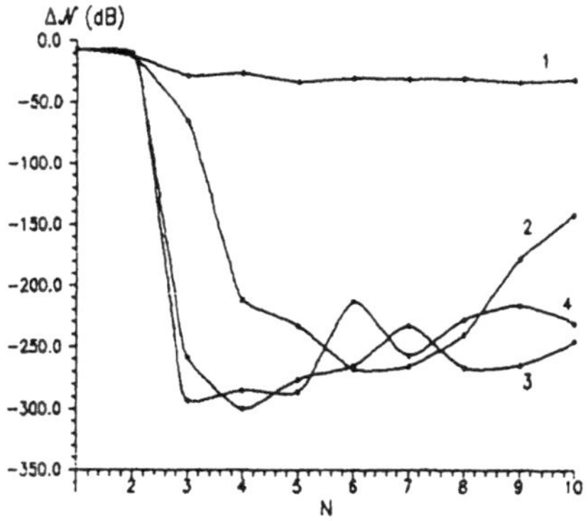

Figure 2

Curves 1–4 in Figure 1 describe the dependence of the value of the power suppressed (in decibels)

$$(\Delta\mathcal{N})\mathrm{d}\,B = 10\lg(\mathcal{N}_{res}/\mathcal{N}_b) \tag{12}$$

on N for four values of the parameter λ equal to 0.5, 0.75, 1, 1.25, respectively. Here $\mathcal{N}_{res}$ is a residual power in the far zone D^+, i.e., the power which is not suppressed in D^+ when using the secondary sources located accordance to our algorithm. Curves 1-4 in Figure 2 describe the dependence of the quantity (12) on N for λ equal to 1.4, 1.5, 1.75 and 2.0.

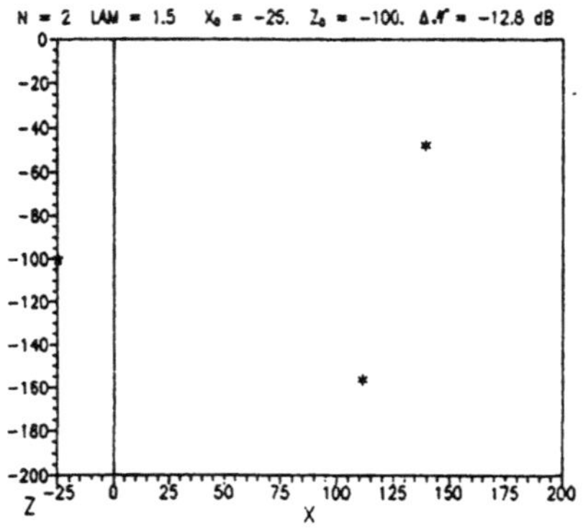

FIGURE 3

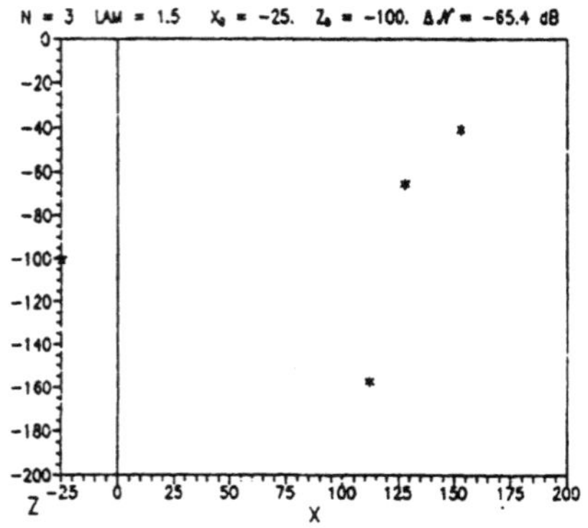

FIGURE 4

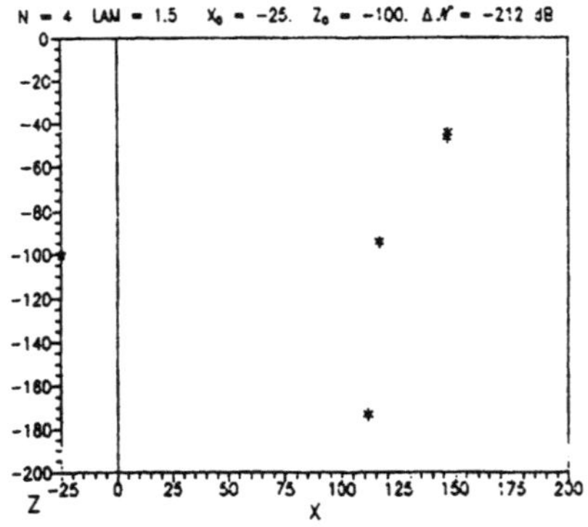

FIGURE 5

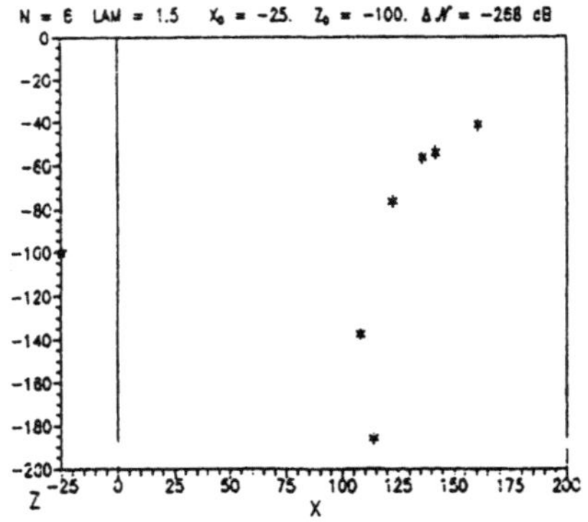

FIGURE 6

In Figures 3-6 we can see the optimal location of the secondary sources (the values of N and λ are specified for each figure as well as the coordinates (x_0, z_0) of the primary source). The quantity $\Delta\mathcal{N}$ denotes the value of the power suppressed by using the secondary sources located as shown in the figures.

Comparing Figures 1 and 2, we see that all curves in Figure 1 are sufficiently smooth, although their "steepness" smoothly increases with the growth of λ. Thus, the change in N from 1 to 10 for the considered values of λ (and, in fact, for all values $\lambda \leq 1.40$) has almost no influence on the value of the power suppressed by the optimal antenna $(\mathbf{Z}, \mathbf{q})$. This implies the low sound suppression efficiency of the antennae $(\mathbf{Z}, \mathbf{q})$ in the far zone D^+ for the case $\lambda < 1.40$.

To the contrary, all the curves in Figure 2 fall steeply down beginning with $N = 2$. In particular, for $\lambda = 1.75$ the quantity $(\Delta\mathcal{N})$ dB, which equals about 6 dB and 12 dB for $N = 1$ and 2, respectively, takes the value of 260 dB for $N = 3$ which means, in fact, the full power suppression of the primary source ($\mathcal{N}_{res} = 10^{-24}$).

Acknowledgement

The work was supported by the Russian Fundamental Research Foundation under Grant 93-013-17399. The author thanks E. Martynenko for the help in conducting numerical experiments.

REFERENCES

Alekseev G. V. (1988), *Mathematical Foundation of Ocean Acoustics.* Vladivostok, Far Eastern State University Press. (in Russian)

Alekseev G. V. and Komarov E. G. (1991), Numerical Simulation of the Sound Radiating Extremum Problems in a Plane Waveguide. *Mathematical Modelling,* 3, no. 12, 52–64. (in Russian)

Alekseev G.V. and Komarov E. G. (1993), On Active Suppression of Sound Fields in Stratified Wave Guides. *Acoust. J.* 39, no. 1, 5–12. (in Russian)

Optical Tomography of Natural and Artificial Optical Disturbances in the Near Terrestrial Space

V.V. ALPATOV, A.V. LIKHACHOV, V.V. PICKALOV, Yu.A. ROMANOVSKY

Institute of Applied Geophysics, Moscow, Russia

Institute of Theoretical and Applied Mechanics, Novosibirsk, Russia

ABSTRACT

The paper presents some evaluations of possibility to use the optical tomographic methods for investigations of natural and artificial irregularities in the near earth space. Results obtained in research of artificial irregularities and auroral airglow are described.

1. Introduction

Recently, of great interest has been an application of new methods for investigating natural and artificial optical irregularities in the near earth environment, atmosphere and near space (NS). We consider as natural the auroral structures, tropical arcs and similar objects. When active experiments in NS are carried out, one can observe artificial optical irregularities such as artificial clouds, artificial auroras etc. The optical irregularities investigations may give important information on natural physical processes in NS and the ones which cause NS pollution by the human activity . Optical irregularity studies may effectively use optical tomography methods and there are now conditions for the appearance of special tomographic systems intended for such objects. In spite of large variety of such irregularities, they have some common features, namely:

- large spatial scales;
- limited number of observation points and observation angles range;
- necessity for three-dimensional distributions of different physical parameters obtained from two-dimensional projection images;
- initial data obtained by different devices: TV, photo cameras, CCD-matrixes and rulers, spectrographs and spectrometers and so on;
- high cost of establishing every additive observation point.

An IAG active experiment program studied a possibility to use optic tomography methods for internal structure reconstruction of natural and artificial objects. In our opinion, the results of specific investigations in this area may be applied for a wide range of optical near space problems. In the paper some results of research in this direction are considered. The objectives of the work were:

1) to study the possibility of optic tomography methods application to determine inner structure of near space natural and artificial optical large scale irregularities;

2) in the case of positive answer, to develop technology for designing local and global tomographic systems intended for near space optical irregularities investigations.

The problem was resolved in some integrated stages:

- analysis of probing radiation propagation in the environment (object);
- selection of the equation describing the relationship between the measured parameters within the object and radiation parameters;
- search for the methods of projection inversion;
- analysis of possible detecting schemes;
- informational analysis;
- selection of reconstruction algorithms,routines and processor for their realization;
- simulation of reconstruction objects, testing of algorithms using model objects;
- development of the reconstruction procedure for the data obtained in field experiments;
- analysis of reconstruction results;
- development of tomographic systems structure.

Each stage displayed problems of different complexity and, apparently, the process is specific for every new object. The results of one series of studies may help to develop technology, facilitate studies of similar objects and to provide an experience in the application of such technology.

Let us briefly show some results obtained for such objects as artificial structures (AS) usually created in active experiments (Alpatov et al., 1985, 1986, 1989, 1991) and for auroral airglow.

2. Results of Research for Artificial Structures

1. For different AS classes, radiation within them is absorbed, scattered and reemitted by internal environment. The relation between the measured radiation and physical parameters may be described by the radiation transport equation (RTE).

2. The form of the equation depends on AS development stages which can be characterized by such an integral parameter as optical thickness. Therefore, different types of AS in different development stages were analyzed using experimental results and different models for this integral parameter evaluation.

3. For AS types and development stages with optical thickness $\tau < 1$, the well known project inversion methods are applicable. When $\tau > 1$, there is no decision in a general case, because it is necessary to apply partial methods or to design them specially. For example, for optically thick barium clouds, a special method of " couple lines" may be

applied, based on the fact that during barium clouds luminosity there are two connected spectral lines, one of which is optically thick and the other is optically thin.

4. Some versions of schemes were analyzed for transmission, emission tomography and tomography of scattering environments. Their merits and faults were evaluated, and their application was classified. Fig. 1 shows table obtained after analyzing different AS types and stages. The table provides a number of conclusions as to which tomographic schemes and inversion methods are preferable or exclusively possible for various types of AS and their development stages.

5. An important stage of investigation consisted in informational analysis, which is especially important for studying such objects as AS and nature irregularities in NS. The informational analysis includes evaluations of the following parameters influence on the accuracy of reconstruction:

- the required number of projections and amount of data in each projection;
- complexity of the studied object;
- possibility to use a priori data;
- informativity of different projections (obtained from various directions);
- the effects of noise;
- the minimum intensity of informative signal;
- the required exposition period and minimum available scales for the reconstruction of irregularities within the object.

The informational analysis revealed requirements on registration devices and on experiment schemes, and provided data to create an object model for computer reconstruction simulation.

6. Informational analysis also showed that the problem of tomographic reconstruction of AS inner distributions belongs to the class of low views and ultra low views tomographic problems. The study also ascertained a possibility to use a priori data on the inner AS parameter distributions, which is especially important in low views tomography. We assume that the best reconstruction algorithms for such tasks are ART, GP, Ment, maximum likelihood algorithms and others in these classes. They are partially realized in $TOPAS - MICRO$ package developed by Pickalov team.

7. An AS model was developed for computer simulation which imitates the stratified stage of AS progress on the basis of a set of different parameter gaussians. Test reconstruction has been made using the modified ART and GP algorithms. Several tasks were solved in the simulation.
There was evaluated, for example, the influence on reconstruction quality of:

- model parameters;
- the number and disposition of projections;
- noises of different intensity.

Fig. 2 presents reconstruction errors dependencies on the iteration number for various number of projections and their total informativity. One can see that for a set of 19 projections the reconstruction error is several times better than that for a set of 3 projections. One can see an essential decrease of the reconstruction error when using projection sets with high total informativity for 5 and 3 projection sets.

8. The research showed the necessity to develop a procedure for object reconstruction from the data obtained in field experiments. The following problems were solved:

- input of projection images into computer (projections were initially registered on video- and photo films);
- removal of geometrical and brightness distortions;
- geometrical fastening of projections to a common point in the space;
- images brightness calibration;
- geometrical and brightness normalization of images (fastening to the same scale and absolute energetic values).

Solution of these complex and difficult problems resulted in a possibility to get images which one can call tomographic projections from experimental images obtained by various devices. Such tomographic projections may be used in reconstructive routines. The procedure described above was applied to 3 images of barium cloud obtained in one of the middle latitude experiments at 170 km. Unfortunately, the projection set was uninformative according to the informativity criterion used in our works: only one projection was situated in the high informativity zone. Images corresponded to optically thin AS development stage, the emission tomography scheme was used. Spatial luminosity distribution was reconstructed which for a given AS development stage is proportional to a spatial distribution of barium ion concentration. Mentioned the convergence of iteration process was noted.

9. The purpose of reconstruction analysis was to check the conformity of the distribution obtained after reconstruction to a priory data and physical concepts of it, and also to formulate some additional requirements to experimental conditions, to algorithms and routines of reconstruction and visualization of its results.
Fig. 3 shows initial projections and the ones obtained after reconstruction. One can see that the general structure features are preserved, though, as was supposed after simulating, the intensity of weaker details essentially decreases. One of the good tests in this case is to determinate post-reconstruction orientation of "invisible" inner details with respect to the vertical z-axis. The point is that it follows from a priori data that at a given AS development stage there exist some structures named "strata" which are oriented preferably along the local magnetic field vector. These strata are similar to some structures in auroral glow. In the experiment carried out it forms 24 degrees to the vertical axis. Fig. 4 shows the form of the inner structure for 50% of the maximum intensity (it is invisible inner structure) in the plane passing through the vertical axis and this structure. One can see that the angle between the vertical axis and this structure, which is similar to strata, is approximately 21 degrees, which, in our opinion, rather well agrees with a priori data and physical presentations. At the same time the analysis of results has shown that it is perfectly insufficient to do reconstruction on a grid of 64x64x64 voxels to obtain the spatial accuracy required in most physical applications.

10. The work allowed us to formulate requirements on devices and experimental methodology including collection of necessary attending information. A version of tomographic system structure was also developed.

3. Results of Auroral Structures Research

The designed method was applied to analyze some practical questions concerning the development of ALIS project (A.Steen, 1990; V.Alpatov et al., 1993). Shortly, it is a system for investigations of auroral airglows which must consists of a number of ground-based stations. These stations will obtain information about aurora in the image form and transmit it to a central point, where information processing must be carried out.

It is very important for such a system to dispose the observation stations in an optimal way and to develop an optimal reconstruction procedure because each station is very expansive and it is supposed that results of tomographic reconstruction may be obtain in real or "pseudoreal" time. In this case informational analysis was used which demonstrated those merits which may be obtain. The informativity criterion of Dr. Kazantsev (Kazantsev, 1992) was used.

We have selected a model consisting of two arcs, each being described by the function f from the Dr. Gustavsson's paper (Gustavsson, 1992):

$$f = Kz^2 \exp\{-az - b(y - c\sin \quad dx)^2\}, \quad z > 70km, \tag{1}$$

where K is an absolute value of a stohastic process, a, b, c, d are any parameters.

Then we assumingly calculated conditions for derivation of parallel informativity projections for various projection directions with respect to the object mass center in the range of polar angles of $0^o - 180^o$ and azimuth angles of $0^o - 360^o$. For the sake of convenience, we used relative informativities and the so-called inhomogeneity index:

$$\gamma = Q_{max}/Q_{min}, \qquad Q_\gamma = Q_i/Q_{min}, \tag{2}$$

where Q_i is the informativity value for the i-th projection, Q_{max} and Q_{min} are the values of the maximum and minimum of informativity.

Fig. 5 presents the distribution of relative informativity for the accepted model of aurora on the Earth surface in the rectangular system of coordinates. Informativity decrease denoting the region of aurora projection on the earth surface.

Then we calculated informativities for 14 observation sites which were supposed to be in the ALIS system. Fig. 6 and 7 show informativity distributions plotted the locations of these sites for two different aurora positions. One can see which sites in which case are in the zone of high and low informativity.

The next step was to compare registration errors at the most and least informative observation sites.

Modeling was carried out in the following way. At first we reconstructed all 14 projections, then 7 ot the most informative and 7 of the least informative ones. In the first set the reconstruction was carried out according to projection numbering from the paper (Steen, 1990), in the second – in the order of decreasing informativity, in the third – in the order of increasing informativity. Three iterations were made for each case. Fig. 8 presents the reconstruction error as a function of the projection number and the iteration number. It is seen that the error in the 2nd set decreases more rapidly as compared to the 1st one until an informative projection appears in the first set. It is seen from 8b that after

3 iterations the reconstruction error for the 2nd set is about 5% worse as compared to the 1st set. A set of 7 non-informative projections produced a 80% error after 3 iterations. How did it really affect the structure reconstruction?

Fig. 9 presents vertical slices of the model region for the accurate and reconstructed solution of all the 3 sets. One can see that even after 3 iterations the model structure displays a good reconstruction both for 14 projections and for the set of 7 informative ones. At the same time, 7 non-informative projections produce an unsatisfactory reconstruction. Fig. 10 shows projections obtained from the model and from the reconstruction using all 3 sets, and confirms the said above. Besides, the modeling has shown possible existence of an optimum distance between the observation sites and the object for the case of cone projection.

4. Possibility of optimization of cone beam projections

The next stage should include the testing of criterion for cone beam projections typical for auroras. Several estimates using the ART algorithm have been recently obtained for regular models.

Fig. 11 shows a scheme of cone beam modeling which produced a notable result. Theoretical concepts suggest that the reconstruction error should monotonously increase as the observation points move away from the object. In our modeling the error has a minimum (see Fig. 12) which varies depending on the model object. The minimum occurs at $S = 5 - 15$, where S is the relative distance between the objects and the receivers. This can be a measure of closeness d (see above) which can be represented as:

$$d \approx (\frac{1}{\sum_1} + \frac{1}{\beta \sum_1}), \tag{3}$$

where $\sum_1$ is a sum of the intersection lengths of rays with the object, $\sum_2$ is a sum of the intersection lengths of rays with the shaped region, β is a coefficient characterizing relative importance of $\sum_2$.

The dashed area in Fig. 11 may be called a "stabilization" region, since the additions occurring in the region in the course of iteration will be less than or equal to zero, which leads to artifact suppression.

5. Conclusions

Thus, the research carried out allowed us to develop a technology for local and global tomographic systems intended for inner structure reconstruction of different natural and artificial optical irregularities in the near space. This technology was tested on different kinds of artificial objects and auroral models intended for ALIS project. The modeling has shown that the informativity criterion Q is applicable for various models and can be used for a priori informativity estimation of the suggested observation sites. Besides, the modeling has shown possible existence of an optimum distance between the observation sites and the object for the case of cone projection.

This can also help to:

a) optimize the observation scheme, which would improve the quality of reconstruction;

b) optimize the reconstruction procedure so as to use the projections in the order of decreasing informativity.

Satisfactory results can be achieved more rapidly, which may be important, for instance, when using the obtained solution in another reconstruction algorithm that requires a good initial approximation. We think that the results of the study may be used when constructing the ALIS .

Acknowledgments

The work was supported partly by Russian Fund of Fundamental Research under grant N 93-05-14137, and partly by the Scientific Research Organization of the Netherlands, NWO, grant No. 713-097.

REFERENCES

Alpatov V.V., Gaydukov V.Yu., Milinevsky G.P. and Romanovsky Yu.A. (1985), On a technique of the artificial clouds parameters determination in the ionosphere at AC computer image processing. In: *Abstracts of V All-USSR Workshop in Dynamic Processes Researches in the Upper Atmosphere of the Earth*, Obninsk, p. 127 (in Russian)

Alpatov V.V., Belotserkovsky M.B., Evtushevsky A.N., Milinevsky G.P. and Romanovsky Yu.A. (1986), Optical Observations of Phenomena During the Expansion of Alkali Metak Hot Clouds in the Earth's Upper Atmosphere. In: *Proceedings of XXVI COSPAR*. Toulouse, France, p. 30.

Alpatov V.V., Bulygin F.V., Levin G.G., Pickalov V.V. and Romanovsky Yu.A. (1989), About a possibility of artificial ionospheric structures researches by optic tomography methods. *Poster report in IV All-USSR computer tomography symposium*, Tashkent. (in Russian)

Alpatov V.V., Levin G.G., Pickalov V.V., Romanovsky Yu.A. (1991), Optical tomography techniques to study geophysical artificial structures. *Proceedings of SPIE.* vol. 1843, Analytical Methods for Optical Tomography, 297–311.

Gustavsson B. (1992), A study of feasible tomographic inversion techniques for ALIS. *IRF Technical Report 039*, April, 1992, ISSN 0284-1738.

Hamaker C., Solmon D.C. (1978), The angles between the null spaces of X-rays. *J. Math. Anal. Appl.* 62, 1–23.

Kazantsev I.G. (1992), Information Content of Projections in Few-Views Tomography. *Proceedings of SPIE.* vol. 1843, Analytical Methods for Optical Tomography, 62–64.

Luenberger D.C. (1969), *Optimization by Vector Space Methods.* Wiley.

Steen A. (1990), A Scientific and Technical Description of ALIS. In: *Proc. NSSR Annual Meeting*, Bolresjo, Norway, 12-14 November, 1990, 153–164.

TYPE OF OBJECT	STAGES OF OBJECT DEVELOPMENT	TIME AFTER INJECT., S	OPTICAL RADIUS	DIMENSIONS, KM	RATE OF SIZE VA-VARIATIONS, KM/SC	Density of emission intensity near detector, erg/cm² s	TOMOGRAPHIC CLASSIFICATION
CLOUDS OF ALCALI METALS	SCATTER RAPID DIFFUSION AND DRIFT SLOW DIFFUSION AND DRIFT	0–2 2–100 >100	> 1 τ_x=0.2–200 τ_x+>>1 τ_x <0.2 τ_x+ >>1 additing 0.001– 1	1–2 2–10 30–50	1–1.5 30–100 M/S 5–100 M/S	$> 1\cdot 10^{-9}$ $>1.\ 10^{-8}$	AVAILABLE SELF-GLOW TRANSMISSION TRANSMISSION EMISSION
JETS	SCATTER and CREATING DIFFUSION AND DRIFT	0–2 0–100 >2	>1 0.001–0.1	0.5 length up 70 0.5–30 length up 70	1–1.5 10–100 M/S drift up 1 m/s	$>1.\ 10^{-8}$ $>1.\ 10^{-8}$	AVAILABLE SELF-GLOW TRANSMISSION EMISSION
AEROSOL CLOUDS	SCATTER DIFFUSION AND DRIFT	0–10 10–1000	>>1 >>1	5–10 10–20	1–1.5 100–15 m/s	$> 1\cdot 10^{-9}$ $> 1\cdot 10^{-9}$	SELF-GLOW in WIDE RANGE TRANSMISSION
NEUTRAL INJECTIONS	SCATTER DIFFUSION	50–60 60–600	< 1 < 1	10–100 >100	0.3–2 KM/s ao 300 M/s	> 40 R ~ 50 R	EMISSION SELF-GLOW in ATMOSPH.EMIS-SOINS

FIGURE 1. Estimates of some characteristics AS for tomographic classification analysis of AS.

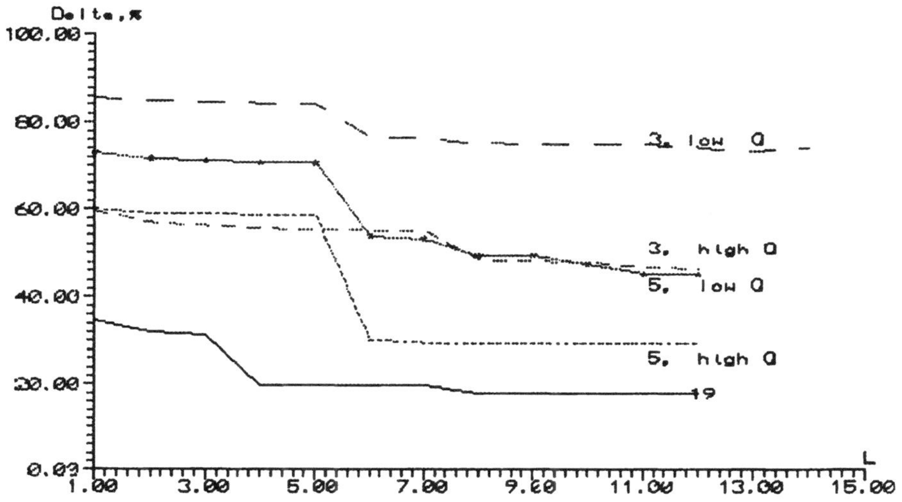

FIGURE 2.

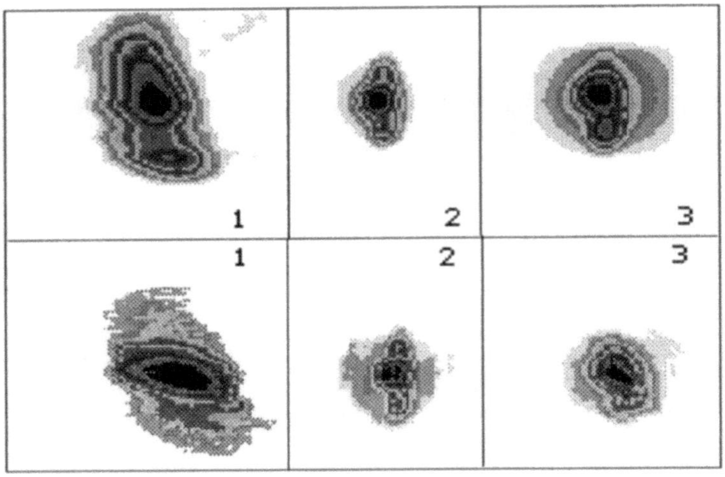

FIGURE 3. Experimental projects (top) and "pseudo projections" on reconstruction distribution (bottom).

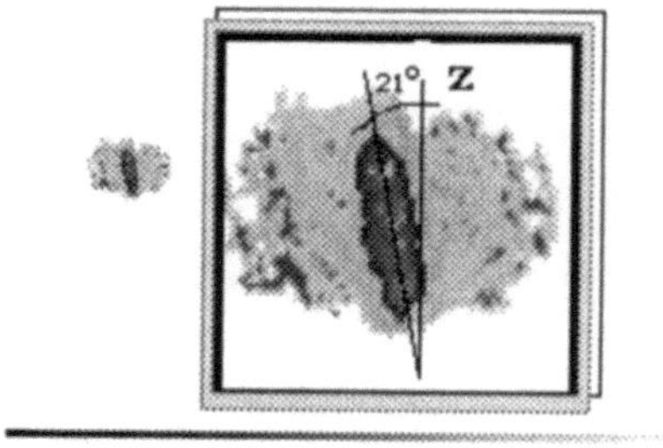

FIGURE 4. Slice of object inner structure.

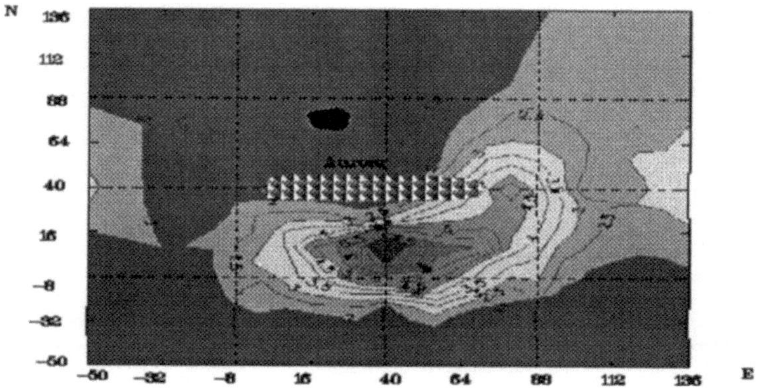

FIGURE 5. Distribution of informative criterium.

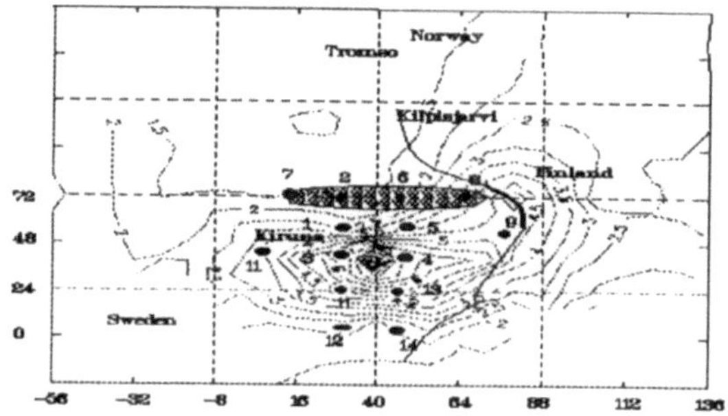

FIGURE 6.

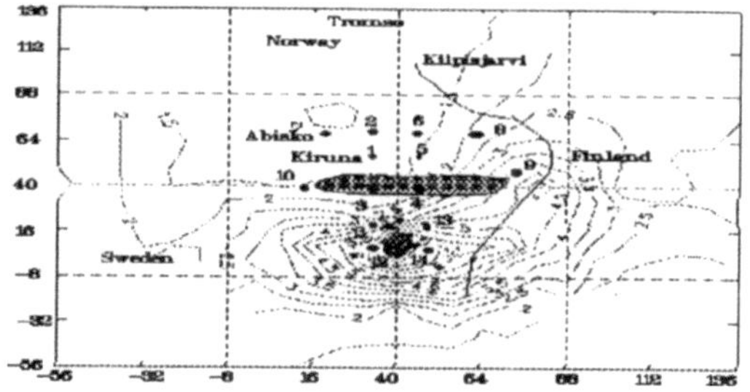

FIGURE 7. Distribution of informative criterium with ALIS stations.

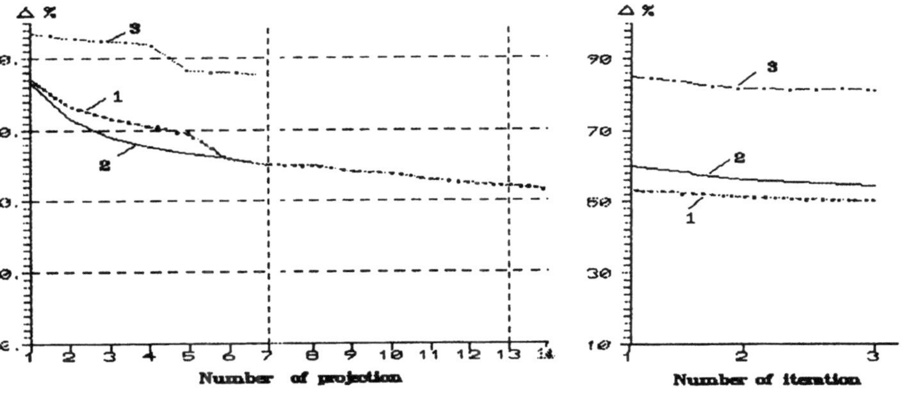

a) b)

FIGURE 8. Dependence of reconstruction error upon: a) number of projection for first iteration; b) number of iteration; 1 – for 14 projections, 2 – for 7 "informative", 3 – for 7 "uninformative".

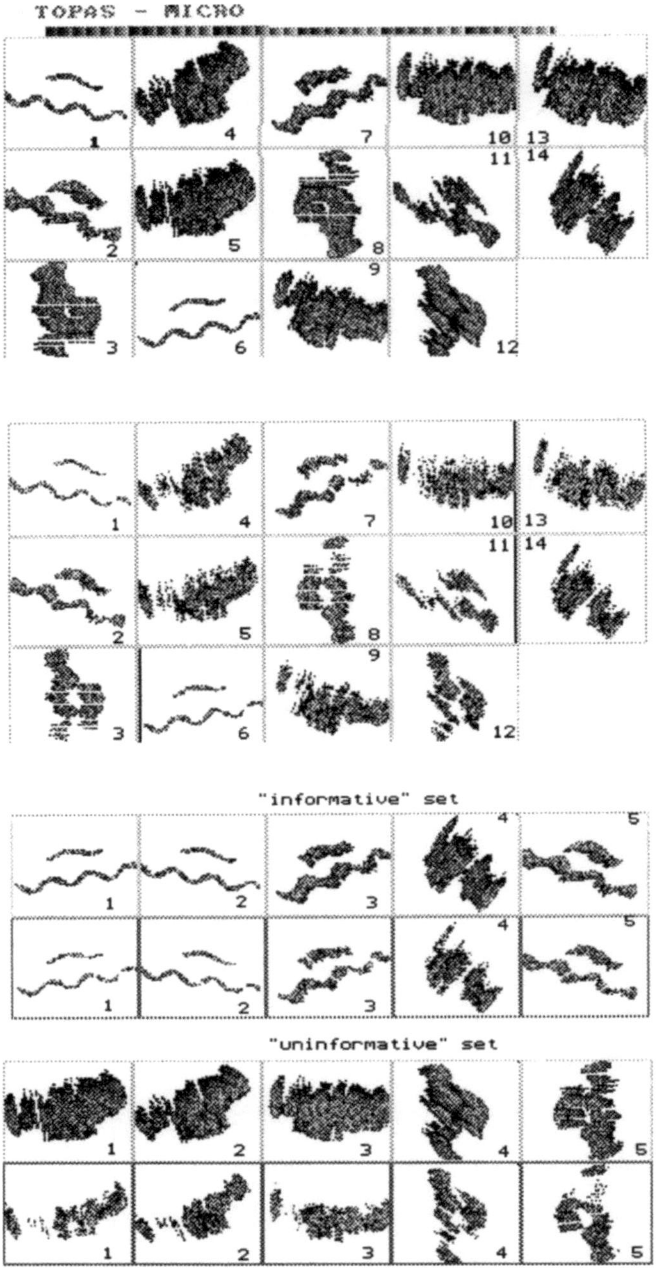

Figure 9. Comparison projections made on model (top) and on reconstructions (bottom) for all 14 ALIS projections and for "informative" and "uninformative" sets.

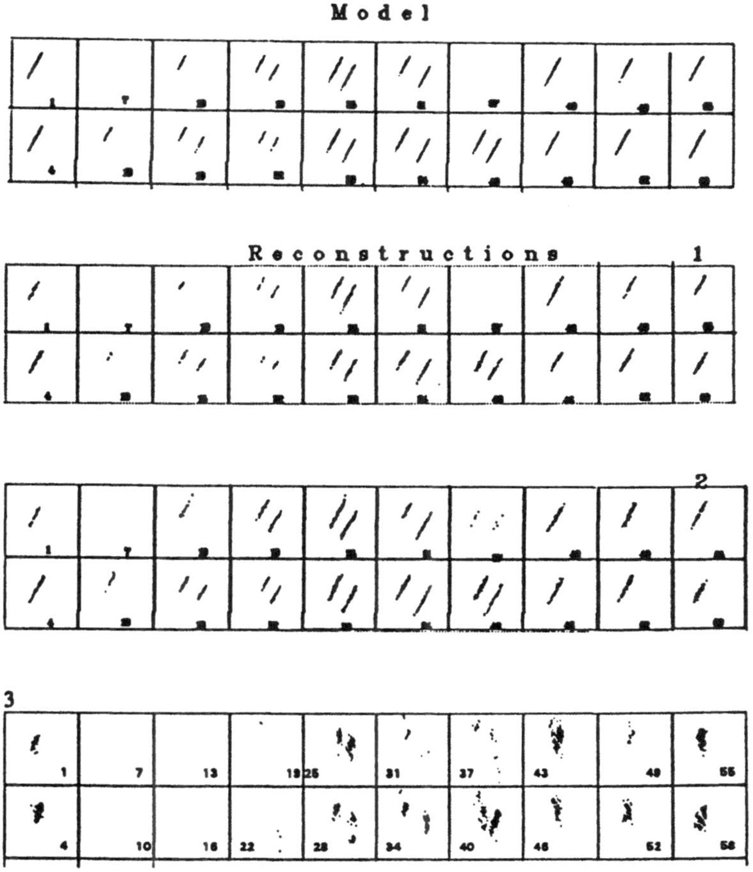

Figure 10. Comparison model and reconstructions slice by slice in a vertical plane: 1) on 14 projections; 2) on 7 "informative" projections; 3) on 7 "uninformative" projections.

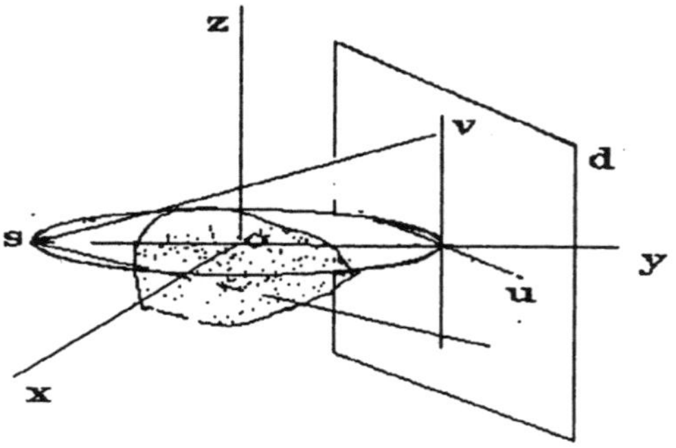

Figure 11. Scheme of cone beam modeling.

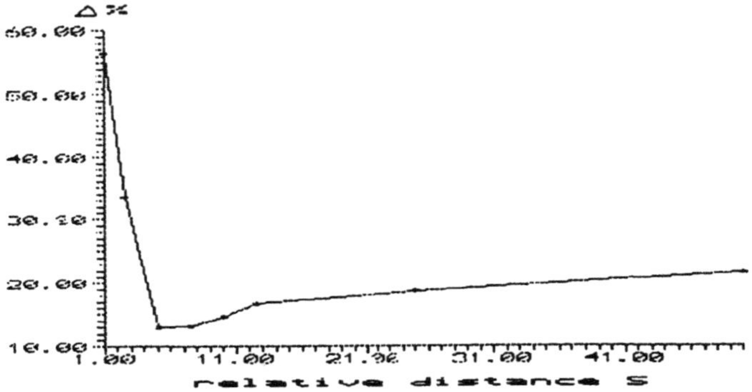

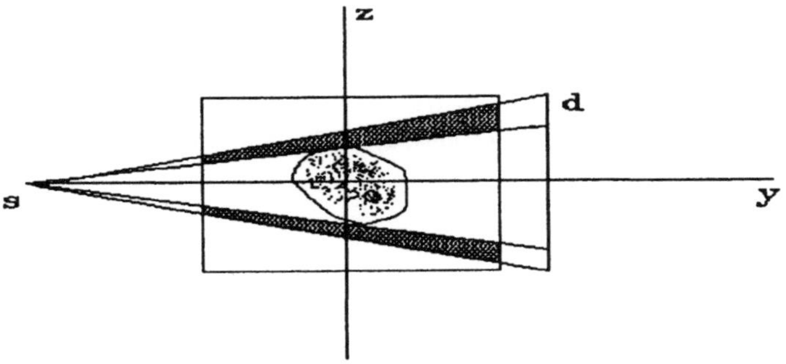

Figure 12. Scheme for explaining of minimum curve.

Some Numerical Experiments in Tomography of Scattering Medium

D. S. ANIKONOV, A. E. KOVTANYUK, I. V. PROKHOROV

Institute of Applied Mathematics of Russians Academy of Sciences, Far East Branch
Radio 7, Vladivostok, 690041, Russia

ABSTRACT

In the paper a tomography problem is considered as an inverse problem for the transport equation. The problem we deal with is to find the coefficient of full attenuation or the discontinuity surfaces of coefficients of the transport equation. The corresponding algorithms are described and examined using two control tests. The results of numerical experiments are given in figures. The suggested methods allow us to solve the tomography problem for an arbitrary scattering and any internal sources.

1. Introduction and Discussion of Problems

We consider the following stationary transport equation

$$(s, \nabla_x \varphi(x,s)) + a(x)\varphi(x,s) = \frac{1}{4\pi}\int\limits_{\Omega} b(x, s\cdot s')\varphi(x,s')ds' + F(x,s), \tag{1}$$

where $x \in G$, G is a convex domain in $\mathbf{R}^3$, $s \in \Omega$, $\Omega = \{s : s \in \mathbf{R}^3, |s| = 1\}$. The function $\varphi(x,s)$ denotes the intensity of the particles flow at the point x, which moves along the s-axis. The function $a(x)$ is the coefficient of full attenuation of the radiation, $b(x, s\cdot s')$ is the indicatrix of scattering and $F(x,s)$ describes the intensity of internal sources.

We denote the length of intersection of the ray $\{x + ts, t > 0\}$ with the domain G by $\sigma(x,s)$. It is clear that, for any $x_0 \in G$, the points $x_0 + \sigma(x_0,s)s$, $x_0 - \sigma(x_0,-s)s$ lie on the boundary of the domain G ($x_0 + \sigma(x_0,s)s \in \partial G$, $x_0 - \sigma(x_0,-s)s \in \partial G$). We denote by Γ^- the set of points $(x,s) \in \partial G \times \Omega$ for which $x = x_0 - \sigma(x_0,-s)s$, $x_0 \in G$, and by Γ^+ the set of points $(x,s) \in \partial G \times \Omega$ for which $x = x_0 + \sigma(x_0,s)s$, $x_0 \in G$.

We add the following boundary conditions to equation (1):

$$\varphi(x,s) = h(x,s), \quad (x,s) \in \Gamma^-; \tag{2}$$

$$\varphi(x,s) = H(x,s), \quad (x,s) \in \Gamma^+. \tag{3}$$

The functions h, H describe the radiation at the boundary of the domain G, $h(x,s)$ is the intensity of the input flow, and $H(x,s)$ is the intensity of the output flow. We here examine the following problem (Anikonov and Prokhorov, 1992).

PROBLEM 1. Determine the coefficient $a(x)$ from equation (1) and the boundary conditions (2),(3) if the functions h, H are given.

This problem has the following connection with tomography. The flow of radiation of a known intensity $h(x,s)$ incidents upon an unknown medium G. Measuring the transient radiation we have to find the index of the medium's internal structure (the function $a(x)$). In particular, the surfaces of discontinuity of the function $a(x)$ are the boundaries between different materials in G. It is clear that cracks and other defects inside objects also may be described in this way.

PROBLEM 2. Determine the discontinuity surfaces of the coefficients a, b, F from equation (1) and the boundary conditions (2),(3) if the function H is known.

The solution of Problem 2 gives less information than that of Problem 1 but it uses less data, too. The function h is not known, but it is supposed to be smooth enough.

Usually the scattering of particles in the medium is taken into account partially, if at all (Frey and Tsui 1990, 1991). We can suggest some methods for solving the tomography problem in the general case, i.e., for an arbitrary scattering and any internal sources.

2. Investigation of the Problems

Assume that the functions $a(x)$, $b(x, s\cdot s')$, $F(x,s)$ belong to the classes $C^1(G_i)$, $C^1(G_i \times [-1,1])$, $C^1(G_i \times \Omega)$, respectively, where $\bar{G} = \bigcup_{i=1}^{p} \bar{G}_i$, $G_i \bigcap G_j = 0$, $i \neq j$, and the domain boundary G_i is piecewise smooth surface of the class C^2.

Let the vector $s \in \Omega$ be characterized by spherical angles: the latitude θ, $0 \leq \theta \leq \pi$, and the longitude γ, $0 \leq \gamma < 2\pi$, i.e. $s = s(\theta, \gamma)$. The angle $\theta = \frac{\pi}{2}$ corresponds to the horizontal plane. We suppose

$$\lim_{\epsilon\to 0}\left| h\left(x, s(\frac{\pi}{2}+\epsilon,\gamma)\right) - h\left(x, s(\frac{\pi}{2}-\epsilon,\gamma)\right)\right| > 0, \quad (x,s)\in\Gamma^-, \tag{4}$$

$$\begin{aligned}\int_0^{\sigma(x,-s(\frac{\pi}{2},\gamma))} a\left(x-\tau s(\frac{\pi}{2},\gamma)\right) d\tau = -\ln\lim_{\epsilon\to 0}\left[H\left(x, s(\frac{\pi}{2}+\epsilon,\gamma)\right) - H\left(x, s(\frac{\pi}{2}-\epsilon,\gamma)\right)\right]^{-1} \\ \times\left[h\left(x-\sigma(x,-s(\frac{\pi}{2}+\epsilon,\gamma))s(\frac{\pi}{2}+\epsilon,\gamma), s(\frac{\pi}{2}+\epsilon,\gamma)\right)\right. \\ \left. - h\left(x-\sigma(x,-s(\frac{\pi}{2}-\epsilon,\gamma))s(\frac{\pi}{2}-\epsilon,\gamma), s(\frac{\pi}{2}-\epsilon,\gamma)\right)\right], \quad (x,s)\in\Gamma^+.\end{aligned} \tag{5}$$

Let the function h depend on a parameter μ, i.e., $h = h(x,s,\mu)$, $\mu \in (0,1)$, and for $s = (s_1, s_2, s_3)$, $(x,s)\in\Gamma^-$, the condition

$$\left|\frac{\partial h}{\partial \mu}(x,s,\mu)\right| \longrightarrow \infty, \quad s_3, \mu \longrightarrow 0 \tag{6}$$

holds. Then

$$\int_0^{\sigma(x,-s_0)} a(x-\tau s_0) d\tau = \ln\lim_{\mu\to 0}\frac{h'_\mu(x-\sigma(x,-s_0)s_0, s_0, \mu)}{H'_\mu(s, s_0, \mu)}, \tag{7}$$

$$(x,s)\in\Gamma^+, \quad s_0 = (s_1, s_2, 0).$$

We can determine the function $a(x)$ using in (5) and (7) the inverse Radon transform and therefore solve Problem 1.

If the coefficients of equation (1) are discontinuous for $x \in L$ (L is a surface in G) and $h(x,s) \in C^1(\partial G \times \Omega)$ then, except for some special cases, the function

$$J(x,s) = \sum_{i=1}^{3} \left\{ \left| \frac{\partial H(x,s)}{\partial x_i} \right| + \left| \frac{\partial H(x,s)}{\partial s_i} \right| \right\}, \qquad (x,s) \in \Gamma^+,$$

is infinite if and only if the ray $(x - ts, t > 0)$ is tangent to the surface L. Using this effect, we construct an algorithm for finding L. This algorithm can be used in ecology for detection the radioactive waste products in a medium. We may apply this method in the case when the intensity of unknown internal sources is not greater than the radioactive background intensity of the surrounding medium.

3. Numerical Results

For examination the above-stated methods we used the following tests.

Test 1.

$$G = \{x:\ x \in R^3,\ |x| < 1\}, \quad \bar{G} = \bigcup_{i=1}^{3} \bar{G}_i, \quad G_1 = \{x:\ x \in R^3,\ 0.7 < |x| < 1\},$$

$$G_2 = \{x:\ x \in R^3,\ 0.5 < |x| < 0.7\}, \quad G_3 = \{x:\ x \in R^3,\ |x| < 0.5\}.$$

Let $F = 0$, $a = a_i$, $b = b_i$ for $(x,s) \in G_i \times \Omega$, $i = 1,2,3$, and $a_1 = a_3 = 4$, $a_2 = 10$, $b_i = 0.3a_i$, $i = 1,2,3$. This is to say that the number of free runs of particles in the medium amounts up to 17. The intensity of the input flow of radiation, according to (4), is the following

$$h(x,s) = \begin{cases} 1 - [(\theta - \frac{\pi}{2})/\frac{\pi}{2}]^\alpha, & \theta \geq \frac{\pi}{2}, \\ 0, & \theta < \frac{\pi}{2}, \end{cases} \qquad a = 0.01.$$

Test 2.

$$G = \{x:\ x \in R^3,\ |x| < 1\}, \quad \bar{G} = \bigcup_{i=1}^{3} \bar{G}_i, \quad G_1 = \{x:\ x \in R^3,\ 0.9 < |x| < 1\},$$

$$G_2 = \{x:\ x \in R^3,\ 0.2 < |x| < 0.9\}, \quad G_3 = \{x:\ x \in R^3,\ |x| < 0.2\}.$$

$a_1 = a_3 = 10$, $a_2 = 3$, $b_i = 0.3a_i$, $i = 1,2,3$. This means that the number of free runs of particles in the medium may be as much as 15. The intensity of the input flow of radiation, according to (6), is

$$h(x,s,\mu) = \left| 1 - \left[\mu + \frac{5}{8} \left| \theta - \frac{\pi}{2} \right| \right]^\alpha \right|, \quad a = 0.01.$$

In testing the intensity of the output flow of radiation was approximated by 12 terms of the Neumann series. The terms were calculated by means of the Monte Carlo method. For calculating the intensity of the output flow at a point, we simulated 20000 trajectories and it took 3 minutes on PC AT Intel 386/40. We used formula (5) in Test 1 and formula (7) in Test 2 for 20 points $(x,s) \in \Gamma^+$. In the centrally symmetric case this allowed us to obtain the values of optical thickness on 1600 lines on the section $x_3 = 0$. Using the method of convolution and back projection, we reconstructed the values of $a(x)$ on the plane $x_3 = 0$.

The exact and reconstructed values of $a(x)$ are given in Figures 1, 2 for Test 1, and in Figures 3, 4 for Test 2, correspondingly.

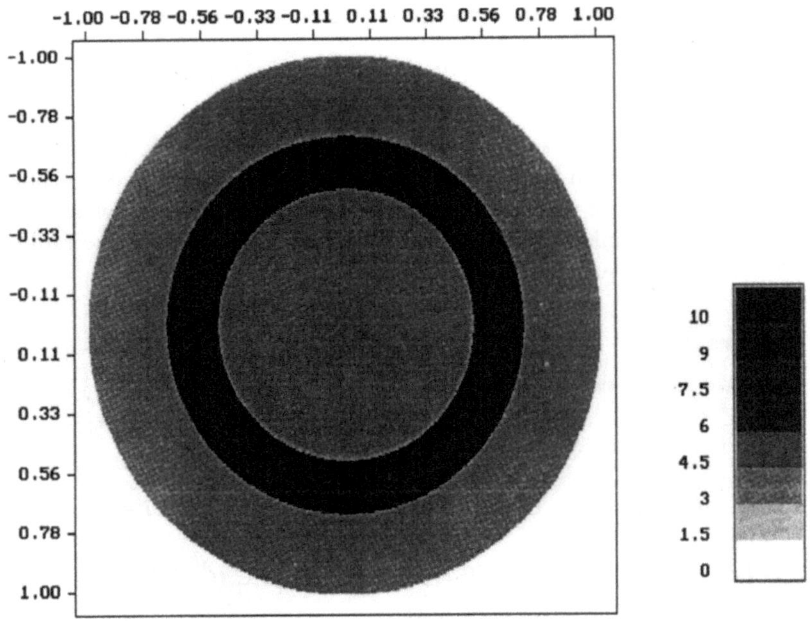

FIGURE 1. Test 1. Exact image of internal structure of the medium G.

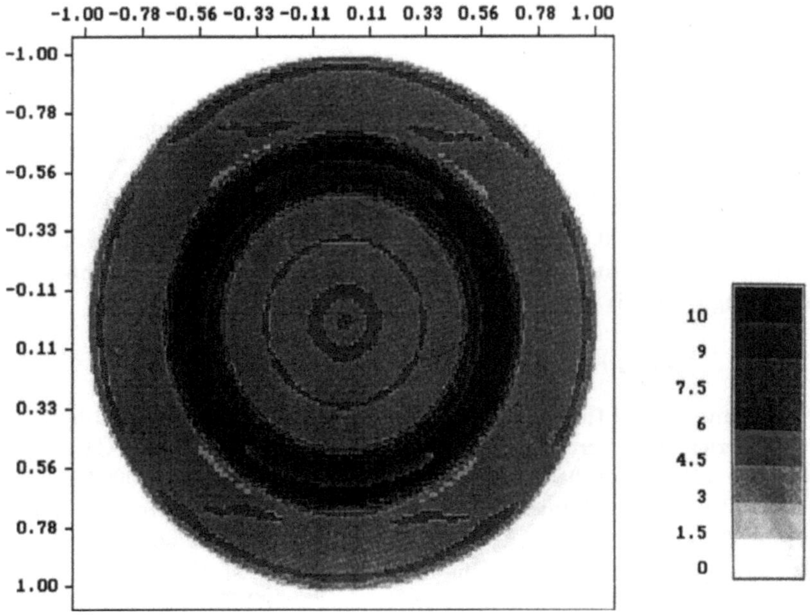

FIGURE 2. Test 1. Reconstructed image of internal structure of the medium G.

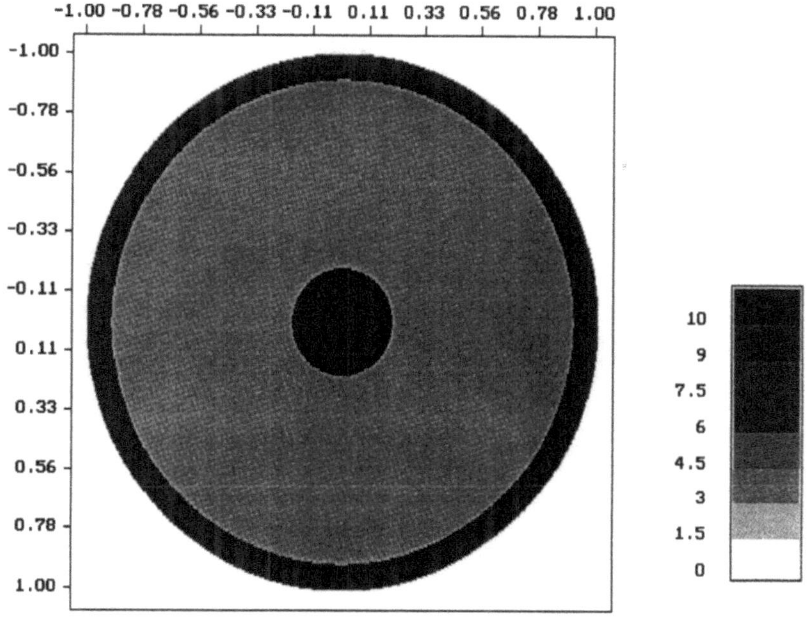

FIGURE 3. Test 2. Exact image of internal structure of the medium G'.

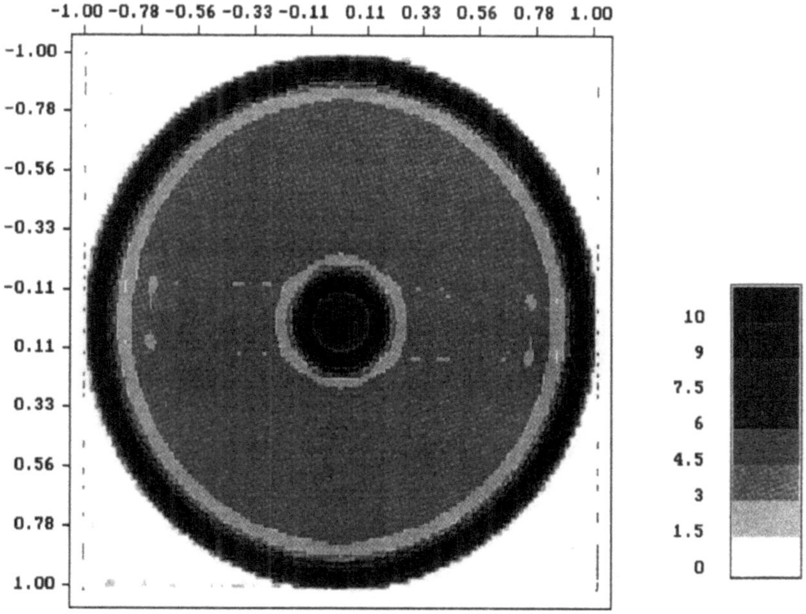

FIGURE 4. Test 2. Reconstructed image of internal structure the of medium G.

REFERENCES

Anikonov, D.S. and Prokhorov, I.V. (1992), Determination of coefficients of the transport equation with energy and angular features of the outside intensity. *Doklady Academii Nauk SSSR* **327**, no. 2, 205–207. (in Russian)

Frey, E.C. and Tsui, B.M.W. (1990), Parametrization of the scatter response function in SPECT imaging using Monte Carlo simulation. *IEEE Trans. Nucl. Sci.*, **37**, no. 3, 1308–1315.

Frey, E.C. and Tsui, B.M.W. (1991), Special properties of the scatter response function in SPECT. *IEEE Trans. Nucl. Sci.*, **38**, no. 2, 789–794.

Integral Geometry

Yu. E. ANIKONOV

Institute of Mathematics
Novosibirsk, Russia

ABSTRACT

The problems of integral geometry are considered. The methods and results concerning this type of problems are presented. The existence and uniqueness theorems, some inversion formulas are given as well.

1. Introduction

The problem of determining unknown objects (functions, differential forms, tensor field) known their integral on manifolds (curves, surfaces) are called the problems of the integral geometry.

The problems of the integral geometry are connected with differential equations, inverse problems, group representations and one can use them in the geophysics, astronomy, medicine.

The first results here were received by Funk (1916), Radon (1917). They founded the integral geometry and discovered the inversion formulas.

Later the new problems of integral geometry were set, many theorems of the uniqueness and existence were proved, the new inversion formulas were discovered and the new applications were given.

The problems of integral geometry were studied in the works of Blaschke (1916), John (1955), Gelfand et al. (1966, 1969, 1980), Gelfand and Goncharov (1987), Kostelyane and Reshethyak (1954), Helgason (1980), Semyanstyi (1960, 1961, 1966), Kirillov (1961), Solman (1976), Lavrent'ev and Romanov (1966), Lavrent'ev (1970), Lavrent'ev and Bukhgeim (1973), Lavrent'ev (1986), Romanov (1967, 1978, 1987), Anikonov and Lavrent'ev (1967), Anikonov (1969a, 1969b, 1978, 1978b, 1983), Anikonov and Romanov (1979), Anikonov and Pestov (1989), Anikonov and Stepanov (1991), Anikonov and Shneiberg (1991), Bukhgeim (1972, 1979, 1983), Mukhometov (1977), Blagoveshchenskii (1986), Pestov (1985), Pestov and Sharafutdinov (1987), Sharafutdinov (1989), Goncharov (1988), Palamodov and Denisjuk (1988).

Applications and information were given in the papers by Tikhonov et al. (1987), Natterer (1986). One general problem of the integral geometry is to find the function

$$f(x) \in C^\infty(R^n), \quad \sup_{x \in R^n} (1 + |x|^m) \left| \frac{\partial^\alpha f}{\partial x_1^{\alpha_1}, \dots, \partial x_n^{\alpha_n}} \right| < \infty, \quad \forall m \geq 0$$

if we know the function

$$\hat{f}(y) = \int_{B(y)} f(x) \mathrm{d}\mu, \quad y \in R^n,$$

where $B(y) \subset R^n$ is a smooth manifold, $y \in R^n$, μ is a measure on $B(y)$ (Gelfand et al., 1966). The Radon transform is

$$\hat{f}(\omega, p) = \int\limits_{(\omega,x)=p} f(x) \mathrm{d}\,\mu, \quad y = (\omega, p),$$

$$B(y) = \{x : (\omega, x) = p\}, \quad \omega \in R^n, \quad |\omega| = 1, \quad p \in R, \quad x \in R^n.$$

In this paper we shall discuss some methods in the integral geometry especially in the curve case and some applications.

2. Inversion Formulas

The inversion formulas in problems of the integral geometry play very important role in the theory and applications.

Theorem 1. *Radon (1917), Gelfand et al. (1966), Helgason (1980). Let a function $\hat{f}(\omega, p)$ be the Radon transform*

$$\hat{f}(\omega, p) = \int\limits_{(x,\omega)=p} f(x) \mathrm{d}\,\nu.$$

Then the function $\hat{f}(\omega, p)$, $\omega \in \Omega$, $\Omega = \{\omega, \ |\omega| = 1\}$, $p \in R$, satisfies the conditions

1. $\hat{f}(\omega, p) \in C^\infty(\Omega \times R)$;

2.

$$\sup_{p \in R}(1 + |p|^m) \left| \frac{\partial^\alpha \hat{f}}{\partial p^\alpha} \right| < \infty, \quad \forall m \geq 0, \quad \alpha \geq 0;$$

3. $\int_R \hat{f}(\omega, p) p^k \mathrm{d}\,p$ is a homogeneous polynomial of the degree k depending on $\omega = (\omega_1, \ldots, \omega_n)$ and the inversion Radon formula holds

$$f(x) = \frac{(-4\pi)^{(n-1)/2}\Gamma(\frac{n}{2})}{\Gamma(\frac{1}{2})} \Delta^{\frac{n-1}{2}} \int\limits_{\Omega} \hat{f}(\omega, (\omega, x)) \mathrm{d}\,\omega$$

where $\Delta^{(n-1)/2}$ is the Laplace operator of the degree $(n-1)/2$.

The other similar formulas are contained in the papers by Semyanistyi (1960, 1961, 1966), Gelfand et al. (1980).

If $n = 2$, $x_1 = x$, $x_2 = y$, $x^2 + y^2 \leq 1$

$$\hat{f}(\phi, p) = \int\limits_{x \cos\phi + y \sin\phi = p} f(x, y) \mathrm{d}\,l, \quad 0 \leq \phi \leq 2\pi, \quad -1 \leq p \leq 1$$

then the different inversion formulas may be written on the basis of theorem 1.

$$f(x, y) = \frac{1}{(2\pi)^2} \Delta \int\limits_0^\pi \mathrm{d}\,\phi \int\limits_{-1}^1 \hat{f}(\phi, p) \ln|p - rn| \,\mathrm{d}\,p$$

$$= \frac{1}{(2\pi)^2} \int\limits_0^\pi \mathrm{d}\,\phi \int\limits_{-1}^1 \frac{\partial^2 \hat{f}(\phi, p)}{\partial p^2} \ln|p - rn| \,\mathrm{d}\,p$$

$$= -\frac{1}{(2\pi)^2}\int_0^{\pi} \mathrm{d}\phi \int_{-1}^{1} \frac{\partial \hat{f}(\phi,p)}{\partial p}\frac{1}{p-rn}\mathrm{d}\,p$$

$$= \frac{1}{(2\pi)^2}\int_0^{\pi} \mathrm{d}\phi \int_{-1}^{1} \hat{f}(\phi,p)\frac{1}{(p-rn)^2}\mathrm{d}\,p$$

$$= \frac{1}{(4\pi)^2}\int_{-\infty}^{\infty}\int_{-\infty}^{\infty}\left[\int_{-1}^{1} \hat{f}(\mathrm{arctg}\,\frac{\omega_1}{\omega_2},t)\,\mathrm{e}^{\mathrm{i}\,t\sqrt{\omega_1^2+\omega_2^2}}\mathrm{d}\,t\right]\mathrm{e}^{\mathrm{i}\,\omega_1 x+\mathrm{i}\,\omega_2 y}\mathrm{d}\,\omega_1\,\mathrm{d}\,\omega_2,$$

$$rn = x\cos\phi + y\sin\phi.$$

3. The Other Formulas

In this section we consider the problem of the integral geometry to find the function $f(x)$, $x \in R^3$ such that

$$\hat{f}(\gamma) = \int_{\gamma} f(x)\mathrm{d}\,l,$$

where γ is a line passing through a smooth curve $\Gamma \subset R^3$. The function $f(\gamma)$ is supposed to be known for all lines such that $\gamma \bigcap \Gamma \neq \emptyset$.

Theorem 2. *(Kirillov, 1961) A function*

$$f(x) \in C^{\infty}(R^3), \quad \sup_{x\in R^3}(1+|x|^m)\left|\frac{\partial^{\alpha} f}{\partial x_1^{\alpha_1},\ldots,\partial x_n^{\alpha_n}}\right| < \infty, \quad \forall m \geq 0, \;\; \alpha \geq 0$$

is restored by $\hat{f}(\gamma)$ and the inversion formulas hold if and only if

$$\{x:\,(x,\omega)=p\}\bigcap\Gamma \neq \emptyset$$

for almost all (ω,p), $|\omega|=1$, $p\in R$.

Tuy (1983) received the inversion formulae for the functions with the compact support when the curve is bounded provided the almost every hyperplane intersecting the support of the unknown function meets the curve Γ transversely in some point. Finch (1985) investigated the analogous inversion procedure for the class of the curves which did not satisfy the hypotheses of Tuy.

Blagoveshcheskii (1986) received the inversion formulae when the curve Γ is a circle in R^3. He received the essentially new formulae. The uniqueness theorems for the integral geometry problems when unknown functions have a compact support were received by Anikonov (1978), Helgason (1980), Leahy et al. (1979).

4. Integral Geometry on the Sphere

The problems of the integral geometry on the sphere were considered in the papers of Blaschke (1916), Semyanistyi (1960, 1961, 1966), Helgason (1980). The unknown functions were supposed to be even on the sphere.

Let n, p be the point of the unit sphere S in R^3 with the center at the origin, (n,p) be the inner product $n\in S$, $p \in S$, $\Gamma(n)=\{p\in S,\;\;(n,p)=0\}$, $S(n)=\{p\in S,\;\;(n,p)\geq 0\}$, $\triangle$ be the Laplace-Beltrami operator, a, b be the constants.

One of the integral geometry problems is to determine the function $f(n) \in C^\infty(S)$, provided the function $\hat{f}(p)$ being known. $p \in S$,

$$\hat{f} = 2a \int\limits_{\Gamma(p)} f(n) \mathrm{d}\,\Gamma + b \iint\limits_{S(n)} (\Delta + 2) f(n) \mathrm{d}\,S,$$

the function $f(p)$ is not necessary even.

Theorem 3. *(Anikonov and Stepanov, 1991) Let $a \neq 0$, $b \neq 0$, $a + b \neq 0$. The inversion formula holds*

$$\hat{f}(n) = \frac{1}{4\pi} \iint\limits_{S} \Phi(p)(1 - (n,p)) \ln(1 - (n,p)) \mathrm{d}\,S,$$

$$\Phi(n) = -\frac{1}{8\pi^2} \frac{\mathrm{d}}{\mathrm{d}\,t} \iint\limits_{(n,p)^2 > t} \frac{(\Delta + 2)\phi_1(p)(n,p) \mathrm{d}\,S}{\sqrt{(n,p)^2 - t}} \Bigg|_{t=0}$$

$$-\frac{1}{4\pi^2 b} \iint\limits_{S(n)} \frac{\Delta \phi^-(p) \mathrm{d}\,S}{(n,p)}, \quad \phi_1 = \frac{1}{a}\phi^+ - \frac{b}{4\pi a(a+b)} \iint\limits_{S} \phi \mathrm{d}\,S,$$

$$\phi^\pm = \frac{\phi(p) \pm \phi(-p)}{2}.$$

5. Equation for Function $\mu(\phi)$

In this section we consider the generalized problem of the integral geometry, namely, we seek for two functions $f(x) \in C^\infty(|x| \leq 1)$, $x = (x_1, x_2)$ and $\mu(\phi) \in C^\infty([0, 2\pi])$, given the function

$$\hat{f}(\phi, p) = \int\limits_{x \cos\phi + y \sin\phi = p} f(x)\, \mathrm{e}^{\,\mu(\phi)(-x_1 \sin\phi + x_2 \cos\phi)} \mathrm{d}\,l.$$

It should be noticed that if the function $\mu(\phi)$ is known, the problem is linear and the inverse formula for the function $f(x)$ holds (Kurchment, 1989).

Theorem 4. *(Anikonov and Shneiberg, 1991) The function $\mu(\phi)$, $0 \leq \phi \leq 2\pi$ solves the differential equation*

$$\int\limits_{-1}^{1} \left[\frac{\partial \hat{f}(\phi,p)}{\partial \phi} + \mu(\phi) p \hat{f}(\phi,p) \right] \mathrm{e}^{\mu'(\phi)p} \mathrm{d}\,p = 0, \quad \frac{\mathrm{d}\,\mu}{\mathrm{d}\,\phi} = \mu.$$

The other similar results are contained in the papers of Anikonov (1983), Anikonov and Shneiberg (1991).

In particular, the uniqueness class of two-dimensional integral geometry problem consists of the smooth functions $f(x_1, x_2, \phi)$ such that

$$\left(\frac{\partial}{\partial x_1} \sin\phi - \frac{\partial}{\partial x_2} \cos\phi \right) \frac{\partial f}{\partial \phi} = 0.$$

6. Support Theorems

Let $\hat{f}(\mathrm{P})$ be the Radon transform

$$\hat{f}(\mathrm{P}) = \int_{\mathrm{P}} f(x) \mathrm{d}\mu,$$

P be a plane.

Theorem 5. *(Helgason, 1980). Let the following conditions hold*

1. $f(x) \in C(R^n)$,
2. $|x^n|\,|f(x)| \leq \infty, \forall k \geq 0, x \in R^n$,
3. $\exists A > 0$ *such that* $\hat{f}(\mathrm{P}) = 0$ *for all* $\mathrm{P} : \rho(0, \mathrm{P}) \geq A$, *where* ρ *is a distance in* R^n.

Then $f(x) = 0$, $|x| > A$.

Theorem 6. *(Logvinenko, 1988). Let the following conditions hold*

1. $f(x) \in C^\infty(R^n)$, $f(x) = o(|x|^{-k})$, $\forall k > 0$, $x \to \infty$,
2. *the function*

$$\hat{f}(\omega, p) = \int_{(\omega,x)=p} f(x) \mathrm{d}\mu$$

satisfies the condition $\left|\hat{f}\right| \leq c_\omega \mathrm{e}^{-\varepsilon_\omega p}$, $c_\omega > 0$, $\varepsilon_\omega > 0$ *are the constants,*

3. $\exists A > 0$ *and a set* $e \in \Omega$, $\mu e \neq 0$ *such that* $\hat{f}(\omega, p) = 0$, $\omega \in e$, $|p| > A$.

Then supp $f(x)$ *is a compact set.*

7. Inversion Formulas for Tensor Fields

Let $f(x, \xi) = \sum_{i=0}^{m} f_{i_1 \ldots i_m} \xi_{i_1} \cdots \xi_{i_m}$, $x \in R^n$, $|\xi| = 1$, $m \geq 0$ be a tensor field such that

1. $f(x, \xi) \in C^\infty(R^n \times |\xi| = 1)$,
2. $$\sup_{x \in R^n} (1 + |x|^k) \left| \frac{\partial^\alpha f}{\partial x_1^{\alpha_1}, \ldots, \partial x_n^{\alpha_n}} \right| < \infty, \quad \forall k \geq 0, \ \alpha_i \geq 0.$$

We consider here the problem of the integral geometry to find $f(x, \xi)$ if we known the function

$$\hat{f}(x, \xi) = \int_{-\infty}^{\infty} f(x + t\xi, \xi) \mathrm{d}t.$$

Theorem 7. *(Sharafutdinov 1989) If the conditions 1,2 hold, then*

3. $f = \tilde{f} + \mathrm{d}v$, $\delta\tilde{f} = 0$,
4. $$\tilde{f} = (-\Delta)^{1/2} \left[\sum_{k=0}^{[m/2]} c_k (i - \Delta^{-1} d^2)^k j^k \right] \mu^m \hat{f}$$

c_k *are constants,* $i, \Delta, \mathrm{d}, j, \mu$ *are the operators defined in the paper of Sharafutdinov (1989).*

8. Integral Geometry Problem Uniqueness of the Solution

Consider the two-dimensional problem of to finding a smooth function $f(r,\phi)$, $0 \le r \le 1$, $0 \le \phi \le 2\pi$ from the function

$$\hat{f}(\rho,\alpha) = \sum_{j=1}^{2} \int_{\rho} f(r,\phi_j)\sqrt{1+\left(r\frac{\partial \phi_j}{\partial r}\right)^2}\, dr, \quad 0 \le \rho \le 1, \quad 0 \le \alpha \le 2\pi,$$

$$\phi_j = \alpha - (-1)^{-j}\sqrt{r - \rho\psi_j(r,\rho)}, \quad \psi \in C^1(0 < \rho \le r \le 1), \quad \psi_1(\rho,\rho) = \psi_2(\rho,\rho).$$

This problem is equivalent to the problem of integral geometry when curves have group's properties (Romanov, 1967).

Theorem 8. *(Romanov, 1967) If $\hat{f}(\rho,\alpha) = 0$, $0 \le \rho \le 1$, $0 \le \alpha \le 2\pi$, then $f(r,\phi) = 0$, $0 \le r \le 1$, $0 \le \phi \le 2\pi$.*

The generalizations and applications of this result may be found in the book by Romanov (1987). In particular, using the inversion formulae one can to prove the uniqueness theorem when manifolds are not plane or sphere.

9. Uniqueness of the Solution to the Integral Geometry Problems in the Class of Analytic Functions

Let the smooth manifolds $B(z,p)$, $z \in R^n$, $p > 0$ satisfy the condition

$$B(z,p) \subset \Omega^+(z,p) = \{(x,y) : x \in R^n,\ y > 0,\ |x-z|^2 + y^2 < p^2,\ n \ge 1\}.$$

Theorem 9. *(Anikonov, 1978) Let the function $f(x,y) \in C^\infty(x \in R^n, y \ge 0)$ be analytic in the variable y in the domain $x \in R^n$, $y \ge 0$. If*

$$\int_{B(z,p)} f(x,y)\, d\mu = 0, \quad z \in R^n, \quad p > 0,$$

Then $f(x,y) = 0$, $x \in R^n$, $y \ge 0$.

10. Differential Identities and Integral Geometry

The differential identities use are well for proving the uniqueness, stability, and existence theorems of the solution in integral geometry. Here we consider only the two-dimensional case. Let M, $\tilde{M}$ be the differential operators

$$MW = \left(\frac{\partial W}{\partial x} - c\right)\cos\theta + \left(\frac{\partial W}{\partial y} - d\right)\sin\theta,$$

$$\tilde{M}W = \left(\frac{\partial W}{\partial x} - c\right)\sin\theta + \left(\frac{\partial W}{\partial y} - d\right)\cos\theta,$$

where $W(x,y,z)$, $\theta(x,y,z)$, $c(x,y)$ $(\partial c/\partial z = 0)$, $d(x,y)$ $(\partial d/\partial z = 0)$ are functions in C^2 in the domain $x^2 + y^2 \le 1$, $0 \le z \le 2\pi$.

Theorem 10. *(Anikonov, 1978) The following identity holds*

$$\tilde{M}W\frac{\partial}{\partial z}MW - MW\frac{\partial}{\partial z}\tilde{M}W = -\frac{\partial\theta}{\partial z}\left[\left(\frac{\partial W}{\partial x} - c\right)^2 + \left(\frac{\partial W}{\partial y} - d\right)^2\right]$$

$$+\frac{\partial}{\partial y}\left(\frac{\partial W}{\partial x}\frac{\partial W}{\partial z}\right) + \frac{\partial}{\partial x}\left(\frac{\partial W}{\partial y}\frac{\partial W}{\partial z}\right) - \frac{\partial}{\partial z}\left(c\frac{\partial W}{\partial y} - d\frac{\partial W}{\partial x}\right).$$

In the case when $c = 0$ and $d = 0$ the identity was obtained by Mukhometov (1977). Let $\lambda(x,y)$, $\mu(x,y)$, $a(x,y)$, $b(x,y)$, $c(x,y)$, and $d(x,y)$ be smooth functions in the domain $x^2 + y^2 \leq 1$, and

$$\gamma(z,t) = \{x(s;z,t),\ y(s;z,t),\quad x|_{s=0} = \cos z,\ \ y|_{s=0} = \sin z\}$$

be a set of smooth extremals of the functional

$$\int_\gamma \left[a\mathrm{d}x + b\mathrm{d}y + \lambda\sqrt{\mathrm{d}x^2 + \mathrm{d}y^2}\right],$$

which joints the points $(\cos z, \sin z)$ and $(\cos t, \sin t)$. Assume that the set γ is regular. In this section we consider the non-linear problem of integral geometry of find the functions λ, μ, a, b, c, and d if the functions

$$\hat{f}(z,t) = \int_{\gamma(z,t)} \left[c\mathrm{d}x + d\mathrm{d}y + \mu\sqrt{\mathrm{d}x^2 + \mathrm{d}y^2}\right],$$

$$\theta(z,t) = \mathrm{arctg}\ \frac{y'_s}{x'_s}\bigg|_{s=0}$$

are known in the domain $0 \leq z \leq 2\pi$, $0 \leq t \leq 2\pi$. Theorem 10 implies the following theorem.

Theorem 11. *(Anikonov, 1978) Let the functions λ, μ be known for (x,y) such that $x^2 + y^2 = 1$. Then the functions $\lambda(x,y)$, $\mu(x,y)$, $\partial a/\partial y - \partial b/\partial x$, and $\partial c/\partial y - \partial d/\partial x$ are uniquely determined from the functions $\hat{f}(z,t)$ and $\theta(z,t)$.*

The problems in multi-dimensional cases were considered by Romanov (1978, 1987) and Pestov (1985). Pestov was the first to use the curvature of Riemannian spaces in the identities.

11. Integral Geometry and Evolution Equations

In the case of line integral the integral geometry problems reduce to Cauchy problems for evolution equations (Anikonov, 1969a), (Anikonov and Lavrent'ev, 1967). After solving these problems, unknown functions can be found by formulae.

The two-dimensional problem of integral geometry reduced to the inverse problem for the transport equation

$$\frac{\partial W}{\partial x}\cos\phi + \frac{\partial W}{\partial y}\sin\phi + K(x,y,\phi)\frac{\partial W}{\partial \phi} = f(x,y)$$

$$W|_{y=0} = \hat{f}(x,\phi),\quad x \in R,\quad 0 \leq \phi \leq \pi,\quad y \geq 0,$$

where the functions $W(x,y,\phi)$ and $f(x,y)$ are unknown. Information is given by the function $\hat{f}(x,\phi)$. The function $K(x,y,\phi)$ is the curvature.

Theorem 12. *(Anikonov, 1978) Let a smooth function $u(x,y,z)$ be the solution to the Cauchy problem*

$$\frac{\partial u}{\partial y} = -\int_0^1 \frac{\partial}{\partial p}\left[\frac{\partial u(x,y,p)}{\partial x}\sqrt{1-p^2} + K\frac{\partial u(x,y,p)}{\partial p}(1-p^2)\right]_{p=z\eta} \mathrm{d}\eta \equiv Au,$$

$$u|_{y=0} = \hat{f}(x, \arcsin z).$$

Then,

$$f(x,y) = \left[\frac{\partial u}{\partial x} + K(x,y,0)\frac{\partial u}{\partial z}\right]_{z=0}.$$

REMARK. If the function K does not depend on y, we have

$$u(x,y,z) = \mathrm{e}^{yA}\hat{f}.$$

12. On the Solvability of a Problem of Integral Geometry

We consider here the question on solvability of a problem of integral geometry in the case when the integration curves are geodesics of a fixed analytic metrics. Let

$$\mathrm{d}s^2 = B^2(x,y)(\mathrm{d}x^2 + \mathrm{d}y^2) \tag{1}$$

be an analytic metrics defined on the (x,y)-plane. Henceforth, we suppose that $B > 0$ and $\partial B/\partial y < 0$. We denote by $\tilde{\gamma}(\xi,\phi)$ the geodesic of the metrics (1) emanating from the point $(\xi, 0)$ at an angle ϕ, and let $\gamma(\xi,\phi)$ be the part of $\tilde{\gamma}(\xi,\phi)$ lying in the half-plane $y \geq 0$. The problem we consider is to find the function $f(x,y)$ in the region $y \geq 0$ if the function

$$\hat{f}(\xi,\phi) = \int_{\gamma(\xi,\phi)} f(x,y)\sqrt{\mathrm{d}x^2 + \mathrm{d}y^2}$$

is given for $-\delta \leq \xi \leq \delta$ and $0 \leq \phi \leq \alpha$.

We start with giving preliminaries. Denote by $\eta(\xi,\phi)$, $\eta > \xi$, a point of the line $y = 0$ that belongs to the geodesic $\gamma(\xi,\phi)$, and let $\theta(\xi,\phi)$ be the angle between $\gamma(\xi,\phi)$ and the x-axis at the point $(\eta, 0)$. Set $a(\xi,\phi) = B(\xi,0)\sin\phi$, $b(\xi,\phi) = B(\eta(\xi,\phi),0)\sin\phi$. Assuming

$$t = \int_\gamma \frac{\sqrt{\mathrm{d}x^2 + \mathrm{d}y^2}}{B}$$

the equations for geodesics of the metrics (1) become

$$\frac{\mathrm{d}^2x}{\mathrm{d}t^2} = B\frac{\partial B}{\partial x}, \quad \frac{\mathrm{d}^2y}{\mathrm{d}t^2} = B\frac{\partial B}{\partial y}, \quad x'^2 + y'^2 = B^2. \tag{2}$$

It is easily seen that any first analytic integral of (2) can be written in the form

$$\tilde{r} = u(x,y,y')x' + v(x,y,y')$$

where $u(x,y,y')$ is an even analytic function of y', and $v(x,y,y')$ is an odd analytic function of y'.

In the following we shall consider real analytic functions in a neighborhood of the origin of the real Euclidean space R^2.

Let E be the set of all analytic functions $g(x,z)$, $x \in R$, $z \in R$. We denote by T the set of analytic functions $r(x,z)$ generated by the first integrals of system (2); namely, if

$$\tilde{r} = u(x,y,y')x' + v(x,y,y')$$

is a first integral of (2), then

$$r(x,z) = u(x,0,z) + v(x,0,z).$$

Let E/T be the quotient space of the equivalence classes: $q_1 \sim g_2$, $g_1 - g_2 \in T$. Let A be the set of the functions $\hat{f}(\xi,\phi)$

$$\hat{f}(\xi,\phi) = \frac{1}{2}\left[g(\eta,b)(1 + B(\eta,0)\cos\phi) - q(\eta,-b)\times\right.$$

$$\times (1 - B(\eta,0)\cos\theta)] - \frac{1}{2}\left[g(\xi,a)(1 + B(\xi,0)\cos\phi)\right.$$

$$\left. - g(\xi,-a)(1 - B(\xi,0))\cos\phi\right],$$

where η, b, a, θ are the defined above functions, $q \in E$. Denote such a representation of the functions $f(\xi,\phi)$ by Ag and write $\hat{f} = Ag$.

We denote by Dg, $g \in E$, the analytic function $f(x,y)$ of the variables (x,y) defined as follows: we consider the system of the equations

$$\frac{\partial u_k}{\partial y} = -\frac{\partial v_k}{\partial x} - 2(k+1)B\frac{\partial B}{\partial y}u_{k+1}$$

$$\frac{\partial v_k}{\partial y} = \frac{\partial u_k}{\partial x} - u_{k+1}B\frac{\partial B}{\partial x} - \frac{\partial u_{k+1}}{\partial x}B^2 - (2k+3)v_k\frac{\partial B}{\partial y}$$

with the Cauchy data

$$u_k|_{y=0} = u_k^0(x), \quad v_k|_{y=0} = v_k^0(x)$$

such that

$$\sum_{k=0}^{\infty}\left(u_k^0 z^{2k} + v_k^0 z^{2k+1}\right) = g(x,z), \quad g \in E.$$

If this problem has an analytic solution u_k, v_k, $k = 1,2,\ldots$, then by the definition

$$Dg = u_0\frac{\partial B}{\partial x} + v_0\frac{\partial B}{\partial y} + \frac{\partial u_0}{\partial x}B.$$

Theorem 13. *(Anikonov 1978a) A necessary and sufficient condition for the solvability of the integral geometry problem in the class of analytic functions $f(x,y)$ is that the function*

$$\hat{f}(\xi,\phi) = \int\limits_{\gamma(\xi,\phi)} f(x,y)\sqrt{dx^2 + dy^2}$$

can be represented in the form $\hat{f} = Ag$. *Moreover, for any function* $g(x, z)$ *the operation* Dg *is defined, and*

$$Ag = \int_{\gamma(\xi,\phi)} Dg\sqrt{\mathrm{d}\,x^2 + \mathrm{d}\,y^2}$$

2. $Ag_1 = Ag_2$ *if and only if the equivalence classes of* E/T *corresponding to the functions* g_1 *and* g_2 *coincide.*

3. *If*

$$\int_{\gamma} f_1\sqrt{\mathrm{d}\,x^2 + \mathrm{d}\,y^2} = \int_{\gamma} f_2\sqrt{\mathrm{d}\,x^2 + \mathrm{d}\,y^2}$$

if f_i, $i = 1, 2$, *are analytic functions, then* $f_1(x, y) = f_2(x, y)$.

13. Integral Geometry and the Structure of Riemannian Spaces

Let M be an n-dimensional $n \geq 2$ simply-connected compact Riemannian manifold of the class C^∞ with a metric tensor $g = (g_{ij})$ and a strictly convex boundary ∂M.

We introduce the following notation: R_{ijkl} is the curvature tensor of the metric g, T_x is the tangent space at x, ΩM is the tangent bundle of unit vectors, $\Omega M = \{(x, \xi) : x \in M,\ \xi \in T_x\}$,

$$|\xi| = \sqrt{g_{ij}(x)\xi^i\xi^j}$$

(here and below summation from 1 to n over repeated subscripts and superscripts is understood),

$$\Omega\partial M = \{(x, \xi) : x \in \partial M,\ \xi \in T_x,\ |\xi| = 1,\ (\xi, v(\xi)) \leq 0\},$$

where $v(x)$ is the outward normal to ∂M at the point x, and $\gamma_{x,\xi}(t)$ is the geodesic defined by the initial data $\gamma_{x,\xi}(0) = x$, $\dot{\gamma}_{x,\xi}(0) = \xi$.

The set of the smooth (class C^∞) covariant symmetric tensor fields of valency m, $m \geq j$ on M and ∂M is denoted by $S_m M$ and $S_m\partial M$ respectively. Define the operation of symmetric covariant differentiation $\mathrm{d}\,S_m M \to S_{m+j}M$, $\mathrm{d} = \sigma\nabla$, where σ is the symmetric operator and ∇ is the operator of the covariant differentiation with respect to the metric g. Henceforth we shall assume that M is the dispersing manifold: any geodesic of it leaves a compactum $K \subset M$, $K \bigcap \partial M = \emptyset$. We define the function $t^0(x, \xi)$, $(x, \xi) \in \Omega^- M$ as the length of the geodesic $\gamma_{x,\xi}$, $(x, \xi) \in \Omega^- M$, and call it the hodograph of the metric g. By the assumptions we have made about M, the hodograph $t^0(x, \xi)$ is well-defined on $\Omega^- M$.

Suppose that $f = f_{i_1\ldots i_m}$ and $f \in S_m M$.

Consider the integrals along the geodesic joining points of ∂M

$$\hat{f}(x, \xi) = \int_0^{t^0(x,\xi)} f_{i_1\ldots i_m}(\gamma_{x,\xi}(t))\dot{\gamma}^{i_1}_{x,\xi}\cdots\dot{\gamma}^{i_m}_{x,\xi}\mathrm{d}\,t \tag{3}$$

Theorem 14. *(Pestov and Sharafutdinov, 1987). Let* M *be a compact dispersing manifold with a strictly convex boundary* ∂M, *and suppose that the sectional curvatures of* M *are nonpositive, that is at each point* $x \in M$, $R_{ijkl}\xi^i\eta^j\xi^k\eta^l \leq 0$ *for any* $\xi, \eta \in T_x$. *Then the integral (3) vanishes when* $(x, \xi) \in \Omega^-\partial M$ *if and only if*

$$f = \mathrm{d}\,v, \quad v \in S_{m-1}M, \quad v|_{\partial M} = 0, \quad m \geq 1$$

$$f = 0, \quad m = 0.$$

It turns out that such theorems of the integral geometry enable us to determine the structure of the Riemannian space from the integral information.

Theorem 15. *(Anikonov and Pestov, 1989) Suppose that the conditions of theorem 14 are satisfied. Then the equations of geodesics admit a first integral*

$$J(\gamma_{x,\xi}(t), \dot{\gamma}_{x,\xi}(t)) = u_{i_1 \dots i_m} \dot{\gamma}^{i_1}_{x,\xi} \dots \dot{\gamma}^{i_m}_{x,\xi},$$

where $u = (u_1, u_2, \dots, u_m) \in S_m M$ *if and only if there is a field* $u^0 \in S_m \partial M$ *such that*

$$u^0_{i_1 \dots i_m}(x)\xi^{i_1} \dots \xi^{i_m} = u^0_{i_1 \dots i_m}\left(\gamma_{x,\xi}(t^0(x,\xi))\right) \times$$

$$\times \dot{\gamma}^{i_1}_{x,\xi}(t^0(x,\xi)) \dots \dot{\gamma}^{i_m}_{x,\xi}(t^0(x,\xi)), \quad (x,\xi) \in \Omega\partial M.$$

14. Some Applications

We consider a multi-dimensional inverse kinematic problem and obtain necessary and sufficient conditions in order that some differential relations for unknown functions are held.

The proof is based on the theorems of the integral geometry. One formulation of the multi-dimensional inverse kinematic problem consists in the following: in the ball $\omega = \{x : |x| \leq 1\}$ of the real Euclidean space of variables $x = (x_1, \dots, x_n)$, $n \geq 2$ we consider a 3 times continuously differentiable metric

$$ds^2 = \lambda^2(x)\,|dx|^2, \tag{4}$$

such that any two points $p \in \omega$, $q \in \omega$ can be joined by a unique geodesic $\gamma(p,q)$ of the metric (4); the function

$$W(p,q) = \int_{\gamma(p,q)} \lambda(x)\,|dx|, \quad |p| = 1, \quad |q| = 1,$$

assumed to be known, and it is required to find $\lambda(x)$ for $|x| \leq 1$.

This problem has an important physical interpretation in the seismology. The function $\lambda(x) = 1/v(x)$ is the inverse to the propagation speed of a perturbation. The function $W(p,q)$ is the time of this propagation along the geodesic $\gamma(p,q)$. The formulated problem consists in determining the velocity $v(x)$ on the basis of the times $W(p,q)$, $|p| = 1$, $|q| = 1$.

Some result on the uniqueness, stability and existence of the multi-dimensional kinematic problems solutions and of the integral geometry closely related to the inverse kinematic problem are contained in the papers of Anikonov (1978), Romanov (1987). The constructive techniques for solving these problems are particularly interesting in applications. Here, necessary and sufficient conditions are presented in order that the differential relations for the known function $W(p,q)$, $|p| = 1$, $|q| = 1$ and the unknown function $\lambda(x)$, $|x| \leq 1$ are simultaneously satisfied. We proceed to formulate the results. Suppose that the components $u_k(x)$, $k = 1, 2, \dots, n$ of the vector-valued function $u(x)$ satisfy the system of equations

$$\frac{\partial u_k}{\partial x_j} = -\frac{\partial u_j}{\partial x_k}, \quad k \neq j, \quad \frac{\partial u_k}{\partial x_k} = \frac{\partial u_j}{\partial x_j}, \tag{5}$$

In the case $n = 2$ this is the system of the Cauchy-Riemann equations; for $n > 2$ its general solution is given by

$$u(x) = Ax - B|x|^2 + 2x(B,x) + c$$

where A is a constant matrix with the elements A_{jk} such that $A_{jk} = -A_{kj}$, $k \neq j$, $A_{jj} = A_0$, B and C are the constant vectors, and (B,x) is the inner product. There is a considerable difference between the classes of vector-valued functions $u(x)$ for $n = 2$ and $n > 2$.

Let $u_k(x) = (u_{k_1}, \dots, u_{k_n})$ and $v_k(x) = (v_{k_1}, \dots, v_{k_n})$ be vector-valued functions satisfying (5) for any k. We set

$$a_{ij} = \frac{1}{2}\sum_{k=1}^{n}(u_{k_j}v_{k_i} + u_{k_i}v_{k_j})$$

and let

$$Q = \sum_{i,j=1}^{n} a_{ij}x_i'x_j'$$

Along any geodesic γ we have

$$\frac{dQ}{dt} = \sum_{i=1}^{n}\left(\sum_{k=1}^{n} v_{ki}\tilde{L}_k^u\lambda + u_{ki}\tilde{L}_k^v\lambda\right)x_i', \quad |x'| = \lambda$$

where

$$\tilde{L}_k^u\lambda = \lambda(x)(\operatorname{grad}\lambda, u_k) + \frac{\lambda^2}{n}\sum_{i=1}^{n}\frac{\partial u_{ki}}{\partial x_i}$$

$$\tilde{L}_k^v\lambda = \lambda(x)(\operatorname{grad}\lambda, v_k) + \frac{\lambda^2}{n}\sum_{i=1}^{n}\frac{\partial v_{ki}}{\partial x_i}$$

Consider the differential expression

$$Lw = \int_{k=1}^{n}\left[\left(u_k(q), \frac{\partial w}{\partial q}\right)\left(v_k(q), \frac{\partial w}{\partial q}\right) - \left(u_k(p), \frac{\partial w}{\partial p}\right)\left(v_k(p), \frac{\partial w}{\partial p}\right)\right]$$

where

$$w(p,q) = \int_{\gamma(p,q)} \lambda(x)\,|dx|, \quad \frac{\partial w}{\partial q} = \left(\frac{\partial w}{\partial q_1}, \dots, \frac{\partial w}{\partial q_n}\right), \quad \frac{\partial w}{\partial p} = \left(\frac{\partial w}{\partial p_1}, \dots, \frac{\partial w}{\partial p_n}\right),$$

and $u_k(x)$ and $v_k(x)$ are vector-valued functions satisfying (5).

Theorem 16. *(Anikonov and Pestov, 1990) In order that*

$$Lw = \phi(q) - \phi(p), \quad |q| = 1, \quad |p| = 1$$

for some continuously differentiable function $\phi(x)$, $|x| \leq 1$ it is necessary and sufficient that

$$\frac{\partial \phi}{\partial x_i} = \sum_{k=1}^{n}\left(v_{ki}\tilde{L}_k^u\lambda + u_{ki}\tilde{L}_k^v\lambda\right), \quad i = 1, 2, \dots, n.$$

To formulate another result in the inverse kinematic problem we suppose that $u(x)$ satisfies (5) and has no more restrictions. Consider the differential operators Lw, $\tilde{L}\lambda$ such that

$$Lw = \left(\frac{\partial w}{\partial q}, u(q)\right) + \left(\frac{\partial w}{\partial p}, u(p)\right) + \beta w,$$

$$\tilde{L}\lambda = \left(\frac{\partial \lambda}{\partial x}, u(x)\right) + \frac{1}{n} \operatorname{sp} \frac{\partial u}{\partial x} + \beta\lambda,$$

where β is a constant, sp $\partial u/\partial x$ is the trace of the matrix $\partial u_k/\partial x_j$, $k, j = 1, 2, \ldots, n$.

Now define M setting Mw be equal to the following determinant of the order $n+2$

$$Mw = \begin{vmatrix} 0 & 0 & \frac{\partial Lw}{\partial q_1} & & \frac{\partial Lw}{\partial q_n} \\ 0 & 0 & \frac{q_1}{|q|} & & \frac{q_n}{|q|} \\ \frac{\partial Lw}{\partial p_1} & \frac{p_1}{|p|} & & & \\ & & & \frac{\partial^2 Lw}{\partial p \partial q} & \\ \frac{\partial Lw}{\partial p_n} & \frac{p_n}{|p|} & & & \end{vmatrix}$$

where $\partial^2 Lw/\partial p \partial q$ is a matrix of the order n.

Let $\Gamma(t)$ denote the gamma function, $\mathrm{d}\,p$, $\mathrm{d}\,q$ denote the surface area elements of the sphere $\partial\omega$ at the points p and q respectively, and let

$$\|F\| = \sup |F(q)|, \quad q \in D$$

Theorem 17. *(Anikonov, 1978). The following inequalities hold*

$$\left\| \frac{\tilde{L}\lambda}{\lambda} \right\|_{C(\omega)} \geq \left\| \frac{Lw}{w} \right\|_{C(\partial\omega \times \partial\omega)},$$

$$\int\limits_{\omega} \lambda^{n-2} \left(\tilde{L}\lambda\right)^2 \mathrm{d}\,x \leq \frac{\Gamma(n/2)}{(n-1)2\pi^{n/2}} \int\limits_{\partial\omega} \int\limits_{\partial\omega} Mw \,\mathrm{d}\,p\,\mathrm{d}\,q.$$

From theorem 17 we obtain the table

$\bar{u} = (0,0,1) = k$	$v = F(x_1, x_2)$
$\bar{u} = [k, x]$	$v = F(x_1^2 + x_2^2, x_3)$
$\bar{u} = x$	$v = \|x\| F\left(\frac{x}{\|x\|}\right)$
$\bar{u} = a_0 x + [k, x]$	$v = x_3 F\left(\frac{\sqrt{x_1^2 + x_2^2}}{x_3}, \sqrt{x_1^2 + x_2^2}\, \mathrm{e}^{-a_0 \operatorname{arctg}(x_2/x_1)}\right)$
$\bar{u} = \alpha k + [k, x]$	$v = F\left(\sqrt{x_1^2 + x_2^2}, x_3 - \alpha \operatorname{arctg} \frac{x_2}{x_1}\right)$
$\bar{u} = -k \|x\|^2 + 2x(k, x)$	$v = \|x\|^2 F\left(\frac{x_1}{\|x\|^2}, \frac{x_2}{\|x\|^2}\right)$
$\bar{u} = \alpha_0 - k \|x\|^2 + 2x(k, x)$	$v = \sqrt{x_1^2 + x_2^2} F\left(\frac{x_2}{x_1}, \frac{\alpha_0 x_3 + \|x\|^2}{\sqrt{x_1^2 + x_2^2}}\right)$
$\bar{u} = -k \|x\|^2 + 2x(k, x) + \alpha k$	$v = \sqrt{x_1^2 + x_2^2} F\left(\frac{x_2}{x_1}, \frac{\|x\|^2 + \alpha}{\sqrt{x_1^2 + x_2^2}}\right)$

where $\bar{u} = \bar{u}(x)$, $n = 3$, $v(x) = 1/\lambda(x)$, F is arbitrary function.

15. One Inverse Kinematic Problem

In this section we consider the certain inverse kinematic problem.

Let R^n be the n-dimensional Euclidean space of the variables $x = (x_1, x_2, \dots, x_n)$, $n = 2, 3, \dots$, and Q be the ball $|x| \le R$, with the boundary ∂Q. Let B be an m-dimensional differentiable manifold imbedded in the ball Q with the local coordinates $u = (u_1, u_2, \dots, u_m)$ and the corresponding local parametrization $x = x(u)$, $|u| \le 1$, $m \le n$.

Consider the Riemannian metric

$$\mathrm{d}\, s^2 = \lambda^2(x)(\mathrm{d}\, x_1^2 + \ldots + \mathrm{d}\, x_n^2) = \lambda^2(x)\, |\mathrm{d}\, x|^2$$

on the ball Q and let p denote a point of the sphere ∂Q. $\mathrm{d}\, s^2$ induces a Riemannian metric

$$\mathrm{d}\, \bar{s}^2 = \sum_{i=1}^{m} \sum_{j=1}^{m} \lambda_{ij}(u) \mathrm{d}\, u_i \mathrm{d}\, u_j$$

on the manifold B. Here we set

$$\mathrm{d}\, x_i = \sum_{j=1}^{m} \frac{\partial x_i}{\partial u_j} \mathrm{d}\, u_j$$

in the equality $\mathrm{d}\, s^2 = \lambda^2\, |\mathrm{d}\, x|^2$

If some integral information linked with the manifold B is given on the surface of the sphere ∂Q, then the problem of investigating this manifold B with induced metric

$$\mathrm{d}\, \bar{s}^2 = \sum_{i=1}^{m} \sum_{j=1}^{m} \lambda_{ij}(u) \mathrm{d}\, u_i \mathrm{d}\, u_j$$

is of interest for geophysics. Such an information may be given as a function

$$\int_{\gamma(x(u),p)} \lambda\, |x| + a(x(u)) = g(u, p)$$

of the variables u and p, where $\gamma(x(u), p)$ is a geodesic of the metric $\mathrm{d}\, s^2 = \lambda^2\, |\mathrm{d}\, x|^2$ joining the points $x(u) \in B$ and $p \in \partial Q$. Here the functions $a = a(x)$ and $\lambda = \lambda(x)$, as well as the parametrization $x = x(u)$ are unknown a priori. The given function $g(u, p)$ may, in general, be not even continuous.

Before we formulate the results of this problem, let us give a physical interpretation of the terms above. The ball Q is the Earth, the manifold $B \subset Q$ is the set of earthquake hypocenters, $1/\lambda(x) = v(x)$ is the speed of the solid waves propagation in the Earth, $a(x)$ is the starting time of an earthquake at the hypocenter $x = x(u)$. The given function $g(u, p)$ is the "indexed" times of the arrival of the disturbance on the surface of the Earth, registered by seismometers, where u is the "index" of the earthquake, and p is the "index" of the seismometer.

Theorem 18. *Let the function $\lambda(x)$ be twice differentiable and suppose that through the every pair of points x_1 and x_2, in the ball Q there passes a unique geodesic $\gamma(x_1, x_2)$ of the metric $\mathrm{d}\, s^2 = \lambda^2\, |x|^2$.*

Then the function $R(u,p_1,p_2)=g(u,p_1)-g(u,p_2)$ *of the variables* (u,p_1,p_2), $|u|\le 1$, $p_i\in\partial Q$, *is continuously differentiable and the following formula holds*

$$\mathrm{d}\,\bar{s}^2=\sum_{i=1}^{m}\sum_{j=1}^{m}\lambda_{ij}(u)\mathrm{d}\,u_i\mathrm{d}\,u_j=\frac{1}{4}\left[\max_{p_1,p_2}\sum_{j=1}^{m}\frac{\partial R}{\partial u_j}\right]^2$$

From theorem 18 it is clear that the intrinsic geometry of the Riemannian manifold B is determined by the function $g(u,p)$. This leads to corollaries of an interest for applications.

Corollary 1. *Let the dimension of the space* R^n *by three, and* B *be a closed convex level surface of a function* $\lambda=\lambda(x)$ *that is,* $B=\{\lambda(x)=c\}$. *Then the surface* B *is uniquely determined by the function* $g(u,p)$ *to within the similarity and position in the ball* Q.

Corollary 2. *Let* $m=n$ *and let* B *be a simply-connected region with a smooth boundary; then both the function* $\lambda=\lambda(x)$, $x\in B$, *and the region* B *are uniquely determined by the function* $g(u,p)$ *to within the conformal transformations of the region* B.

Given additional hypotheses on B and the function $a(x)$ we may somewhat strengthen theorem 18, namely

Theorem 19. *Let* B *be a simply-connected* n*-dimensional region with a smooth boundary and let the function* $a(x)=0$, $x\in B$.

Then the elements of the matrix $c_{ij}(u)$ *inverse to* $\lambda_{ij}(u)$ *satisfy the relationship*

$$\sum_{i=1}^{n}\sum_{j=1}^{n}\frac{\partial g}{\partial u_i}\frac{\partial g}{\partial u_j}c_{ij}(u)=1$$

for all $p\in\partial Q$ *and, therefore, are uniquely determined by the set* $g^k(u)=g(u,p_k)$ *of values of the function* $g(u,p)$ *at the points* $p=p_k=\partial Q$ *where the determinant* $(\partial g^k/\partial u_i)(\partial g^k/\partial u_j)$ *is nonzero,* $k=1,2,\ldots,n(n+1)$, $i,j=1,2,\ldots,n$. *Here the function* $\lambda(x)$, $x\in B$ *and the region* B *are unique to within the conformal maps.*

REFERENCES

Anikonov Yu.E. (1969a), *On Certain Problems of Calculus of Variations and Integral Geometry*, Preprint, Computer Center, Novosibirsk.

Anikonov Yu.E. (1969b), Inversion formulae for problems of kinematic seismic and integral geometry, In: *Mathematical Problems of Geophysics* no. 1, Computer Center, SO AN SSSR, Novosibirsk, 41–48.

Anikonov Yu.E. (1978a), On solvability of problem in integral geometry, *Math. VSSR Sb.*, **30**, 239–247.

Anikonov Yu.E. (1978b), *Some Methods for Study of Multi-Dimensional Inverse Problems for Differential Equations*, Nauka, Navosibirsk.

Anikonov Yu.E. (1982), On problems of integral geometry, In: *Questions of Correctness of Inverse Problems in Mathematical Problems*, Computer Center, SO AN SSSR, Novosibirsk, 20–29.

Anikonov Yu.E. (1983), On uniqueness classes of solution for problems of integral geometry, In: *Methods of Investigation of Incorrect Problems of Mathematical Physics*, Computer Center, SO AN SSSR, Novosibirsk, 16–17.

Anikonov Yu.E. and Lavrent'ev M. M. (1967), A certain class of problems in integral geometry, *Dokl. Akad. Nauk SSSR*, **176**, 1240–1241.

Anikonov Yu.E. and Pestov L. N. (1990), Integral geometry and structure of Riemannian spaces, *Sov. Math. Dokl.*, **40**, 90–93.

Anikonov Yu.E. and Stepanov V. N. (1991), Inversion formula in one problems of integral geometry, *Dokl. Akad. Nauk SSSR*, **316**, 93–95.

Anikonov Yu.E. and Romanov V. G. (1983), On uniqueness form of degree one form its integral along geodesics, In: *Incorrect Mathematical Problems and Problems of Geophysics*, Computer Center, SO AN SSSR, Novosibirsk, 22–27.

Anikonov Yu.E. and Shneiberg I. Yu. (1991), Radon transformation with variable absorption, *Dokl. Akad. Nauk SSSR*, **318**, 265–267.

Blagoveshchenskii A.S. (1986), Reconstruction of a function from known its integral taken along linear manifolds, *Math. Notes*, **39**, 457–461.

Blaschke W. (1916), *Kreis und Kugel*, Veit, Leipsig.

Bukhgeim A.L. (1972), On some problems of integral geometry, *Siberian Math. J.*, **13**, 23–29.

Bukhgeim A.L. (1979), On the solvability of a certain problems of integral geometry, In: *Incorrect Mathematical Problems and Problems of Geophysics*, Computer Center, SO AN SSSR, Novosibirsk, 37–46.

Bukhgeim A.L. (1983), *Volterra Equations and Inverse Problems*, Nauka, Novosibirsk.

Finch D. V. (1985), Cone-beam reconstruction with sources on curve, *SIAM J. Appl. Math.*, **45**, 665–673.

Funk P. (1916), Über eine geometrische Anwendung and Abelschen Integralgleiching, *Math. Ann.*, **77**, 125–129.

Gelfand I. M. and Goncharov A. B. (1987), Recovery of a compactly supported function by its integrals over lines intersecting given set of points in space, *Sov. Math. Dokl.*, **34**, 373–376.

Gelfand I. M., Gindikin S. G., and Graev M. I. (1980), Integral geometry in affine and projective spaces, In: *Current Problems in Mathematics*, **16**, Vsesojuz Inst. Nauch i Tehc. Informacii, Moscow, 53–226.

Gelfand I. M., Graev M. I., and Vilenkin N. Ja. (1966), Generalized functions, In: *Integral Geometry and Representation Theory*, **5**, Academic Press, New-York–London.

Gelfand I. M., Graev M. I., and Shapiro Z. Ja. (1969), Differential forms and integral geometry, *Functional Anal. i Prilozhen*, **3**, 24–40.

Goncharov A. B. (1988), Methods of integral geometry and recovering function with compact support from its projections in unknown directions, *Acta Applicande Mathematicae*, **11**, 213–222.

Helgason S. (1980), *The Radon Transform*, Birkhaser Verlag, Boston.

John F. (1955), *Plane Waves and Spherical Means Applied to Partial Differential Equations*, Interscience Publishers, New-York–London.

Kirillov A. A. (1961), On a problem of I .M. Gelfand, *Sov. Math. Dokl.*, **2**, 268–269.

Kostelyanets P. O. and Reshetnyak Yu. G. (1954), Determination of completely additive function by its values on half-spaces, *Uspekhi Mat. Nauk*, **9**, 135–140.

Kuchment P. (1989), *Paley-Wiener Theory for Exponential Radon Transform on the Plane*, Lecture on the Joint Summer Research conference on Integral Geometry, Arcafa (CAL), USA.

Lavrent'ev M. M. and Bukhgeim A. L. (1973), On a class of problems of integral geometry, *Soviet Math. Dokl.*, **14**, 750–760.

Lavrent'ev M. M. and Romanov V. G. (1966), Three linearized inverse problems for hyperbolic equation, *Soviet Math. Dokl.*, **7**, 1650–1652.

Lavrent'ev M. M., Romanov V. G., and Shishatskii S. G. (1986), *Ill-Posed Problems in Mathematical Physics and Analysis Translations of Mathematical Monographs, 64*, American Mathematical Society, Providence R. I.

Lavrent'ev M. M., Romanov V. G., and Vasiliev V. G. (1970), Multi-dimensional inverse problems for differential equations, In: *Lecture Notes in Mathematics*, no. 167, Springer-Verlag, Berlin–New-York.

Leahy J. V., Smith K.T., and Solmon D. C. (1979), *Uniqueness, Nonuniqueness and Inversion in the X-ray and Radon Problems*, Proc. International Symposium on Ill-Posed Problems, Newark, D. E.

Logvinenko V. N. (1988), Multi-dimensional analysis theorem of Kaztrait and Planjerely-Polia and their applications, In: *Complex Analysis*, Krasnoyarsk, 90–98.

Mukhometov R. G. (1977), The problem of reconstructing a two-dimensional Riemannian method and integral geometry, *Soviet Math. Dokl.*, **18**, 27–31.

Natterer F. (1986), *The Mathematics of Computerized Tomography*, Teubner and John Willy sons, Stuttgart.

Palamodov V. P. and Denisjuk A. S. (1988), Inversion de la transformation de Radon d'apres des donnes incompletes, *C. R. Acad. Sci. Paris*, **307**, 181–183.

Pestov L. N. (1985), The problem of integral geometry for a symmetric m-form and its applications in kinematics, In: *Inverse Problems of Mathematical Physics*, Computer Center, SO AN SSSR, Novosibirsk, 105–119.

Pestov L. N. and Sharafutdinov V. A. (1987), Integral geometry of tensor field on manifold of negative curvature, *Soviet Math. Dokl.*, **295**, no. 6, 305–307.

Radon J. (1917), Über die Bestimmiung von Functionen durch ihre Integral werse längs gewisser Mannigfaltigkeiten, *Math-Nat*, **69**, 262–277.

Romanov V. G. (1967), Function reconstructing by means of integral along family of curves, *Siberian Math. J.*, **8**, 1206–1208.

Romanov V. G. (1978), Integral geometry on geodesics of isotropic Riemannian metric, *Soviet Math. Dokl.*, **19**, 847–851.

Romanov V. G. (1987), *Inverse Problem of Mathematical Physics*, VNU Science Press, Utrecht.

Semyanistyi V. I. (1960), On some integral transformations in Euclidean space, *Dokl. Akad. Nauk SSSR*, **134**, 536–539.

Semyanistyi V. I. (1961), Homogeneous functions and some problems of integral geometry in space of constant curvature, *Soviet Math. Dokl.*, **2**, 59–62.

Semyanistyi V. I. (1966), Certain problems of integral geometry in Euclidean and non-Euclidean spaces, *Trudy Sem. Vektor Tensor. Anal.*, **13**, 244–302.

Sharafutdinov V. A. (1989), Radial transformation of symmetric tensor fields, *Soviet Math. Dokl.*, **39**, 82–85.

Solman D. C. (1976), X-ray transformation, *J. Math. Anal. Appl.*, **56**, 61–83.

Tikhonov A. N., Arsenin V. Ya., and Timonov A. A. (1987), *Mathematical Problems in Computerized Tomography*, Nauka, Moscow.

Tuy H. K. (1983), An inverse formula for cone-beam reconstruction, *SIAM J. Appl. Math.*, **43**, 546–552.

Inversion Formulae for One Class of Multidimensional Integral Volterra Equations of the First Kind

A. APARTSYN

Siberian Energy Institute SB RAS
Irkutsk, Russia

ABSTRACT

The paper presents inversion formulae for multidimensional integral Volterra equations of the first kind appearing in mathematical modelling of nonlinear dynamic systems by the Volterra series.

In the recent years the area of application of computed tomography ideas and methods has substantially been extended. In a number of fields of applied mathematics, some traditional approaches have been interpreted tomographically and enriched by new more efficient algorithms and programs. In this paper we consider such an approach typical for the identification theory of dynamic objects for which active test experiments can be conducted.

1. As is known, the Volterra series are widely used in modelling nonlinear dynamic systems, which is based on the Frecher theorem, continual analog of the Wierstrass theorem on approximation of continuous functions by polynomials.

Let scalar time functions $x(t)$ and $y(t)$ be the input and output signals of a stationary dynamic system, respectively. Then the continuous dependence of $y(t)$ on $x(t)$ is represented as the integro-power Volterra series

$$y(t) = \sum_{m=1}^{\infty} \int_0^t \cdots \int_0^t K_m(s_1, \ldots, s_m) \prod_{k=1}^{m} x(t - s_k) \mathrm{d}s_k, \quad 0 \le t \le T, \tag{1}$$

and a partial sum of (1), with a suitable choice of the number of terms q, can be used to construct high speed simulation models, training simulators, etc. The problem of identifying the Volterra kernels K_m, $m = 1, \ldots, q$, symmetrical in all their arguments (transient characteristics of the system) should be solved in advance. It is the problem on which we focus our attention.

In the linear case when series (1) consists of one term

$$y(t) = \int_0^t K_1(s) x(t - s) \mathrm{d}s, \quad 0 \le t \le T, \tag{2}$$

one pair of the input-output is enough to identify $K_1(t)$. For example, if

$$x(t) = e(t), \quad t \in [0, T], \tag{3}$$

where $e(t)$ is the Heaviside function, then

$$K_1(t) = y(t), \quad t \in [0,T]. \tag{4}$$

To generalize (4) to the m - dimensional case, we introduce a $(m-1)$ - parametric family of the test signals

$$x_{\omega_1 \cdots \omega_{m-1}}(t) = e(t) + 2\sum_{k=1}^{m-2}(-1)^k e(t - \sum_{j=1}^{k}\omega_j)$$

$$+(-1)^{m-1}e(t - \sum_{j=1}^{m-1}\omega_j), \quad t \in [0,T], \quad 0 \le \omega_j \le T, \quad j = 1,\ldots,m-1. \tag{5}$$

Then the problem of identifying the kernel $K_m(s_1,\ldots,s_m)$, due to its symmetry, is reduced to the linear m -dimensional integral Volterra equation of the first kind

$$VK_m \equiv \sum_{j_1+\cdots+j_{m-1}=m} (-1)^{\sum_{k=1}^{[(m-1)/2]} j_{2k}} \frac{m!}{j_1!\ldots j_{m-1}!} V_{j_1\cdots j_{m-1}} K_m = f_m, \tag{6}$$

where the integral operator $V_{j_1\cdots j_{m-1}}$ includes j_1 - times integration of K_m from $t-(\omega_1 + \cdots + \omega_{m-1})$ to $t-(\omega_1+\cdots+\omega_{m-2})$, and $f_m \equiv f(t,\omega_1,\ldots,\omega_{m-1})$ is a linear combination of responses $y_{\omega_1\ldots\omega_{m-1}}(t)$ of the dynamic system to the signals of type (5). In computed tomography the right-hand side of (6) plays a role of projection data.

The cases $m=2$ and $m=3$ are of particular interest for applications. From (6) it follows that

$$\int_{t-\omega_1}^{t}\int_{t-\omega_1}^{t} K_2(s_1,s_2)\mathrm{d}s_1\mathrm{d}s_2 = f_2(t,\omega_1), \quad 0 \le \omega_1 \le t \le T, \quad \text{for } m=2, \tag{7}$$

and

$$(V_{30} - 3V_{21} + 3V_{12} - V_{03})K_3 = f_3(t,\omega_1,\omega_2) \quad \text{for } m=3, \tag{8}$$

where

$$V_{30}K_3 = \int_{t-\omega_1}^{t}\int_{t-\omega_1}^{t}\int_{t-\omega_1}^{t} K_3(s_1,s_2,s_3)\mathrm{d}s_1\mathrm{d}s_2\mathrm{d}s_3; \tag{9}$$

$$V_{21}K_3 = \int_{t-\omega_1}^{t}\int_{t-\omega_1}^{t}\int_{t-\omega_1-\omega_2}^{t-\omega_1} K_3(s_1,s_2,s_3)\mathrm{d}s_1\mathrm{d}s_2\mathrm{d}s_3; \tag{10}$$

$$V_{12}K_3 = \int_{t-\omega_1}^{t}\int_{t-\omega_1-\omega_2}^{t-\omega_1}\int_{t-\omega_1-\omega_2}^{t-\omega_1} K_3(s_1,s_2,s_3)\mathrm{d}s_1\mathrm{d}s_2\mathrm{d}s_3; \tag{11}$$

$$V_{03}K_3 = \int_{t-\omega_1-\omega_2}^{t-\omega_1}\int_{t-\omega_1-\omega_2}^{t-\omega_1}\int_{t-\omega_1-\omega_2}^{t-\omega_1} K_3(s_1,s_2,s_3)\mathrm{d}s_1\mathrm{d}s_2\mathrm{d}s_3; \tag{12}$$

$$0 \le \omega_1 + \omega_2 \le t \le T; \quad \omega_1,\omega_2 \ge 0.$$

If f_m in (6) belongs to the range of values of the operator V, then the solution of (6) is unique in the class of continuous functions symmetrical in all arguments and it is determined by the recurrence relation

$$K_m(t, t-\omega_1, \ldots t-\omega_1-\ldots-\omega_{m-1}) = \mathrm{D}_m f_m, \tag{13}$$

$$\mathrm{D}_m = \frac{(-1)^{m-1}}{2m} \frac{\partial}{\partial \omega_{m-1}} \left[\hat{\mathrm{D}}_{m-1} - \mathrm{D}_{m-1}\right], \quad m = 3, 4, \dots; \tag{14}$$

$$\hat{\mathrm{D}}_m = \frac{(-1)^{m-1}}{2m} \frac{\partial}{\partial \omega_{m-1}} \left[\hat{\mathrm{D}}_{m-1} - \mathrm{D}_{m-1}\right], \tag{15}$$

$$\mathrm{D}_1 = (f'_{m_t} + f'_{m_{\omega_1}})/2, \quad \mathrm{D}_2 = \frac{\partial}{\partial \omega_1} \mathrm{D}_1, \quad \mathrm{D}_2 = \frac{\partial}{\partial \omega_2} \mathrm{D}_1. \tag{16}$$

In particular,

$$K_2(t, t-\omega_1) = \frac{1}{2} \frac{\partial}{\partial \omega_1} \left(\frac{\partial}{\partial t} + \frac{\partial}{\partial \omega_1} \right) f_2 \quad (m = 2); \tag{17}$$

$$K_3(t, t-\omega_1, t-\omega_1-\omega_2) = \frac{1}{12} \left(\frac{\partial^2}{\partial \omega_2{}^2} - \frac{\partial^2}{\partial \omega_1 \partial \omega_2} \right) \left(\frac{\partial}{\partial t} + \frac{\partial}{\partial \omega_1} \right) f_3 \quad (m = 3); \tag{18}$$

$$K_4(t, t-\omega_1, t-\omega_1-\omega_2, t-\omega_1-\omega_2-\omega_3) = \frac{1}{96} \left(\frac{\partial^3}{\partial \omega_2{}^2 \partial \omega_3} - \frac{\partial^3}{\partial \omega_1 \partial \omega_2 \partial \omega_3} \right.$$

$$\left. - \frac{\partial^3}{\partial \omega_2 \partial \omega_3^2} + \frac{\partial^3}{\partial \omega_1 \partial \omega_3^2} \right) \left(\frac{\partial}{\partial t} + \frac{\partial}{\partial \omega_1} \right) f_4 \quad (m = 4). \tag{19}$$

2. So far we have assumed that $x(t)$ is a scalar time function. The situation becomes more complicated if the input signal is a vector-function $x(t) = (x_1(t), \dots, x_n(t))$. In this case the truncated Volterra series which involves transient characteristics of the orders not exceeding q takes the form

$$y(t) = \sum_{\nu=1}^{q} \sum_{1 \le i_1 \le \dots \le i_\nu \le n} \int_0^t \int_0^t K_{i_1 \dots i_\nu}(s_1, \dots, s_\nu) x_{i_1}(t - s_1) \dots x_{i_\nu}(t - s_\nu) \mathrm{d}s_1 \dots \mathrm{d}s_\nu,$$

$$t \in [0, T]. \tag{20}$$

So the Volterra kernels characterizing the system sensitivity to simultaneous change in several components of $x(t)$ should be identified in addition. A special feature of the new identification problems is that these kernels may be non-symmetrical in some variables, which requires the corresponding increase in a number of responses to the test disturbances.

Let us illustrate it by an example of a quadratic model ($q = 2$) with two inputs $x_1(t)$ and $x_2(t)$. Denote by $y_{\omega_1 \cdot \omega_2}(t)$ a system response to the signal

$$x_1(t) = e(t) - e(t - \omega_1), x_2 = e(t) - e(t - \omega_2), \quad 0 \le \omega_1, \omega_2 \le T.$$

With regard to (20) it is not hard to see that

$$\int_{t-\omega}^{t} \left\{ \int_{t-\omega}^{t} K_{12}(s_1, s_2) \mathrm{d}s_1 \right\} \mathrm{d}s_2 = \overset{(1)}{f}(t, \omega) \equiv y_{\omega,T}(t) - y_{\omega,0}(t) - y_{0,T}(t), \tag{21}$$

$$\int_{t-\omega}^{t} \left\{ \int_{t-\omega}^{t} K_{12}(s_1, s_2) \mathrm{d}s_2 \right\} \mathrm{d}s_1 = \overset{(2)}{f}(t, \omega) \equiv y_{T,\omega}(t) - y_{0,\omega}(t) - y_{T,0}(t), \tag{22}$$

where

$$0 \le \omega \le t \le T.$$

If $\overset{(1)}{f}, \overset{(2)}{f}$, are chosen so that system (21), (22) admits a solution in the class of functions continuous on $\Omega_2 = \{s_1, s_2/0 \le s_1, s_2 \le T\}$, then this solution is unique in this class and determined by the formulae

$$K_{12}(t-\omega, t) = \overset{(1)}{f''}_{t\omega} + \overset{(1)}{f''}_{\omega^2}, \tag{23}$$

$$K_{12}(t, t-\omega) = \overset{(2)}{f''}_{t\omega} + \overset{(2)}{f''}_{\omega^2}, \tag{24}$$

where

$$0 \le \omega \le t \le T.$$

Though (23), (24) differ from (17) only in the multiplier 1/2, the function f_2 in (17) has another physical sense.

3. Let us proceed now to the numerical solution of equation (6). As is known (see, for example, Apartsyn, 1987) the straightforward discretization of integral Volterra equations of the first kind can generate a self-regularizing algorithm that produces a stable approximate solution provided the discretization step is coordinated with the error in initial data. The discretization applied to (6) leads to the numerical differentiation formulae. Their convergence and self-regularizing properties for the scalar case are studied in Apartsyn (1992). Let us consider some features of the vector case by the example of system (21),(22). Approximating the integrals in (21), (22), we come to a system of linear algebraic equations (SLAE) of the type

$$h^2 \sum_{l=1}^{i} \sum_{m=i-j+1}^{i} K^h_{l-1/2,m-1/2} = \overset{(1)}{f^h}_{i,j}, \tag{25}$$

$$h^2 \sum_{l=i-j+1}^{i} \sum_{m=1}^{i} K^h_{l-1/2,m-1/2} = \overset{(2)}{f^h}_{i,j}, \tag{26}$$

where $i = 1, \dots, N; j = 1, \dots, i; Nh = T; K^h_{l-1/2,m-1/2}$ is a mesh approximation of the kernel $K_{12}(s_1, s_2)$ at the point $((l-1/2)h, (m-1/2)h)$; $\overset{(1)}{f^h}_{i,j}, (\overset{(2)}{f^h}_{i,j})$ is a value of the function $\overset{(1)}{f}(t,\omega), (\overset{(2)}{f}(t,\omega))$ at the point (ih, jh).

Formally, SLAE (25), (26) is overdetermined since it consists of $N(N+1)$ equations and N^2 unknowns. However, as $\overset{(1)}{f}(t,t) = \overset{(2)}{f}(t,t)$ for every $t \in [0,T]$, we have to exclude N equations corresponding to the right-hand side of $\overset{(1)}{f}_{i,j}($ or $\overset{(2)}{f}_{i,j}), i = 1, \dots, N$. The unique solution of (25), (26) can easily be shown to be determined by the formulae

$$K^h_{i-j+1/2,i-1/2} = \frac{\overset{(1)}{f}{}^h_{i,j} - \overset{(1)}{f^h}_{i,j-1} - \overset{(1)}{f^h}_{i-1,j-1} + \overset{(1)}{f^h}_{i-1,j-2}}{h^2}, \quad i = 2, \dots, N, \quad j = 2, \dots, i, \tag{27}$$

$$K^h_{i-1/2,i-j+1/2} = \frac{\overset{(2)}{f^h}_{i,j} - \overset{(2)}{f^h}_{i,j-1} - \overset{(2)}{f^h}_{i-1,j-1} + \overset{(2)}{f^h}_{i-1,j-2}}{h^2}, \quad i = 2, \dots, N, \quad j = 2, \dots, i, \tag{28}$$

$$K^h_{i-1/2,i-1/2} = \frac{\overset{(1)}{f^h}_{i-1,j-1} + \overset{(1)}{f^h}_{i,1} + \overset{(1)}{f^h}_{i,1} - \overset{(1)}{f^h}_{i,i}}{h^2}, \quad i = 1, \dots, N. \tag{29}$$

Formulae (27), (28) are difference analogues of (25), (26) for the inner points of the domains $\Omega 1 = \{0 \leq s_1 \leq s_2 \leq T\}$ and $\Omega 2 = \{0 \leq s_2 \leq s_1 \leq T\}$, respectively, and use the standard four-point computational molecule providing the approximation accuracy of the order of h^2 .

It can be shown that equations (29) approximating the kernel K_{12} on the diagonal $s_1 = s_2$ and including the values of both $\overset{(1)}{f}$ and $\overset{(2)}{f}$ have also the local truncation error $O(h^2)$.

A complex of mathematical models for dynamics of thermal power plants is recently developed in the Siberian Energy Institute of Siberian Branch of RAS on the basis of the approach presented.

REFERENCES

Apartsyn A. S. (1987), Discretization methods for regularization of some integral equations of the first kind. In: *Methods of numerical analysis and optimization.* Novosibirsk: Nauka, 263–291. (in Russian)

Apartsyn A. S. (1992), *On some identification method for nonlinear dynamic systems.* On engineering mathematics and applications (ISEMA) - 92. Shenzhen, China, 288–291.

The Cauchy Problem for A–Harmonic Functions [1]

E. V. ARBUZOV and A. L. BUKHGEIM

Institute of Mathematics of the Siberian Branch of the Russian Academy of Science, Novosibirsk, Russia

ABSTRACT

In the article we consider a Cauchy problem for A-harmonic functions being equivalent to a Radon problem with incomplete data. The theorem of uniqueness, the criterion for solvability and the formula of Carleman type for a solution of this problem are proved.

1. Introduction

In the article by Bukhgeim (1991) it is shown that the problem of reconstructing a function $a(x)$ in some domain $\Omega \subset \mathbf{R}^2$ from integrals along all straight lines crossing some section S on the boundary of the domain (the Radon problem with incomplete data) turns out to be equivalent to the Cauchy problem for some operator Δ_A with Cauchy data given on S. Namely, the function $a(x)$ is found with the help of the formula

$$a(x) = 2\left((E+A)^{-1}\frac{\partial}{\partial x_1}f(x)\right)_1,$$

where $f(x) = (f_0(x), f_1(x), f_2(x), \ldots)$ is a function with values in the space l_2, $A = -U^*$ is the shift operator defined by the formula

$$U^* : (f_0, f_1, f_2, \ldots) \to (f_1, f_2, \ldots),$$

and the function $f(x)$ solves the following Cauchy problem:

$$\Delta_A f(x) \equiv (E-A)^2\frac{\partial^2}{\partial x_1^2}f(x) + (E+A)^2\frac{\partial^2}{\partial x_2^2}f(x) = 0, \quad x \in \Omega,$$

$$f(x)\mid_S = f_0(x),$$

$$\frac{\partial}{\partial \nu_A}f(x)\mid_S \equiv \left(\nu_1(x)(E-A)^2\frac{\partial}{\partial x_1}f(x) + \nu_2(x)(E+A)^2\frac{\partial}{\partial x_2}f(x)\right)\mid_S = f_1(x). \qquad (1.1)$$

Here $f_0(x)$ and $f_1(x)$ are given functions on S and $(\nu_1(x), \nu_2(x))$ is the unit vector of the external normal at the point $x \in S$.

In this paper we consider the following Cauchy problem: let X be a Hilbert space, $A : X \to X$ be a linear continuous operator, $\|A\| < 1$; Ω be a bounded simply connected

[1] The work was supported by the Russian Foundation of Fundamental Research under the grant No. 93-01-01-753.

domain in $\mathbf{R}^2$, S be a smooth closed curve in the domain Ω dividing it into two connected components Ω^+ and Ω^- and orientated as the boundary of Ω^-. It is required to find conditions under which there is a solution of problem (1.1) in the domain Ω^-.

We denote by $h_A(\Omega; X)$ the subspace of the Hilbert space $L_2(\Omega; X)$ consisting of functions $f \in C^2(\Omega; X)$ which satisfy the equation $\Delta_A f(x) = 0$ for $x \in \Omega$.

Using the functions $f_0(x)$, $f_1(x)$ for $x \in \Omega \setminus S$, we define the function

$$F(x) = \int\limits_S \left(\frac{\partial}{\partial \nu_A} \Gamma(x-y) f_0(y) - \Gamma(x-y) f_1(y) \right) ds, \tag{1.2}$$

where $\Gamma(x)$ is the fundamental solution of the operator Δ_A defined in Bukhgeim (1991).

The main results of the article are the following: the criterion for solvability of problem (1.1) and the formula of Carleman type for a solution of this problem. These results generalize those obtained by Shlapunov (1992) for the classical Laplace operator. The following theorems are their precise formulations.

Theorem 1.1. *Let Y be a Hilbert space dense in X and compactly embedded into X; $\|A\| < 1$; $S \in C^2$, $f_0(x) \in C^1(S, X)$, $f_1(x) \in C(S, X)$; and in Ω^+ there is a subdomain Ω' with a piecewise smooth boundary which satisfies the following conditions: the complement $\Omega \setminus \Omega'$ cannot be represented in the form $\Omega \setminus \Omega' = F \cup K$, where $F \cap K = \emptyset$, $K \neq \emptyset$ is a compact set and F is a closed set in Ω.*

Then

1. *there exists a system of functions $b_k(x), k = 1, 2, 3, \ldots$, which represents an orthonormal basis in $h_A(\Omega; X)$ and an orthogonal basis in $h_A(\Omega'; X)$:*

2. *Problem (1.1) is solvable if and only if the series $\sum\limits_{k=1}^{\infty} |c_k|^2$ converges, where c_k are the coefficients in the Fourier expansion of the function $F(x)$ with respect to the basis $b_k(x)$ of the space $h_A(\Omega'; X)$:*

$$c_k = \frac{< F, b_k >_{L_2(\Omega';X)}}{\|b_k\|_{L_2(\Omega';X)}}.$$

Moreover, if the solution exists then it is unique and is given by the formula

$$f(x) = F(x) - \sum_{k=1}^{\infty} c_k b_k(x), \quad x \in \Omega^-.$$

Theorem 1.2. *On the assumptions of Theorem 1.1 every function $f \in C^1(\Omega^- \cup S; X)$ which is A-harmonic in Ω^- can be represented with the help of the following formula: $\forall x \in \Omega^-$*

$$f(x) = \lim_{m \to \infty} \int\limits_S \left(\frac{\partial}{\partial \nu_A} G_m(x,y) f(y) - G_m(x,y) \frac{\partial}{\partial \nu_A} f(y) \right) ds,$$

where $G_m(x,y)$ is the operator function $G_m(x,y) : X \to X$ defined by the formula

$$G_m(x,y)u = \Gamma(x-y)u - \sum_{k=1}^{m} \frac{b_k(x)}{\|b_k\|_{L_2(\Omega';X)}} < \Gamma(x-y)u\,,\ b_k(x) >_{L_2(\Omega';X)}, \tag{1.3}$$

for $u \in X$. The limit is attained uniformly on compact subsets in Ω^-.

Now we describe briefly the contents of the article.

In Item 2 the definition and some properties of A-harmonic functions are given.

In Item 3 the existence of the system of functions $b_k(x)$ with the properties specified in Theorem 1.1 is proved.

In Items 4–5 we prove the criterion for solvability of problem (1.1) and the Carleman formula adduced in Theorem 1.2.

2. Definition and Some Properties of A–harmonic Functions

Let X be a complexification of a real Hilbert space X': $X = X' \oplus \mathrm{i}\, X'$ (for $u \in X$ $u = v + \mathrm{i}\, w$, where $v, w \in X'$); $\mathcal{L}(\mathcal{X})$ is the algebra of linear continuous operators. Let Ω be an open connected set in $C(\sim \mathbf{R}^2)$. For a function $u : \Omega \to X, u \in C^1(\Omega; X)$, we denote

$$\partial u = \frac{\partial u}{\partial z} = \frac{1}{2}\left(\frac{\partial u}{\partial x_1} + \frac{1}{\mathrm{i}}\frac{\partial u}{\partial x_2}\right); \quad \bar{\partial} u = \frac{\partial u}{\partial \bar{z}} = \frac{1}{2}\left(\frac{\partial u}{\partial x_1} - \frac{1}{\mathrm{i}}\frac{\partial u}{\partial x_2}\right);$$

$$\partial_A u = \partial u - A\bar{\partial} u, \quad \bar{\partial}_A u = \bar{\partial} u - A\partial u.$$

DEFINITION 2.1. A function $u \in C^1(\Omega; X)$ is called *A-analytic in Ω* if $\bar{\partial}_A u(z) = 0$ for all $z \in \Omega$.

We suppose that the operator A is real and $\|A\| < 1$. Let $\Omega \subset \mathbf{R}^2$.

DEFINITION 2.2. A function $v \in C^2(\Omega; X)$ is called *A-harmonic in Ω* if $\Delta_A v(x) = 0$ for all $x \in \Omega$.

REMARKI. f $u = v + \imath w$ and the domain Ω is simply connected, then the function u is A-analytic in $\Omega \subset \mathbf{R}^2$ if and only if the functions v, w are A–harmonic in Ω; moreover, the following relations hold:

$$(E - A)v_{x_1} = (E + A)w_{x_2},$$
$$(E + A)v_{x_2} = -(E - A)w_{x_1}. \tag{2.1}$$

This remark is proved in Bukhgeim(1991).

For $z \in C, z \neq 0$, we define the operator

$$B(z) = (z + \bar{z}A)^{-1},$$

and for $x \in \mathbf{R}^2$, $x \neq 0$, we define the operator function

$$\Gamma(x) = \frac{1}{2\pi}(E - A^2)^{-1} \ln \left(x_1^2(E + A)^2 + x_2^2(E - A)^2\right)^{\frac{1}{2}}$$

being the fundamental solution of the operator Δ_A.

As is shown in Bukhgeim(1991), $\Gamma(x) \in \mathcal{L}(\mathcal{X})$ at $x \neq 0$, and for all $e \in X$ the function $B(z)e$ is A–analytic and $\Gamma(x)e$ is A-harmonic.

With the help of these operators the A–analogues of the Cauchy and Green formulae are obtained in Bukhgeim (1991).

Theorem 2.3.

1. *Let $u \in C^1(\Omega; X)$, then for all $\zeta \in \Omega$*

$$u(\zeta) = \frac{1}{2\pi \mathrm{i}} \left(\int_{\partial\Omega} (\mathrm{d}\, z + A \mathrm{d}\, \bar{z}) B(z - \zeta) u(z) + \int_{\Omega} B(z - \zeta) \bar{\partial}_A u(z) \mathrm{d}\, z \wedge \mathrm{d}\, \bar{z} \right)$$

and, if $u(z)$ is A-analytic in Ω,

$$u(\zeta) = \frac{1}{2\pi \mathrm{i}} \int_{\partial\Omega} (\mathrm{d}\, z + A \mathrm{d}\, \bar{z}) B(z - \zeta) u(z).$$

2. *Let $f \in C^2(\Omega; X) \cap C^1(\bar{\Omega}; X)$, then for all $x \in \Omega$*

$$f(x) = \int_{\partial\Omega} \left(\frac{\partial \Gamma}{\partial \nu_A}(x - y) f(y) - \Gamma(x - y) \frac{\partial f}{\partial \nu_A}(y) \right) \mathrm{d}\, s + \int_{\Omega} \Gamma(x - y) \Delta_A f(y) \mathrm{d}\, y$$

and if $f(x)$ is A–harmonic in Ω,

$$f(x) = \int_{\partial\Omega} \left(\frac{\partial \Gamma}{\partial \nu_A}(x - y) f(y) - \Gamma(x - y) \frac{\partial f}{\partial \nu_A}(y) \right) \mathrm{d}\, s.$$

Other properties of A-harmonic functions are formulated in the following theorems.

Theorem 2.4. *Let Ω be a simply connected domain in $\mathbf{R}^2$ and $\|A\| < 1$. Then for every compact set $K \subset \Omega$ and every neighborhood $K' \subset \Omega$ of the set K there are constants c_α such that the following inequality holds for function $f \in h_A(\Omega; X)$:*

$$\|\partial^\alpha f\|_{C(K,X)} \leq c_\alpha \|f\|_{L_1(K';X)} .$$

Here α is multi–index.

Corollary 2.5. *Every A–harmonic function belongs to $C^\infty(\Omega; X)$.*

Corollary 2.6. *If a sequence of functions f_j A-harmonic in Ω converges uniformly to some function f on compact subsets in Ω , then f is also A–harmonic in Ω.*

Theorem 2.7. (uniqueness of an A-harmonic extension). *Let $\|A\| < 1$, Ω be a simply connected domain in $\mathbf{R}^2$, f be an A–harmonic in Ω function such that for all multi–indices α, $|\alpha| \geq 0$, $\partial^\alpha f(x) = 0$ for some $x \in \Omega$. Then $f(x) = 0$ in Ω.*

Theorem 2.8. (an analogue of the Runge theorem). *Let $\Omega_1 \subset \Omega_2$ be open sets in $\mathbf{R}^2$ which satisfy the following assumption: the complement $\Omega_2 \setminus \Omega_1$ is not representable in the form $\Omega_2 \setminus \Omega_1 = F \cup K$, where $F \cap K = \emptyset, K \neq \emptyset$ is a compact set and F is a closed set in Ω_2; Ω_1 is a simply connected set; and let $\|A\| < 1$.*

Then for every function f A–harmonic in Ω_1 there exists a sequence of A–harmonic functions f_j in Ω_2 converging uniformly to f on the compact set $K \subset \Omega_1$.

At the beginning, using the technique employed for usual analytic functions (for example see Hörmander, 1965), we obtain these results for A–analytic functions. Thereafter our statements are easily obtained from Remark.

Theorem 2.9. *Let Ω be a simply connected domain in $\mathbf{R}^2$, $f(x) \in C^1(\bar{\Omega}; X)$ and S be a smooth curve in Ω dividing it into two connected components Ω^- and Ω^+; and let $\|A\| < 1$.*

If the function $f(x)$ is A-harmonic in Ω^- and Ω^+, then $f(x)$ is A–harmonic in Ω.

PROOF. In Bukhgeim (1991) the following statement is proved: a continuous function $u(z)$ is A–analytic in some open set $\Omega \subset \mathbf{C}$ if and only if for every simply connected domain $\Omega' \subset \Omega$ with a piecewise–smooth boundary

$$\int_{\partial\Omega'} (\mathrm{d}\, z + A \mathrm{d}\, \bar{z}) u(z) = 0.$$

Let us consider the function $u = f + \mathrm{i}\, g$, where g is connected with f according to formulae (2.1). If the function u is A–analytic in Ω, then the function f is A-harmonic in Ω, because the domain Ω is simply connected. For every domain $\Omega' \subset \Omega$ we have

$$\int_{\partial\Omega'} (\mathrm{d}\, z + A \mathrm{d}\, \bar{z}) u(z) = \sum_i \int_{\partial\Omega_i'^+} (\mathrm{d}\, z + A \mathrm{d}\, \bar{z}) u(z) + \sum_j \int_{\partial\Omega_j'^-} (\mathrm{d}\, z + A \mathrm{d}\, \bar{z}) u(z),$$

where $\Omega_i'^+$ is a simply connected component of the set $\Omega'^+ = \Omega' \cap \Omega^+$ and $\Omega_j'^-$ is a simply connected component of the set $\Omega'^- = \Omega' \cap \Omega^-$. But the function $u(z)$ is A-analytic in $\Omega_i'^+$ or in $\Omega_j'^-$; therefore, all the integrals in the right-hand side are equal to zero, i.e.,

$$\int_{\partial\Omega'} (\mathrm{d}\, z + A \mathrm{d}\, \bar{z}) u(z) = 0$$

for all $\Omega' \subset \Omega$. It means that the function $u(z)$ is A-analytic in Ω.

Theorem 2.10. (an analogue of the Stieltjes-Vitali theorem). *Let Ω be a simply connected domain in $\mathbf{R}^2$, $\|A\| < 1$, Y be a Hilbert space dense in X and compactly embedded into X; $f_n(x)$, $n = 1, 2, 3, \ldots$ be a sequence of functions such that $f_n(x) \in h_A(\Omega; X) \cap C(\Omega; Y)$ and for every compact subset $K \subset \Omega$*

$$\|f_n\|_{C(K;Y)} \leq c\,(K).$$

Then there exists a subsequence $f_{n_j}(x)$ converging to some function $f \in h_A(\Omega; X)$ uniformly on every compact subset in Ω.

PROOF. From Theorem 2.4 we infer that for every compact set $K \subset \Omega$ we have

$$\left\|f_n(x^1) - f_n(x^2)\right\|_Y \leq c_1 \left|x^1 - x^2\right|, \quad x^1, x^2 \in K,$$

where the constant c_1 does not depend on n.

Let e_k, $k = 1, 2, 3, \ldots$ be an orthonormal basis in Y and let

$$v_n^k(x) = < f_n(x), e_k >_Y .$$

Then $v_n^k(x) \in C(\Omega)$, and for every $k = 1, 2, 3 \ldots$ $\left|v_n^k(x)\right| < c(K)$ at $x \in K$; moreover, $\left|v_n^k(x^1) - v_n^k(x^2)\right| < c_1 \left|x^1 - x^2\right|$ at $x^1, x^2 \in K$. Therefore, for every $k = 1, 2, 3, \ldots$ there exists a subsequence $v_{n_j}^k(x)$ and a function $v^k(x) \in C(\Omega)$ such that

$$\sup_K \left|v_{n_j}^k(x) - v^k(x)\right| \to 0, \quad j \to \infty.$$

We denote

$$f(x) = \sum_{k=1}^{\infty} v^k(x) e_k,$$

$$f_{n_j}(x) = \sum_{k=1}^{\infty} v^k_{n_j}(x) e_k.$$

It can easily be shown that for every $x \in K$ the subsequence $f_{n_j}(x)$ converges weakly to $f(x)$ in the space Y. But Y is compactly embedded into X; therefore, for every $x \in K$ $f_{n_j}(x)$ converges to $f(x)$ in the norm of the space X, furthermore, $\{f_{n_j}(x)\}$ converges uniformly to $f(x)$ on the compact subset $K \subset \Omega$. Using Corollary 2.6, we obtain our statement.

3. Bases with Double Orthogonality

In this section we shall prove the existence of a system of functions meeting the conditions described in Theorem 1.1. Namely, the following theorem formulates conditions under which there exist functions $b_k(x)$ forming an orthonormal basis of $h_A(\Omega; X)$ and an orthogonal basis of $h_A(\Omega'; X)$, where Ω' is some subdomain of Ω^+. Such a system is called *a basis with double orthogonality.* This theorem is analogous to the theorem by Shlapunov and Tarkhanov (1990).

Theorem 3.1. *Let Y be a Hilbert space dense in X and compactly embedded into X, Ω' be a simply connected subdomain of Ω^+ with a piecewise-smooth boundary and such that the complement $\Omega \setminus \Omega'$ cannot be represented in the form $\Omega \setminus \Omega' = F \cup K$, where $F \cap K = \emptyset$, $K \neq \emptyset$ is a compact set, and F is a closed set in Ω. Then there exists a basis with double orthogonality.*

We shall use in the proof results by Krasichkov (1986).

Theorem 3.2. *Let H be a Hilbert space; B be a subspace of H; P be the operator of orthogonal projection on B; and $D : H \to H$ be some self-adjoint linear operator. Assume the operator $PD|_B$: $B \to B$ to be one-to-one and completely continuous. Then there exists a system φ_j, $j = 1, 2, 3, \ldots$, enjoying the following properties:*

1. *the system φ_j is complete in B;*
2. *the system φ_j is orthonormal in H;*
3. $< D\varphi_i, \varphi_j >= 0$ *for* $i \neq j$.

This system is the system of eigenvectors of the operator PD and there are no other systems with such properties.

In our case we consider $L_2(\Omega; X)$ as H and $h_A(\Omega; X)$ as B.

Before we define the operator D we recall that the space Y can be considered as the domain of definition of some unbounded self-adjoint positive linear operator $\Lambda : Y \to X$ (for example, see Lions and Magenes, 1986), such that for all $u, v \in Y$ we have

$$< u, v >_Y = < \Lambda u, \Lambda v >_X; \quad \|u\|_Y = \|\Lambda u\|_X;$$

and $\Lambda^{-1} : X \to Y \subset X$ is a completely continuous and self-adjoint operator as an operator acting from X to X.

Now we define the operator D by the formula

$$Df(x) = \Lambda^{-1}\chi_{\Omega'}(x)f(x), \quad f(x) \in L_2(\Omega; X),$$

where $\chi_{\Omega'}(x)$ is the characteristic function of the subset Ω'.

It is clear that the operator $D : L_2(\Omega; X) \to L_2(\Omega; X)$ is a self-adjoint linear operator.

If the operator $PD\ |_{h_A(\Omega;X)}: h_A(\Omega; X) \to h_A(\Omega; X)$ is one-to-one and completely continuous then from Theorem 3.2 we shall obtain the existence of a system of functions $b_k(x)$ being the orthonormal basis of the space $h_A(\Omega; X)$ and the orthogonal system of functions in the space $h_A(\Omega'; X)$. If, in addition, this system is complete in $h_A(\Omega', X)$, then it will be the orthogonal basis of $h_A(\Omega', X)$.

Therefore, to prove the theorem, we have to show that

1. the operator $PD\ |_{h_A(\Omega;X)}: h_A(\Omega; X) \to h_A(\Omega; X)$ is completely continuous;
2. the operator $PD\ |_{h_A(\Omega;X)}$ is a one-to-one operator;
3. $h_A(\Omega; X)$ is dense in $h_A(\Omega'; X)$.

The third condition ensues immediately from Theorem 2.8.

The second condition readily follows from Theorem 2.7. Assume that $PDf = 0$ for some function $f(x) \in h_A(\Omega; X)$, then $Df \in h_A(\Omega; X)^{\perp}$ or $< f, Df >_{L_2(\Omega;X)} = 0$. But we have

$$< f, Df >_{L_2(\Omega;X)} = < f, \Lambda^{-1}\chi_{\Omega'} f >_{L_2(\Omega;X)} = < f, \Lambda^{-1} f >_{L_2(\Omega';X)} = 0.$$

Since the operator Λ^{-1} is positive, self-adjoint and completely continuous, we obtain $\|f\|_{L_2(\Omega';X)} = 0$, but $f \in h_A(\Omega; X)$ and, in particular, $f \in C^2(\Omega; X)$; it means that $f = 0$ at $x \in \Omega'$. Using Theorem 2.7 (Ω is a simply connected domain), we obtain $f = 0$ in Ω.

To check the first condition, we note that the operator P is a continuous linear operator. Hence, the operator $PD\ |_{h_A(\Omega;X)}: h_A(\Omega; X) \to h_A(\Omega; X)$ is completely continuous if the operator $D\ |_{h_A(\Omega;X)}: L_2(\Omega; X) \to L_2(\Omega; X)$ is completely continuous. Therefore, it suffices to show that the operator $D\ |_{h_A(\Omega;X)}$ maps a bounded set into a compact set.

Let us consider a sequence of functions $f_j(x) \in h_A(\Omega; X)$ such that $\|f_j\|_{L_2(\Omega;X)} \le c < \infty$ for all j and let $F_j(x) = Df_j(x)$.

Our statement will be proved if we show that there exists a subsequence $F_{j_k}(x)$ and a function $F(x)$ such that

$$\|F_{j_k} - F\|_{L_2(\Omega:X)} \to 0, \quad k \to \infty.$$

At $x \in \Omega$ we have

$$\|F_j(x)\|_Y = \|\Lambda Df_j(x)\|_X = \|\chi_{\Omega'}(x)f_j(x)\|_X .$$

Therefore, at $x \in \Omega'$ we, by Theorem 2.4, obtain

$$\|F_j(x)\|_Y = \|f_j(x)\|_X \le c_0 \|f_j\|_{L_2(\Omega;X)} \le c_0 c = c',$$

$$\left\|F_j(x^1) - F_j(x^2)\right\|_Y \le c_1 \left|x^1 - x^2\right| \|f_j\|_{L_2(\Omega;x)} \le c'' \left|x^1 - x^2\right|, \quad x^1, x^2 \in \Omega'.$$

It is clear that $F_j(x) \in C(\Omega'; Y) \cap L_2(\Omega; Y)$ and for every compact set $K \subset \Omega'$ we have

$$\|F_j\|_{C(K;Y)} \le c_1 \quad \forall j.$$

Therefore, reasoning as in the proof of Theorem 2.10, we can show that there exists a subsequence $F_{j_k}(x)$ converging to a function $F'(x) \in C(\Omega'; Y)$ weakly in Y and such that $\{F_{j_k}(x)\}$ converges to $F'(x)$ in the norm of X. Hence, for every compact set $K \subset \Omega'$ $\|F_{j_k} - F'\|_{C(K;X)} \to 0$ as $k \to \infty$ and $\|F_{j_k} - F'\|_{L_2(K,X)} \to 0$ as $k \to \infty$.

Now for all $\varepsilon > 0$ we can find a compact set $K_\varepsilon \subset \Omega'$ and a number j_ε such that for all $j_k > j_\varepsilon$

$$\|F_{j_k} - F'\|_{L_2(K_\varepsilon;X)} < \frac{\varepsilon}{2}$$

and

$$\|F_{j_k} - F'\|_{L_2(\Omega' - K_\varepsilon;X)} < \frac{\varepsilon}{2}.$$

Therefore, we obtain $\|F_{j_k} - F'\|_{L_2(\Omega';X)} \to 0$ as $k \to \infty$.

If we put $F = \chi_{\Omega'} F'$, then $\|F_{j_k} - F\|_{L_2(\Omega;X)} \to 0$ as $k \to \infty$.

Hence, the operator $D\,|_{h_A(\Omega;X)}$: $L_2(\Omega; X) \to L_2(\Omega; X)$ is completely continuous and the first condition is checked. So our theorem is proved.

4. Criterion for Solvability of Problem 1.1

Now we return to our problem. Let us consider the following fundamental solution of Δ_A :

$$\Gamma(x) = \frac{1}{2\pi}(E - A^2)^{-1} \ln \left(x_1^2(E + A)^2 + x_2^2(E - A)^2\right)^{\frac{1}{2}},$$

here $\|A\| < 1$ as usual in this paper. Since

$$x_1^2(E + A)^2 + x_2^2(E - A)^2 = |x|^2 (E - \mu(x)A)(E - \bar{\mu}(x)A),$$

where

$$\mu(x) = -\frac{x_1^2 - x_2^2}{|x|^2} + \imath\sqrt{1 - \left(\frac{x_1^2 - x_2^2}{|x|^2}\right)^2},$$

then

$$\Gamma(x) = \frac{1}{2\pi}(E - A^2)^{-1} \left(\ln|x| + \frac{1}{2}\ln(E - \mu(x)A) + \frac{1}{2}\ln(E - \bar{\mu}(x)A)\right).$$

It is easy to show that

$$\frac{\partial}{\partial \nu_A}\Gamma(x) = \frac{\nu \cdot x}{2\pi\,|x|^2}(E - \mu(x)A)^{-1}(E - \bar{\mu}(x)A)^{-1}(E - A^2),$$

where $\nu = \nu(x)$ is the unit vector of the outward normal at the point x. The operator $(E - \mu(x)A)^{-1}$ exists because $|\mu| = 1$ and $\|A\| < 1$; therefore,

$$(E - \mu(x)A)^{-1} = \sum_{k=0}^{\infty}(\mu(x)A)^k,$$

$$\left\|(E - \mu(x)A)^{-1}\right\| \le \frac{1}{1 - \|A\|}.$$

At first we intend to prove some auxiliary results.

Lemma 4.1. *Let $S \in C^1$, $f_0 \in C^1(S; X)$ and $f_1 \in C(S; X)$. Then there exists a function $f \in C^1(\mathbf{R}^2; X)$ such that, for $x \in S$,*

$$f(x) = f_0(x), \quad \frac{\partial}{\partial \nu_A} f(x) = f_1(x).$$

To prove the lemma, we apply the Whitney theorem:

Theorem 4.2. *Let K be a compact set in $\mathbf{R}^n$, u_α, $|\alpha| \le k$, be continuous functions on K with values in a Banach space X. For $|\alpha| \le k$ we define*

$$U_\alpha(x,y) = \left\| u_\alpha(x) - \sum_{|\beta|\le k-|\alpha|} u_{\alpha+\beta}(y)\frac{(x-y)^\beta}{\beta!} \right\| \cdot |x-y|^{|\alpha|-k}, \quad x,y \in K, \quad x \ne y;$$

$$U_\alpha(x,x) = 0 \quad \text{at} \quad x \in K.$$

If all the functions $U_\alpha, |\alpha| \le k$, are continuous on $K \times K$, then there exists a function $v \in C^k(\mathbf{R}^n; X)$ such that

$$\partial^\alpha v(x) = u_\alpha(x), \quad x \in K, \quad |\alpha| \le k.$$

Under these conditions we can choose a function $v(x)$ such that

$$\sum_{|\alpha|\le k} \sup \|\partial^\alpha v\| \le C(K) \left(\sum_{|\alpha|\le k} \sup_{K\times K} U_\alpha + \sum_{|\alpha|\le k} \sup_K \|u_\alpha\| \right).$$

This theorem is a obvious generalization of the Whitney theorem for usual functions presented in Hörmander (1983).

PROOF. of Lemma 4.1 In our case $k = 1, K = S$. The functions u_α on S are defined by the following formulae:

$$u_0(x) = f_0(x);$$

$$u_{1,0}(x) = [\nu_1^2(E-A)^2 + \nu_2^2(E+A)^2]^{-1}(\nu_1 f_1 - \nu_2(E+A)^2\frac{\partial}{\partial\nu_\perp}f_0)(x);$$

$$u_{0,1}(x) = [\nu_1^2(E-A)^2 + \nu_2^2(E+A)^2]^{-1}(\nu_2 f_1 + \nu_1(E-A)^2\frac{\partial}{\partial\nu_\perp}f_0)(x);$$

where $(\nu_1(x), \nu_2(x))$ is the unit vector of the outward normal at the point $x \in S$ and

$$\frac{\partial}{\partial\nu_\perp}f_0(x) = -\nu_2(x)\frac{\partial}{\partial x_1}f_0(x) + \nu_1(x)\frac{\partial}{\partial x_2}f_0(x)$$

is the tangent derivative at the point $x \in S$.

All functions $u_\alpha(x), |\alpha| \le 1$, are continuous on S, because $f_1(x) \in C(S;X), f_0(x) \in C^1(S;X), S \in C^2$. Note that

$$B(y) = \left[\nu_1^2(E-A)^2 + \nu_2^2(E+A)^2\right]^{-1}(y) \in \mathcal{L}(\mathcal{X}) \quad \dagger \in \mathcal{S},$$

and

$$\|B\| \le \frac{1}{(1-\|A\|)^2}.$$

Let us consider the functions

$$U_\alpha(x,y) = \left\| u_\alpha(x) - \sum_{|\beta|\le k-|\alpha|} u_{\alpha+\beta}(y)\frac{(x-y)^\beta}{\beta!} \right\| \cdot |x-y|^{|\alpha|-k}.$$

We have

$$U_0(x,y) = \frac{1}{|x-y|}\left\|u_0(x) - (u_0(y) + u_{1,0}(y)(x_1-y_1) + u_{0,1}(y)(x_2-y_2))\right\|;$$

$$U_{1,0}(x,y) = \|u_{1,0}(x) - u_{1,0}(y)\| ;$$
$$U_{0,1}(x,y) = \|u_{0,1}(x) - u_{0,1}(y)\| , \quad x, y \in S, \quad x \neq y,$$

and $U_\alpha(x,x) = 0$ at $x \in S$.

The continuity of the functions $U_{1,0}(x,y)$ and $U_{0,1}(x,y)$ on $S \times S$ follows from that of the functions u_α on S.

We shall show that the function $U_0(x,y)$ is continuous on $S \times S$ as well. Let us consider the numerator of this function. In accordance with the definition of u_α, it is equal to

$$\bar{U}_0(x,y) = f_0(x) - f_0(y) - B\left(\nu_1 f_1 - \nu_2(E+A)^2 \frac{\partial}{\partial \nu_\perp} f_0\right)(y)(x_1 - y_1) -$$
$$B\left(\nu_2 f_1 + \nu_1(E-A)^2 \frac{\partial}{\partial \nu_\perp} f_0\right)(y)(x_2 - y_2).$$

In other words, we have

$$\bar{U}_0(x,y) = f_0(x) - f_0(y) +$$
$$B\nu_2(y)(E+A)^2 \frac{\partial}{\partial \nu_\perp} f_0(y)(x_1 - y_1) - B\nu_1(y)(E-A)^2 \frac{\partial}{\partial \nu_\perp} f_0(y)(x_2 - y_2) -$$
$$(\nu_1(y)(x_1 - y_1) + \nu_2(y)(x_2 - y_2))Bf_1(y). \tag{4.1}$$

Since for $x, y \in S$

$$\frac{x_1 - y_1}{|x-y|} \to \nu_2(x),$$
$$\frac{x_2 - y_2}{|x-y|} \to -\nu_1(x),$$

as $y \to x$, then

$$\frac{\nu_1(y)(x_1 - y_1) + \nu_2(y)(x_2 - y_2)}{|x-y|} \to < \nu(x), \nu_\perp(x) >= 0.$$

Therefore, the last term in (4.1) is $o(|x-y|)$.

Let the curve S be given in the parametric form:

$$x = x(t) = (x_1(t), x_2(t)), \quad t \in [t_0, t_1],$$

and $y = x(t')$ for some $t' \in [t_0, t_1]$. Then the unit vector of the normal is given by the formula

$$\left(\frac{\dot{x}_2(t)}{|\dot{x}(t)|}, -\frac{\dot{x}_1(t)}{|\dot{x}(t)|}\right).$$

Moreover,

$$x - y = \dot{x}(t')(t - t') + o(|t - t'|);$$
$$|x - y| = |\dot{x}(t')| \; |t - t'| + o(|t - t'|);$$
$$x_1 - y_1 = -\nu_2(y)\,|\dot{x}(t')|\,(t - t') + o(|t - t'|);$$
$$x_2 - y_2 = \nu_1(y)\,|\dot{x}(t')|\,(t - t') + o(|t - t'|).$$

Hence,

$$\bar{U}_0(x,y) = f_0(x(t)) - f_0(x(t')) -$$

$$B\nu_2^2(y)(E+A)^2\frac{\partial}{\partial\nu_\perp}f_0(x(t'))\,|\dot{x}(t')|\,(t-t') - B\nu_1^2(y)(E-A)^2\frac{\partial}{\partial\nu_\perp}f_0(x(t'))\,|\dot{x}(t')|\,(t-t')+$$

$$B[\nu_2(y)(E+A)^2-\nu_1(y)(E-A)^2]\frac{\partial}{\partial\nu_\perp}f_0(x(t'))\,\mathrm{o}\,(|t-t'|)+\mathrm{o}\,(|x-y|)=$$

$$f_0(x(t))-f_0(x(t'))-BB^{-1}(y)\frac{\partial}{\partial\nu_\perp}f_0(x(t'))\,|\dot{x}(t')|\,(t-t')+\mathrm{o}\,(|t-t'|)=$$

$$f_0(x(t))-f_0(x(t'))-|\dot{x}(t')|\left[\frac{\dot{x}_1(t')}{|\dot{x}(t')|}\frac{\partial}{\partial x_1}f_0(x(t'))+\frac{\dot{x}_2(t')}{|\dot{x}(t')|}\frac{\partial}{\partial x_2}f_0(x(t'))\right](t-t')+\mathrm{o}\,(|t-t'|)=$$

$$f_0(x(t))-f_0(x(t'))-\frac{\mathrm{d}}{\mathrm{d}\,t}f_0(x(t'))(t-t')+\mathrm{o}\,(|t-t'|)=\mathrm{o}\,(x-y).$$

It means that the function $U_0(x,y)$ is continuous on $S\times S$. Therefore, from Theorem 4.2 we obtain the existence of a function $f\in C^1(\mathbf{R}^2;X)$ such that

$$\partial^\alpha f(x)=u_\alpha(x),\quad |\alpha|\le 1, x\in S.$$

In other words, for $x\in S$ we have

$$f(x)=u_0(x)=f_0(x)$$

and

$$\frac{\partial}{\partial\nu_A}f(x)=\nu_1(x)(E-A)^2\frac{\partial}{\partial x_1}f(x)+\nu_2(x)(E+A)^2\frac{\partial}{\partial x_2}f(x)=$$

$$\nu_1(x)(E-A)^2u_{1,0}(x)+\nu_2(x)(E+A)^2u_{0,1}(x)=$$

$$B(x)\Bigg[\nu_1(x)(E-A)^2\nu_1(x)f_1(x)-\nu_1(x)(E-A)^2\nu_2(x)(E+A)^2\frac{\partial}{\partial\nu_\perp}f_0(x)$$

$$+\nu_2(x)(E+A)^2\nu_2(x)f_1(x)+\nu_2(x)(E+A)^2\nu_1(x)(E-A)^2\frac{\partial}{\partial\nu_\perp}f_0(x)\Bigg]=$$

$$B(x)B^{-1}(x)f_1(x)=f_1(x).$$

Thus, the function $f(x)$ is the required one.

Let $x_0\in S$, and $\nu(x_0)$ be the unit vector of the outward normal to the domain Ω^- at the point x_0. Using the methods suggested by Tarkhanov (1991) and Shlapunov (1992) and , we can obtain the following result:

Lemma 4.3. *Let $S\in C^2, f_0\in C^1(S;X)$ and $f_1\in C(S;X)$. Then*

1. *there exists a function $f\in C^1(\mathbf{R}^2;X)$ such that, for $x\in S$,*

$$f(x)=f_0(x),\quad \frac{\partial}{\partial\nu_A}f(x)=f_1(x);$$

2. *it holds that*

$$\lim_{\varepsilon\to 0}(\partial^\alpha F(x_0-\varepsilon\nu(x_0))-\partial^\alpha F(x_0+\varepsilon\nu(x_0)))=\partial^\alpha f(x_0),$$

where $|\alpha|\le 1$ and the function $F(x)$ is defined by formula (1.2);

3. *the function $F^+(x)=F(x)\,|_{\Omega^+}$ has a smooth extension onto the set $\Omega^+\cup S$ if and only if the function $F^-(x)=F(x)\,|_{\Omega^-}$ has a smooth extension onto the set $\Omega^-\cup S$.*

The first statement has been proved in Lemma 4.2, the second and the third are generalizations of the results obtained by Tarkhanov (1991) (Lemma 29.5) and Shlapunov (1992) (Lemma 1.1) respectively, and can be proved in a similar way.

Now we turn to proving the main results of this article.

Theorem 4.4. *Let $S \in C^2, f_0 \in C^1(S;X)$ and $f_1 \in C(S;X)$. Then problem (1.1) is solvable if and only if the function $F^+(x)$ has an A-harmonic extension from Ω^+ to Ω^-.*

PROOF. a) Assume $f(x)$ to be a solution of problem (1.1) and consider the function $\Phi(x) = F^+(x)$ for $x \in \Omega^+$ and $\Phi(x) = F^-(x) - f(x)$ for $x \in \Omega^-$.

Then for an arbitrary connected piece $S_1 \subset S$ there exists $\Omega_1 \subset \Omega^-$ such that $S_1 \subset \partial\Omega_1$ and the function $f \in C^1(\bar{\Omega}_1; X)$ is A-harmonic in Ω_1. Using the analogue of the Green formula for A-harmonic functions, we can write

$$f(x) = \int_{\partial\Omega_1} \left(\frac{\partial\Gamma}{\partial\nu_A}(x-y)f(y) - \Gamma(x-y)\frac{\partial f}{\partial\nu_A}(y) \right) \mathrm{d}\, s.$$

For $x \in \Omega^-$ we have

$$\Phi(x) = F^-(x) - f(x) = \int_{S\setminus S_1} \left(\frac{\partial\Gamma}{\partial\nu_A}(x-y)f_0(y) - \Gamma(x-y)f_1(y) \right) \mathrm{d}\, s$$

$$- \int_{\partial\Omega_1\setminus S_1} \left(\frac{\partial\Gamma}{\partial\nu_A}(x-y)f(y) - \Gamma(x-y)\frac{\partial f}{\partial\nu_A}(y) \right) \mathrm{d}\, s.$$

The terms in the right-hand side are A-harmonic functions in a neighborhood of S_1, and this equality implies that the function $F^-(x)$ has a smooth extension onto the set $\Omega^- \cup S_1$. But S_1 is chosen arbitrary; therefore, $F^-(x)$ has a smooth extension onto the set $\Omega^- \cup S$. It means that the function $F^+(x)$ has a smooth extension onto the set $\Omega^+ \cup S$ as well. Therefore, the restriction of $\Phi(x)$ onto the sets $\Omega^\pm$ continues smoothly onto the sets $\Omega^+ \cup S$ and $\Omega^- \cup S$. From the second statement of Lemma 4.3 we obtain, for an arbitrary point $x_0 \in S$, that

$$\lim_{\varepsilon\to 0} (\partial^\alpha\Phi(x_0 - \varepsilon\nu(x_0)) - \partial^\alpha\Phi(x_0 + \varepsilon\nu(x_0))) = 0$$

for $|\alpha| < 1$. Thus, the function $\Phi(x)$ continues smoothly onto the domain Ω and is an A-harmonic in Ω^+ and Ω^-. Hence, from Theorem 2.9, we infer that $\Phi(x)$ is A-harmonic in Ω. So, $\Phi(x)$ is an A-harmonic extension of $F^+(x)$.

b) Let $F^+(x)$ have an A-harmonic extension $\Phi(x)$ in Ω. Then, from Lemma 4.3, we conclude that $F^-(x)$ continues smoothly onto $\Omega^- \cup S$.

Denote $f(x) = F^-(x) - \Phi(x)$ for $x \in \Omega^-$. Then $\Delta_A f(x) = 0$ in Ω^-; therefore,

$$f(x) = \int_{\partial\Omega^-} \left(\frac{\partial\Gamma}{\partial\nu_A}(x-y)f(y) - \Gamma(x-y)\frac{\partial f}{\partial\nu_A}(y) \right) \mathrm{d}\, s.$$

On the other hand,

$$f(x) = \int_S \left(\frac{\partial\Gamma}{\partial\nu_A}(x-y)f_0(y) - \Gamma(x-y)f_1(y) \right) \mathrm{d}\, s - \Phi(x).$$

From Lemma 4.3 we get

$$\lim_{\varepsilon\to 0}(f(x_0-\varepsilon\nu(x_0))=\lim_{\varepsilon\to 0}\left(F^-(x_0-\varepsilon\nu(x_0))-\Phi(x_0-\varepsilon\nu(x_0))\right)=$$

$$\lim_{\varepsilon\to 0}F^-(x_0-\varepsilon\nu(x_0))-\Phi(x_0)=$$

$$\lim_{\varepsilon\to 0}(F^-(x_0-\varepsilon\nu(x_0))-F^+(x_0+\varepsilon\nu(x_0)))=f(x_0),$$

since $\Phi^- |_S=\Phi^+ |_S=F^+ |_S$. Similarly,

$$\frac{\partial}{\partial\nu_A}f(x_0)=f_1(x_0).$$

Consequently, the function $f(x)=F^-(x)-\Phi(x)$ is a solution of our problem.

Theorem 4.5. *Let $S\in C^2, f_0\in C^1(S;X)$ and $f_1\in C(S;X)$. If a solution of problem (1.1) exists, then it is unique.*

PROOF. If $f(x)$ is a solution of Problem 1.1, then the function

$$\Phi(x)=F^-(x)-f(x)\quad x\in\Omega^-\cup S$$

is an A-harmonic extension of the function $F^+(x)$ from Ω^+ onto Ω. Therefore, if $f_0=f_1\equiv 0$ and $f(x)\neq 0$ for $x\in\Omega^-$ is a solution, then we infer that $F^+(x)=F^-(x)\equiv 0$ and $\Phi(x)=-f(x)\neq 0$ for $x\in\Omega^-$ is an A-harmonic function in Ω^-. But it contradicts the uniqueness of an A–harmonic extension.

5. Criterion for Solvability of Problem 1.1 in Terms of Bases with Double Orthogonality

Let the conditions described in Theorem 3.1 be satisfied, then there exist functions $b_k(x)$ forming an orthonormal basis of $h_A(\Omega;X)$ and an orthogonal basis of $h_A(\Omega';X)$. Consider

$$c_k=\frac{< F^+,b_k >_{L_2(\Omega';X)}}{\|b_k\|_{L_2(\Omega';X)}}=$$

$$\frac{1}{\|b_k\|_{L_2(\Omega';X)}}\int_S\left(<\frac{\partial\Gamma}{\partial\nu_A}(x-y)f_0(y),b_k(x)>_{L_2(\Omega';X)}-<\Gamma(x-y)f_1(y),b_k(x)>_{L_2(\Omega';X)}\right)ds,$$

i.e., the coefficient of $F^+(x)$ in the expansion in the basis of $h_A(\Omega';X)$. The following theorem is true.

Theorem 5.1. *Let $S\in C^2, f_0\in C^1(S;X), f_1\in C(S;X)$, and all the conditions of Theorem 3.1 be satisfied. Then problem (1.1) is solvable if and only if the series $\sum_{k=1}^{\infty}|c_k|^2$ converges.*

PROOF. a) Let $f(x)$ be a solution of problem (1.1), then from Theorem 4.4 we conclude that $F^+(x)$ continues A–harmonically from Ω^+ onto Ω. The A–harmonic extension of $F^+(x)$ is equal to

$$\Phi(x)=F^-(x)-f(x),\quad x\in\Omega^-\cup S.$$

Since $F^+(x) \in h_A(\Omega^+; X)$ and $F^-(x) \in h_A(\Omega^-; X)$, we have $\Phi(x) \in h_A(\Omega; X)$. Therefore, $\Phi(x) = \sum_{k=1}^{\infty} c_k b_k(x)$ and the series converges in the norm of the space $L_2(\Omega; X)$ and, from the Bessel's inequality, we conclude

$$\sum_{k=1}^{\infty} |c_k|^2 \leq \|\Phi\|^2_{L_2(\Omega;x)} < \infty.$$

b) Let $\sum_{k=1}^{\infty} |c_k|^2$ converge, then, by the Riesz–Fisher theorem, there exists a function $\Phi(x) \in h_A(\Omega; X)$ such that $\Phi(x) = \sum_{k=1}^{\infty} c_k b_k(x)$. But in this case the function $\Phi(x)$ is an A-harmonic extension of the function $F^+(x)$ from Ω^+ onto Ω. And, from Theorem 4.4 we infer that the function $f(x) = F^-(x) - \Phi(x)$, for $x \in \Omega^-$, is a solution of the problem.

The results of Theorems 3.1, 4.5, and 5.1 imply Theorem 1.1.

6. The Carleman Formula

In this item the proof of Theorem 1.2 is given.

Let $f(x)$ be an A-harmonic function in Ω^-, and let $f \in C^1(\Omega^- \cup S; X) \cap C^2(\Omega^-; X)$. So f is a solution of Problem 1.1; therefore, the function $F^+(x)$ has an A–harmonic extension $\Phi(x)$ in Ω, and $\bar{f} = F^- - \Phi$ is a solution as well. It is clear that $\bar{f} = f$ in $\Omega^- \cup S$.

The function $\Phi(x)$ belongs to $h_A(\Omega; X)$; therefore,

$$f(x) = F^-(x) - \Phi(x) = F^-(x) - \sum_{k=1}^{\infty} c_k b_k(x) = F^-(x) - \lim_{m\to\infty} \sum_{k=1}^{m} c_k b_k(x),$$

where the limit is understood in the norm of the space $L_2(\Omega^-; X)$. From Theorem 2.4 it follows that the series converges in the norm of the space $C(K; X)$ for every compact set $K \subset \Omega^-$.

We obtain, for $x \in \Omega^-$,

$$f(x) = \int_S \left(\frac{\partial \Gamma}{\partial \nu_A}(x-y) f(y) - \Gamma(x-y) \frac{\partial}{\partial \nu_A} f(y) \right) \mathrm{d}\, s -$$

$$\lim_{m\to\infty} \sum_{k=1}^{m} \frac{b_k(x)}{\|b_k\|_{L_2(\Omega';X)}} \int_S \Big(< \frac{\partial \Gamma}{\partial \nu_A}(x-y) f(y), b_k(x) >_{L_2(\Omega';X)}$$

$$- < \Gamma(x-y) \frac{\partial f}{\partial \nu_A}(y), b_k(x) >_{L_2(\Omega';X)} \Big) \mathrm{d}\, s =$$

$$\lim_{m\to\infty} \int_S \Bigg[\left(\frac{\partial \Gamma}{\partial \nu_A}(x-y) - \sum_{k=1}^{m} \frac{b_k(x)}{\|b_k\|_{L_2(\Omega';X)}} < \frac{\partial \Gamma}{\partial \nu_A}(x-y)\bullet, b_k(x) >_{L_2(\Omega';X)} \right) f(y) -$$

$$\left(\Gamma(x-y) - \sum_{k=1}^{m} \frac{b_k(x)}{\|b_k\|_{L_2(\Omega';X)}} < \Gamma(x-y)\bullet, b_k(x) >_{L_2(\Omega';X))} \right) \frac{\partial f}{\partial \nu_A}(y) \Bigg] \mathrm{d}\, s =$$

$$\lim_{m\to\infty} \int_S \left(\frac{\partial}{\partial \nu_A} G_m(x,y) f(y) - G_m(x,y) \frac{\partial f}{\partial \nu_A}(y) \right) \mathrm{d}\, s,$$

where $G_m(x, y)$ is defined by formula (1.3).

In conclusion we give an example of the spaces X, Y and the operator Λ^{-1} satisfying the conditions of Theorem 1.1.

Let us consider the scale of Hilbert spaces $l_{2,s}$ for $s \in (0,1)$ with norms

$$\|u\|_s^2 = \sum_{j=0}^{\infty} s^{-2j} |u_j|^2,$$

where $u = (u_0, u_1, u_2, ...)$. If the operator A is the shift operator defined in Introduction, then

$$\|Au\|_s^2 = \sum_{j=0}^{\infty} s^{-2j} |u_{j+1}|^2 s^{-2} s^2 \leq s^2 \|u\|_s^2 .$$

Thus, $\|A\| \leq s < 1$ in the space $l_{2,s}$, whereas $\|A\| = 1$ in l_2. This is the main requirement on the operator A in this article.

Let us consider the space $Y = l_{2,ss'}$ with the norm

$$\|u\|_{ss'}^2 = \sum_{j=0}^{\infty} (ss')^{-2j} |u_j|^2, \quad s' \in (0,1).$$

We have $\|u\|_s \leq \|u\|_{ss'}$; therefore, Y is embedded into X and the embedding operator is completely continuous. Indeed, if $\|u_n\|_Y < c \quad \forall n$, we have

a). $\|u_n\|_X < c \quad \forall n;$

b). $\forall \varepsilon > 0 \quad \exists N(\varepsilon)$ such that for all $k > N(\varepsilon)$

$$\sum_{j=k}^{\infty} s^{-2j} |u_j|^2 = \sum_{j=k}^{\infty} (ss')^{-2j} s'^{2j} |u_j|^2 < s'^{2k} \|u\|_{ss'}^2 < s'^{2k} c^2 < \varepsilon.$$

The operator Λ^{-1} is defined by the following rule:

$$\Lambda^{-1} : (u_0, u_1, u_2, \ldots) \to (u_0, s' u_1, s'^2 u_2, \ldots),$$

Then

$$\left\|\Lambda^{-1} u\right\|_Y^2 = \sum_{j=0}^{\infty} (ss')^{-2j} \left|s'^j u_j\right|^2 = \sum_{j=0}^{\infty} s^{-2j} |u_j|^2 = \|u\|_X^2 .$$

Consequently, $\Lambda^{-1} : X \to Y$ is the operator required in Theorem 3.1.

REFERENCES

Bukhgeim A. L. (1991), Inversion Formulae in Inverse Problems. Supplement to the book of Lavrent'ev M. M. and Savel'ev L. Yu. *Linear Operators and Ill-Posed Problems,* Nauka, Moscow, 276-320, (in Russian)

Hörmander L. (1966), *An Introduction to Complex Analysis in Several Variables.* Princeton. N.J., a.o., D. Van Nostrand Co.

Hörmander L. (1983), *The Analysis of Linear Partial Differential Operators. Distribution Theory and Fourier Analysis.* Berlin a.o. Springer.

Krasichkov I. F. (1968), Systems of Functions with the Properties of the Double Orthogonality. *Mathematical Reviews,* vol. 4, no. 5, 551-556, (in Russian)

Lions J. L. and Magenes E. (1968), *Problemes aux limites non homogenes et applications,* Dunod. Paris,vol. 1.

Shlapunov A. A. (1992), About Cauchy Problem for Laplace Equation. *Siberian Math. J.*, vol. 33, no. 3, 205-215, (in Russian)

Shlapunov A. A. and Tarkhanov N. N. (1990), *To Cauchy Problem for Holomorphic Functions of the Lebesgue Class in Domain.* Krasnoyarsk, Preprint no. 52, The L.V.Kirensky Inst. of physics, (in Russian)

Tarkhanov N. N. (1991), *The Loran Series for Solutions of Elliptic Systems.* Nauka, Novosibirsk (in Russian)

Simultaneous Determination of Sounding Signal and Wave Propagation Velocity in a Vertically-Inhomogeneous Medium [1]

Alexander V. AVDEEV

Novosibirsk Computing Center
Prospect Lavrentieva 6, Novosibirsk 630090, Russia

ABSTRACT

The problem of simultaneous determination of the sounding signal form and the wave propagation velocity in a vertically-inhomogeneous medium is considered in the article.

It should be noted that usually the form of the sounding signal is either unknown or given only approximately, although its accurate estimate is required for practical solution of many inverse problems.

We propose an interactive numerical algorithm which determines simultaneously both the wave propagation velocity and the sounding signal form in a vertically-inhomogeneous medium. This algorithm is based on an optimization approach, i.e., on minimization of the data misfit functional of observed and calculated data.

Results of numerical experiments are given to illustrate the efficiency of the method.

Introduction

In the article we consider simultaneous determination of the sounding signal form and the wave propagation velocity in a vertically-inhomogeneous medium. The solution of this problem is sought as the minimum point of a data misfit functional. The data misfit functional is defined as quadratic deviation of the wave field that is recorded at the free surface, from the field calculated for "experimental" values of velocity and input signal. Advantages of this approach over other numerical methods are its universality and the capability to take into account a priori information on the structure of the medium under study at each step of the process of functional minimization. The approach is described in many papers including those by Alekseev et al. (1993) and Kolb et al. (1986).

It should be noted that usually the form of the sounding signal is either unknown or given only approximately, although its accurate estimate is required in solving many inverse problems.

Similar formulations of these problems were studied theoretically by Blagoveshchenskii (1970) and Lavrent'ev et al. (1982). Unique solvability of the given problem to some finite depth was shown and uniqueness theorem for the linearized problem was proved in these publications.

More complete bibliography on numerical solution of the problems of simultaneous determination of the medium structure and the input signal form is given by Carrion

[1]The work was supported by the Russian Foundation of Fundamental Research under the grant No. 93-05-8638.

et al. (1990). Of particular interest is the article by Tarantola (1984) in which least-squares inversion is proposed for solving problems of this type. However, this work does not contain any numerical results.

In the present article we describe an interactive numerical algorithm which determines simultaneously the wave propagation velocity and the sounding signal form in a vertically-inhomogeneous medium. This algorithm is based on an optimization approach, i.e., on minimization of the data misfit functional of observed and calculated data. Results of numerical experiments are given to illustrate the efficiency of the method.

1. Wave propagation in a vertically-inhomogeneous medium is described by the following initial boundary value problem:

$$\frac{\partial^2 U}{\partial t^2} = c_*^2(z)\Delta U, \tag{1}$$

$$U\,|_{t<0} \equiv 0, \tag{2}$$

$$\frac{\partial U}{\partial z}\,|_{z=0} = f(t)g(x,y), \tag{3}$$

where $c_*(z)$ is the piecewise-continuous function characterizing the wave propagation velocity in the medium, $f(t)$ and $g(x,y)$ describe, correspondingly, influence in time and space distribution of sources on the free surface. Let $z_1, \dots, z_n$ be boundaries of layers with the velocities $c_1, \dots, c_n$. For this model it is necessary to add to relations (1)–(3) the conditions at $z = z_i$:

$$[U]_{z=z_i} = [\frac{dU}{dz}]_{z=z_i} = 0, \quad i = 1, \dots, N.$$

Let us formulate a problem inverse to problem (1)–(3).

The inverse problem is to determine the functions $c(z)$ and $f(t)$ from the information

$$U\,|_{z=0} = U_0(x,y,t); \quad 0 < t < T; \quad (x,y) \in S \subset \{(x,y,z) : z = 0\} \tag{4}$$

on the oscillation regime of the observation surface $z = 0$.

The solution of this problem will be sought as the minimum point of the data misfit functional

$$\Phi[c^2(z)] = \int_0^T dt \int_S \{U_0(x,y,t) - B[c^2(z), f(t)](x,y,t)\}^2 dxdy, \tag{5}$$

where $B[c^2(z), f(t)]$ is the operator transforming the functions $c(z)$ and $f(t)$ into the solution of the direct problem (1)–(3) at $z = 0$ (the wave field on the set S).

2. The semi-analytical method described by Fatianov and Mikhailenko (1988), was chosen for solving the direct problem (1)–(3).

Following Fatianov and Mikhailenko (1988) the initial problem (1)–(4) can be rewritten in the following form:

$$\frac{d^2 U}{dz^2} = \nu^2 U, \tag{6}$$

$$\frac{dU}{dz}\,|_{z=0} = F(\omega), \tag{7}$$

where

$$\nu^2 = k^2 - \omega^2/c^2(z),$$

$$F(\omega) = \int_0^T f(t)\exp(-\mathrm{i}\,\omega t)\mathrm{d}t,$$

and the uniqueness of its solution needs the principle of limiting absorption to be fulfilled, i.e.,

$$U(z,k,\omega) = \lim_{\varepsilon\to+0} U(z,k,\omega - \mathrm{i}\,\varepsilon),$$

where $U(z,k,\omega-\mathrm{i}\,\varepsilon) \to 0$ for $z \to \infty$.

As was shown by Alekseev et al. (1993), realization of a search for the minimum point of the data misfit functional in the frequency range (k,ω) enables us to reduce considerably computational resources necessary for the multiple solution of the direct problem. Moreover, it is possible to make detailed analysis of spectra of the wave fields at each step of calculations.

In this connection, assuming that the recording time T and the size of the observation domain S on the free surface are sufficiently large, we rewrite functional (5) in an equivalent form:

$$\Phi[n^2(z)] = \int_{\omega_1}^{\omega_2} \mathrm{d}\omega \int_{k_1}^{k_2} \mid U_0(k,\omega) - \hat{B}[n^2(z),F(\omega)](k,\omega)\mid^2 k\mathrm{d}k, \tag{8}$$

where (ω_1,ω_2) and (k_1,k_2) are the time and the space frequency ranges determined by the spectral composition of the sounding signal $F(\omega)$ and by the size of the observation domain S. The operator $\hat{B}[n^2(z),F(\omega)]$ transforms the functions $n^2(z) = c^{-2}(z)$ ($c(z)$ is a current wave propagation velocity) and $F(\omega)$ (the spectrum of the sounding signal) into the solution of the direct problem (6) - (7) at $z = 0$.

3. The set of functions $\{n^2(z), F(\omega)\}$, where

$$0 < \delta \le n^2(z) \le N < \infty, \quad \lim_{z\to\infty} n^2(z) = n_0^2,$$
$$n^2(z) - n_0^2 \in L_2(0,\infty) \cap C^1(0,\infty),$$
$$F(\omega) \in L_2(\omega_1,\omega_2),$$

will be called the set of models M.

The image of the set M under the action of the operator $\hat{B}$ will be called the set of acceptable data D ($D = \hat{B}(M)$).

After introducing on M and D scalar products, the data misfit functional (8) can be rewritten in the following form:

$$\Phi[n^2(z),F(\omega)] = (U_0 - \hat{B}[n^2(z),F(\omega)], U_0 - \hat{B}[n^2(z),F(\omega)])_D.$$

Using the conjugate Green function method described by McGillivray and Oldenbourgh (1990), we obtain expressions for the gradients of the data misfit functional (8) over velocity and impulse:

$$\nabla_n^2\Phi[n^2(z),F(\omega)](\xi) = -2\,\mathrm{Re}\int_{\omega_1}^{\omega_2}(\omega+\mathrm{i}\,\varepsilon)^2\bar{F}(\omega)\mathrm{d}\omega \tag{9}$$
$$\times \int_{k_1}^{k_2}\left[U_0(k,\omega) - \hat{B}[n^2(z),F(\omega)](k,\omega)\right]\bar{G}^2(\xi,k,\omega)k\mathrm{d}k,$$

$$\nabla_F \Phi[n^2(z), F(\omega)](\xi) = -2\,\mathrm{Re} \int\limits_{k_1}^{k_2} \left[U_0(k,\omega) - \hat{B}[n^2(z), F(\omega)](k,\omega)\right] \tag{10}$$

$$\times G(\xi, k, \omega) k \mathrm{d}k - 2\,\mathrm{i}\,\mathrm{Im} \int\limits_{k_1}^{k_2} \left[U_0(k,\omega) - \hat{B}[n^2(z), F(\omega)](k,\omega)\right]$$

$$\times G(\xi, k, \omega) k \mathrm{d}k,$$

where $G(\xi, k, \omega)$ is the solution of the problem (6)–(7) for $F(\omega) = 1$.

4. First-order optimization methods of descent were used to make a search for the minimum point of the data misfit functional, namely:

a) the method of steepest descent (for the impulse spectrum):

$$F_{j+1}(\omega) = F_j(\omega) - \alpha_j \nabla_F \Phi[n^2(z), F_j(\omega)];$$
$$\alpha_j = \arg\min_{a\geq 0} \Phi\left[F_j(\omega) - \alpha \nabla_F \Phi[n^2(z), F_j(\omega)]\right];$$

b) the method of conjugate gradients (for velocity):

$$n^2_{j+1}(z) = n^2_j(z) - \alpha_j P_j(z),$$
$$\alpha_j = \arg\min_{a\geq 0} \Phi\left[n^2_j(z) - \alpha P_j(z)\right],$$
$$P_0(z) = \nabla^2_n \Phi\left[n^2_0(z)\right],$$
$$P_j(z) = \nabla^2_n \Phi\left[n^2_j(z)\right] - \beta_j P_j(z),\ j \geq 1,$$
$$\beta_j = \left(\nabla^2_n \Phi\left[n^2_j(z)\right], \nabla^2_n \Phi\left[n^2_{j-1}(z)\right] - \nabla^2_n \Phi\left[n^2_j(z)\right]\right).$$

The step α_j in both methods was chosen using the method of "golden section".

These methods were found to be very efficient for a wide range of problems.

Program package GEOMIN elaborated by Avdeev et al. (1992) was used for the calculations. Program blocks were written so as to organize the iterative process of simultaneous determination of the sounding signal form and the wave propagation velocity in a vertically-inhomogeneous medium.

The chosen velocity model of the medium is shown in Fig. 1. The impulse in the spectral domain (Fig. 3-4) was given in the following form:

$$F(\omega) = \left[\exp\left(-\left(\tfrac{\omega - 2\pi f}{\pi f}\right)^2\right) + \exp\left(-\left(\tfrac{\omega + 2\pi f}{\pi f}\right)^2\right)\right] \tag{11}$$
$$\times \exp(-\mathrm{i}\, 1.75\omega/f).$$

The initial approximation of the impulse was chosen to be zero (Fig. 2–3), and that of the velocity was taken as a linear function (Fig. 1).

All calculations were made for the time frequency range from 10 to 40 Hz, typical for seismic prospecting.

Following Alekseev et al. (1993), space frequency ranges were chosen so as to provide the best convergence of the iterative process to the minimum point of the data misfit functional.

The first approximations to impulse and to velocity are shown in Fig. 2–3, and Fig. 4, respectively. The final results of calculations are given in Fig. 5-7.

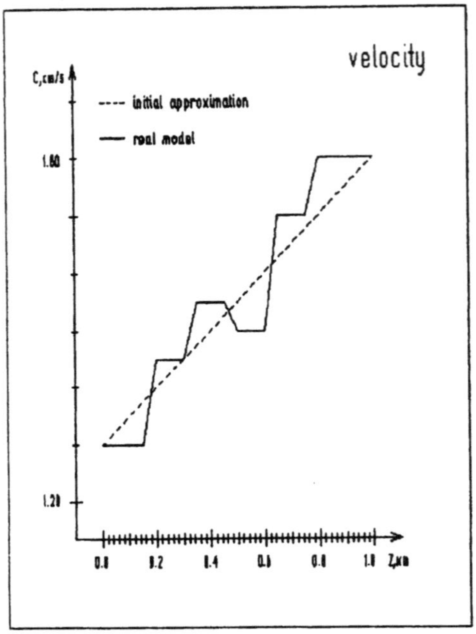

FIGURE 1.

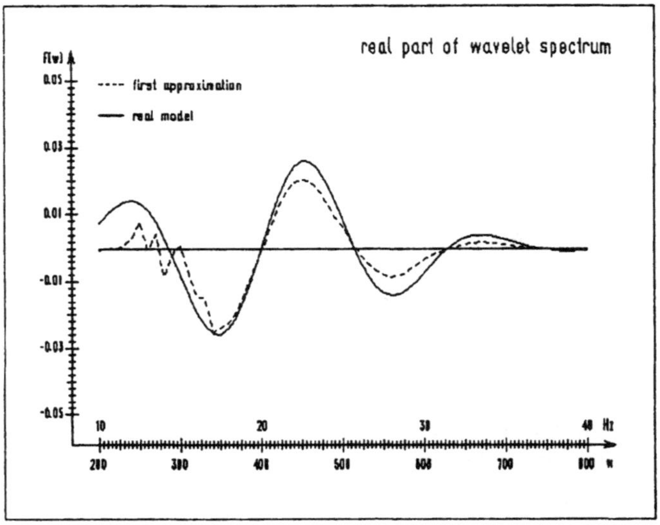

FIGURE 2.

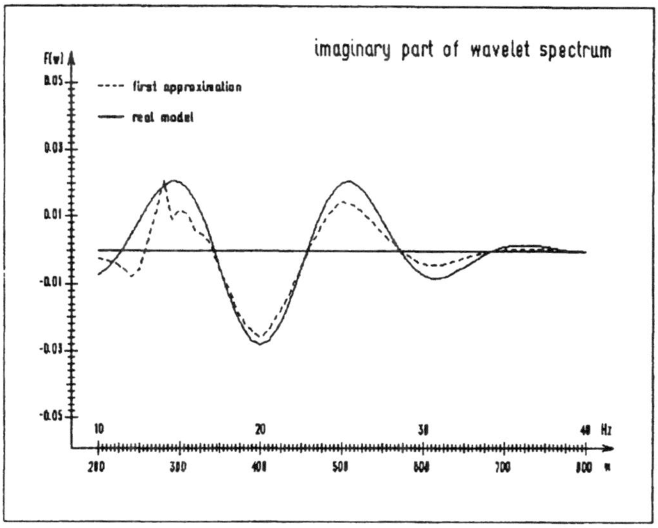

FIGURE 3.

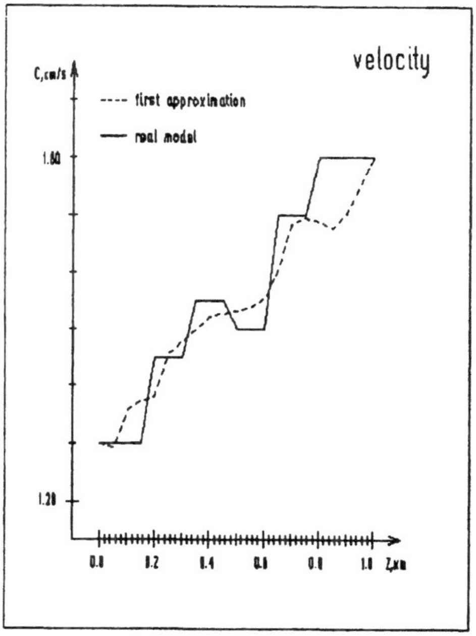

FIGURE 4.

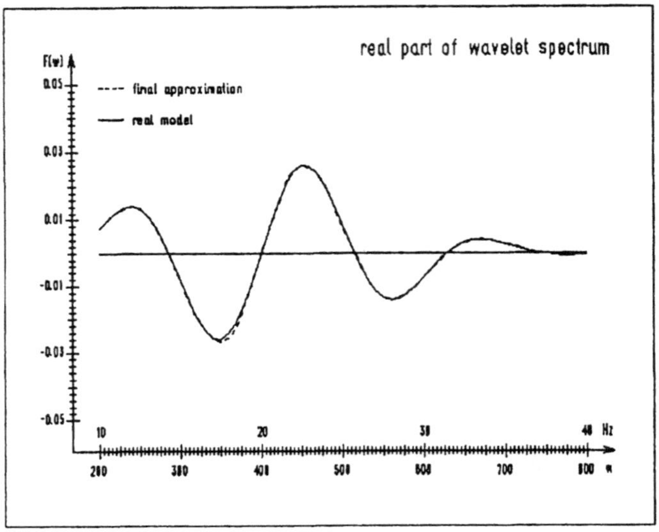

FIGURE 5.

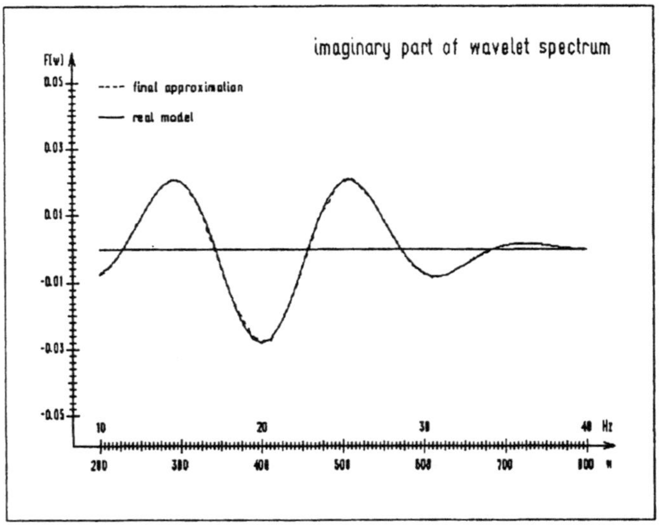

FIGURE 6.

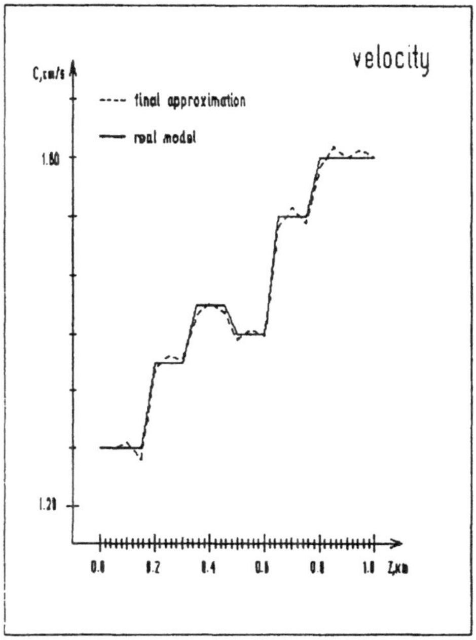

FIGURE 7.

REFERENCES

Alekseev A. S., Avdeev A. V., Fatianov A. G. and Cheverda V. A. (1993), Wave processes in vertically-inhomogeneous media: a new strategy for a velocity inversion. *Inverse Problems*, vol. 9, no. 3, 367–390.

Avdeev A. V., Vinogradov S. P., Derji N. M. and Cheverda V. A. (1992), Program package realizing a closed cycle of modeling of wave processes in vertically-inhomogeneous media on the base of an optimization approach. *Abstracts of All-Union Conference "Conditionally Well-Posed Problems of Mathematical Physics and Analysis"* (Novosibirsk), 139. (in Russian)

Blagoveshchenskii A. S. (1970), Inverse problems for the wave equation with an unknown source wavelet. In: *Problems of Mathematical Physics*, 4, Leningrad: Nauka, 27–39. (in Russian)

Carrion P. M., Sacramento S. dos S. and Pestana R. da C. (1990), Source wavelet and its angular spectrum from plane-wave seismograms. *Geophysics*, 55(8), 1026–1034.

Fatianov A. G. and Mikhailenko B. G. (1988), Methods for calculating non-stationary wave fields in non-elastic stratified inhomogeneous media. *Dokl. AN SSSR*, 301(4), 834–39. (in Russian)

Kolb P., Collino F. and Lailly R. (1986), Prestack inversion of a 1-D medium. *Proc. Inst. Electr. Electron. Eng.*, 74(4), 498–508.

Lavrent'ev M. M., Reznitckaya K. G. and Yakhno V. G. (1982), *One-dimensional Inverse Problems of Mathematical Physics.* Novosibirsk: Nauka, 1–88.

McGillivray P. R. and Oldenbourgh D. W. (1990), Methods for calculating Frechet derivatives and sensitivities for the non-linear inverse problems: a comparative study. *Geophysical Prospecting*, 38(5), 499–524.

Tarantola A. (1984), Inversion of seismic reflection data in the acoustic approximation. *Geophysics*, 49(8), 1259–1266.

Vector Tomography of Plasmas Using Faraday Rotation

A. BALANDIN[1], G. FUCHS, V. PICKALOV[2], J. RAPP and H. SOLTWISCH

Institit für Plasmaphysik ASS–Euratom, Forschungszentrum Jülich GmbH,
D–52425 Jülich Germany

[1]Siberian Energy Institute, 664033 Irkutsk Russia

[2]Institute of Theor. and Appl. Mech. 630090 Novosibirsk Russia

ABSTRACT

A tomographic method widely based on analytical techniques and working for arbitrary cross–sections is presented. The case is simulated for Faraday rotation on tokamaks with circular boundary in the poloidal plane. It allows in a straight–forward manner to derive the poloidal magnetic flux function Ψ from measurements of the Faraday rotation angle. The described method relies on the premises that the plasma density is known and that it depends on the flux function only.

1. Introduction

The poloidal magnetic flux function is the key function for most physical processes in tokamaks. Unfortunately this function is hard to calculate, because the Shafranov equation (Shafranov, 1966) does not provide a way to determine it uniquely.
Most measurements from the plasma core in tokamaks are based on methods, which give line integrated data (Sauthoff and von Goeler, 1979; Myers and Levin, 1978) and require some inversion procedure (Natterer, 1986).
The Ψ function is closely related to the vector functions of the poloidal magnetic field $\vec{B}_p$ and the current density $\vec{j}$. We derive a relation between a generalized flux function and the Faraday angle α by which the electric vector of a polarized electromagnetic wave is rotated, when the wave is passed through a magnetized plasma (De Marco and Segre, 1972; Soltwisch, 1992).

2. Theory

The angle α at which the direction of the electric vector of a polarized electromagnetic wave is changed when it passes through a plasma immersed in a magnetic field is given by Soltwisch (1992).

$$\alpha(p,\vartheta) = \gamma \int\limits_{L(p,\vartheta)} n_e(x,y)\vec{B}(x,y)\cdot \mathrm{d}\vec{l}, \tag{1}$$

where γ is a constant, n_e the electron density, $\vec{B}$ the magnetic field strength and $\mathrm{d}\vec{l}$ the line element along a line of sight L (cf. Fig. 1).

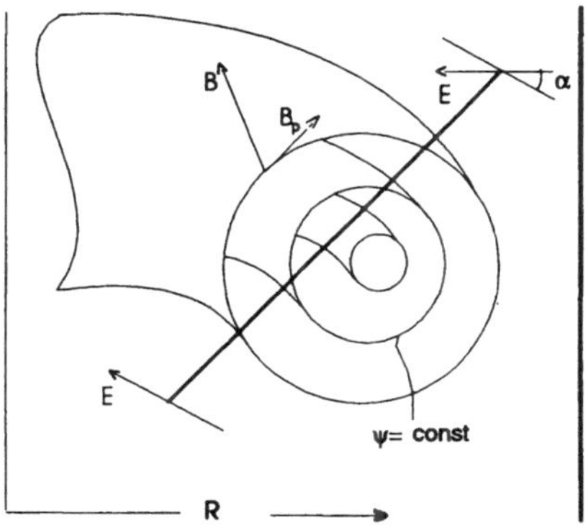

Figure 1: The figure explains the parameters used throughout the paper and shows schematically the isocontours of Ψ on TEXTOR.

We want to apply our method to a tokamak, where – neglecting the diamagnetic effect – the toroidal component of $\vec{B}$ is known from the coil configuration and we have to evaluate the poloidal component of $\vec{B}$ only. $\vec{B}_p$ can be expressed by a vector potential $\vec{A}$. Because $\vec{B}_p$ lies in the poloidal plane, $\vec{A}$ has a toroidal component A_φ only and we can derive $\vec{B}_p$ from the poloidal magnetic flux Ψ in the following manner:

$$\vec{B} = \nabla \times \vec{A} \tag{2}$$

$$\Psi = A_\varphi \cdot R \tag{3}$$

$$R = R_0 + x \tag{4}$$

We introduce a generalized vector potential $Q(x, y)$:

$$\gamma\, n_e(x,y)\vec{B}_p = \nabla \times (Q\,\vec{e}_\varphi) \tag{5}$$

$$= \frac{\partial Q}{\partial y}\vec{e}_x - \frac{\partial Q}{\partial x}\vec{e}_y \tag{6}$$

Using

$$\mathrm{d}\vec{l} = \{-\sin\vartheta, \cos\vartheta\}\mathrm{d}l = \vec{e}_l \mathrm{d}l \tag{7}$$

and equ. 6 we may rewrite equ. 1 as

$$\alpha(p,\vartheta) = -\sin(\vartheta)\int \frac{\partial Q}{\partial y}\,\mathrm{d}l - \cos(\vartheta)\int \frac{\partial Q}{\partial x}\,\mathrm{d}l \tag{8}$$

If the coordinate system is rotated by the angle ϑ, i.e. such that the new variable x' is in the direction of p, we get:

$$\alpha(x',\vartheta) = -\int \frac{\partial Q}{\partial x'}\mathrm{d}y' = -\frac{\partial}{\partial x'}\int Q\,\mathrm{d}y' \tag{9}$$

Rewriting this into

$$\alpha(x',\vartheta) = -\frac{\mathrm{d}}{\mathrm{d}x'}\mathcal{R}\{Q(x',y')\}, \tag{10}$$

$$\mathcal{R}\{Q(x',y')\} = -\int \alpha(x',\vartheta)\,\mathrm{d}x', \tag{11}$$

$$Q(x,y) = -\mathcal{R}^{-1}\left\{\int \alpha(p,\vartheta)\,\mathrm{d}p\right\}, \tag{12}$$

where we used $\mathcal{R}$ for the Radon operator and $\mathcal{R}^{-1}$ for its inverse, we get a simple relation between the measured α and the generalized vectorpotential Q. After some algebra we get for the Ψ function

$$\Psi(x,y) = R(x)\int_{\sqrt{a^2-x^2}}^{y} B_x(x,y)\,\mathrm{d}y, \tag{13}$$

$$\Psi(x,y) = R(x)\int_{\sqrt{a^2-x^2}}^{y} \frac{\partial Q}{\partial y}\frac{\mathrm{d}y}{\gamma n_e(x,y)}. \tag{14}$$

It should be mentioned, that relations similar to equ. 10–12 have been used for quite different diagnostics as schlieren- and speckle- tomography (Faris and Byer, 1987; Okoshi and Nishimura, 1981).

3. Computer Simulations

In order to demonstrate the usefulness of the tomographic method we choose a model function Q

$$s_i = \sin \beta_i \quad c_i = \cos \beta_i \tag{15}$$

$$Q(x,y) = \sum_{i=1}^{4} k_i \exp\left(-\frac{[(x-x_i)c_i + (y-y_i)s_i]^2}{a_i^2/4\ln 2} - \frac{[(x-x_i)s_i + (y-y_i)c_i]^2}{b_i^2/4\ln 2}\right) \tag{16}$$

with the different constants given in Table 1. The functions Q and ∇Q are shown in Fig. 2.

i	k_i	x_i	y_i	a_i	b_i	β_i
1	1.0	0	0	0.65	0.65	0
2	-0.97	0	0	0.6	0.58	$-\pi/12$
3	0.12	0.04	0	0.7	0.6	0
4	0.12	-0.04	0	0.7	0.6	0

Table 1:

For the reconstruction we used a set of 12 viewing directions equally spaced in angle with $0 \leq \vartheta < \pi$ and with 33 beams each. From equ. 10 we calculated the Faraday angle α analytically. To the α we added 3% of random normal noise. Afterwards we used these data as input for the inversion according (for numerical algorithms see Pickalov and Melnikova, 1984)

$$Q(x,y) = -\frac{1}{2\pi^2}\int_0^\pi d\vartheta \int_{-\infty}^{\infty} \frac{\alpha_\vartheta(p)}{p-p_0} dp \tag{17}$$

with $p_0 = x\cos\vartheta + y\sin\vartheta$.
The reconstructed tomograms Q and $|\nabla Q|$ are shown in Fig. 4b, 4d respectively. The average error of the reconstruction is 8.5%. To calculate the derivative we used biquadratic interpolation.
To reconstruct $\vec{B}_p$ requires that n_e is known, which in practice usually is derived from another measurement. In most practical cases one has at the boundary $n_e = 0$. However, $\vec{B}_p$ can either be measured there by other diagnostics or physically meaningful boundary conditions can be set.
The method presented here may possibly be extended to other vector tomographic problems e.g. determination of velocity fields (Koslover and McWilliams, 1986; Zhang et al., 1990). Tomographic measurements of the Faraday rotation angle on tokamaks have been done up to now with 1 viewing direction only, the inversion does therefore require additional assumptions.

Acknowledgement

This work was funded by the Deutsche Forschungsgemeinschaft and the Russian Academy of Science.

REFERENCES

De Marco F., Segre S.E. (1972), The polarization of an e.m. wave propagating in a plasma with magnetic shear. The measurement of poloidal magnetic field in tokamak. *Plasma Phys.*, vol. 14, 245–252.

Faris G.W., Byer R.L. (1987), Beam-deflection optical tomography. *Optics Letters*, vol. 12, no. 2, 72–74.

Koslover R., McWilliams (1986), Measurement of multidimensional ion velocity distributions by optical tomography. *Rev. Sci. Instr.*, vol. 57, no. 10, 2441–2448.

Myers B.R., Levin M.A. (1978), Two-dimensional spectral line emission reconstruction as a plasma diagnostic tool. *Rev. Sci. Instrum.*, vol. 49, no. 5, 610–616.

Natterer F. (1986), *The Mathematics of Computerized Tomography.* Teubner, Stuttgart, 222 p.

Okoshi T., Nishimura M. (1981), Measurement of axially nonsymmetrical refractive-index distributions of optical fiber performs by a triangular mask method. *Appl. Opt.*, vol. 20, no. 14, 2407–2411.

Pickalov V.V. and Melnikova T.S. (1984), Tomographic measurements of plasma temperature fields. *Beitr. Plasmaphys.* vol. 24, 1984, 417–430.

Sauthoff N.R., von Goeler S. (1979), Techniques for the reconstruction of two-dimensional images from projections. *IEEE Trans. Plasma Sci.*, vol. 7, no. 3, 141–147.

Shafranov V.D. (1966), Plasma Equilibrium in a Magnetic Field, *Rev. Plasma Phys.*, vol. 2, 103. Consultants Bureau New York.

Soltwisch H. (1992) Current density measurements in tokamak devices. *Plasma Phys. and Contr. Fusion*, vol. 34, no. 12, 1669–1698.

Zhang Y., Coplan M.A., Moore J.H., Berenstein C.A. (1990), Computerized tomographic imaging for space plasma physics. *J. Appl. Phys.*, vol. 68, no. 11, 5883–5889.

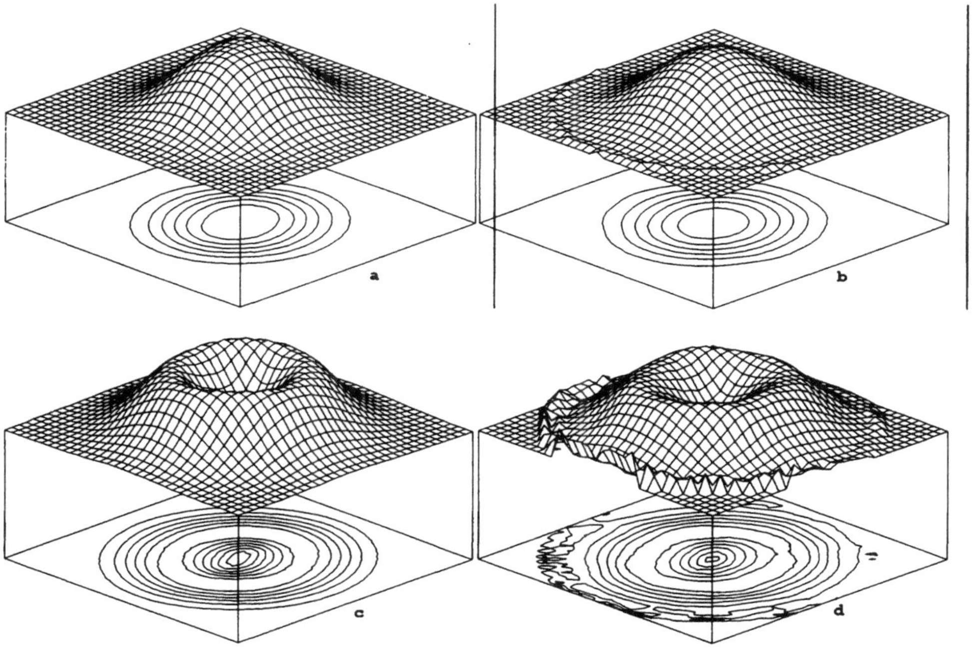

Figure 2: a) the model function Q and b) its reconstructed tomogram; c) the exact $|\nabla Q|$ and d) its reconstructed tomogram.

A Variational Approach to Non-Linear Backprojection

V .A .BARANOV

Research Institute of Introscopy, Tomsk, Russia

ABSTRACT

The problem of tomographic reconstruction from non-negative projection data is considered. The concept of non-linear backprojections (NLBP) as some estimators of presumably non-negative solution of integral projection equations is introduced. The minimal projection method is demonstrated to be a solution of minimization problems with constraints for norms of discrepancy functions on the class of NLBP-estimates. A comparison of the classical backprojection with the minimal projection method is performed. Interrelations between NLBP and POCS (projections onto convex sets) methods are revealed. The results of computer simulation and tomographic reconstruction are depicted.

Non-linear reconstruction techniques are being increasingly used in a present-day tomography and image processing (Gordon et al., 1970; Gordon, 1974; Frieden, 1972; Wernecke and d'Addario, 1977; Minerbo, 1979). There are several points supporting this line of investigation, at least one of them is obvious and lies in fact that non-linear estimates of integral projection equations solutions may be more exact than linear ones and require little or no additional computational cost. What is more, there exist important problems, such as a 3D tomography from a small number of views (Gordon et al., 1970; Minerbo, 1979 Gull and Daniell, 1978; Hanson, 1982), for which the approaches based on the "linear philosophy" and linear estimates are unsuitable.

Nowadays, almost for all intents and purposes 3D retrieval techniques are more preferable than 2D ones. At the same time, 3D reconstruction is computationally intense and the primary source of troubles with 3D tomography consists in the peculiarities of the well-known 3D reconstruction algorithms — "pseudo-universal" ones — requiring excessive computational times even with the use of modern powerful array processors. However, just the basic mathematical concept predetermines to a great extent a hardware configuration of future imaging system. When designing such a system, one should carefully classify possible lines of its application and strive for an optimum balance between conflicting requirements to reduce the cost and to widen the application area. Under these conditions the use of non-linear reconstruction from a small number of views is almost inevitable, the more so as the number of projections is the main parameter determining the system complexity and therefore its hardware cost.

Most commonly, non-linear techniques elaborated to date have not been adequately studied and substantiated as compared with linear ones. As a rule, they are statistical in nature and sometimes, in actual fact, heuristic procedures. Among them methods based on variational principles of statistics (like the maximum entropy and the maximum likelihood (Gordon et al., 1970; Gordon, 1974; Frieden, 1972; Wernecke and d'Addario, 1977; Minerbo, 1979) are most justified and mathematically correct. So, when investigating sparse-data image recovery, such techniques as well as Bayesian methods (Hanson and Wecksung, 1983) are a major focus of interest.

Now it is commonplace that the inverse problem from limited data is unsolvable. In other words, because the problem is ill-conditioned, a unique solution does not exist and one may hope at most to find some "pseudo-solution". In these circumstances variational principles may be used for stabilization of solution and play a similar part as regularization procedures in the theory of ill-posed problems (Tikhonov and Arsenin, 1977). Typically, such an approach leads to iterative techniques (Gordon et al., 1970; Minerbo, 1979), which being flexible and convenient enough for an account of a priory and a posteriory data, place, however, considerable demands on computer speed, memory and disk space, demands which become pressing in 3D case. More often than not, computation times are not feasible for NDT applications, to say nothing of inadequacy of such techniques for on-line testing and for imaging at real time rates. The grave drawback of iterative techniques is an impossibility to reconstruct any given cross-section irrespective to others. These disadvantages retard important applications sending us in search of alternative ways of tomographic imaging or at least of some efficient non-linear estimates to improve existent recovery methods.

In this connection it is pertinent to distinguish a fresh approach to the backprojection method which belongs to most old and yet conceptually attractive ones in tomography.

Consider the problem of tomographic reconstruction of 3D object from its 2D projections. For definiteness sake a registration geometry is supposed the same as in classical tomosynthesis. The test-object is spaced between a source plane and a registration plane which are parallel to one another. The position of an arbitrary point $\mathbf{r}$ is specified by the Cartesian coordinates x, y, z ($\mathbf{r} \equiv (x, y, z)$ is a radius vector of a point). The z-axis is perpendicular to the registration plane, so x, y are the plane coordinates. The origin of coordinates is on the registration plane ($z = 0$). N of fixed positions of the radiation source on the source plane ($z = F$, where F is a film-focus distance) correspond to N of projections on the registration plane. The position of an arbitrary cross-sectional plane is specified by the parameter z ($z < F$) — its "depth".

If a function $t(\mathbf{r})$ describes the internal structure of the object under investigation, it is bound to be a solution of N integral equations system

$$\int t(\mathbf{r}) dL_n(\mathbf{r}, \mathbf{R}) = P_n(\mathbf{R}), \quad (\forall n, n = 1 \rightarrow N; \forall \mathbf{R}, z = 0). \tag{1}$$

Here $P_n(\mathbf{R}) = P_n(\mathbf{r})$ are beam-sums incidental to the point $\mathbf{r}$ (in other words, corresponding to beams passing from the n-th source position through an arbitrary point of test-object with coordinates x, y, z). Although projections are two-dimensional, beam-sums are defined in the whole volume of interest ($0 < z < F$) and may be treated as functions of x, y, z. The integration in (1) is performed along the beams. (One may suppose that all the points $\mathbf{R}$ in (1) belong to some manifold, say, the cross-section $z = 0$).

In parallel with this we consider another (generalized and simplified) formulation of the problem. Let us assume that R and C are two ordered finite sets (with the cardinal numbers M and L, respectively) of some objects $\mu_m \in R$ ($\forall m, m = 1 \rightarrow M$) and $\sigma_l \in C$ ($\forall l, l = 1 \rightarrow L$), with each of them a real non-negative number is associated. We suppose that the set C contains all available measured data (to a greater or lesser extent degraded by noise) and the set R describes the unknown test-object to be reconstructed.

Elements of the set R may be treated as components of a vector μ in some M-dimensional space $\mathcal{L}_R$ and, similarly, elements of C are components of a vector σ in an L-dimensional space $\mathcal{L}_C$. (To be more specific, μ belongs to the convex subset of $\mathcal{L}_R$ — its first quadrant. Similarly, σ is in the first quadrant of $\mathcal{L}_C$). Data components $\sigma_l \in C$ can be matched one-to-one with some subsets of R ($r_l \subset R$) — samples from R. It is supposed that $\sigma_l = \phi(r_l)$ where $\phi(r)$ is an additive measure defined as $\phi(r) \equiv \sum_m \mu_m \quad (\forall \mu_m, \mu_m \in r)$ and the argument r is an arbitrary subset of R.

Consider a matrix I_{lm}, whose components are values of characteristic functions ("arithmetized" Boolean variables): $I_{lm} = 1$ if $\mu_m \in r_l$ and 0 otherwise. Then a linear matrix equation may be derived, which describes projection from $\mathcal{L}_R$ into the sample space $\mathcal{L}_C$:

$$\sum_{m=1}^{M} I_{lm}\mu_m = \sigma_l, \quad (\forall l, l = 1 \rightarrow L). \tag{2}$$

Matrix rows and columns can be matched one-to-one with elements from potency classes of R and C, respectively ($l \leftrightarrow r_l$, $m \leftrightarrow c_m \subset C, c_m$ is a list of data elements incidental to μ_m). We assume that $M \gg L$ (i.e., the data C are essentially incomplete). For overwhelming majority of matrix columns all I_{lm} are zero, so "null-space" (Louis, 1981) exists. Practically, M may be very large. Basically, in the limiting case $M \rightarrow \infty$ we may regard R as a countably infinite linearly ordered set.

We call I_{lm} the incidence matrix. Its destination is to describe interrelations between (presumably geometric) objects from two basic classes 1) "pixels" represented by columns of I_{lm}, 2) "beams" or maybe "spots" which are represented by rows of I_{lm}. In this case the data in the set C may be interpreted as "beam-sums" or "spot-sums". So, in the context of this scheme, all information connected with registration geometry is built into I_{lm} and one may forget geometric peculiarities of the problem.

(For brevity sake we call two varieties of problem formulations based on (1) and (2), respectively, as Scheme 1 (Sch. 1) and Scheme 2 (Sch. 2). It is easy to express results obtained with the use of Sch. 2 in terms of Sch. 1 and vice versa. One may regard Sch. 1 as a pragmatic problem formulation and Sch. 2 as some theoretical tool for its investigation.)

Equations (1) as well as (2) may be solved with diverse additional suppositions in order to stabilize the solution using straightforward or iterative techniques. In doing so, various estimates $T(\mathbf{r})$ of the solution $t(\mathbf{r})$ are commonly used, and among them the use of backprojection estimate, i.e., the sum (or normalized sum) of beam-sums

$$T\{x,y,z\} = \frac{1}{N}\sum_{n=1}^{N} P_n(x,y,z) \tag{3}$$

is a standard practice. If we add, according tp the backprojection technique, to each "pixel" all available (i.e., incidental) "beam-sums", we can obtain (3) in terms of Sch. 2:

$$\mu_m = \sum_{l=1}^{L} I_{lm}\, \sigma_l. \tag{4}$$

So, the inverse problem in the framework of backprojection approach is solved using the transposed incidence matrix (in place of inverse or pseudoinverse one). It goes without saying that such a substitution can be considered as nothing more than a rough approximation but, on the other hand, the inversion formula (4) is presumed to be initial material to construct other more exact estimations. The most popular CT reconstruction algorithms are based on diverse improvements of the basic "pre-method" (3) at the expense of filtering (in spatial or frequency domain) at the stages of preprocessing or postprocessing or both. These methods are not without merit but share a common grave disadvantage — their inadequacy in the case when the number of views is small.

Nevertheless, it should be realized that there exists another way for generalization of the "pre-method" (3), namely, the use of non-linear estimates of the type

$$T = T\{x, y, z\} = f(P_1, P_2, \dots, P_N) \tag{5}$$

which we call estimates of non-linear backprojection (the NLBP-estimates). In general case f is a non-linear function of N non-negative variables (beam-sums $P_1, P_2, \dots, P_N$ incidental to x, y, z), which is non-negative and symmetric (i.e., invariant with respect to any permutation of its arguments). From the standpoint of Sch. 2, the NLBP-estimate may be expressed as $\nu_m = f(c_m)$. (Most commonly, we have for 3D tomography

$$\sum_{m=1}^{M} I_{lm} = K = \text{const}\,, \ \sum_{l=1}^{L} I_{lm} = N = \text{const}\,, \tag{6}$$

where K is the number of layers and N is the number of sources. Therefore, as in the case of (5), f is a function of N arguments, although the second equation (6) is violated in the case when the m-th column describes a source position).

In the case $N = 1$, a difference between (3) and (5) does not exist and (5) is simply a propagator $f(P_n) = T\{x, y, z\} = P_n(\mathbf{r}) = P_n(\mathbf{R})$ of the projection values (beam-sums $P_n(\mathbf{R})$) to all the pixels $\mathbf{r}$ incidental to $\mathbf{R}$. The NLBP-estimates may be either analytical or statistical in nature. Normally, we take a statistical approach, so that the function f may be treated as some statistic — a function of a random sample $P_1, P_2, \dots, P_N$. For instance, in the special case of (3) f is linear averaging. In another important (essentially non-linear) special case of (5),

$$T = T\{x, y, z\} = \min(P_1, P_2, \dots, P_N) \tag{7}$$

is the NLBP-estimate widely used in the "minimal projection method" elaborated by the author (Baranov and Chekalin, 1988; Baranov et al., 1989, 1990; Baranov and Janisov, 1986). One of the basic constituents of the NLBP-concept is that (5), in fact, is some non-linear averaging to be used instead of (3), so the class of estimates defined as the subset of (5) by the constraints

$$\min(P_1, P_2, \dots, P_N) \le f \le \max(P_1, P_2, \dots, P_N) \tag{8}$$

is of prime importance. From the conceptual point of view, two types of estimates in class (8) are most interesting: 1) order statistics

$$f = S_k(P_1, P_2, \dots, P_N) \tag{9}$$

(say, the median, in the case $k = N/2$ or $(N+1)/2$, or the minimal projection estimate (7), in the case $k = 1$) and 2) non-linear averaging of the form

$$F(f) = \frac{1}{N} \sum_{n=1}^{N} F(P_n), \tag{10}$$

where $F(P)$ in (10) is a monotone non-negative function. So, the non-linear equation (10) is solvable with respect to f, and the inverse transform F^{-1} exists (e.g. f is the geometric mean for $F(P) = \ln P$, the harmonic mean for $F(P) = 1/P$, etc.). It is more preferable to regard $F(P)$ as bijections $[0,1] \leftrightarrow [0,1]$ (a subset of (10).) – a group of non-linear transformations with composition as the group operation and the identity function $F(P) = P$ as the group unity. In this case the monotonically increasing functions $F(P)$ may be treated as some probability distributions.

The NLBP-estimates, obtained by one way or another, may be equally used either in iterative or straightforward (say, some generalizations of the convolutional backprojection) algorithms. One of the early steps toward the NLBP-approach (5) has been attempted by Pickalov and Sharapova. Their estimate (Pickalov and Sharapova, 1989) is equal to an ordinary backprojection solution (3) in the case when all the beam-sums $P_1, P_2, \ldots, P_N$ are greater than some threshold value P_0 and is 0 otherwise. Numerous NLBP-type algorithms elaborated by the author (Baranov and Chekalin, 1988; Baranov et al., 1989b, 1990; Baranov and Janisov, 1986) for NDT applications were proved to be successful (especially for reconstruction from limited data).

Now when non-linear backprojection methods are well developed, it is necessary to answer the following questions: 1) What are (from the mathematical viewpoint) the reasons for the use of the NLBP-approach? (In particular, why the NLBP-estimates may be more exact than linear ones, say, classical backprojection?). 2) What is the systematic pathway (if exists) to derive non-linear backprojection operators?

It is more realistic to suppose that the NLBP-estimates are connected directly with minimization problems for some functionals (Baranov, 1993) and may be obtained as solutions of these problems with constraints. There exist diverse ways to construct such functionals, and from a conceptual point of view, it is more worthwhile to use variational principles of statistics (such as the minimum chi-squared, the maximum likelihood and especially the maximum entropy) as the essentials for constructions of this kind. This line of research involves also the variational derivation of space filtering procedures for non-linear backprojectors and the variational approach to iterative techniques. However, because this is quite extensive work to be the subject matter for a single paper, we shall restrict our consideration predominantly to some functionals of quadratic and Chebyshev types with the aim to reveal the nature of non-linear backprojection.

Reasoning from this, we transform basic integral equations (1) into an equivalent variational problem. Define the discrepancy function $d(T)$ (depending on $T(\mathbf{r})$ — some estimate of the solution $t(\mathbf{r})$ of (1)) as the difference between the right and the left sides of (1):

$$d(T) = P_n(\mathbf{R}) - \int T(\mathbf{r})dL_n(\mathbf{r},\mathbf{R}). \tag{11}$$

According to the general concept of approximation theory, one can construct a measure of the discrepancy between $T(\mathbf{r})$ and $t(\mathbf{r})$ as a non-negative functional — a norm of the discrepancy function (11):

$$\Psi(T) = \| d(T) \| = \| P_n(\mathbf{R}) - \int T(\mathbf{r})dL_n(\mathbf{r},\mathbf{R}) \| . \tag{12}$$

Supposedly, the functional (12) is bound to reach zero minimum provided $T(\mathbf{r}) = t(\mathbf{r})$ and vice versa. (However, since (11) is a vector in the sample space, (12) is unaffected by null-space components). Starting from variational principles, one should solve an equation $\delta\Psi(T) = 0$ in order to find $T(\mathbf{r})$. In fact, by its very nature (in view of limited data, etc.) this problem should be formulated as minimization with constraints (most closely corresponding to the class of approximation functions $T(\mathbf{r})$), especially with constraints induced by available information about the object of interest. In this connection, it is pertinent to discriminate two possible situations: +) The only additional information is non-negativity of $t(\mathbf{r})$ and $P_n(\mathbf{R}$. (This condition is in accordance with the definition of non-linear backprojection (5)). ++) Apart from "+", there is some information (a priori or deduced from the basic equation (2)) concerned with special features of object-type. It is self-evident that in the case "++" estimations should be more exact than in the case "+". But above all it is necessary to solve the problem with minimal additional information (the case "+").

An estimate $T(\mathbf{r})$ may be, in particular, dependent on additional parameters $\gamma_1, \ldots, \gamma_Q$, whose values are to be determined by minimization:

$$T(\mathbf{r}) = f(\mathbf{r};\ \gamma_1, \ldots, \gamma_Q). \tag{13}$$

So, it is possible to take the NLBP-estimate in the form

$$T(\mathbf{r}) = f(P_1, P_2, \ldots, P_N;\ \gamma_1, \ldots, \gamma_Q) \tag{14}$$

and to find the parameters $\gamma_1, \ldots, \gamma_Q$ by well-known methods, for example, by solving a system of equations $\frac{\partial \Psi}{\partial \gamma_q} = 0 \quad (\forall q, q = 1 \to Q)$. The parameters number Q seems to be small enough to solve the system numerically. Thus defined estimate (14) with the determined $\gamma_1, \ldots, \gamma_Q$ may be used repetitively for the object-type in question as an algorithm constituent. Moreover, this approach offers a possibility to create adaptive-type algorithms, whose parameters depend upon characteristics of the test-object. Of course, non-parametric minimization is also possible. (In this case the value of $T(\mathbf{r})$ for each pixel may be treated as a minimization parameter, so that the number of parameters is equal to the number of pixels).

Fundamentally, two types of norm — the Euclid and the Chebyshev ones — are of essence and should be the focus of attention. Expressed in an explicit form with the use of these norms, the discrepancy functionals (12) are respectively

$$\Psi_{MSQ}(T) = (\| d \|_2)^2 = \int \sum_{n=1}^{N} \{P_n(\mathbf{R}) - \int T(\mathbf{r}) dL_n(\mathbf{r}, \mathbf{R})\}^2 dX dY \tag{15}$$

for the Euclid norm and

$$\Psi_{CHEB}(T) = \| d \|_\infty = \max_{X,Y,n} \mid P_n(\mathbf{R}) - \int T(\mathbf{r}) dL_n(\mathbf{r}, \mathbf{R}) \mid \tag{16}$$

for the Chebyshev one. The integration $(dXdY)$ in (15) is performed over the registration plane $(\mathbf{R} = (X, Y, 0))$ and in the general case over some manifold. The domain of definition of X, Y and n in (16) is the same as in (15).

Consider non-parametric minimization of (15) and (16) on the class of non-linear backprojections (5) with constraints. Assume that data are non-degraded. First we consider some properties of upper and lower estimates of $t(\mathbf{r})$. Because it is more convenient to work in the context of non-geometric approach (Sch. 2), rewrite (15) and (16) respectively as

$$\Psi_{msq}(\nu) = \sum_{l=1}^{L} (\sigma_l - \sum_{m=1}^{M} I_{lm} \nu_m)^2 \tag{17}$$

and

$$\Psi_{cheb}(\nu) = \max_{l=1}^{L} \mid \sigma_l - \sum_{m=1}^{M} I_{lm} \nu_m \mid . \tag{18}$$

The vector ν ($\nu \in \mathcal{L}_R$) in (17) and (18) is some estimation of the solution μ, so the discrepancy vector $d(\nu)$ ($d \in \mathcal{L}_C$) is defined as

$$d(\nu) = \sigma_l - \sum_{m=1}^{M} I_{lm}\nu_m. \tag{19}$$

Suppose that ν_{upp} and ν_{UPP} are two upper estimates of μ such that the inequalities hold:

$$\nu_{UPP,j} \geq \nu_{upp,j} \geq \mu_j, \ (\forall j, j = 1 \rightarrow M) \tag{20}$$

and, respectively, d_{upp} and d_{UPP} are the associated discrepancy vectors (19) for these estimates. For an arbitrary upper estimate, each component of the discrepancy vector is non-positive, $d_l \leq 0$ ($\forall l, l = 1 \rightarrow L$), and we have the sequence of inequalities

$$\nu_{UPP,j} \geq \nu_{upp,j} \Rightarrow \ d_{UPP,l} \leq d_{upp,l} \Rightarrow \mid d_{UPP,l} \mid \geq \mid d_{upp,l} \mid \Rightarrow \tag{21}$$

$$\Rightarrow \max_{l=1}^{L} \mid d_{UPP,l} \mid \geq \max_{l=1}^{L} \mid d_{upp,l} \mid$$

$$\Rightarrow (d_{UPP,l})^2 \geq (d_{upp,l})^2 \Rightarrow \sum_{l=1}^{L}(d_{UPP,l})^2 \geq \sum_{l=1}^{L}(d_{upp,l})^2.$$

Hence,

$$\Psi_{msq}(\nu_{UPP}) \geq \Psi_{msq}(\nu_{upp}), \quad \Psi_{cheb}(\nu_{UPP}) \geq \Psi_{cheb}(\nu_{upp}) \tag{22}$$

Similarly, for two arbitrary lower estimates ν_{low} and ν_{LOW},

$$\nu_{LOW,j} \leq \nu_{low,j} \leq \mu_j, \ (\forall j, j = 1 \rightarrow M), \tag{23}$$

components of discrepancy vectors are non-negative and the sequence of inequalities holds:

$$\nu_{LOW,j} \leq \nu_{low,j} \Rightarrow \ d_{LOW,l} \geq d_{low,l} \Rightarrow \mid d_{LOW,l} \mid \geq \mid d_{low,l} \mid \Rightarrow \tag{24}$$

$$\Rightarrow \max_{l=1}^{L} \mid d_{LOW,l} \mid \geq \max_{l=1}^{L} \mid d_{low,l} \mid$$

$$\Rightarrow (d_{LOW,l})^2 \geq (d_{low,l})^2 \Rightarrow \sum_{l=1}^{L}(d_{LOW,l})^2 \geq \sum_{l=1}^{L}(d_{low,l})^2,$$

and

$$\Psi_{msq}(\nu_{LOW}) \geq \Psi_{msq}(\nu_{low}), \quad \Psi_{cheb}(\nu_{LOW}) \geq \Psi_{cheb}(\nu_{low}) \tag{25}$$

According to (25), $\Psi_{msq}(\nu_{LOW})$ and $\Psi_{cheb}(\nu_{LOW})$ are maximal in the limiting case $\nu_{LOW} = 0$. So, for an arbitrary lower estimate ν_{low}, we obtain

$$\Psi_{msq}(\nu_{low}) \leq (\| \sigma \|_2)^2, \quad \Psi_{cheb}(\nu_{low}) \leq \| \sigma \|_\infty \tag{26}$$

Now, in view of (22), the problem may be refined as non-parametric minimization of (17) and (18) on the class of non-linear backprojections (5),

$$\nu_j = f(c_j), \tag{27}$$

with constraints (8):

$$\min_{l=1}^{L} \sigma_l \leq f(c_j) \leq \max_{l=1}^{L} \sigma_l, \ (\forall l, I_{lj} \neq 0). \tag{28}$$

Because μ_j is a summand of all the incidental beam-sums σ_l, the non-linear backprojector (7)

$$\nu_j = \min_{l=1}^{L} \sigma_l, \ (\forall l, I_{lj} \neq 0) \tag{29}$$

is an upper estimate of μ. And since (28) is a class of upper NLBP-estimates of μ due to (22), the limiting estimate (29) is an exact solution of our minimization problem.

It stands to reason that information built into the basic system of equations (2) is incompletely used in (29) (otherwise (17) and (18) would be equal to zero). Actually, a domain

$$\mu_m \leq \min_{l=1}^{L} \sigma_l, \ (\forall l, I_{lm} \neq 0) \tag{30}$$

determined by backprojector (29) is a solution of the system of $M * L$ linear inequalities

$$I_{lm}\mu_m \leq \sigma_l \ (\forall m, m = 1 \rightarrow M; \ \forall l, l = 1 \rightarrow L), \tag{31}$$

into which (2) is converted.

One can see an intimate relationship between the NLBP-approach elaborated by us and the POCS-methods (projections onto convex sets) (Gubin et al., 1967; Bregman, 1965; Youla, 1978). Each inequality (31) defines some convex domain b_{lm} in the first quadrant of $\mathcal{L}_R$ ($b_{lm} \subset \mathcal{L}_R$). So, the upper estimate (29) defines a convex domain $B(I,\sigma)$ equivalent to (30):

$$B(I,\sigma) = \bigcap_{m=1,l=1}^{M,L} b_{lm}, \ (\forall m, m = 1 \rightarrow M,; \ \forall l, l = 1 \rightarrow L), \tag{32}$$

a rectangular hyperparallelepiped, not counting components of the null-space. Each elementary convex constraint (31) will cause the values of the discrepancy functionals (17) and (18) to decrease (not to increase, at least). In the context of pure non-linear backprojection approach (5) (in accord with "+"), we know nothing about domain (32). However, if additional information (the case "++") may be represented as some convex constraint A, then a convex domain S may be defined as

$$S = A \cap B(I,\sigma) \tag{33}$$

so that $\mu \in S$. An additional portion of information connected with A will cause the values of (17) and (18) to decrease additionally. In fact, diverse suppositions about the test-object are often representable as convex constraints (33), so that many postprocessing procedures may be constructed as convex projections. For example, if a condition A lies in the fact that some lower estimate ν_{lA} of μ exists, then an upper estimate ν_{uA} may be obtained (Baranov and Chekalin, 1991b) so that the condition A may be regarded as a convex constraint of form (33).

(Using the notation of Sch. 1, it is convenient to express an estimate $T_A(\mathbf{r})$ connected with an additional constraint A in form (13) as a difference between (7) and some non-negative function Δ_A, where parameters $\gamma_1, \ldots, \gamma_Q$ are to be determined by minimization with constraints of (15) or (16)

$$T_A(\mathbf{r}) = \min_{n=1}^{N} P_n(\mathbf{r}) - \Delta_A(\mathbf{r}; \ \gamma_1, \ldots, \gamma_Q) \tag{34}$$

so that $T_A(\mathbf{r})$ is to be found in the convex domain (32). If $T_A(\mathbf{r})$ is an upper estimate, then its convex projection onto $B(I,\sigma)$ defines (33). The constraints connected with lower estimates may be also understood as convex ones. So, in the two-cycle iterative method worked out by the author (Baranov and Chekalin, 1991) two sequences of convex constraints related respectively to lower and upper estimates are used to find a non-negative solution).

Let us compare two backprojection operators — the classical backprojector (3,4) and the minimal projection method (7,29) (BP and MP, respectively). We assume that BP is normalized backprojection. First of all, $MP(\sigma) \leq BP(\sigma)$, whence, due to (22),

$$\Psi_{msq}(MP) \leq \Psi_{msq}(BP), \quad \Psi_{cheb}(MP) \leq \Psi_{cheb}(BP). \tag{35}$$

Whereas BP is a linear and additive method such that for an arbitrary λ and two data sets σ_1 and σ_2 we have

$$BP(\lambda\sigma) = \lambda BP(\sigma), \quad BP(\sigma_1 + \sigma_2) = BP(\sigma_1) + BP(\sigma_2), \tag{36}$$

MP is non-linear,

$$MP(\lambda\sigma) = \lambda MP(\sigma)\ (\lambda \geq 0), \quad MP(\sigma_1 + \sigma_2) \geq MP(\sigma_1) + MP(\sigma_2), \tag{37}$$

and multiplicative (from the standpoint of (32)). MP is equal to zero for an arbitrary pixel if only one of beam-sums incidental to this pixel is equal to zero.

The difference between two methods is more readily visualized after consideration of some functionals, which may be interpreted as measures of variation or similarity measures. So, one can derive the classical backprojection (Baranov, 1993) as a solution of a variation problem

$$\delta V(T) = \delta \int \sum_{n=1}^{N} \{P_n(\mathbf{r}) - T(\mathbf{r})\}^2 dxdydz = 0. \tag{38}$$

The integration in (38) is performed over an arbitrary region or the whole domain of definition of $T(\mathbf{r})$. Since the values of $T(\mathbf{r})$ for each pixel are considered as minimization parameters, it is necessary to solve the system of equations $\frac{\partial V}{\partial T(\mathbf{r})} = 0$ $(\forall \mathbf{r})$ and for an arbitrary small domain the equation $\int \sum_{n=1}^{N} \{P_n(\mathbf{r}) - T(\mathbf{r})\} dxdydz = 0$ holds. Thus, the exact solution of (38) is $T\{x, y, z\} = \frac{1}{N} \sum_{n=1}^{N} P_n$.

Let us construct a multiple similarity measure of N arbitrary non-negative functions $P_1(\mathbf{r}), P_2(\mathbf{r}), \ldots, P_N(\mathbf{r})$, $(\mathbf{r} = (x, y, z))$, (for instance, of our N projections $P_n(\mathbf{r}) = P_n(\mathbf{R})$) as

$$\Omega = v\,(\bigcap_{n=1}^{N} \{P_n\}) \tag{39}$$

or a normalized multiple measure as

$$\omega = \frac{v\,(\bigcap_{n=1}^{N} \{P_n\})}{v\,(\bigcup_{n=1}^{N} \{P_n\})}, \tag{40}$$

where $\{P\}$ is a domain $(\tau \leq P(\mathbf{r}))$ in the (x, y, z, τ)-space and v is some additive measure defined in this space, say, a "volume": $v = \int P(\mathbf{r}) dxdydz$.

It is natural to find a non-negative "common part" $T(\mathbf{r})$ of the functions $P_1(\mathbf{r}), P_2(\mathbf{r}), \ldots, P_N(\mathbf{r})$ as a solution of the minimization problem for the similarity measure (39):

$$\Omega(T) = v\,(\bigcap_{n=1}^{N} \{P_n(\mathbf{r}) - T(\mathbf{r})\}) = \min_{T(\mathbf{r})}. \tag{41}$$

One can easily see that the solution of (41) is the minimal projection method (7) and $\Omega(MP) = 0$. It is remarkable that in the context of MP-method an estimate T(**r**) results from some pure geometrical construction

$$\{T(\mathbf{r})\} = \bigcap_{n=1}^{N} \{P_n(\mathbf{r})\} \tag{42}$$

in the 4-dimensional space (or in the 3-dimensional (x, y, τ)-space if a separate cross-section is visualized).

As can be inferred from (38), the classical backprojection is statistical in nature. When considering possible approaches to improve BP-technique, account must be taken of two its aspects: 1) The data are degraded by noise (not necessarily additive). 2) Like many well-known methods, such as MENT (Minerbo, 1979), MART (Gordon et al., 1970), etc., backprojection is a deterministic procedure based on statistical principles. An estimation of an element μ_m in the framework of backprojection approach (classical or modified) is treated as a "local" problem because only N of "incidental" sample sums $\sigma_l = \phi(r_l)$ are used as data (for each of them μ_m is a summand). A classical solution derived by minimum of variance (exactly as in the case of minimization (38)) is the linear averaging (3, 4), so backprojection (4) is used as a classical noise-reduction procedure (deterministic in the case of non-degraded data, with a "noise" caused by other elements of the set R). However, one can see that the preconditions of limit theorems governing the probability convergence of backprojection (4) to a "true" value of μ_m are violated. Strictly speaking, the assumption of "noise" as being equally distributed and stochastically independent for two arbitrary elements is contrary to fact. What is more, the mathematical expectation of "noise" is non-zero. This is why backprojection (3, 4) is a rather poor approximation even if $N \to \infty$.

One way to improve the estimate (3) is to transform the probability measure of "noise" on the basis of (10). The function $F(P)$ is to be found through minimization of (15) or (16) or from the equation

$$\delta\ V_F(T) = \delta \int \sum_{n=1}^{N} \{F(P_n(\mathbf{r})) - F(T(\mathbf{r}))\}^2 dx dy dz = 0 \tag{43}$$

(a non-linear modification of (38)). The values of $F(P)$ play the part of parameters $\gamma_1, \ldots, \gamma_Q$ in (14), and a solution is representable in the form

$$T_F(\mathbf{r}) = F^{-1}\left(\frac{1}{N} \sum_{n=1}^{N} F(P_n(\mathbf{r}))\right). \tag{44}$$

If $F(p)$ is a fixed element in a class of transformations, then non-parametric minimization of $V_F(T)$ yields (44) (in particular, (3) if $F(p) = p$) and

$$V_F = N \int \left(\frac{1}{N} \sum_{n=1}^{N} F^2(P_n(\mathbf{r})) - \left(\frac{1}{N} \sum_{n=1}^{N} F(P_n(\mathbf{r}))\right)^2\right) dx dy dz\ \geq 0. \tag{45}$$

It is easy to see that the MP-method results from (44) as a limiting case. For example, if $F(p) = \frac{1}{p^\eta}$, then

$$\lim_{\eta \to \infty} T_F(\mathbf{r}) = \min_{n=1}^{N} P_n(\mathbf{r}). \tag{46}$$

According to (45), $V_F \to 0$ in the limiting case. Therefore, the minimal projection method (7) is an exact solution of the variational problem (43). Thus, there is no need to use (44) in the case of deterministic reconstruction. However, statistical characteristics of "noise" are of importance, especially in derivation of estimates more exact than MP (i.e., lower than MP) according to (34), the more so in elaboration of probabilistic non-convex-type space filtering procedures for (7).

In the case of noisy and contradictory data, the question of statistical stability of MP is still an open question. Unfortunately, the effect of noise generally is not accountable in the form of convex constraints (33). However, intuition suggests that the use of (44) serves to choose an optimum backprojector "between" MP and BP (between precision and stability), having regard to characteristics of noise. Another possibility to do so is connected with the use of order statistics (9). (The determination of parameters in (44) and (9) is beyond the scope of this paper).

Potentialities of non-linear backprojection techniques will be appreciated more readily after consideration of some examples of cross-sectional reconstruction from 2-dimensional projections.

In Figs. 1.1–1.4 the results of computer simulation are represented. The test-object (i.e., mathematically generated phantom) consists of 8 reciprocally parallel planes (or layers). The information content of any individual layer is the image of a letter (from the set A,B,C,...,H). In Fig. 1.1 four "shadowgrams" are depicted which are the results of "X-raying" of the test-object from four views. In Fig. 1.2 the reconstruction results of 2 layers of the test-object from 4 views are shown: at the left — using the backprojection technique (3) (classical tomosynthesis), at the right — using the minimal projection method (7). In Fig. 1.3 the reconstruction results of the same layers from 8 views are depicted. In Fig. 1.4 the reconstruction results of other 2 layers of the test-object from 4 views are shown.

In the experiment some results of which are represented in Figs. 2.1 and 2.2, a ferro-concrete wall (300 mm thick) contained 7 layers of reinforcement was X-rayed from 8 views using portable betatron with an accelerated electron maximum energy of 6 MeV. The test-object shadow projections were recorded on X-ray films and then picked up by TV-camera and fed into the computer memory. In Fig. 2.1,4 various projections are represented. In Fig 2.2 the reconstruction results of 2 layers from 8 views are depicted (using an algorithm on the basis of estimators (9), which are order statistics (with $k = 2$ at the left and $k = 3$) with some preprocessing).

An immediate practical consequence of the non-linear backprojection approach is that it allows cost-effective and versatile 3D reconstruction techniques (like "non-linear tomosynthesis") (Baranov and Chekalin, 1988, 1990; Baranov et al., 1989b) to be elaborated for NDT. Most commonly, these methods are feasible for reconstruction from a limited number of projections (with computational times appreciably reduced as compared with other non-linear techniques). The object-type most widespread in NDT is "more or less contrast inclusions in more or less uniform (and noisy) background". In these conditions the NLBP-approach often makes possible a remarkable simplification of a reconstruction problem without sacrifice of image quality.

Nowadays, the hardware implementation of nonlinear backprojection methods (with the use of table-driven array processor (Baranov and Chekalin, 1991a)) seems to be the only way to industrial on-line 3D tomographic testing. The same approach can be used to design the real-time X-ray 3D tomographic microscope. These methods can be easily adapted to widely ranging sizes, shapes and densities of test-objects. (The conventional (already existent and available) NDT equipment may be used). NLBP-techniques are equally at home with the use of radiography as well as X-ray television (Baranov et al., 1991), thermoluminescent storage screens (Baranov et al., 1993) or other types of registrator. Apart from NDT, there exist strong possibilities to use this approach in medicine (Baranov et al., 1989) and physics. Some non-tomographic applications of the method are of interest as well.

It should be noted that basic advantages of the classical backprojection approach (3) (first and foremost that it is simple and natural) are not violated by generalization (5), which is an important prerequisite for widespread use of estimates (5) as structural materials for diverse reconstruction procedures.

Non-linear backprojection approach has set us rethinking currently available reconstruction methods and offers the possibility to generalize and improve some of them. The problem (especially its statistical interpretation and such its aspects as elaboration of special non-linear space filtering techniques aimed to object-type, elaboration of iterative and adaptive-type algorithms based on the NLBP-concept) is a valid one for further investigation.

REFERENCES

Baranov V. A. (1993), *Non-Linear Backprojection as a Variation Problem*, Proc. of International symposium on computerized tomography, Novosibirsk, 24.

Baranov V. A. and Chekalin A. S. (1988), Digital system of tomosynthesis for non-destructive testing, *Defectoscopia*,**5**, 30–36.

Baranov V. A. and Chekalin A. S. (1991a), *Table-Driven Processor for 3D-Tomography and Image Processing Non-Linear Techniques Hardware Implementation*, Proc. of 5-th all-union symposium on computed tomography, Zvenigorod, 164–165.

Baranov V. A. and Chekalin A. S. (1991b), *Iterative 3D-Reconstruction Techniques on the Basis of Non-Linear Tomosynthesis Algorithms*, Proc. of 5-th all-union symposium on computed tomography, Zvenigorod, 229–230.

Baranov V. A. and Janisov V. V. (1986), *Iterative Method for Digital Tomosynthesis*, Proc. of 5-th all-union conference "The enhancement of reliability and improvement of quality in aviation engineering" , Andropov, 92–93. (in Russian)

Baranov V. A., Berdonosov V. A., Matveev Yu. V., Temnik A. K., Chakhlov V. L., and Chekalin A. S. (1988), *Cross-Sectional Visualization of Contrast Inclusions in Concrete from Series of Radiographs Obtained with the Use of Portable Betatron*, Proc. of 6-th all-union conference on applied accelerators, Leningrad, 167. (in Russian)

Baranov V. A., Chekalin A. S., and Temnik A. K. (1989a), *Digital Tomosynthesis for Medical Diagnostics*, Proc. of the 9-th all-union conference "The measurements in medicine and their metrology", Moscow, 107.

Baranov V. A., Chekalin A. S., and Temnik A. K. (1989b), *The Synthesis of Tomograms from Series of Shadowgraphs - Modified Algorithms, Software and Hardware*, Proc. of "The 6-th all-union symposium on computed tomography", Tashkent, part 2, 87–88.

Baranov V. A., Chekalin A. S., and Temnik A. K. (1990), *3D-Tomography from Small Number of Views*, Proc. of 12-th all union scientific conference "Physical methods of non-destructive testing", Sverdlovsk, 111–112.

Baranov V .A., Temnik A. K., and Chekalin A. S. (1991), *X-Ray Microtomography on the Basis of Tomosynthesis Techniques: Software and Hardware Implementation*, Proc. of 5-th all-union symposium on computed tomography, Zvenigorod, 241.

Baranov V. A., Temnik A. K., Chakhlov V. L., and Chekalin A. S. (1991), *The 3D-Tomography System on the Basis of Portable Betatron for Heavy Components Testing*, Proc. of 5-th all-union symposium on computed tomography, Zvenigorod, 123.

Baranov V. A., Dmitrieva A. V., Moskalev Yu. A., Petushenko V. F., Temnik A. K., Chakhlov V. L., and Chekalin A. S. (1993), *Tomographic 3D Reconstruction from Small Number of Views with the Use of Thermoluminescent Storage Screens*, Proc. of International symposium on computerized tomography, Novosibirsk, 25.

Frieden B. R. (1972), Restoring with Maximum likelihood and entropy, *J. Opt. Soc. Am.*, **62**, 511–518.

Gordon R. (1974), A tutorial on ART, *IEEE Trans. Nucl. Sci.*, **21**, 78–93.

Gordon R., Bender R., and Herman G.T. (1970), Algebraic reconstruction techniques for three-dimensional electron microscopy and x-ray photography, *J. Theor. Biol.*, **29**, 471–481.

Gull S. F. and Daniell G. J. (1978), Image reconstruction from incomplete and noisy data, *Nature*, **272**, 686–690.

Hanson K. M. (1982), CT reconstruction from limited projection angles, *Proc. PIES*, **347**, 166–173.

Hanson K. M. and Wecksung G. W. (1983), Bayesian approach to limit-angle reconstruction in computed tomography, *J. Opt. Soc. Amer.*, **73**, 1501–1509.

Louis A. K. (1981), Ghosts in tomography – The null space of the Radon transform, *Math. Meth. Appl. Sci.*, **3**, 1–10.

Medoff B. P., Brody W. R., and Macovsci A. (1982), *Image Reconstruction from Limited Data*, Proc. Int. Workshop on Physics and Engineering in Medical Imaging (Pacific Grove, 1982), IEEE Computer Society, Silver Spring, 188–192.

Medoff B. P., Brody W. R., Nassi M., and Macovsci A. (1983), Iterative convolution backprojection algorithm for Image reconstruction from limited data, *J. Opt. Soc. Am.*, **73**, 1493–1500.

Minerbo G. (1979), MENT: A maximum entropy algorithm for reconstructing a source from projection data, *Comput. Graph. Process.*, **10**, 48–68.

Pickalov V. V. and Sharapova N. B. (1990), *Quick Algorithms of Cracks Detection on the Basis of Non-Linear Backprojection*, Proc. of "The 6-th all-union symposium on computed tomography", Tashkent, part 2, 152–153.

Tikhonov A. N. and Arsenin V. Y. (1977), *Solutions of Ill-Posed Problems*, V.H.Winston and Sons, Washington, D.C.

Wernecke S. J. and d'Addario L. R. (1977), Maximum entropy image restoration, *IEEE Trans. Comput.*, **26**, 351–364.

Youla Dante C. (1978), Generalized Image Restoration by the Method of Alternating Orthogonal Projections, *IEEE Trans. on Circuits and Systems*, **25**, 694–702.

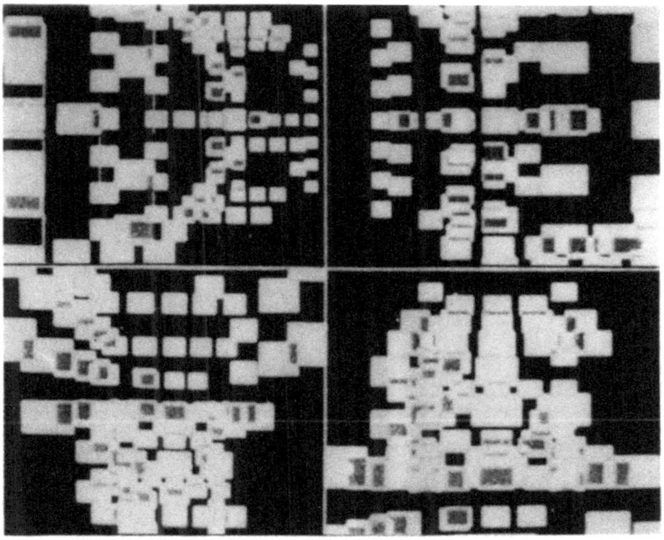

FIGURE 1.1

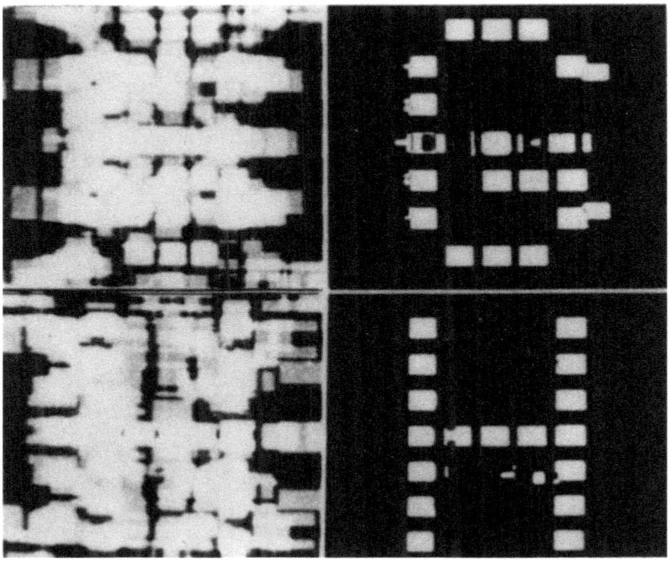

FIGURE 1.2

FIGURE 1.3

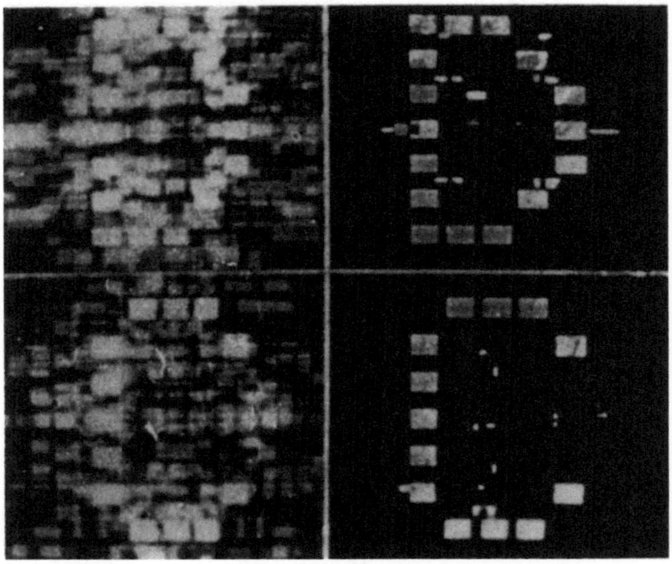

FIGURE 1.4

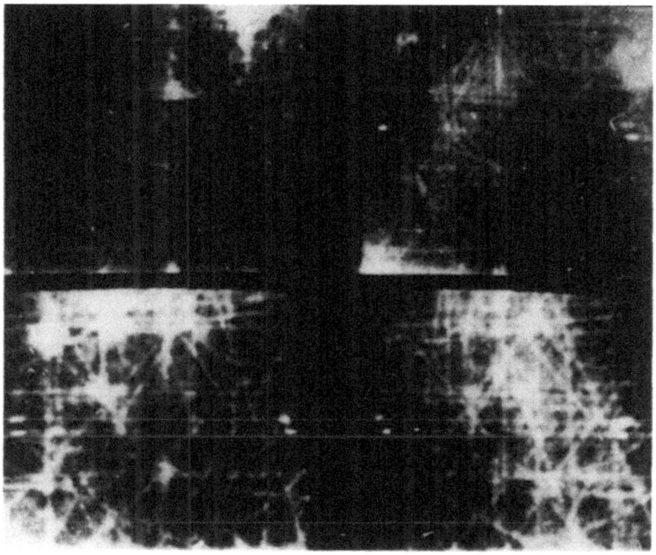

FIGURE 2.1

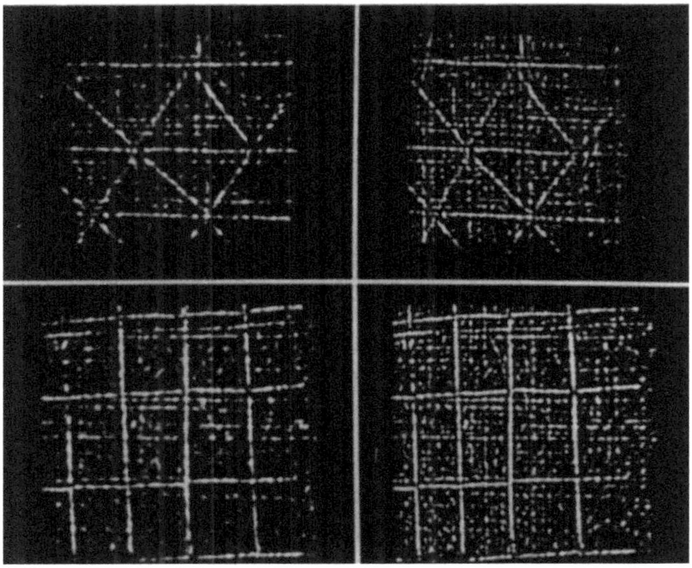

FIGURE 2.2

Three-Dimensional Restoration of Single Photon Emission Computed Tomography Images Using the Kalman-Metz Filter

D. BOULFELFEL, R. M. RANGAYYAN, L. J. HAHN, R. KLOIBER

Department of Electrical and Computer Engineering, University of Calgary; Division of Nuclear Medicine, Foothills Hospital

Calgary, Alberta, Canada T2N 1N4, and T2N 2T9

ABSTRACT

Restoration of single photon emission computed tomography (SPECT) images is usually performed using two-dimensional (2-D) shift-invariant techniques on either projection images (pre-reconstruction restoration) or on SPECT slices (post-reconstruction restoration). This study presents the application of a three-dimensional (3-D) Kalman-Metz filter in restoring SPECT images. This filter consists of a 2-D shift-variant Kalman filter cascaded with a 1-D Metz filter. The 3-D filter is shown to perform better than the 2-D methods because it takes into consideration the inter-slice information of the filtering process and the shift-variance of the blur function. This filter was applied to SPECT images of a resolution phantom and a large, truncated cone phantom containing two types of cold spots: a sphere, and a triangular prism. The filter was also applied to clinical images of the liver and spleen. The images were acquired on an ADAC GENESYS camera. A comparison was performed between results obtained in this study and those obtained by 3-D shift-invariant restoration filters. Quantitative analysis of the 3-D Kalman- Metz restored images performed through measurement of root mean squared (RMS) errors showed a considerable reduction in error over images restored using the other methods.

Introduction

Several methods for restoring single photon emission computed tomography (SPECT) images have been proposed (Boulfelfel et al. 1992a,b,c; Accepted a,b). Most of these methods assume a shift-invariant and stationary SPECT modulation transfer function (MTF). However, many investigators have shown that such assumption will lead to poor results if the SPECT images are large compared to the field of view of the camera or if their distance to the camera varies considerably during the imaging process (Tsui and Jaszczak, 1984; Coleman et al., 1989; Glick et al., 1989). Some of these authors also provided methods to lessen the effects of non-stationarity and shift-variance of the MTF on the restoration algorithm.

Tsui and Jaszczak (1984) studied the effects of the collimator-detector system, sampling bin width, and the backprojection filter on the spatial resolution of SPECT images and concluded that an adequate choice of these parameters achieves a minimal loss in resolution, stationarity, and minimal aliasing during reconstruction. Glick et al. (1989) also studied the effects of the reconstruction process on the anisotropy and the nonstationarity of the MTF and have indicated how these can be reduced by proper selection of the methods used in reconstruction.

Larsson (1980) investigated the effects of using the arithmetic and geometric means of line spread functions (LSF) taken at opposing angular positions in air and water to improve stationarity. He found that both means of opposing LSFs result in a LSF of approximately the same full-width at half-maximum (FWHM) as the LSF at the center of rotation of the camera, although the arithmetic mean shows significant amplitude differences for the different distances. Axelsson et al. (1984) investigated the shape of the LSF resulting from the geometric mean of opposing planar views and concluded that the LSF was approximately stationary in the central two-thirds of their phantom. Msaki et al. (1987) also concluded that the stationarity of the two-dimensional (2-D) MTF improved in their study of the variation in the PSF of the geometric mean of opposing views for source positions both orthogonal and parallel to the collimator axis. Coleman et al. (1989) also used the arithmetic and geometric mean of opposing SPECT views to calculate the 2-D MTF as a function of camera angle and source location and concluded that the geometric mean provides an approximately stationary 2-D MTF except near the edges of the phantom.

We have also evaluated the use of the geometric mean opposing projections in pre-reconstruction restoration, and have found that this pre-processing step leads to improved results (Boulfelfel et al., 1992c). We also studied the 3-D MTF of SPECT imaging and its use in the restoration of SPECT images, and found that the 3-D MTF shows better stationarity and provides better results in restoring large objects because it uses the inter-slice blur, which is ignored by 2-D techniques (Boulfelfel et al., 1992a, Accepted a).

Although, the above techniques reduce the shift-variance of the MTF, they do not totally eliminate it. This shift-variance manifests itself more near the edges of the field of view of the camera. Furthermore, the Poisson noise is nonstationary, while all the proposed techniques assume a stationary noise function. Therefore, an ideal restoration method for SPECT images should support a shift-variant PSF and a nonstationary noise model. We have recently investigate the use of a shift-variant implementation of the 2-D Kalman filter in the restoration of SPECT images (Boulfelfel et al., Accepted b). The Kalman filter, being iterative, allows the use of different MTFs and noise functions for each iteration. However, the 2-D Kalman filter performs only a partial restoration as it does not take into consideration the inter-slice blur. The objective of this study is to use a shift-variant implementation of the 3-D Kalman-Metz filter in post-reconstruction restoration of SPECT images, and to compare the results obtained with those of 3-D shift-invariant restoration filters.

Kalman-Metz Filter

We have shown in a recent study that when planar images are subjected to tomographic reconstruction, the 2-D blur which exists in each projection becomes 3-D in the SPECT bloc of images (Boulfelfel et al. 1992a). Therefore, the 2-D MTF in each planar projection contributes to the formation of a 3-D MTF after reconstruction. The resulting 3-D MTF is not isotropic in the three directions as the SPECT data is ramp-filtered and interpolated in the x-y plane, while no processing is performed along the z-axis (inter-slice direction). Furthermore, while the MTF is shift-variant in the x-y direction, the z-component is shift-invariant as the z-axis of rotation of the camera.

In this study, we propose the use of a 2-D shift-variant Kalman filter to restore the SPECT data in the x-y plane cascaded with a 1-D Metz filter in the inter-slice direction. The 2-D Kalman filter is performed over a block of SPECT images in the spatial domain, while the 1-D Metz filter is performed over the Kalman-filtered data in the spatial-frequency domain after a 1-D Fourier transform in the z-axis. The 3-D Kalman-Metz filter deals with both the shift-variance of the MTF and the inter-slice blur, and should give better results than either 3-D shift-invariant filters or 2-D shift-variant filters.

The proposed 2-D shift-variant Kalman filter used in this study is a modified implementation of the reduced update Kalman filter (RUKF) proposed by Woods and Ingle(1981) and Woods and Radewan(1977), which offers a significant reduction in the total computational load. The modifications that we propose to this implementation are to change the region of support (ROS) from the non-symmetric half-plane (NSHP) to a quarter plane (QP) (Marple, 1987), and to use a 2-D autoregressive (AR) model (Marple, 1987) instead of the Markov model proposed by Woods and Radewan(1977). The AR model is chosen over the Markov model because of its ease of implementation and the availability of techniques to estimate the optimal model order.

The dynamic model of a system at the time sample m is given by (Woods and Radewan, 1977)

$$\mathbf{f}(m,n) = \mathbf{A}\mathbf{f}(m-1,n) + \mathbf{w}(m,n), \tag{1}$$

$$\mathbf{g}(m,n) = \mathbf{H}\mathbf{f}(m,n) + \mathbf{v}(m,n). \tag{2}$$

where $\mathbf{f}(m,n)$ is the signal state matrix, $\mathbf{g}(m,n)$ is the observation (or blurred) image, $\mathbf{v}(m,n)$ is the observation noise, $\mathbf{w}(m,n)$ is the driving noise which generates the signal, $\mathbf{A}$ is the state transition matrix, and $\mathbf{H}$ is the blur matrix. In this model, the observation and driving noise are assumed to be uncorrelated, while, Gaussian, zero-mean noise sources with

$$E[\mathbf{w}(m,n)\mathbf{w}^T(k,l)] = \mathbf{Q}_w \tag{3}$$

and

$$E[\mathbf{v}(m,n)\mathbf{v}^T(k,l)] = \mathbf{Q}_v \tag{4}$$

where $\mathbf{Q}_w$ and $\mathbf{Q}_v$ are the correlation matrices of $\mathbf{w}$ and $\mathbf{v}$ respectively.

Using an AR model for the state transition matrix and rewriting the model in a scalar form leads to the following signal and observation models:

$$f(m,n) = -\sum_k \sum_l a(k,l) f(m-k, n-l) + w(m,n), \tag{5}$$

$$g(m,n) = -\sum_k \sum_l h(k,l) f(m-k, n-l) + v(m,n), \tag{6}$$

the summations are over the chosen ROS. The RUKF scalar equations as developed by Woods and Ingle (1981) are as follows:

$$m \to m+1(\text{ at end of line } m \to 1 \text{ and } n \to n+1);$$

$$\hat{g}_b^{(m,n)}(m,n) = -\sum_k \sum_l a(k,l)\hat{g}_a^{(m-1,n)}(m-k, n-l); \tag{7}$$

$$\hat{g}_a^{(m,n)}(i,j) = \hat{g}_b^{(m,n)}(i,j) + k^{(m,n)}(m-i, n-j)$$

$$\times \left[g(m,n) - \sum_k \sum_l h(k,l)\hat{g}_a^{(m,n)}(m-k, n-l) \right] \tag{8}$$

$$R_b^{(m,n)}(m,n;k,l) = -\sum_o \sum_p a(o,p) R_a^{(m-1,n)}(m-o, n-p; k,l); \tag{9}$$

$$R_b^{(m,n)}(m,n;m,n) = -\sum_k \sum_l a(k,l) R_b^{(m,n)}(m,n;m-k,n-l) + \sigma_w^2; \tag{10}$$

$$k^{(m,n)}(i,j) = \frac{\sum_k \sum_l h(k,l) R_b^{(m,n)}(m-k,n-l;m-i,n-j)}{\sum_k \sum_l \sum_i \sum_j h(k,l)h(i,j) R_b^{(m,n)}(m-k,n-l;m-i,n-j) + \sigma_v^2}; \tag{11}$$

$$R_a^{(m,n)}(i,j;k,l) = R_b^{(m,n)}(i,j;k,l) - k^{(m,n)}(m-i,n-j) \times \sum_o \sum_p h(o,p) R_b^{(m,n)} 9m - o, n-p; k,l); \tag{12}$$

where R is the error covariance matrix, k is the Kalman matrix, $\hat{g}$ is the restored image, the subscripts a and b indicate "after" and "before" updating, the superscripts indicate the iteration, and σ_w^2 and σ_v^2 are the variances of the driving and observation noise. These Kalman filtering equations have the following interpretation. Equation (7) uses the dynamic system model to project the last estimate of the restored matrix forward. Then in (9) and (10) the error covariance matrix is predicted one step forward. Equation (11) is the updated Kalman gain matrix. In equation (8), the estimate is updated using the new observation $g(m,n)$. Finally, equation (12) is the update of the error covariance matrix. Equation (7) to (12) need initial conditions for both $\hat{g}$ and R. However, since SPECT images are located within the intrinsic circle of the image square matrix and the scanning is started from the left corner of the SPECT image, an initial zero matrix can be taken for both $\hat{g}$ and R.

The above entities are computed over a QP window of dimensions MxM, where M is width of the AR matrix or the blur matrix, whichever is wider. The AR matrix width is chosen to be the optimal AR order using the Akaike information criterion (AIC) (Akaike, 1974; Kay, 1987). The blur matrix is truncated to twice the FWHM value of the PSF to contain all significant effects of the blur matrix on the image. The QP window performs a raster scan on the image. The blur function h is chosen to be one of sixteen blur functions depending on the location of the window; the sixteen blur functions map all the different changes in the blur as a function of the position of the object with respect to the center of the image.

The AR model in equation (12) was generated by a 2-D implementation of the Levinson algorithm (Marple, 1987). The algorithm uses the autocorrelation matrix of the window to estimate the AR coefficient and the driving noise variance (σ_w), and is one of the last algorithms described in the literature.

The observation noise may be modeled by a Poisson distribution (Todd-Pokropek and Jarritt, 1982). The noise variance (σ_v^2) is estimate locally for each window at each step of the iterations since the SPECT noise process is non-stationary. σ_v^2 is estimated to be equal to the local total count in the window.

After applying the 2-D Kalman filter on the block of SPECT images, the data is Fourier transformed in the inter-slice direction to restore it using the 1-D Metz filter. The Metz filter is chosen over other restoration filters as its only required apriori information is the inter-slice component of the MTF. Other filters require information about the noise which can not easily be predicted in our case, since the noise component has been modified by the 2-D Kalman filter applied on the SPECT slice. The Metz filter is based on an extension to the inverse filter that would perform noise suppression at high frequencies. The Metz filter is defined as (Metz and Beck, 1974)

$$\hat{F}_m(u) = \frac{[1-(1-H(u)^2)^X]}{H(u)} G(u), \tag{13}$$

where X is factor that controls the extent to which the inverse filter is predominant before the noise suppression feature takes over.

Experimental Work

The MTF model was estimated through a measure of the PSF at different angles, geometric averaging of conjugate projections, and application of the reconstruction algorithm to compute the 3-D PSF. Images of a point source phantom containing 37MBq (1mCi) of Tc-99m solution were obtained using the ADAC GENESYS SPECT system at 16 angular positions using the 512x512-pixel format. Measurements of the point source were performed for sixteen point sources located at distances from zero to the radius of rotation of the camera for each given radius of rotation, using an ADAC GENESYS camera with a low-energy, general-purpose (LEGP) collimator. These measurements were taken in air, as scatter and attenuation effects are not considered in this study. The measured point sources were then reconstructed using the discrete filtered backprojection (DFBP) algorithm to include the reconstruction effects on the PSF. We have shown in a previous study that a choice of sixteen blur functions is enough to map the changes in the blur function (Boulfelfel et al., Accepted b).

The resolution phantom contains 9 pairs of hot spots of diameters 39, 22, 17, 14, 12, 9, 8, 6, and 5 mm (Nuclear Associates, Chapel Hill, North Carolina). The phantom was filled with 1mCi of Tl-201-Chloride. The phantom was held centered at the axis of rotation of the GENESYS camera with the LEGP collimator at distances of 21.7 and 28 cm, respectively, from the camera, and 120 projections of size 128x128 each were acquired over 360°. The SPECT images of different trans-axial slices were reconstructed using the MicroDELTA software.

The cone phantom used in this study was constructed with a truncated cone of length 18.2 cm and end diameters of 41 and 88 mm. Within the truncated cone were inserted a solid sphere of 18 mm diameter and a solid equilateral triangular prism of 100 mm length and 35 mm sides. The cone was then filled with 37MBq (1mCi) Tc-99m solution. 120 planar projections spanning 360° were taken at a collimator distance of 25 cm using the GENESYS camera with a low-energy, general-purpose (LEGP) collimator. The projections were processed through Siemens MicroDELTA software that uses the backprojection algorithm with a ramp filter to obtain SPECT images. All the images used in this study are of size 128x128 pixels.

The liver and spleen imaging procedure at the Foothills Hospital is as follows. Two mCi a Tc-99m-based radiopharmaceutical is given to the patient. After the isotope has accumulated in the liver and spleen, sixty-four 64x64 projections are acquired using the GENESYS camera with the LEGP collimator. The time for each projection is 40 seconds. The projections are acquired over 360° in an elliptical trajectory with an average radius of rotation of 25 cm. Two energy peaks were used in the acquisition of the projections to perform scatter correction using the dual-energy-window subtraction technique. Transverse SPECT images were reconstructed after averaging and correcting for attenuation using the MicroDELTA processor. The Chang (1978) algorithm was used for attenuation correction.

Results

The acquired (original) and the Kalman-Metz restored SPECT images of the resolution and truncated cone phantom are shown in figure 1. The Kalman-Metz restored image were obtained with a 13x13, 7x7, and 11x11 windows respectively which correspond to the optimal orders of the AR model. The Kalman restored images are clearly seen to have superior sharpness as compared with the acquired images. The RMS error (with reference to the known phantom geometry) for the acquired images of the resolution phantom and the truncated cone were 74.22 and 66.21, for the 3-D Metz filtered images 34.12 and 24.51, the 2-D Kalman filtered images 36.71 and 22.41, and the 3-D Kalman-Metz filtered images 29.19 and 18.43. The Kalman-Metz filtered image shows clearly smaller RMS value than the two other restored images.

Figure 2 shows the acquired and 3-D Kalman-Metz restored SPECT images of the liver and spleen. This figure shows clearly that the 3-D Kalman-Metz filter improves the quality of the image.

Discussion

The 3-D shift-variant Kalman-Metz filter performs better than 3-D shift-invariant techniques or 2-D techniques because it takes into consideration both the anomalies introduced by increased blurring of reconstruction of points located farther from the center of rotation of the camera and inter-slice blur. However, the Kalman-Metz filter does not conserve the number of counts after restoration.

The processing time of the Kalman-Mets filter depends on the size of the moving window, and the Kalman equations show that it depends on M^4, which makes a choice of larger windows more expensive. For instance, a Myrias machine using 320 parallel processors and running on UNIX performs Kalman restoration of 128x128 SPECT images in 81 seconds for a 5x5 window, but needs 1912 seconds for a 7x7 window.

Acknowledgement

This project is supported by a Natural Sciences and Engineering Research Council of Canada grant.

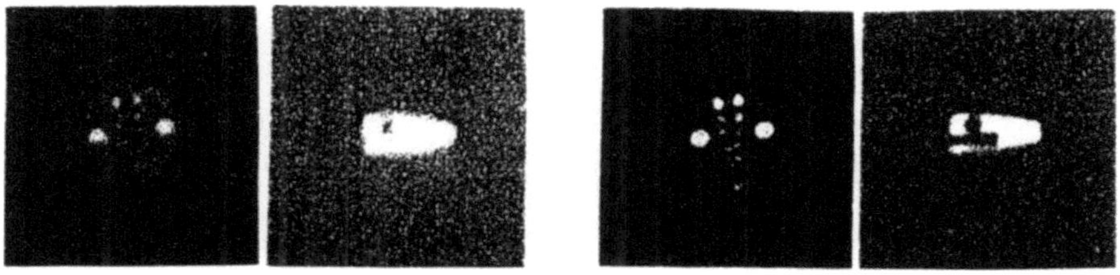

(a) (b)

Figure 1. SPECT images of the resolution and truncated cone phantom: (a) acquired images and (b) Kalman-Metz filter restored images.

(a) (b)

FIGURE 2. SPECT images of Liver and spleen: (a) acquired images and (b) Kalman-Metz filter restored images.

REFERENCES

Akaike H. (1974), A New Look at the Statistical Model Identification. In: *IEEE Trans. Autom. Control* 19, 716–723.

Axelsson B., Msaki P., Israelsson A. (1984), Subtraction of Compton Scattered Photons in Single-Photon Emission Computerized Tomography. In: *J. Nucl. Med.* 25, 490–494.

Boulfelfel D., Rangayyan R. M., Hahn L. J., Kloiber R. (1992a), Three-Dimensional Restoration of Single Photon Emission Computed Tomography Images. In: *Proc. of the 1991 IEEE Symposium Nucl. Sci. and Med. Imag.* 1986–1989.

Boulfelfel D., Rangayyan R. M., Hahn L. J., Kloiber R. (1992b), Pre-Reconstruction Restoration of Myocardial Single Photon Emission Computed Tomography Images. In: *IEEE Trans. Med. Imag.* 11, 336–341.

Boulfelfel D., Rangayyan R. M., Hahn L. J., Kloiber R. (1992c), Use of the Geometric Mean of Opposing Planar Projections in Pre-Reconstruction Restoration of SPECT Images. In: *Phys. Med. Biol.* 37, 1915–1929.

Boulfelfel D., Rangayyan R. M., Hahn L. J., Kloiber R. (Accepted a), Three-Dimensional Restoration of Single Photon Emission Computed Tomography Images. In: *IEEE Trans. Nucl. Sci.*

Boulfelfel D., Rangayyan R. M., Hahn L. J., Kloiber R., Kuduvalli R. (Accepted b), Two-dimensional restoration of single photon emission computed tomography images using the Kalman filter. In: *IEEE Trans. Med. Imag.* 25, 490–494.

Chang L. T. (1978), A Method for Attenuation Correction in Radionuclide Computed Tomography. In: *IEEE Trans. Nucl. Sci.* 25, 638–643.

Coleman M., King M. A., Glick S. J., Knesaurek K., Penney P. C. (1989), Investigation of the Stationarity of the Modulation Transfer Function and the Scatter Fraction in Conjugate View SPECT Restoration Filtering. In: *IEEE Trans. Nucl. Sci.* 36, 969–971.

Glick S. J., King M. A., Penney B. C. (1989), Characterization of the Modulation Transfer Function of Discrete Filtered Backprojection. In: *IEEE Trans. Med. Imag.* 8, 203–213.

Kay S. M. (1987), *Modern Spectral Estimation*, Prentice-Hall, Englewood Cliffs, N.J.

Larsson S. A. (1980), Gamma Camera Emission Tomography. In: *ACTA Radiologica Supplementum 363.*

Marple S. L. (1987), *Digital Spectral Analysis*, Prentice-Hall, Englewood Cliffs, N.J.

Metz C. E., Beck R. M. (1974), Quantitative Effects of Stationary Linear Image Processing on Noise and Resolution of Structure in Radionuclide Images. In: *J. Nuc. Med.* 15, 164–170.

Msaki P., Axelsson B., Dahl C. M., Larsson S. A. (1987), Generalized Scatter Correction Method in SPECT Using Point Scatter Distribution Functions. In: *J. Nucl. Med.* 28, 1861–1869.

Todd-Pokropek A. E., Jarritt P. H. (1982), The Noise Characteristics of SPECT systems. In: *Computed Emission Tomography* 361–389, ed. Ell P. J. and Holman B. L., Oxford University Press, Oxford.

Tsui B. M. W., Jaszczak R. J. (1984), Interactions of Collimation, Sampling and Filtering on SPECT Spatial Resolution. In: *IEEE Tran. Nucl. Sci.* 31, 527–532.

Woods J. W., Ingle V. N. (1981), Kalman Filtering in Two Dimensions: Further Results. In: *IEEE Trans. Acount. Speech Sig. Proc.* 29, 188–196.

Woods J. W., Radewan C. H. (1977), Kalman Filtering in Two Dimensions. In: *IEEE Trans. Inf. Theory* 23, 473–481.

An Extended Kaczmarz's Method for l_p Minimum Norm Solutions

Achiya DAX

Hydrological Service, P.O.B. 6381, Jerusalem 91060, ISRAEL

ABSTRACT

This paper presents a row relaxation method for solving the problem

$$\text{minimize } \|\underline{x}\|_p^p / p$$

$$\text{subject to } A\underline{x} = \underline{b}$$

where $1 < p < \infty$. It is shown that the dual of this problem has the form

$$\text{maximize } \underline{b}^T \underline{y} - \left\|A^T \underline{y}\right\|_q^q / q$$

where $q = p/(p-1)$. Moreover, let $\underline{y}$ solve the dual and let $\underline{x} = (x_1, \ldots, x_n)^T \in \mathbf{R}^n$ solve the primal, then $x_j = \left|\underline{c}_j^T \underline{y}\right|^{q-1} \operatorname{sign}(\underline{c}_j^T \underline{y})$ where $\underline{c}_j$ denotes the j-th column of A. That is, a primal solution is easily retrieved from a dual one. Maximizing the dual objective function by changing one variable at a time results in an iterative scheme that resembles Kaczmarz's method. This feature makes the new scheme suitable for problems in which A is large, sparse and unstructured. Numerical experiments illustrate the ability of the proposed method to handle very large problems of this kind.

1. Introduction

This paper presents a row relaxation method for solving the problem

$$\text{minimize } P(\underline{x}) = \|\underline{x}\|_p^p / p$$

$$\text{subject to } A\underline{x} = \underline{b} \tag{1.1}$$

where $1 < p < \infty$ is a real number, A is a real $m \times n$ matrix, $\underline{b} = (b_1, \ldots, b_m) \in \mathbf{R}^m$, and $\underline{x} = (x_1, \ldots, x_n) \in \mathbf{R}^n$ denotes vector of unknowns. For the sake of clarity we mention that

$$\|\underline{x}\|_p = \left(\sum_{j=1}^{n} |x_j|^p\right)^{1/p}.$$

Hence an alternative way to express our objective function is

$$P(\underline{x}) = \left(\sum_{j=1}^{n} |x_j|^p\right) / p.$$

It is assumed throughout the paper that the system of linear equations $A\underline{x} = \underline{b}$ is solvable. This ensures the existence and the uniqueness of a point $\underline{x}^* \in \mathbf{R}^n$ that solves (1.1). The rows of A are denoted by $\underline{a}_i^T$, $i = 1, \ldots, m$, while the columns of A are denoted as $\underline{c}_j$, $j = 1, \ldots, n$. The dual index of p is denoted by q. That is, $q = p/(p-1)$ and $1/p + 1/q = 1$.

A need for calculating minimum norm solutions arises in optimal control problems and other applications. See, for example, Luenberger (1969), Cadzow (1971, 1973), and Kolev (1975). The term "row relaxation" method refers to an iterative scheme whose basic iteration is composed of a sweep along the rows of A. This type of method is suitable for problems in which A is large sparse and unstructured. In particular it was proved to be a useful tool for handling large linear systems that arise in the field of image reconstruction from projections. In this field it is possible to avoid storing A. Instead the nonzero entries of the i-th row are newly generated from the experimental data at each iteration. For this reason these methods are also called "row generation" methods or "row action" methods. Perhaps the most famous method of this kind is the method of Kaczmarz (1937), which was rediscovered under the name ART (Algebraic Reconstruction Technique) by Gordon et al. (1970) and Herman et al. (1973). For a detailed discussion of row relaxation methods and the special environment that characterizes their use see, for example, Censor (1981).

The motivation behind the proposed method lies in the following observation. The dual of (1.1) has the form

$$\text{maximize } D(\underline{y}) = \underline{b}^T \underline{y} - \left\| A^T \underline{y} \right\|_q^q / q, \tag{1.2}$$

and if $\underline{y}$ solves the dual then the coordinates of the primal solution satisfy

$$x_j = \left| \underline{c}_j^T \underline{y} \right|^{q-1} \operatorname{sign}(\underline{c}_j^T \underline{y}). \tag{1.3}$$

In other words, the primal solution is easily obtained from a dual one. The proof of these assertions is given in the next section. Maximizing the dual objective function by changing one variable at a time results in a row relaxation method that resembles Kaczmarz's method. In fact, when $p = 2$ the two methods coincide. Therefore the new method can be viewed as extended from of Kaczmarz's method.The details of the proposed scheme are explained in Section 3 while Section 4 discusses its convergence properties. Finally, in Section 5, we provide numerical experiments that demonstrate the feasibility of our ideas.

2. Duality Relations

This section investigates the duality relations between (1.1) and (1.2). For this purpose we introduce the following notations. Let

$$\underline{z}(\underline{y}) = \left(z_1(\underline{y}), \ldots, z_n(\underline{y}) \right)^T \in \mathbf{R}^n$$

be a vector function whose j -th component is

$$z_j(\underline{y}) = \left| \underline{c}_j^T \underline{y} \right|^{q-1} \operatorname{sign}(\underline{c}_j^T \underline{y}).$$

Then $A\underline{z}(y)$ is the gradient vector of $\left\| A^T \underline{y} \right\|_q^q / q$ and

$$\underline{y}^T A \underline{z}(\underline{y}) = \left\| A^T \underline{y} \right\|_q^q = \left\| \underline{z}(\underline{y}) \right\|_p^p = \left\| A^T \underline{y} \right\|_q^q / q + \left\| \underline{z}(\underline{y}) \right\|_p^p / p \tag{2.1}$$

Similarly we use

$$\underline{g}(\underline{x}) = (g_1(\underline{x}), \ldots, g_n(\underline{x}))^T \in \mathbf{R}^n$$

to denote the gradient vector of $P(\underline{x})$. That is,

$$g_j(\underline{x}) = |x_j|^{p-1} \operatorname{sign}(x_j).$$

Now one can verify that $\underline{g}(\underline{x})$ and $\underline{z}(\underline{y})$ are related by the equality

$$\underline{g}(\underline{z}(\underline{y})) = A^T \underline{y}, \tag{2.2}$$

which means that $A^T\underline{y}$ is the gradient vector of $P(\underline{x})$ at the point $\underline{x} = \underline{z}(\underline{y})$.

Theorem 1. *Let $\underline{x}^*$ solve (1.1), then there exists $\underline{y}^* \in \mathbf{R}^m$ such that $\underline{x}^* = \underline{z}(\underline{y}^*)$ and $\underline{y}^*$ solves (1.2). Conversely, let $\underline{y}^*$ solve (1.2) then the vector $\underline{x}^* = \underline{z}(\underline{y}^*)$ is the unique solution of (1.1). In both cases we have*

$$D(\underline{y}^*) = P(\underline{x}^*),$$

and the classical primal-dual inequality

$$D(\underline{y}) \leq P(\underline{x})$$

holds for all $\underline{x} \in \{\underline{u} \mid A\underline{u} = \underline{b}\}$ and $\underline{y} \in \mathbf{R}^m$.

PROOF. Recall that a point $\underline{x}^* \in \mathbf{R}^n$ solves(1.1) if and only if there exists a point $\underline{y} \in \mathbf{R}^m$ such that

$$A\underline{x}^* = \underline{b} \quad \text{and} \quad \underline{g}(\underline{x}^*) = A^T\underline{y}^* \tag{2.3}$$

(These are the well-known Kuhn-Tucker optimality conditions). Hence by comparing the equalities $\underline{g}(\underline{x}^*) = A^T\underline{y}$ and $\underline{g}(\underline{z}(\underline{y}^*)) = A^T\underline{y}^*$ we deduce that $\underline{x}^* = \underline{z}(\underline{y}^*)$. Consequently $\underline{y}^*$ satisfies

$$A\underline{z}(\underline{y}^*) = \underline{b} \tag{2.4}$$

which is the optimality condition of (1.2). We have seen therefore, that $\underline{x}^*$ solves (1.1) then the corresponding Lagrange vector, $\underline{y}^*$, solves (1.2). Conversely, let $\underline{y}^*$ solve (1.2) then the optimality condition (2.4) holds and the vector $\underline{x}^* = \underline{z}(\underline{y}^*)$ satisfies (2.3), which proves that $\underline{x}^*$ solve (1.1). In both cases the pair $\underline{x}^*$ and $\underline{y}^*$ satisfies the relations $\underline{x}^* = \underline{z}(\underline{y}^*)$ and

$$(\underline{y}^*)^T(A\underline{x}^* - \underline{b}) = (\underline{y}^*)^T(A\underline{z}(\underline{y}^*) - \underline{b}) = 0.$$

Thus by applying (2.1) we obtain that

$$\underline{b}^T\underline{y}^* = (\underline{y}^*)^T A\underline{z}(\underline{y}^*) = \left\|\underline{z}(\underline{y}^*)\right\|_p^p /p + \left\|A^T\underline{y}^*\right\|_q^q /q,$$

and

$$\underline{b}^T\underline{y}^* - \left\|A^T\underline{y}^*\right\|_q^q /q = \left\|\underline{z}(\underline{y}^*)\right\|_p^p /p = \|\underline{x}^*\|_p^p /p.$$

It is also worthwhile to mention that Theorem 1 can be extended to include inequality constraints. In this case dual variables that correspond to inequalities are restricted to stay nonnegative (see Dax, 1993). In particular, when $p = 2$ one obtains that the dual of the quadratic programming problem

$$\text{minimize } \frac{1}{2} \|\underline{x}\|_2^2$$

$$\text{subject to } A\underline{x} \geq \underline{b}, \tag{2.5}$$

has the form

$$\text{maximize } \underline{b}^T\underline{y} - \left\|A^T\underline{y}\right\|_2^2/2$$

$$\text{subject to } \underline{y} \geq \underline{o} \tag{2.6}$$

Maximizing the objective function of (2.6) by changing one variable at a time, while keeping the dual variables nonnegative, results in the method of Lent and Censor (1980). This suggests that a similar modification may enable the current method to handle inequality constraints. The implementation of this idea is left, however, outsize the scope of this paper.

3. A Row Relaxation Method

In this section we describe a row relaxation method for calculating a minimizer of the function

$$F(\underline{y}) = \left\|A^T\underline{y}\right\|_q^q/q - \underline{b}^T\underline{y}.$$

The basic iteration of the proposed algorithm is composed of m steps, where the i-th step, $i = 1, \ldots, m$, considers the i-th row of A. Let $\underline{y} = (y_1, \ldots, y_m)^T$ denote the current estimate of the solution at the beginning of the i-th step, and let

$$\underline{r} = A^T\underline{y}$$

denote the corresponding "residual" vector. Then at the i-th step y_i alone is changed in an attempt to reduce the objective function value, and all the other variables are kept fixed. Let $y_i + \hat{\theta}$ denote the new value of y_i. Then $\hat{\theta}$ is determined by a "line search" procedure that is aimed at minimizing the one parameter function

$$f(\theta) = F(\underline{y} + \theta\underline{e}_i) = \|\theta\underline{a}_i + \underline{r}\|_q^q/q - \theta b_i - \underline{b}^T\underline{y},$$

where $\underline{e}_i$ denotes the i-th column of the $m \times m$ unit matrix.

The success of the line search procedure depends on a proper choice of its starting point, θ_0. If $q \geq 2$ then the Newton's step-length

$$\theta_0 = -f'(0)/f''(0)$$

seems to be a natural starting point. However, in order to avoid a division by zero, in practice θ_0 defined by the rule

$$\theta_0 = -f'(0)/\max\{\varepsilon, f''(0)\} \tag{3.1}$$

where ε is a small positive constant.

If $1 < q < 2$ then it is also necessary to guard against the possibility that $f''(0)$ is undefined or "too large". Let M_i denote the subset of $\{1, 2, \ldots, m\}$ that contains the indices of the non-zero components of $\underline{a}_i$, and let α_i and ρ_j, $j \in M_i$, denote the corresponding components of $\underline{a}_i$ and $\underline{r}$, respectively. Then minimizing $f(\theta)$ is equivalent to minimizing the function

$$\sum_{j \in M_i} |\theta\alpha_j + \rho_j|^q/q - \theta b_i,$$

and differentiating $f(\theta)$ gives

$$f'(\theta) = \sum_{j\in M_i} \alpha_j \, |\theta\alpha_j + \rho_j|^{q-1} \operatorname{sign}(\theta\alpha_j + \rho_j) - b_i,$$

$$f'(0) = \sum_{j\in M_i} \alpha_j \, |\rho_j|^{q-1} \operatorname{sign}(\rho_j) - b_i,$$

$$f''(\theta) = (q-1) \sum_{j\in M_i} \alpha_j^2 \, |\theta\alpha_j + \rho_j|^{q-2},$$

and

$$f''(0) = (q-1) \sum_{j\in M_i} \alpha_j^2 \, |\rho_j|^{q-2}.$$

Consequently when $1 < q < 2$ the second derivative of $f(\theta)$ does not exist at points θ in which there exists an index $j \in M_i$ such that $\theta\alpha_j + \rho_j = 0$; while $f''(0)$ is undefined or "too large" whenever one of the relevant residuals, ρ_j, happens to be "too small". One way to overcome this difficulty is to replace (3.1) by the rule

$$\theta_0 = -f'(0)/\max\left\{\varepsilon,\ (q-1)\sum_{j\in M_i} \alpha_j^2\omega_j\right\} \tag{3.2}$$

where

$$\omega_j = |\rho_j|^{q-1} / \max\{\varepsilon,\ |\rho_j|\}.$$

This way $f''(0)$ is replaced by a term which cannot be "too small" or "too large".

Now that the computation of θ_0 has been resolved, we proceed to consider the question of how to complete the line search. Let θ^* denote the unique minimizer of $f(\theta)$. If $\underline{y}$ is close to a minimizer of $F(\underline{y})$ then θ^* is expected to be small and the Newton's steplength, θ_0, is likely to provide a good estimate of θ^*. On the other hand, when $\underline{y}$ lies far from a minimizer of $F(\underline{y})$ there is no need to compute an accurate estimate of θ^*. Hence again θ_0 is likely to provide a reasonable solution. However, in both cases it is necessary to ensure a sufficient decrease in the objective function value. This can be achieved by shortening the steplength until we reach a point $\hat{\theta}$ such that $f'(\hat{\theta})$ has the same sign as $f'(0)$. The details of the proposed line search algorithm are summarized below.

<u>The i-th step of the basic iteration</u>

a) <u>Computing an initial steplength θ_0</u>: If $f'(0) = 0$ skip to the next step. In this case $\hat{\theta} = 0$ is a minimizer of $f(\theta)$. Otherwise θ_0 is computed from (3.1) or (3.2) according to the value of q.

b) <u>Decreasing the steplength when necessary</u>: Set $\hat{\theta}$ to be the first point in the sequence

$$\theta_l = \theta_0/2^l, \quad l = 0, 1, 2, \ldots,$$

such that

$$f'(0)f'(\theta_l) \geq 0.$$

c) <u>Updating $\underline{r}$</u>: Set the new value of $\underline{r}$ to be $\underline{r} + \hat{\theta}\underline{a}_i$.

Since $f(\theta)$ is a strictly convex function, the restriction $f' \neq 0$ indicates that $\theta^* \neq 0$ and that $f(\theta)$ is monotonic decreasing in the interval whose end points are 0 and θ^*. This property ensures that the iterative process of substep b) is finite and that θ^* lies in the above interval. In other words,

$$0 < \left|\hat{\theta}\right| \leq |\theta^*|$$

and

$$f(\hat{\theta}) < f(\theta^*).$$

Note that the algorithm does not actually use or update $\underline{y}$. Hence, the only working space which is required by the algorithm is the n-vector $\underline{r}$, and the only data which are needed during the i-th step are b_i and the non-zero components of $\underline{a}_i$. These features make the proposed method suitable for solving very large problems.

4. Convergence Analysis

This section investigates the convergence properties of the proposed algorithm. We shall start by introducing some necessary notation. Let $\underline{y}^{(k,i)}$ denote the value of $\underline{y}$ at the beginning of the i-th step of the k-th iteration, $i = 1, \ldots, m$, $k = 1, 2, 3, \ldots$. Then $\underline{y}^{(k,i+1)}$ is obtained from $\underline{y}^{(k,i)}$ by the rule

$$\underline{y}^{(k,i+1)} = \underline{y}^{(k,i)} + \hat{\theta}\underline{e}_i,$$

where $\underline{e}_i$ denotes the i-th column of the $m \times m$ unit matrix, and $\hat{\theta}$ is obtained by the line-search procedure described in the previous section. The corresponding values of $F(\underline{y})$, $\underline{r}$ and $\underline{x}$ are denoted by $F^{(k,i)} = F(\underline{y}^{(k,i)})$, $\underline{r}^{(k,i)} = A^T\underline{y}^{(k,i)}$, and $\underline{x}^{(k,i)} = \underline{z}(\underline{y}^{(k,i)})$, respectively. Note that each $\underline{y}^{(k,i)}$ has a unique decomposition of the form $\underline{y}^{(k,i)} = \hat{\underline{y}}^{(k,i)} + \tilde{\underline{y}}^{(k,i)}$ where $\hat{\underline{y}}^{(k,i)} \in$ Range(A) and $\tilde{\underline{y}}^{(k,i)} \in$ Null(A^T).

The line search algorithm ensures that the sequence $\{F^{(k,i)}\}$ is monotonic decreasing. Therefore, since $F(\underline{y})$ has a minimizer, this sequence is bounded from below and

$$\lim_{k\to\infty} \left(F^{(k,i)} - F^{(k,i+1)}\right) = 0. \tag{4.1}$$

Let $\underline{x}^*$ denote the solution of (1.1). Then $F(\underline{x})$ can be rewritten in the form

$$F(\underline{y}) = \left\|A^T\underline{y}\right\|_q^q / q - (\underline{x}^*)^T A^T \underline{y},$$

and the Hölder inequality

$$\left|(\underline{x}^*)^T A^T \underline{y}\right| \leq \|\underline{x}^*\|_p \left\|A^T\underline{y}\right\|_q$$

gives

$$F(\underline{y}) \geq \left\|A^T\underline{y}\right\|_q^q / q - \|\underline{x}^*\|_p \left\|A^T\underline{y}\right\|_q = \left\|A^T\underline{y}\right\|_q \left(\left\|A^T\underline{y}\right\|_q^{q-1} / q - \|\underline{x}^*\|_p\right).$$

Hence the fact that sequence $\{F^{(k,i)}\}$ is bounded from above $F^{(1,1)}$ implies that the sequences $\{\underline{r}^{(k,i)}\}$, $\{\underline{x}^{(k,i)}\}$, $\{A\underline{x}^{(k,i)} - \underline{b}\}$ and $\{\hat{\underline{y}}^{(k,i)}\}$ are bounded. Consequently each of these sequences has at least one cluster points.

Observe that $A\underline{x}^{(k,i)} - \underline{b}$ is the gradient vector of $F(\underline{y})$ at the point $\hat{\underline{y}}^{(k,i)}$. Therefore the sequence $\{\underline{x}^{(k,i)}\}$ converges to $\underline{x}^*$ if and only if

$$\lim_{k\to\infty} A\underline{x}^{(k,i)} = \underline{b}. \tag{4.2}$$

Recall also the i-th component of $A\underline{x}^{(k,i)} - \underline{b}$ equals $f'(0)$ and that $F(\underline{y})$ is continuously differentiable. Hence in order to ensure (4.2) it is sufficient to verify that

$$\lim_{k\to\infty} f'(0) = 0 \tag{4.3}$$

and

$$\lim_{k\to\infty} \hat{\theta} = 0 \tag{4.4}$$

Theorem 2. *If $q > 2$ the sequence $\{\underline{x}^{(k,i)}\}$ converges to $\underline{x}^*$.*

PROOF. The first part of the proof establishes (4.3). Assume for simplicity that $f'(0) < 0$. (The other possibility, $f'(0) > 0$, is treated in a similar way.) Then the special structure of $f''(\theta)$ and the fact that the sequence $\{\underline{r}^{(k,i)}\}$ is bounded imply the existence of a positive constant, γ, such that

$$f''(\theta) \le \gamma \tag{4.5}$$

for all $0 \le \theta \le \theta^*$. (Recall that θ^* denotes the unique minimizer of $f(\theta)$.) Consequently the one parameter quadratic function

$$h(\theta) = f(0) + \theta f'(0) + \frac{1}{2}\theta^2\gamma$$

satisfies

$$f(\theta) \le h(\theta)$$

for all $0 \le \theta \le \theta^*$. The minimizer of $h(\theta)$ lies at the point

$$\eta = -f'(0)/\gamma,$$

and this point satisfies both $0 < \eta \le \theta^*$ and $\eta \le \theta_0$. On the other hand the line search procedure ensures that

$$\eta/2 < \hat{\theta} \le \theta^*.$$

This leads to

$$h(\eta/2) \ge f(\eta/2) \ge f(\hat{\theta}),$$

where the right inequality is due to the fact that $f(\theta)$ is monotonic decreasing in the interval $[0, \theta^*]$. Now the inequalities

$$f(0) - f(\hat{\theta}) \ge f(0) - f(\eta/2) \ge f(0) - h(\eta/2) = h(0) - h(\eta/2) = (3/8)\left(f'(0)\right)^2/\gamma$$

give

$$F^{(k,i)} - F^{(k,i+1)} \ge (3/8)\left(f'(0)\right)^2/\gamma,$$

and limit (4.1) results in (4.3).

The second part of the proof establishes (4.4). Since $|\hat{\theta}| \le |\theta_0|$ it is sufficient to show that

$$\lim_{k\to\infty} \theta_0 = 0.$$

Let μ denote the number of non-zero components in $\underline{a}_i$ and assume for simplicity that $M_i = \{1, 2, \ldots, \mu\}$. Let the vectors $\underline{u} = (u_1 = \ldots = u_\mu)^T$ and $\underline{v} = (v_1 = \ldots = v_\mu)^T$ be defined by the equalities

$$u_i = \alpha_i^2 |\rho_i|^{q-2} \quad \text{and} \quad v_i = \rho_i/\alpha_i, \quad i = 1, \ldots, \mu.$$

Then, since $f'(0) \neq 0$, $f'' = (q-1)\|\underline{u}\|_1 \neq 0$ and

$$f'(0)/f''(0) = \underline{u}^T\underline{v}/\Big((q-1)\|\underline{u}\|_1\Big).$$

Hence by combining (3.1) with the Hölder inequality $\left|\underline{u}^T\underline{v}\right| \leq \|\underline{u}\|_1 \|\underline{v}\|_\infty$ we obtain

$$|\theta_0| \leq |f'(0)/f''(0)| \leq \|\underline{v}\|_\infty/(q-1) \leq \|\underline{v}\|_\infty \leq \rho_{\max}/\alpha_{\min}, \tag{4.6}$$

where

$$\alpha_{\min} = \min\{|\alpha_j| \mid j \in M_i\}$$

and

$$\rho_{\max} = \max\{|\rho_j| \mid j \in M_i\}.$$

From this point the proof continues by contradiction. Assume that there exists a positive constant, δ, such that the inequality

$$\delta \leq |\theta_0| \tag{4.7}$$

holds for infinitely large values of k. Then (4.6) and (4.7) give

$$\delta\alpha_{\min} \leq \rho_{\max},$$

while the inequalities

$$f''(0) \geq (q-1)\sum_{j\in M_i}|p|^{q-2}\alpha_{\min}^2 \geq \rho_{\max}^{q-2}\alpha_{\min}^2 \geq (\delta\alpha_{\min})^{q-2}\alpha_{\min}^2 = \delta^{q-2}\alpha_{\min}^q$$

show that $f''(0)$ exceeds a certain positive constant. Therefore the limit (4.3) and the inequality $|\theta_0| \leq |f'(0)/f''(0)|$ imply (4.5), which contradicts our assumption.

5. Numerical Results

In this section we provide the results of some preliminary experiments with the new method. The test problems have the form (1.1) where A has $m = l(l+1)/2$ rows and $n = 40000$ columns. The first l row of A are generated in the following way. Each row has only four nonzero elements that have random locations. That is, the column indices of these elements are random integers between 1 and 40000. (The random numbers generator is of uniform distribution). All the nonzero elements in the first l rows have the same value, 1. The other $l(l-1)/2$ rows are defined as $\underline{a}_i^T + \underline{a}_j^T$, $1 \leq i < j \leq l$. The elements of the right hand side vector, $\underline{b}$, are defined by the rule $b_i = 4$ for $i = 1, \ldots, l$, and $b_i = 8$ for $i = l+1, \ldots, m$. This ensures that the point $(1, 1, \ldots, 1)^T$ solves our system. The values of l that have been used in our experiments are $l =$ 100, 200, 300, 400 and 500; giving $m =$ 5050, 20100, 45150, 80200 and 125250, respectively. Each of the five linear systems have been solved with various values of p (see Tables 1 and 2). The method of Kaczmarz (i.e., $p = 2$) was started from the origin point $\underline{r}_0 = A^T\underline{y}_0 = (0, 0, \ldots, 0)^T \in \mathbf{R}^n$. Then the solution point for a given value of p has been used as a starting point for the next upward (downward) value of p. The solution of each problem was carried out by performing only four iterations of the proposed method. (There was no need in performing more iterations.) Recall that the gradient vector of $F(\underline{y})$ has the form $\underline{b} - A\underline{z}(\underline{y})$. Hence the rate of convergence of our algorithm can be assessed by recording the corresponding values of $\left\|\underline{b} - A\underline{z}(\underline{y})\right\|_2^2$. Thus for each pair of values of p and m the tables contain two figures: The upper figure provides the value of $\left\|\underline{b} - A\underline{z}(\underline{y})\right\|_2^2$ at the starting point, while the lower figure gives the value of this quantity at the solution point (i.e., after four iterations).

The results of our experiments are quite encouraging. We see that the method retains a fast rate of convergence when p moves away from 2. This gives hope that the range of problems which can be handled efficiently by the new method is similar to that of Kaczmarz's method. Note that although the convergence theory is somewhat deficient when $p > 2$, in practice the rate of convergence is not necessary slower in this case.

Table 1: Results for upward values of p.

	m = 5050	m = 20100	m = 45150	m = 80200	m = 125250
p =2.0	3.2 E+5 7.0 E-11	1.3 E+6 2.4 E-10	2.9 E+6 1.9 E-9	5.1 E+6 1.3 E-8	8.0 E+6 8.8 E-8
p =2.25	4.8 E-1 1.1 E-13	2.3 E+0 4.3 E-10	6.5 E+0 4.1 E-10	4.2 E+1 3.6 E-9	8.0 E+1 5.5 E-9
p =2.5	1.4 E-1 1.0 E-16	7.0 E-1 3.7 E-10	1.9 E+0 1.3 E-10	1.3 E+1 1.7 E-9	2.4 E+1 4.0 E-9
p =3.0	1.2 E-1 3.2 E-11	6.0 E-1 3.4 E-10	1.6 E+0 5.6 E-10	1.1 E+1 2.1 E-9	2.0 E+1 6.7 E-9
p =4.0	4.9 E-2 2.9 E-11	2.6 E-1 1.6 E-10	6.9 E-1 8.1 E-10	4.4 E+0 2.3 E-9	8.2 E+0 4.8 E-9
p =6.0	1.1 E-2 1.1 E-13	5.6 E-2 2.6 E-10	1.5 E-1 5.9 E-10	9.5 E-1 3.2 E-9	1.8 E+0 4.6 E-9
p =11.0	1.4 E-3 1.0 E-16	7.6 E-3 3.4 E-10	2.0 E-2 1.1 E-9	1.3 E-1 2.3 E-9	2.4 E-1 3.7 E-9
p =21.0	8.7 E-5 1.3 E-10	4.6 E-4 1.5 E-10	1.2 E-3 5.5 E-11	7.6 E-3 3.3 E-9	1.4 E-2 6.6 E-9

Table 2: Results for downward values of p.

	$m = 5050$	$m = 20100$	$m = 45150$	$m = 80200$	$m = 125250$
p =2.0	3.2 E+5	1.3 E+6	2.9 E+6	5.1 E+6	8.0 E+6
	7.0 E-11	2.4 E-10	1.9 E-9	1.3 E-8	8.8 E-8
p =1.8	2.0 E+0	9.3 E+0	2.6 E+1	1.8 E+2	3.3 E+2
	4.1 E-12	1.5 E-10	5.0 E-10	1.7 E-9	4.9 E-9
p =5/3	3.0 E+0	1.4 E+1	3.9 E+1	2.7 E+2	5.0 E+2
	1.1 E-13	2.3 E-10	2.0 E-10	2.4 E-9	4.7 E-9
p =1.5	2.3 E+1	1.0 E+2	3.0 E+2	2.0 E+3	3.8 E+3
	1.0 E-10	2.6 E-10	1.1 E-9	4.8 E-9	2.7 E-8
p =4/3	2.3 E+2	9.2 E+2	2.8 E+3	1.9 E+4	3.7 E+4
	1.7 E-10	4.0 E-10	7.9 E-10	3.0 E-5	1.1 E-4
p =1.2	2.5 E+3	8.7 E+3	2.9 E+4	1.9 E+5	3.8 E+5
	2.8 E-10	3.1 E-10	7.4 E-6	1.3 E+0	1.5 E+0
p =1.1	3.7 E+4	1.1 E+5	4.0 E+5	2.5 E+6	5.1 E+6
	1.7 E-2	1.6 E-2	1.1 E-1	1.7 E+0	4.0 E+0
p =1.05	5.8 E+4	1.8 E+5	6.3 E+5	3.8 E+6	8.0 E+6
	4.0 E-7	1.3 E-7	3.5 E-6	2.8 E-5	9.0 E-5

REFERENCES

Cadzow J. A. (1971), Algorithm for the Minimum–Effort Control Problem, *IEEE Trans. Automatic Control*, vol. AC-16, no. 1, 60–63.

Cadzow J. A. (1973), A Finite Algorithm for Minimum 1.. Solution to a System of Consistent Linear Equations, *SIAM J. Numer. Anal.*, no. 10, 607–617.

Censor Y. (1981), Row–Action Methods for Huge and Sparse Systems and Their Applications, *SIAM Review*, no. 23, 444–466.

Dax A. (1993), On Minimum Norm Solutions, *J. of Optim. Theory and its Applic.*, no. 76, 183–193.

Gordon R., Bender R., Herman G. T. (1970), Algebraic Reconstruction Techniques (ART) for Three Dimensional Electron Microscopy and X–Ray Photography, *J. Theoret. Biol.*, no. 29, 471–481.

Herman G. T., Lent A., and Rowland S. W. (1973), ART: Mathematics and Applications, *J. Theoret. Biol.*, no. 42, 1–32.

Kaczmarz S. (1937), Angenaherte Auflosung Von Systemen Linearer Gleichungen, *Bull. Acad. Polon. Sci.* Lett. A, no. 35, 355–357.

Kolev L. (1975), Iterative Algorithm for the Minimum Fuel and Minimum Amplitude for Linear Discrete Systems, *Int. J. Control*, no. 21, 779–784.

Lent A., and Censor Y. (1980), Extension of Hildreth's Row–Action Method for Quadratic Programming, *Siam J. Cont. Optim.*, no. 18, 444–454.

Luenberger D. G. (1969), *Optimization by Vector Space Methods*, John Wiley, New York.

On Explicit Formulae for Inverting the X-Ray Transform

A. S. DENISJUK

Krasnoyarsk State Pedagogical Institute

Prosp. Marksa 100, Krasnoyarsk, Russia

ABSTRACT

The problem we consider is to find an explicit formula for inverting the X-ray transform in $\mathbf{R}^n$ after the data known for some $n-$dimensional family of lines provided that only a finite number of integrations and differentiations may be used. We consider also the problem of describing all the families for which there exist such formulae. We define a new class of families which is referred to as the class of quasi-hyperbolic complexes of lines. An inversion formula is obtained for the simplest example of a quasi-hyperbolic complex, namely, for the set of lines in $\mathbf{R}^5$, tangent to a given three-dimensional surface.

1. We consider the problem of finding an explicit formula for reconstruction of a function $f(x)$, $x \in \mathbf{R}^n \subset \mathbf{R}^n\mathbf{P}^n$, after its X-ray transform

$$g(y) = \int_y f(x)ds.$$

Here $y \in G$ is a line in $\mathbf{R}^n\mathbf{P}^n$, ds is the Lebesgue's measure on the line y, G is the set of all lines in $\mathbf{R}^n\mathbf{P}^n$. (When talking about geometrical properties of lines, it is convenient to deal with a projective space, that is why we fix an immersion $\mathbf{R}^n \subset \mathbf{R}^n\mathbf{P}^n$. The X-ray transform is defined in the usual affine way).

We assume that $g(y)$ is known only for $y \in K$, where $K \subset G$ is an n-dimensional submanifold. Following the paper by Gelfand and Graev (1968), we will call smooth submanifolds $K \subset G$, $\dim K \geq n-1$, the complexes of lines. The second part of the problem is to describe all $n-$dimensional complexes of lines for which there exists such a formula.

An analogous problem in the complex space $\mathbf{C}^n$ was solved in the paper by Gelfand and Graev (1968) for local inversion formulae. The n-dimensional complexes of lines in $\mathbf{C}^n$ for which there exists a local inversion formula were said to be *admissible complexes*. In the real space the inversion formula cannot be local. We consider "simple" inversion formulae.

Following V.P. Palamodov, a formula will be called *simple* if it inverts explicitly the X-ray transform and involves only a finite number of integrations and differentiations. In particular, a simple formula should include neither analytic continuation nor non-explicit solution of an integral equation.

2. Geometrical properties of the complexes of lines are of great importance for reconstruction. We will remind some definitions.

DEFINITION 1. Let $K \subset G$, $\dim K = n - 1 + r$, $(r \geq 0)$, $y \in K$. A point $x_0 \in y$ is called *critical* if

$$R(x_0) = \dim(T_y K \cap T_y G_{x_0}) > r;$$

the number $k(x_0) = R(x_0) - r$ is said to be the *multiplicity* of the critical point x_0. The line $y \in K$ is said to be *critical* if it contains an infinite number of critical points.

Here G_{x_0} denotes the submanifold of lines in $\mathbf{R}^n\,\mathbf{P}^n$, which pass through the point x_0. Note that $\dim G_{x_0} = n - 1$, whence in general position $k(x) = 0$.

Every non-critical line contains no more than $\operatorname{codim} K$ of critical points counting with their multiplicities.

DEFINITION 2. A complex of lines K is called *hyperbolic* if every non-critical line $y \in K$ contains the maximal number, i.e., $\operatorname{codim} K$, of critical points counting with their multiplicities.

REMARK 3. The concept of multiplicity of a critical point given in Definition 1 is the same as in the paper by Denisjuk (1990) and differs from the one used in the paper by Gelfand and Graev (1968). If we define the multiplicity as in the paper by Malyusz (1981), then, for n-dimensional complexes of lines in $\mathbf{C}^n$, the class of complexes described in Definition 2 will coincide with the class of admissible complexes of lines.

Consider some instances of hyperbolic complexes of lines (Denisjuk, 1990). In $\mathbf{R}\,\mathbf{P}^3$ a three-dimensional hyperbolic complex is either of the following families:

(1) a family of lines intersecting a given curve $\gamma \subset \mathbf{R}\,\mathbf{P}^3$,

(2) a family of lines tangent to a given surface $W \subset \mathbf{R}\,\mathbf{P}.^3$

For the former family, various formulae for inverting the X-ray transform are known (Kirillov, 1961; Tuy, 1983; Grangeat, 1987; etc. — see the paper by Palamodov (1990) for a more complete bibliography).

As the latter family, a reconstruction formula was obtained by the author in collaboration with V.P. Palamodov (Denisjuk, 1990, 1993; Palamodov 1990).

In $\mathbf{R}^n\,\mathbf{P}^n$ any hyperbolic complex of "general position" can be represented as an intersection of *elementary* ones having only one critical point. The elementary hyperbolic complex is either of the following sets:

(1) a family of lines intersecting a given submanifold $M \subset \mathbf{R}^n\,\mathbf{P}^n$, $0 < \dim M < n - 1$,

(2) a family of lines tangent to a given submanifold $W \subset \mathbf{R}^n\,\mathbf{P}^n$, $\operatorname{codim} W = 1$.

It turns out that for any n-dimensional hyperbolic complex of lines in $\mathbf{R}^n\,\mathbf{P}^n$ one can obtain simple reconstruction formulae for the X-ray transform if an additional global completeness condition holds. It has been done in Denisjuk (1990, 1993) for elementary complexes. In general case reconstruction formulae have been found by Palamodov (1992).

3. Now we recall the completeness condition, which is necessary for existence of a simple inversion formula (Palamodov, 1989).

DEFINITION 4. Let $K \subset G$ be a complex of lines, $\dim K = n$. The complex K is said to satisfy the *completeness condition* at a point $(x,\omega) \in T^* \mathbf{R}^n \mathbf{P}^n$ if the range of the mapping

$$\Pi_{x,\omega} : K \times \mathbf{RP} \to \overset{n}{\mathbf{R}} \mathbf{P}^n \times \mathbf{R}$$
$$(y,t) \mapsto (y(t), \langle \omega, \dot{y}(t) \rangle)$$

equals $n+1$ at a point (y_0, t_0) such that $y_0(t_0) = x$, $\omega \perp y_0$.

Here $y(t)$ is the parametrization of the line y.

Proposition 5. *(Palamodov, 1992) If a complex K admits a simple inversion formula for functions f, $\operatorname{supp} f \subset B \subset \mathbf{R}^n$, then it satisfies the completeness condition at almost any points $(x,\omega) \in T^*B$.*

REMARK 6. The concept of critical point is closely connected with the completeness condition: a point $x \in y \in K$ is critical for K if and only if the completeness condition fails for any $\omega \in T^*_x \mathbf{R}^n \mathbf{P}^n$ such that $\omega \perp y$.

4. Now we are to define a new class of complexes of lines for which a simple reconstruction formula can be derived.

Let H be the set of all hyperplanes in $\mathbf{R}^n \mathbf{P}^n$.

DEFINITION 7. Let $y \in K$, $K \subset G$, $\dim K = n + r, r > 0$, $y \subset h \in H$. A point $x_0 \in y$ is called *$h-$critical* for the complex K if

$$R_h(x_0) = \dim(T_y K \cap T_y G_{x_0,h}) > r;$$

the number $k(x_0) = R_h(x_0) - r$ is said to be the *multiplicity* of the $h-$critical point x_0. The line $y \in K$ is *$h-$critical* if it contains an infinite number of $h-$critical points.

Here $G_{x_0,h}$ is the set of lines passing through the point x_0 and belonging to h.

If x_0 is a critical point, it will be $h-$critical for any plane $h \in H$ such that $y \subset h$.

Proposition 8. *A point $x_0 \in y \in K$, $\dim K = n$, is $h-$critical if and only if the completeness condition fails for $\omega \in T^*_{x_0} \mathbf{R}^n \mathbf{P}^n$ such that $\omega \perp h$.*

We shall consider not only the complex K but also a family $K_{\mathbf{C}}$ of complexifications of lines from K. $K_{\mathbf{C}}$ is a submanifold in $G_{\mathbf{C}}$, the manifold of lines in $\mathbf{C}^n$. Denote by $H_{\mathbf{C}}$ the manifold of all hyperplanes in $\mathbf{C}^n$.

DEFINITION 9. Let $h \in H_{\mathbf{C}}$. A line $y \in K$ is said to have a *complex $h-$critical point* if it's complexification $y \otimes \mathbf{C}$ has an $h-$critical point. A line $y \in K$ is called $h-$critical if it contains an infinite number of $h-$critical points.

Proposition 10. *Let $h \in H_{\mathbf{C}}$. Unless a line $y \in K \subset G$ is $h-$critical, it contains no more than $\operatorname{codim} K$ of $h-$critical points counting with their multiplicities.*

DEFINITION 11. A complex $K \subset G$ is said to be *quasi-hyperbolic* if all it's $h-$critical points, belonging to non-$h-$critical lines, are real and critical.

REMARK 12. If we demand the number of critical points of any non-h-critical line to be maximum, then the class just defined will coincide with the class of hyperbolic complexes.

REMARK 13. If we neglect the requirement for critical points to be real, then the class described in Definition 11 will coincide with the class of admissible complexes. It can be proved using the following characteristic property of admissible complexes of lines.

Theorem 14. *(Gelfand and Graev, 1968) A complex K is admissible if and only if, for any non-critical line $y_0 \in K$, the plane tangent to a cone K_x, which is formed by points of lines closed to y_0 and intersecting it at a non-critical point $x \in y$, does not depend on x.*

Consider some instances of quasi-hyperbolic complexes of lines. Any hyperbolic complex is quasi-hyperbolic. Other instances of quasi-hyperbolic complexes can be found in considering intersections of elementary quasi-hyperbolic complexes that contain only one critical point. For example, the following sets are elementary quasi-hyperbolic complexes:

(1) the set of lines in $\mathbf{R}^n$ intersecting a given submanifold $M \subset \mathbf{R}^n$, $0 \leq \dim M \leq n-2$;

(2) the set of lines in $\mathbf{R}^n$ tangent to a given submanifold $W \subset \mathbf{R}^n$, $\frac{n}{2} \leq \dim W \leq n-1$.

In the latter case there are instances of quasi-hyperbolic complexes of lines that are not hyperbolic (when $\dim W < n-1$).

5. Turning to inversion formulae, we shall restrict our consideration to the family of lines in $\mathbf{R}^5$ tangent to a fixed surface $W \subset \mathbf{R}^5$ — it is the simplest example of a quasi-hyperbolic complex which is not hyperbolic. In the general case, inversion formulae can be constructed similarly.

Theorem 15. *Let $W \subset \mathbf{R}^5$ be a smooth three-dimensional surface, K_W be the set of lines tangent to W, $B \subset \mathbf{R}^5$ be a compact set, H be a hyperplane $\langle \omega, x\rangle = p$. Suppose that the intersection $H \cap W$ contains an open two-dimensional submanifold W_H, which satisfies the following conditions:*

1. *$\Pi_H^{-1}(B \cap H)$ is a compact set, where the mapping Π_H,*

$$\Pi_H : S(W_H) \times \mathbf{R} \to H((w,\xi),t) \mapsto w + \xi t,$$

is the parametrization of lines in K_W. Here $S(W_H)$ is the restriction to W_H of a bundle of unit vectors tangent to W, $w \in W$, ξ is a unit tangent vector.

2. *Let $S_H(W_H) \subset S(W_H)$ be a subbundle of unit tangent vectors parallel to the plane H. We suppose that the δ-degree of restriction of the map Π_H to $S_H(W_H) \times \mathbf{R}$ is constant and non-zero. (We suppose that orientations on $S_H(W_H) \times \mathbf{R}$ and H are chosen in some way and fixed).*

Then there exists a first-order differential operator L_H such that for any function f, $\operatorname{supp} f \subset B$,

$$\frac{\mathrm{d}^3}{\mathrm{d}p^3}\,\mathrm{R}_4 f(\omega,p) = \frac{-1}{\delta} \int\limits_{S_H(W_H)} L_H^3 g(w,\xi)\omega_H,$$

where

$$\mathrm{R}_4 f(w,p) = \int\limits_{\langle\omega,x\rangle=p} f(x)\mathrm{d}S \quad \textit{is the Radon transform,}$$

$$g(w,\xi) = \int f(w+\xi t)\mathrm{d}t \quad \textit{is the X-ray transform,}$$

and ω_H is the volume form on $S_H(W_H)$ transferred by the map

$$\Pi|_{S_H(W_H)\times\{1\}} \to \mathbf{R}^5$$

$$(w, \xi) \mapsto w + \xi.$$

PROOF. Choose local coordinates $\tau = (\tau_1, \tau_2, \tau_3)$ on W so that τ_1 and τ_2 are coordinates in W_H, whereas τ_3 is a transversal coordinate. The X-ray transform will be written as follows:

$$g(\tau, \varphi, \lambda) = \int f(w(\tau) + t(w_1 \cos\varphi + w_2 \sin\varphi + \lambda w_3))dt,$$

where $w(\tau)$ is a local parametrization of the surface W, $w_j = \frac{dw}{d\tau_j}$, $j = 1, 2, 3$. It is evident that τ, φ, λ can be taken as local coordinates on K_W.

We denote by Ω_H the volume form on the plane H. Then for the Radon transform we obtain the formula

$$\mathrm{R}_4 f(w, p) = \int_H f(x)\Omega_H$$

$$= \frac{-1}{\delta} \int_{S_H(W_H)} \int f(w(\tau_1, \tau_2, 0) + t(w_1 \cos\varphi + w_2 \sin\varphi))t^3 dt \wedge \omega_H.$$

The operator L_H is sought in the form

$$L_H = \frac{1}{\langle \omega, w_3 \rangle} \frac{\partial}{\partial \lambda} + \alpha \frac{\partial}{\partial \varphi},$$

where the coefficient $\alpha = \alpha(\tau, \varphi, \lambda)$ will be defined below.

Differentiate $g(\tau, \varphi, \lambda)$:

$$\frac{\partial g}{\partial \varphi}(\tau_1, \tau_2, 0, \varphi, \lambda) = \int \langle f', -w_1 \sin\varphi + w_2 \cos\varphi \rangle t\, dt,$$

$$\frac{\partial g}{\partial \lambda}(\tau_1, \tau_2, 0, \varphi, \lambda) = \int \langle f', w_3 \rangle t\, dt.$$

So,

$$L_H\, g = \int \langle f', \frac{1}{\langle \omega, w_3 \rangle} w_3 + \alpha(-w_1 \sin\varphi + w_2 \cos\varphi)) t \rangle t\, dt$$

$$= \int \langle f', V_1 + \alpha V_2 \rangle t\, dt,$$

where V_1 and V_2 are the corresponding vector fields.

Now we have to use the following property of the Radon transform: for any vector field V and any function $f(x)$

$$\mathrm{R}(Vf)(\omega, p) = -\mathrm{R}(\mathrm{div}\ V)(\omega, p) + \frac{d}{dp} \mathrm{R}(\langle \omega, V \rangle f)(\omega, p).$$

In our case $\langle \omega, V_1 + \alpha V_2 \rangle \equiv 1$, and we can choose α so that

$$\mathrm{div}\,(V_1 + \alpha V_2) = 0. \tag{1}$$

Indeed, equation (1) written in the coordinates $(\tau_1,\ \tau_2,\ \varphi,\ \lambda,\ t)$ becomes

$$\mathrm{div}\left(\frac{1}{t}\frac{1}{\langle \omega, V_1 \rangle}\frac{\partial}{\partial \lambda} + \frac{\alpha}{t}\frac{\partial}{\partial \varphi}\right) \equiv \frac{1}{\sqrt{G}}\frac{\partial}{\partial \lambda}\left(\frac{1}{t}\frac{1}{\langle \omega, V_1 \rangle}\sqrt{G}\right) + \frac{1}{\sqrt{G}}\frac{\partial}{\partial \varphi}\left(\frac{\alpha}{t}\sqrt{G}\right) = 0,$$

where $G = G(\tau_1, \tau_2, \varphi, \lambda, t)$ is the determinant of the metric tensor on $S(W_H) \times \mathbf{R}$ transferred from the Euclidean metrics on $\mathbf{R}^5$ by the map Π_H. From the last equation the coefficient α can be expressed by the explicit formula:

$$\alpha(\tau_1, \tau_2, \varphi, \lambda) = \frac{-1}{\sqrt{G(\tau_1, \tau_2, \varphi, \lambda, t)}} \int_0^{\varphi} \frac{\partial}{\partial \lambda} \frac{\sqrt{G(\tau_1, \tau_2, \theta, \lambda, t)}}{\langle \omega, V_1 \rangle} d\theta$$

$$= \frac{-1}{\sqrt{G_H(\tau_1, \tau_2, \varphi, \lambda)}} \int_0^{\varphi} \frac{\partial}{\partial \lambda} \frac{\sqrt{G_H(\tau_1, \tau_2, \theta, \lambda)}}{\langle \omega, V_1 \rangle} d\theta,$$

where $G_H(\tau_1, \tau_2, \theta, \lambda)$ is the determinant of the metric tensor on $S(W_H)$ which is induced from the Euclidean metrics in $\mathbf{R}^5$ by the map

$$\Pi_H|_{S(W_H)\times\{1\}} : (w, \xi) \mapsto w + \xi.$$

The last equation holds because of homogeneity of G in t.

Applying to g the operator L_H three times and integrating over $S_H(W_H)$, we will obtain the desired formula. The theorem has been proved.

An analogous theorem can be proved similarly for the complex of lines in $\mathbf{R}^{2k-1}$, which are tangent to a given k-dimensional surface ($k \geq 4$).

Corollary 16. *If the conditions of Theorem 15 hold for all hyperplanes in $\mathbf{R}^5$ which intersect B, then for the complex of lines K_W there exists a simple formula for reconstructing the X-ray transform of every function $f(x)$, $\operatorname{supp} f \subset B$.*

PROOF. If the corollary conditions hold, we can determine the integrand in the well-known inversion formula for the Radon transform:

$$f(x) = \frac{1}{2(2\pi)^4} \int_{S^4} \frac{d}{dp}\left(\frac{d^3}{dp^3} R_4 f\right)(\omega, \langle \omega, x \rangle)\, d\omega,$$

where S^4 is a unit sphere, and $d\omega$ is a volume element on it.

6. In conclusion we formulate a hypothesis about existence of simple inversion formulae for the X-ray transform.

Hypotheses 17. *For any admissible complex of lines there exists a simple formula for inversion of the X-ray transform (with an additional global completeness assumption similar to that of Theorem 15).*

To our knowledge, no inversion formulae have been found for non-admissible complexes of lines. The following proposition shows that non-admissible complexes have *a lot of points* where the necessary completeness condition fails (see Definition 4). It lowers chances for the existence of such formulae.

Proposition 18. *Let $K \subset G$ be an $n-$dimensional complex of lines which is not admissible, $y \in K$. Then for any point $x \in y$ there exists a hyperplane $h \in H$ such that $y \subset K$ and the point x is $h-$critical (therefore, the completeness condition fails at the point $(x, \omega) \in T^*\mathbf{R}^n\mathbf{P}^n$, $\omega \perp h$).*

REFERENCES

Denisjuk A. S. (1990), The local structure of hyperbolic complexes of curves. *Russian Math. Surveys*, vol. 45, no. 5, 225–226.

Denisjuk A. S. (1991), *An Investigation on Integral Geometry in Real Space.* Thesis, Moscow State University, Dep. Mech. Math.

Denisjuk A. S. (1993), Inversion of generalized Radon transform. *Transl. of Mathematical Monographs.* (to appear)

Gelfand I. M. and Graev M. I. (1968), Complexes of lines in $\mathbf{C}^n$ *Functional Anal. Appl.*, vol. 2, 219–229.

Grangeat P. (1987), *Analyse d'un Systéme d'imagerie 3D par Reconstruction à Partir de Radiographies X en Gèomètrie Conique.* These de doctorat, Grenoble.

Kirillov A. A. (1961), On a problem of I. M. Gel'fand. *Sov. Math. Dokl.*, vol. 2, 268–269.

Malyusz K. (1979), The structure of admissible complexes of lines in $\mathbf{CP}^n$. *Trans. Moscow Math. Soc.*, 1981 vol. 1, 195–226.

Palamodov V. P. (1989), Problems of data completeness and of inversion of X-ray transform in space. *Thes of IV All-Union Symp. on Computer Tomography*, Novosibirsk, vol. 1, 30–31. (in Russian)

Palamodov V. P. (1990), An inversion formula for three-dimensional ray transform. *Mathematical Problems of Tomography. Proceedings Oberwolfach.* Springer Lecture Notes in Math. (G. T. Herman, A. K. Louis, F. Natterer, eds.), vol. 1497, 53–62.

Palamodov V. P. (1992), *Radon transform on real algebraic varieties*, September 1–5, 1992. Vienna.

Tuy H. K. (1983), An inversion formula for cone-beam reconstruction. *SIAM J. Appl. Math*, vol. 43, 546–552.

Algebraic Reconstructions from Efficient Sampling Schemes in Tomography

L. DESBAT

TIMC - IMAG, Faculté de Médecine
38706 La Tronche FRANCE

ABSTRACT

Interlaced sampling have been proved to be twice as much efficient as the usual standard parallel schemes in tomography (in fact interlaced schemes are optimal parallel sampling). Recent results on efficient sampling have been obtained for fan beam geometries and for tomography with few data. Thus efficient sampling plays nowadays an important role in the design of medical and industrial scanners. Reconstruction algorithms for interlaced schemes, based on the filtered backprojection are well studied. Recent developments on optimal algorithms allowed to obtain very good results, even in the case of few data. In the present work, we study the reconstructions from interlaced data using classical algebraic algorithms and optimal regularization methods. Both SIRT and optimal algorithms are considered. The Stochastic *GCV* is used for the regularization parameter choice. Comparisons with reconstructions from standard schemes are provided. Numerical experiments are done both on simulated and real data. Applications of these algorithms for the reconstruction of the catalist density of pipes in circulating gasoil craking plants are presented.

1. Introducton

Tomography is a well known and wide used tool in the medical field (see Cormack, 1978; Herman, 1980) The purpose is to reconstruct an attenuation function f (in fact essentially the density function) from a finite number of its line integral, see Figure 1:

$$y_{p,q} = Rf(\phi_p, s_q) = \int_{\langle\theta_p,x\rangle_{\mathbb{R}^2}=s_q} f(x)\mathrm{d}x, \qquad \begin{cases} p = 0,\dots,P-1 \\ q = 1,\dots,Q \end{cases}, \tag{1}$$

where $\theta_p = (\cos\phi_p, \sin\phi_p)^t$. For some years, tomography is also used for the control and study of some industrial process (see for example Martin et al., 1992). Because of all the industrial constraints, process tomography often yields few data. A critical issue is then to take the most information on the studied function from the small amount of data. Pure algebraic approach, like in Kasantsev (1991) or in Desbat and Turlier (1991), can produce interesting results, in particularly when having some a priori information, but are far to be general and easy to use.

Efficient sampling of the Radon transform (Rattey and Lindgren, 1981; Natterer, 1986) is a general way to choose the best measurement geometry in the lack of a priori information. Interlaced sampling, first introduced by Cormack (1978), have been proved to be optimal in tomography. A standard scheme is just an equidistant sampling in the

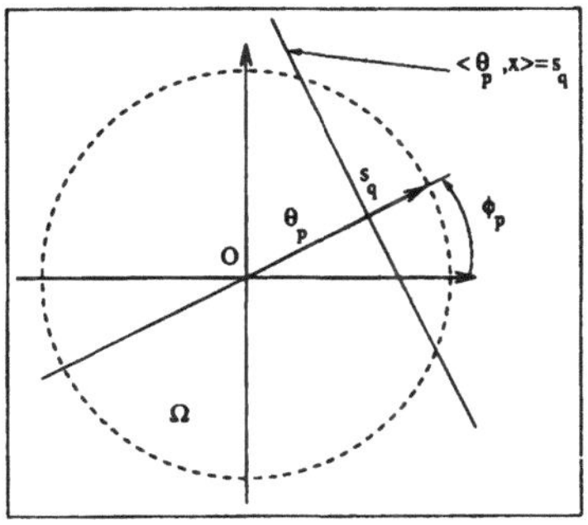

Figure 1: Measurement geometry.

Radon space (ϕ, s) in both directions. The corresponding interlaced sampling consist of *half* of the standard points taken in an interlaced way, i.e. the even translations for the even directions and the odd translations for the odd directions.

Generalizations of sampling theorems Faridani (1990), see also Faridani (1993), allows to consider new efficient sampling schemes essentially derived from the interlaced ones, see (Faridani, 1990) (Desbat, 1993) for the study of interlaced sampling perturbations. The question of optimal sampling is crucial in the scanner design, see Natterer (1993) for results in the case of fan beam geometries.

The usual algorithms have to be adapted when using efficient sampling in order to get the best results. The challenge is to obtain from interlaced data the same results as in the corresponding standard scheme (having twice as much data). Algorithms have been developed for the analytical approach, i.e. discretization of an inverse of the Radon transform such as filtered backprojection (see Faridani (1990) for references). Klaverkamp (1991) proposed a very elegant way to reconstruct from integral over strips in an optimal algebraic approach and thus to get in interlaced sampling almost the same result as in the corresponding standard scheme. Note that the Cormack's original interlaced scheme does not lead to so good results because its strip wide is twice too large. A not so elegant but more stable and easy to used algorithm was been proposed in Desbat and Turlier (1992).

In this report, we consider algorithm of type SIRT (Simultaneous Iterative Reconstruction Technique) for solving problem in the classical image estimation approach through pixel expansions. Our aim is to obtain almost the same results from interlaced data as from the corresponding standard scheme. We present in the first part some facts about efficient sampling and interlaced schemes in tomography. We then introduce algebraic algorithms for the reconstruction from interlaced data. Then we propose numerical experiments on both simulated and real data.

2. Interlaced Sampling

2.1 *Efficient sampling*

In tomography we have to produce sampling of the Radon transform of a function. The goal of efficient sampling is to produce "good" estimation of the unknown function with a minimal number of measurement. Let us first denote $\hat{f}$ the Fourier transform of *a* function $f \in \mathcal{S}(\mathbb{R}^n)$:

$$\hat{f}(\xi) = \int_{\mathbb{R}^n} f(x) e^{-i\langle x,\xi\rangle} \mathrm{d}x$$

Most of the results obtained are based on the Petersen and Middleton sampling theorem, because it allows us to take in account the geometrical properties of the Fourier transform support of the function to be sampled.

Theorem 1. *Petersen and Middleton (Petersen and Middleton, 1962; Natterer, 1986). Let $f \in \mathcal{S}(\mathbb{R}^n)$ and suppose that the support of $\hat{f}$ is a subset of $K \subset \mathbb{R}^n$. Let W be a non-singular real (n,n)-matrix, if the sets $K + 2\pi(W^{-1})^t k$, $k \in \mathbb{Z}^n$ are mutually disjoint then*

$$f(x) = S_W f \stackrel{\text{def}}{=} \sum_{k \in \mathbb{Z}^n} f(Wk) g(x - Wk),$$

where $g(y) = (2\pi)^{-n/2} |\det W| \hat{\chi}_K(-y)$, with χ_K the characteristic function of the set K.

Restrictions have to done for a function to be reconstructed from a finite number of measures. In tomography, the function is supposed to be b-band-limited, i.e. it's Fourier transform vanishes outside a ball of radius b. It has been shown (Rattey and Lindgren, 1981; Natterer, 1986) that the Fourier transform of Rf (viewed as a function on $T \times \mathbb{R}$, with $T = \mathbb{R}/2\pi\mathbb{Z}$) is small outside the set:

$$K(\vartheta, b) = \{(\nu, \sigma) \in \mathbb{Z} \times \mathbb{R} : |\sigma| < b, \ |\nu| < \max(|\sigma|/\vartheta, (1/\vartheta - 1)b)\}, \tag{2}$$

where $0 < \vartheta < 1$ (arbitrary close to 1) , provided f is essentially b-band-limited.

Thus an adapted version of Theorem 1,see Natterer (1986), can now be used for sampling the Radon transform of a b-band-limited function with $K = K(\vartheta, b)$ leading to sampling conditions for various schemes.

In the standard scheme, the Radon transform is sampled on the points $(\theta_p, s_q) = (\pi p/P, qh), k = 0, ..., 2P-1; q \in \mathbb{Z}$ such that $|qh| \leq 1$, i.e. on the lattice $\{W_S l, l \in A_W\} \cap T \times [-1, 1]$ where:

$$W_S = \begin{pmatrix} \pi/P & 0 \\ 0 & h \end{pmatrix}, \quad P \in \mathbb{N}, \quad h > 0.$$

The sampling conditions (to satisfy "the sets $K(\vartheta, b) + 2\pi(W_S^{-1})^t k$, $k \in \mathbb{Z}^n$ are mutually disjoint"): $P \geq b/\vartheta$ and $h \leq \pi/b$, are given in Natterer (1986). Remark that $h = 2/Q$, thus the optimal relationship between P and Q is here: P is the first integer larger than $\pi/2\,Q$.

The Radon transform can be correctly sampled in the corresponding interlaced scheme $\{W_I k, k \in A_W\} \cap T \times [-1, 1]$, where

$$W_I = \begin{pmatrix} 2\pi/P & -\pi/P \\ 0 & h \end{pmatrix},$$

with essentially half a the standard points (with the same sampling conditions on P and h). Remark that P and Q must have a different parity. Thus in section , Q is fixed in the following way: $Q = \lfloor 2/\pi\, P \rfloor$; if $\mathrm{mod}(Q, 2) = \mathrm{mod}(P, 2)$ then $Q = Q - 1$.

2.2 *Generalization*

Some generalizations have been proposed in order give new efficient sampling. For that purpose, a sampling theorem based on too caorse lattices is provided in Faridani (1990). In this case the set K is overlapped by the sets $K + 2\pi(W^{-1})^t k$, $k \in \mathbb{Z}^n$.

This builds a partition $K = \bigcup_{l=1}^{L} K_l$. K_l is the part exactly overlapped by the sets $K + 2\pi(W^{-1})^t k_{l,m}$, $m = 0, \ldots, M_l - 1$. More precisely:

$$\forall l \in \{1, \ldots, L\}, \quad \exists \{k_{l,0}, \ldots, k_{l,M_l-1}\} \subset \mathbb{Z}^n, \quad 1 \leq M_l < \infty$$

such that

$$\forall \xi \in K_l, \quad k \in \mathbb{Z}^n : \left(\xi - 2\pi \left(W^{-1}\right)^t k\right) \in K \Leftrightarrow k \in \{k_{l,0}, \ldots, k_{l,M_l-1}\}$$

M_l is the number of overlappings of the set K_l (from the sets $K + 2\pi \left(W^{-1}\right)^t k, k \in \{k_{l,0}, \ldots, k_{l,M_l-1}\}$). It turns out that $M = \max_{l=1}^{L} M_l$ is the minimum number of lattices required for a correct sampling.

The M shifted sets $\Lambda_r = \{\alpha_r + Wk, k \in \mathbb{Z}^n\}, r = 1, \ldots, M$, produced a correct sampling if the shifts $\alpha_r \in \mathbb{R}^n$ are chosen such that $\forall l$, $l = 1, \ldots, L$, the following linear system:

$$\begin{array}{c} \sum_{r=0}^{M-1} \beta_r^l = 1 \\ \sum_{r=0}^{M-1} \beta_r^l e^{-2i\pi((W^{-1}\alpha_r), k_{l,j})_{\mathbb{R}^2}} = 0, \;\; \text{for } j = 1, \ldots, M_l - 1 \end{array} \tag{3}$$

can be solved for the $\beta_r^l \in \mathbb{C}, r = 0, \ldots, M-1$. Under these conditions we have:

$$S_W f(x) = \sum_{r=0}^{M-1} \sum_{k \in \mathbb{Z}^n} f(\alpha_r + Wk) g_r(x - \alpha_r - Wk) \tag{4}$$

where $g_r(y) = (2\pi)^{-n/2} |\det W| \sum_{l=1}^{L} \beta_r^l \hat{\chi}_{K_l}(-y)$.

In Desbat (1993) sampling on coarse grids are considered with lattices generated by the matrix

$$W_{N,M} = \begin{pmatrix} 2N\pi/P & 0 \\ 0 & 2Mh \end{pmatrix}. \tag{5}$$

It has been shown that only $2NM$ grids generated by (5) are necessary (maximum number of overlapping in Faridani's theorem) for having a correct sampling. The correctness and the optimality of schemes mixing parallel and fanbeam sampling have been proved. The central point here is that interlaced sampling can be produced with lattices generated by the matrix (5), with the shifts $\alpha_{t,r}$ given in the next proposition and we have

Proposition 1. *(Desbat, 1993) If*

$$\alpha_{t,r} = \begin{cases} (2t\pi/P, rh)^t & r \text{ even} \\ ((2t+1)\pi/P, rh)^t & r \text{ odd} \end{cases} \quad t = 0, \ldots, N-1, \; r = 0, \ldots, 2M-1$$

then the matrix of the systems (3) do not depend on l, $\forall l = 1, \ldots, L$. Let us denote this matrix $\Omega_{N,2M}$, then $1/\sqrt{2NM}\Omega_{N,2M}$ is unitary.

Thus, perturbations of interlaced sampling can be easily considered (cf solvability of (3) in Faridani's theorem).

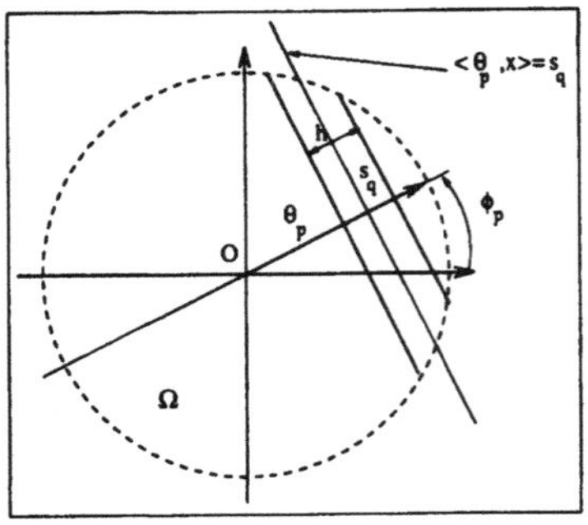

Figure 2: Measurements geometry: strip $L_{p,q}$

3. Algorithms

The way to deal directly with interlaced sampling is not obvious. The easiest way is probably to interpolate the lacking points in the standard scheme from the interlaced ones and to applied classical algorithms. However, filtered backprojection algorithms (or adaptation of the classical algorithms) have been proposed for reconstructions directly from the interlaced points (see Faridani (1990) and references therein). Algorithms to deal with interlaced sampling in the direct algebraic approach are rather new.

3.1 *Optimal algorithm*

The problem here is to compute the minimal norm least squares solution of the tomography problem with finite number of data in Sobolev spaces, see Natterer (1987). In our very classical approach the unknown function f is supposed to be in $L^2(\Omega)$, where Ω is the unit disk.

3.1.1 **Standard data.** The measurements $\mathbf{y}$ are supposed to be integrals over strips of the unknown function f:

$$\mathbf{y} = Rf \text{ where } \mathbf{y}_{p,q} = \int\limits_{L_{p,q}} f(x)\mathrm{d}x \;\; p = 0, \ldots, P-1; \;\; q = 1, \ldots, Q, \tag{6}$$

where $L_{p,q}$ is the strip of width h centered on s_q orthogonal to the line generated with the unit vector $\theta_p = (\cos\phi_p, \sin\phi_p)$, see Figure 2.

$$x \in L_{p,q} \Leftrightarrow s_q - h/2 < \langle\theta_p, x\rangle \leq s_q + h/2$$

The usual starting point for estimating f from (6) is to compute $f^\dagger$ the minimal norm least squares solution in $L^2(\Omega)$:

$$f^\dagger = R^\dagger \mathbf{y} = R^*(RR^*)^\dagger \mathbf{y}, \tag{7}$$

where $R^\dagger$ is the usual generalized inverse of R, R^* is the adjoint of R; namely $R^*\mathbf{y} = \sum\limits_{p,q} \mathbf{y}_{p,q} l_{p,q}$, $RR^*_{p,q,p',q'} = \langle l_{p,q}, l_{p',q'}\rangle$ where the $l_{p,q}$ are the strip characteristic functions defined by:

$$l_{p,q}(x) = \mathbb{1}_{L_{p,q}}(x) = \begin{cases} 1 & \text{if } x \in L_{p,q} \\ 0 & \text{else.} \end{cases}$$

The computation of (7) is generally very instable, so we have to introduce regularization. The following low pass filter is also frequently used:

$$f_\tau = R^*\mathbf{f}_\tau \text{ with } \mathbf{f}_\tau = (RR^*)^{\tau\dagger}\mathbf{y}, \tag{8}$$

where

$$(RR^*)^\tau = UD^\tau U^t \text{ with } D^\tau{}_{ii} = \begin{cases} D_{ii} & \text{if } D_{ii} \geq \tau \max_j D_{jj} \\ 0 & \text{otherwise} \end{cases}, \tau > 0.$$

(the D_{ii} are the eigenvalues of RR^*). f_τ is called, in the following, the optimal algorithm solution in standard schemes.

3.1.2 **Interlaced data.** There are two interlaced schemes underlying a standard one:

$$\mathbf{y}^i_{p,q} = \begin{cases} \mathbf{y}_{p,2q-1} & \text{if } p \text{ odd }, q = 1, \ldots, \lceil Q/2 \rceil \\ \mathbf{y}_{p,2q} & \text{if } p \text{ even }, q = 1, \ldots, \lfloor Q/2 \rfloor \end{cases}$$

is the first one, it's complementary in the standard scheme being the second one. Klaverkamp (1991) proposed the following algorithm based on an fine analysis of the SVD of the Radon transform. Let

$$\mathbf{y}'_{2p+1,2q-1} = \mathbf{y}^i_{2p+1,q}, \mathbf{y}'_{2p,2q} = \mathbf{y}^i_{p,q}, \mathbf{y}'_{2p,2q-1} = \mathbf{y}'_{2p,2q-1} = 0,$$

i.e., the non-measured data are replaced by 0. Using the standard matrix, and solving the problem with a truncated generalized inverse (8), yields essentially the same result as from the standard scheme:

$$f^{iK}{}_\tau = R^*\mathbf{f}^{iK}_\tau \text{ with } \mathbf{f}^{iK}_\tau = (RR^*)^{\tau\dagger}\mathbf{y}',$$

τ must have the right value, see Klaverkamp (1991): τ has to be a bit less than the smallest eigenvalue of the first block corresponding to the zero frequency in the block-diagonalization. In this case, very good results are obtained. They show that the approximation space $\mathcal{L}(l_i)_{i=1,\ldots,PQ}$ has to be same as those of the standard scheme in order to obtain almost the same reconstruction. This idea leads to the following formulation of the problem:

$$\min_{f \in \mathrm{ra}(R^*)} ||R^i f - \mathbf{y}^i||^2$$

where, R^i represents the interlaced measurement mapping:

$$\begin{array}{rccc} R^i: & L^2(\Omega) & \rightarrow & \mathbb{R}^{PQ/2} \\ & f & \rightarrow & \mathbf{y}^i \end{array}$$

We compute a truncated SVD and provide the solution:

$$f^{iD}_\tau = R^*\mathbf{f}^{iD}_\tau \text{ with } \mathbf{f}^{iD}_\tau = (R^iR^*)^{\tau\dagger}\mathbf{y}^i,$$

f^{iD}_τ can be very fast computed and is not so sensitive to the truncation parameter as f^{iK}_τ. In the following, we denote f^{iD}_τ as "optimal algorithm" solution in the interlaced case. In both standard and interlaced case, τ is chosen as in Klaverkamp (1991).

3.2 *SIRT for image reconstruction*

In this approach, the approximation space is given by the user (in contrary to the previous approach). A very common approximation space in image reconstruction is a pixel expansion. The function f is approximated by an image represented by a vector $\mathbf{f} \in \mathbb{R}^{n^2}$: $f(x) = \sum_{j=1}^{n^2} f_j p_j(x)$, where $p_j(x)$ is the characteristic function of the jth pixel in the image. We generally compute a minimal norm least squares solution:

$$\min_{\mathbf{f}\in\mathbb{R}^2} ||R\mathbf{f} - \mathbf{y}||^2, \tag{9}$$

where R is a $PQ \times n^2$ matrix in a standard scheme (respectively a $PQ/2 \times n^2$ matrix in the interlaced scheme), with $R_{(p-1)*Q+q,j} = \int_{\langle\theta_p,x\rangle_{L^2(\Omega)}=s_q} p_j(x)dx$. (in fact, for the simplification of the computations we approximate $R_{(p-1)*Q+q,j}$ by $\mathbb{1}_\Omega(x_j) \int_{\langle\theta_p,x\rangle_{L^2}=s_q} p_j(x)dx$ where x_j is the central point of the pixel number j). The problem (9) is generally ill-posed (the ill-posedness roughly increases with the pixel number) and has to be regularized (Tikhonov, 1963).

In order to produce formulations working for the interlaced scheme, we try to satisfy the "b-band-limited" condition. Let $\hat{\mathbf{f}}$ be the 2-dimensional Discrete Fourier Transform (DFT) coefficients of the vector image $\mathbf{f}$. f must be b-band-limited, i.e., $\text{supp}(\hat{f}) \subset \mathbb{D}(O, b)$, the centered disk of radius b. We first replace the problem (9) by

$$\min_{\mathbf{f}\in\mathbb{R}^2, \text{supp}(\hat{\mathbf{f}})\subset\mathbb{D}(O,b)} ||R\mathbf{f} - \mathbf{y}||^2$$

which is replaced by the problem

$$\min_{\mathbf{u}\in\mathbb{R}^2} ||R\hat{P}_b\mathbf{u} - \mathbf{y}||^2$$

where $\hat{P}_b = F^{-1}P_bF$ ($b = .95 * (Q-1)/2$ in practice), F being the 2-dimensional Discrete Fourier Transform and P_b the projector on the centered disk of radius b, i.e.

$$(P_b\mathbf{f})_{(j-1)*n+i} \left(= (P_b\mathbf{f})_{i,j}\right) = \begin{cases} \mathbf{f}_{(j-1)*n+i} & \text{if } \sqrt{(i-n/2)^2 + (j-n/2)^2} \le b \\ 0 & \text{else} \end{cases}.$$

Remark: the condition $\sqrt{(i-n/2)^2 + (j-n/2)^2} \le b$ has to be adapted to the ordering of the DFT that you use but means that only the coefficients corresponding to "low" frequencies are kept). We then use a classical regularization for the previous problem

$$\min_{\mathbf{u}\in\mathbb{R}^2} ||R\hat{P}_b\mathbf{u} - \mathbf{y}||^2 + \tau||\mathbf{u}||^2. \tag{10}$$

The solution u_τ of (10) yields $f_b = \sum_{i=1}^{n^2} u_{\tau i} p_i$ which is called in the following algebraic "b-band-limited" solution. In the numerical applications, we fixed $b = .95 * (Q-1)/2$.

An other way (mathematically different) to interpret the expression "b-band-limited" is to say that the solution is smooth enough. Thus, we also introduce the following classical regularization of the problem (9)

$$\min_{\mathbf{u}\in\mathbb{R}^2} ||R\mathbf{f} - \mathbf{y}||^2 + \tau||\Delta\mathbf{f}||^2 \tag{11}$$

where Δ is the usual Laplacian matrix

$$(\Delta \mathbf{f})_{i,j} = 4\mathbf{f}_{i,j} - (\mathbf{f}_{i-1,j} + \mathbf{f} + i + 1, j + \mathbf{f}_{i,j-1} + \mathbf{f}_{i,j+1}).$$

Our operator differs only on the edges from the classical Laplacian matrix: for example, the points on the corner are transformed in the following way $(\Delta \mathbf{f})_{1,1} = 2\mathbf{f}_{1,1} - (\mathbf{f}_{1,2} + \mathbf{f}_{2,1})$ and the other points not on the corner like $(\Delta \mathbf{f})_{2,n} = 2\mathbf{f}_{2,n} - (\mathbf{f}_{1,n} + \mathbf{f}_{2,n-1} + \mathbf{f}_{3,n})$. In the following the solution of (11) will be called R4 regularized solution.

3.3 *Regularization parameter estimation*

In regularization methods like the two previous ones, the regularization parameter τ estimation is crucial. As the tomographic problem is not too severely ill conditioned (Natterer, 1977) we decide to minimized the Generalized Cross Validation for choosing τ (Desbat, 1989; Desbat and Girard, 1989; Golub et al., 1979)

$$GCV(\tau) = \frac{(1/PQ)||(I - A_\tau)\mathbf{y}||^2}{((1/PQ)\,\mathrm{trace}\,(I - A_\tau))^2}$$

where A_τ is defined by $\mathbf{y}_\tau = A_\tau \mathbf{y} = R\mathbf{f}_\tau$, $\mathbf{f}_\tau$ being the regularized solution associated to the parameter τ. The computation of $A_\tau \mathbf{v}$ is quite easy because it essentially involves the resolution of a linear system composed by sparse matrices (R, R^*, I are sparse), and Fourier transform for (10). Thus the most costly operation in the estimation of $GCV(\tau)$ is the trace of A_τ. Girard (1991) proposed to replace the GCV by the following estimation:

$$SGCV(\tau) = \frac{(1/PQ)||(I - A_\tau)\mathbf{y}||^2}{(1 - \frac{\mathbf{w}^t A_\tau \mathbf{w}}{\mathbf{w}^t \mathbf{w}})^2},$$

where $\mathbf{w}$ is a normal noise $\mathcal{N}(0, I)$. The following reconstructions are provided with the parameter τ minimizing the Stochastic GCV: $SGCV(\tau)$.

4. Numerical Experiments

4.1 *Simulated data*

In Figure 3, we present the phantom that we have used for testing the different algorithms. Note that the data are produced with integral over strips, thus the pixel image approach of SIRT algorithms is a bit disadvantaged.

In Figure 4, we present contour lines of reconstructions from $P = 20 \times Q = 11$ data with our three algebraic algorithms from standard and interlaced data. Respective perspectives are provided in Figure 5. We can first remark that all reconstructions are similar and good (see Figure 5). Reconstructions from optimal algorithm in L^2 are indeed not smooth as can be seen in Figure 4. Contour lines are very sensitive to small differences in the images. Thus we can do this second remark: interlaced and standard data yield in each method essentially the same reconstruction. The minimization of $SGCV$ allows us to choose efficiently the right smoothing parameter (acting as a "cut off frequency") from both interlaced and standard data. See the similarities of "b-band-limited" reconstructions and "R4 regularized" ones.

The same remarks can be done for reconstruction from $P = 30$ and $Q = 19$ geometries, i.e. 570 data (respectively 285) in the standard (respectively interlaced) geometry: see

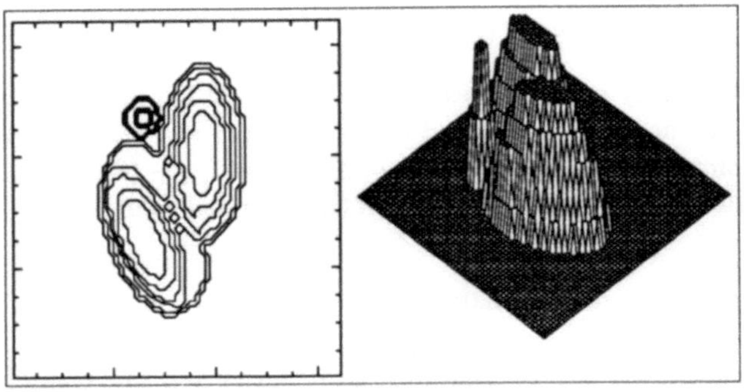

Figure 3: Phantom. Left: contour lines. Right: perspective.

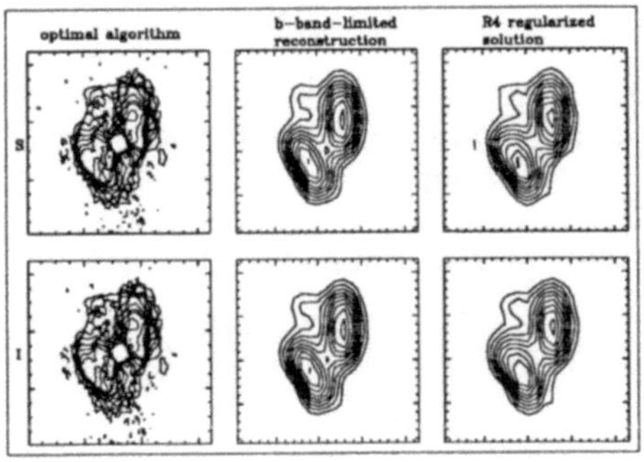

Figure 4: Contour lines of the reconstructions with $P = 20, Q = 2/h = 11$. S (first line): Standard data, I (second line): Interlaced data.

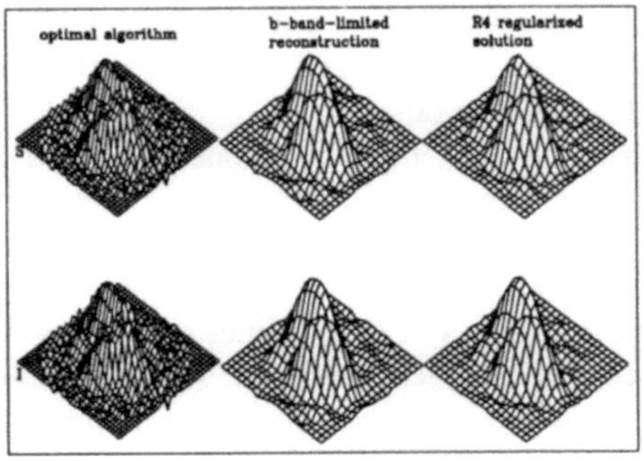

Figure 5: Perspectives of the reconstructions with $P = 20, Q = 2/h = 11$.

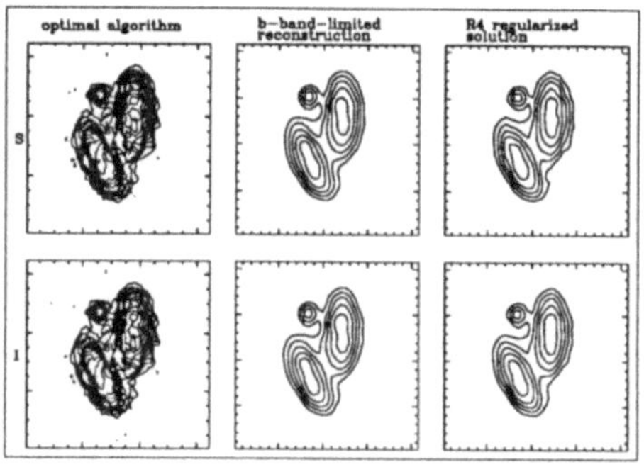

Figure 6: Contour lines of the reconstructions with $P = 30, Q = 2/h = 19$.

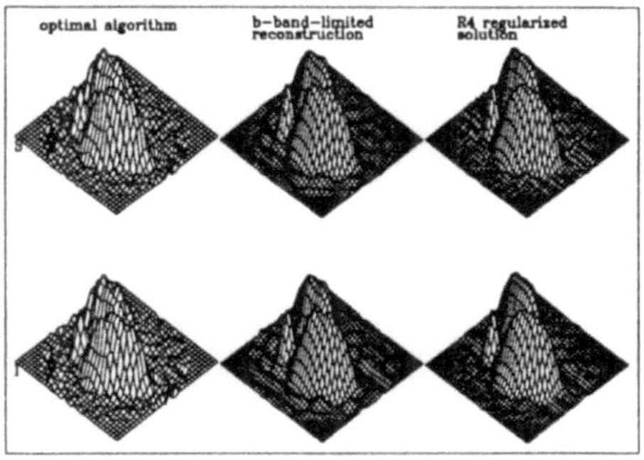

Figure 7: Perspectives of the reconstructions with $P = 30, Q = 2/h = 19$.

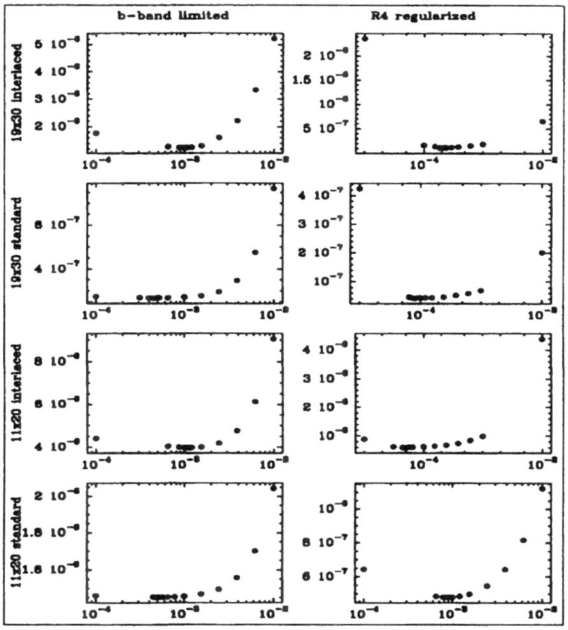

Figure 8: Minimization of the $SGCV(\tau)$ for the SIRT algorithms.

Figure 6 and 7. The points corresponding to the minimization of $SGCV(\tau)$ are presented in Figure 8.

We also have done experiments on noisy data. We added, to the previous data, a centered Gaussian noise with a standard deviation of 5% of the maximum of the data. We can see that the noise affects a bit more the optimal algorithm and the b-band-limited reconstructions. As can be seen in Figures 9 and 10, the R4 regularized solution seems to have adapted its smoothness to the noisy data thanks to the $SGCV(\tau)$ in both standard and interlaced case, see Figure 11.

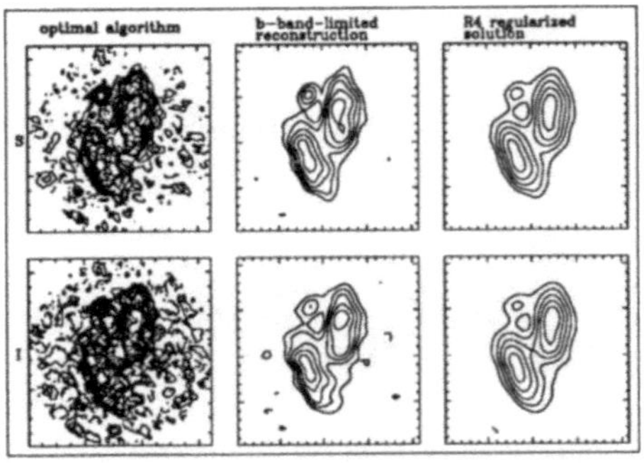

Figure 9: Contour lines of the reconstructions with noisy data, $P = 30, Q = 2/h = 19$.

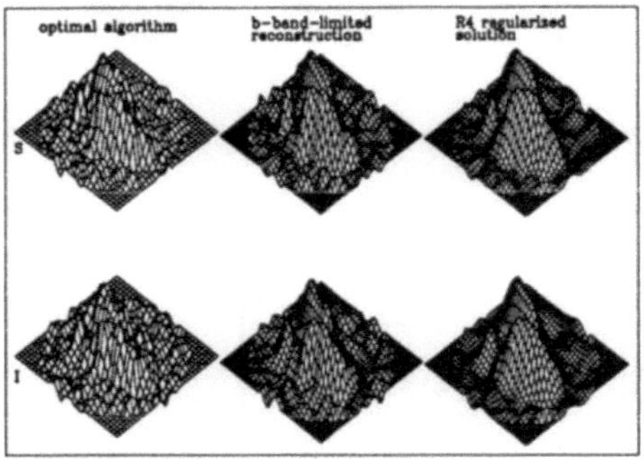

Figure 10: Perspectives of the reconstructions with noisy data, $P = 30, Q = 2/h = 19$.

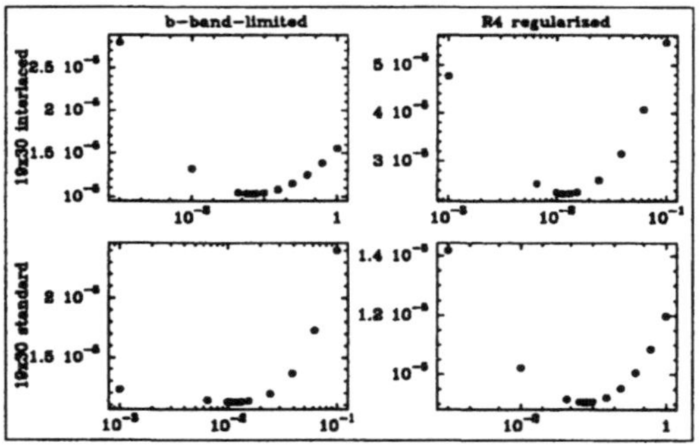

Figure 11: Minimization of the $SGCV(\tau)$ for the SIRT algorithms in the case of noisy data.

4.2 *Real data*

We present here reconstruction from data published in Desbat and Turlier (1993). These data come from γ-tomography of reactors in petroleum raffinery. The aim is to reconstruct a catalyst density in a gasoil cracking reactor (its repartition affects the efficiency of the reaction). The major problem here is that we have to produce reconstruction from relative few data. We present, in Figure 12, reconstructions from a standard geometry with $P = 18$ and $Q = 10$. In 13, we show a "b-band-limited" reconstruction with b a bit too large: $b = 5$. Instabilities can be seen in the reconstruction. Remark that measures of the empty pipe have been first done and subtracted to the measures that we used. Thus we reconstruct only the catalyst repartition. The high density near the tubes is physically good explained. We propose reconstructions of this pipe from the same amount of data in an interlaced geometry with $P = 24$ $Q = 15$. Note that the resolution is improved. The $SGCV$ adapts very well the R4 regularization parameter for these real interlaced data. The reconstructions have to be compared to those of Figure 12 having the same number of data and an obvious lower resolution.

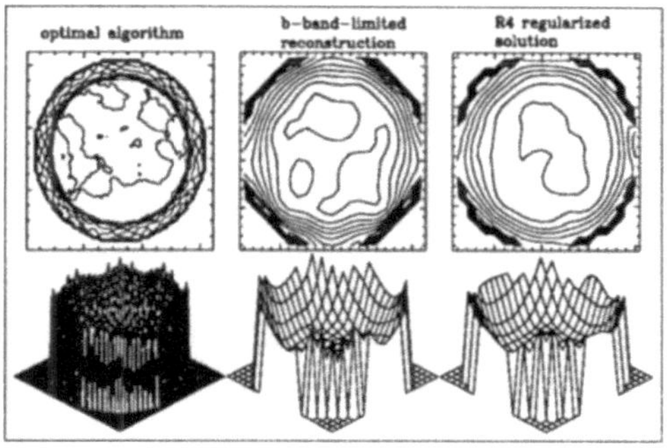

Figure 12: Contour lines and perspectives of the reconstructions with $P = 18, Q = 10$ from standard measurement of a reactor.

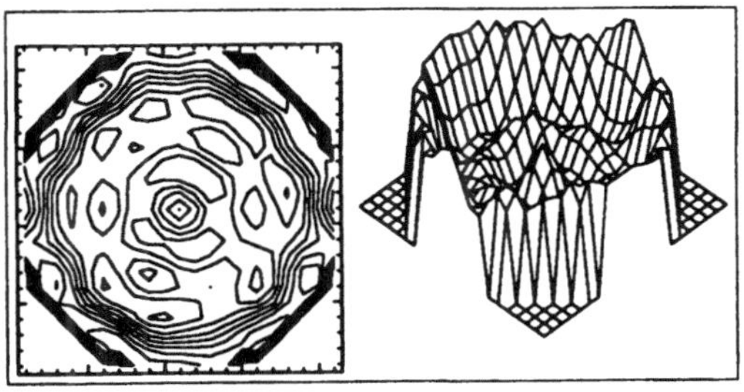

Figure 13: "b-band-limited reconstruction with $b = 5$ with $P = 18, Q = 10$ from the same standard real measurement of a reactor. Left: contour lines. Right: perspective.

5. Conclusion

The interlaced schemes contain, *with twice less data*, essentially the same information as the corresponding standard schemes. We provided algebraic algorithms performing reconstructions almost identical from standard and interlaced schemes. In the SIRT approach, the key point is to reconstruct "b-band-limited" pixel images. A SIRT algorithm

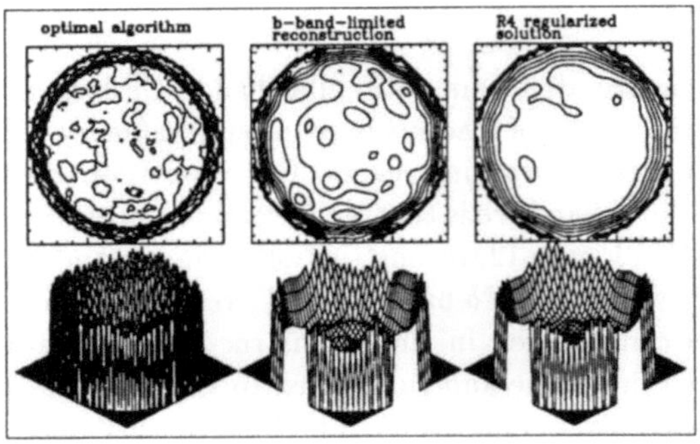

Figure 14: Contour lines and perspectives of the reconstructions with $p = 24, Q = 15$. from interlaced measurements of a reactor.

with a Laplacian like regularization operator has been proposed. The $SGCV(\tau)$ has demonstrated its ability to choose efficiently the smoothness of the solution from both interlaced and standard schemes, with noisy or unnoisy data, even from relatively few data. Other numerical experiments should be done in other sampling schemes such as efficient fan-beam or perturbed interlaced schemes.

REFERENCES

Cormack A.M. (1978), Sampling the Radon transform with beam of finite width, *Phys. Med. Biol.*, vol. 23, no. 6, 1141–1148.

Desbat L. (1989), Parameter Regularization Estimation, In: C. Jaschek and F. Murtagh, editors, *Errors, Bias and Uncertainties in Astronomy*, Cambridge University Press, 313–316.

Desbat L. (1993), Efficient sampling on coarse grids in tomography, *Inverse Problems*, vol. 9, 251–269,

Desbat L. and Girard D. (1989), The "minimum reconstruction-error" choice of regularization parameters: some effective methods and their application to deconvolution problems, report, LMC, D. U. BP 53 X, 38041 Grenoble (FRANCE).

Desbat L. and Turlier P. (1991), Measurement Optimization in Tomography, In: R. Vichnevtsky and J.J.H. Miller,editors, *IMACS'91, 13th World Congress on Computation and Applied Mathematics*, vol. 3, 1523–1534.

Desbat L. and Turlier P. (1992), Efficient reconstruction with few data in industrial tomography, In: M.S. Beck and E. Campogrande and M. Morris and R.A. Williams and R.C. Waterfall, editors, *ECAPT92*, Computational Mechanics Publications, 285–294.

Desbat L. and Turlier P. (1993), Efficient reconstruction with few data in industrial tomography, In: *ECAPT93*.

Faridani A. (1990), An application of a multidimensional sampling theorem to computed tomography, In: *AMS-IMS-SIAM Conference on Integral Geometry and Tomography*, Contemporary Mathematics, vol. 113, 65–80.

Faridani A. (1993), A generalized sampling theorem for locally compact abelian groups, Preprint, Oregon State University, Departement of Mathematics, Corvalis OR 97331, U.S.A.

Girard D. A. (1991), Asymptotic optimality of the fast randomized versions of GCV and C_L in ridge regression and regularization, *Ann. of Stat.*, vol. 19, no. 4, 1950–1963.

Golub G., Health M. and Wahba G. (1979), Generalized Cross Validation for choosing a good ridge parameter, *Technometrics*, vol. 21, 215–224.

Herman G.T. (1980), *Image Reconstruction from Projections*, New York: Academic.

Kasantsev I. (1991), Information content of projections, *Inverse Problems*, vol. 7, 887–898.

Klaverkamp W. (1991), *Tomographische Bildrekonstruktion mit direkten algebraischen Verfahren.*, Ph.D. thesis, Inst. f. Num. Math. Münster.

Martin M.P., Turlier P., Bernard J.R. and G. Wild (1992), Gas and solid behavior in cracking circulating fluidized beds, *Powder Technology*, vol. 70, 249–258.

Natterer F. (1977), The finite element method for ill-posed problems, *RAIRO MA2N*, vol. 11, no. 3, 271–278.

Natterer F. (1986), *The Mathematics of Computerized Tomography*, Wiley.

Natterer F. (1987), Efficient implementation of optimal algorithms in Computerized Tomography, *Math. Meth. in the Appl. Sci.*, vol. 8, no. 6.

Natterer F. (1993), Sampling in fan-beam tomography, *SIAM J. Appl. Math.*, vol. 53, no. 2, 358–380.

Petersen D.P. and Middleton D. (1962), Sampling and reconstruction of wawenumber-limited functions in N-dimensional euclidean space, *Inf. Control*, vol. 5, 279–323.

Rattey P.A. and Lindgren A.G. (1981), Sampling the 2-D Radon transform, *IEEE Trans. ASSP*, vol. 29, 994–1002.

Tikhonov A.N. (1963), Regularization of incorrectly Posed Problems, *Akad. Nauk. SSSR*, vol. 153, 49–52.

An Approach to 2-D Ray Tracing Intended for Non-Linear Seismic Tomography

P. G. DITMAR

Scientific and Production Firm "Development Of Geophysical Software" ("DOGS")

P.O.Box 194, St.Petersburg 197198, Russia

ABSTRACT

Several years ago, a new method for seismic ray tracing was proposed, based on preliminary time field computation. Once computed, the time field is used to determine all the raypaths originated at the given source. Rays are traced in the backward direction, from a receiver toward the source, along the time field anti-gradient.

A new modification of this method is proposed. It is assumed that a 2-D region under consideration is divided into cells of constant velocity by a rectangular net. The time field is explicitly computed at the grid nodes as well as at some intermediate points located at cell boundaries. The time field at other points of a boundary is determined by an interpolation. Computation is based on the graph theory. The obtained set of times is then used for the purpose of ray-tracing. Such an approach allows one to achieve any pre-determined computational accuracy, provided the discretization errors are not considered.

Some features of a computational algorithm developed on the base of the proposed approach are discussed. It is stressed that particular attention must be paid to computations at the vicinity of the source. A case is also taken into consideration where sources and/or receivers slightly deviate from the given plane. The algorithm is tested with several model examples. It is clearly shown that the proposed approach allows one to compute just first arrivals, whatever complicated a medium would be.

Introduction

Seismic tomography as a method of seismic data interpretation has become widely used nowadays. Many new results have been obtained on its base, both in global seismology and in seismic exploration. In most cases, however, the interpretation was performed in rather brute way, being based on too simplified assumptions about seismic raypaths. Namely, the raypaths have been usually assumed either corresponding to a velocity model that depends on the depth only, or simply straight. Meantime, it is well-known that the quality of tomographic recovery is unacceptable if the true velocity model differs from that used for ray tracing by more than 10–15 per cent (Dines and Lytle, 1979; Ditmar, 1990).

Till the very last time, the main reason to use simplified assumptions about the raypaths was the absence of tracing algorithms sufficiently meeting the requirements of seismic tomography. Among these requirements are: high performance; stable work with arbitrary velocity patterns; ability to determine the raypaths corresponding just to the first arrivals; possibility to utilize easily cell parametrization of a medium, traditional for seismic tomography. Shooting and bending, conventional algorithms of ray tracing, hardly satisfy these requirements (Moser, 1991; Schneider et al., 1992).

Fortunately, an alternate ray tracing technique, based on preliminary time field computation, was proposed recently. It consists of two main steps: 1) computation of the time field corresponding to a given source; 2) ray tracing itself, which is performed in the backward direction, from a receiver toward the source, along the direction of the time field anti-gradient. Such technique meets the requirements of seismic tomography much better. In particular, it is very efficient potentially because the time field computation, which is the most time-consuming step, is performed only once for each source; the following tracing of rays emitted from this source takes considerably short time.

Such a method was first proposed by Nakanishi and Yamaguchi (1986) and then developed by Moser (1991). An asset of their version is the application of the graph theory. Thanks to it, the computed raypaths correspond just to the first arrivals, whatever complicated a medium would be. What about drawbacks, the main of them is associated with the assumption that a raypath consists of straight segments whose ends are placed at discrete points. As a consequence, an acceptable accuracy of computation may be achieved only if the number of these points is large enough, which results in low performance.

Another modification of the method was also proposed, based on the finite-difference time field computation (Vidale, 1988, 1990; van Trier and Symes, 1991). For the sake of completeness, Pilipenko (1979) and Reshef and Cosloff (1986) should also be mentioned here, who considered the finite-difference time field computation too but did not mention ray tracing.

Unfortunately, traditional finite-difference schemes fail in complicated media with sharp velocity contrast. That's why a number of papers appeared recently, whose authors, explicitly or not, tried to combine the graph theory and the finite-difference approach (Asming and Buyanov, 1990; Podvin and Lecomte, 1991; Qin et al., 1992).

The present study is devoted to further development of the ray tracing method outlined above. A new approach is proposed, that allows one to achieve any pre-determined computational accuracy, provided that the discretization errors are not considered. Some features of an algorithm developed within the framework of the approach are described. Several model tests are considered providing the possibility to evaluate the accuracy and the reliability of our algorithm.

Theory

Preface

FORMULATION OF THE PROBLEM. Let a simply connected domain Ω be characterized by a velocity distribution $V(x,y)$. Let it be also known that a source is situated at a point $s \in \Omega$ with coordinates (x_s, y_s), whereas a receiver is located at a point $r \in \Omega$ with coordinates (x_r, y_r). Then the problem of ray tracing is to find such a path $L_{sr} \in \Omega$ that the traveltime

$$T = \int\limits_{L_{sr}} \frac{dl}{V(x,y)} \tag{1}$$

is minimum. In the case when such a path is not unique, it is enough to find one of possible paths. For brevity, we shall call such a path as "the shortest path connecting the points s and r". Besides, the variable T as a function of the receiver position will be referred to as "the time field".

The following statement should help us to solve the problem.

Statement 1. *Let a point r be located inside a closed contour $K \in \Omega$, and the time field is known for each point $k \in K$ with coordinates (x_k, y_k), whereas a point s is situated outside the contour K. Then:*

1. *The time field at the point r is equal to*

$$T(x_r, y_r) = \min_{k \in K} \left[T(x_k, y_k) + \int\limits_{\hat{L}_{kr}} \frac{dl}{V(x,y)} \right], \qquad (2)$$

where $\hat{L}_{kr}$ is the shortest path connecting the points k and r and including no points common with the contour K, except for the point k.

2. $\hat{L}_{kr} \in L_{sr}$.

The proof of this statement is evident.

Note 1. *Statement 1 remains valid when the contour K is not closed but its ends belong to the boundary of the domain Ω.*

Note 2. *In the case when both of the points s and r are located inside the contour K, the time field at the point r is equal to the lesser of two values: the first value is determined by equality (2), whereas the second one is equal to*

$$\hat{T}(x_r, y_r) = \int\limits_{\hat{L}_{sr}} \frac{dl}{V(x,y)}, \qquad (3)$$

where $\hat{L}_{sr}$ is the shortest path connecting the points s and r and including no points common with the contour K. If the second value is less, then $L_{sr} = \hat{L}_{sr}$.

It should be pointed out further that as far as we consider the problem from the viewpoint of seismic tomography, it is worth assuming that the domain Ω is rectangular and is divided into cells of constant velocity by a rectangular net. In this case, as will soon become evident, to trace a ray, it is enough to know the time field only at the points situated at the cell boundaries.

General conception

To begin with, we shall consider a hypothetical case when the time field is explicitly calculated in the whole infinite set of points belonging to the cell boundaries. In such a case, a procedure for solving the problem might be as follows.

The first step of ray tracing, i.e. the time field computation, starts from an initialization. In the process of initialization, the time field at the points located at the boundaries of the cell, the source lies within, is calculated . It is taken into account that the corresponding shortest paths may be either straight, or two-segmented (Figure 1a). All the other points under consideration are assigned infinite values.

The following computation consists in a succession of iterations with the following operations on each:

1) All the points located at the corners of cells, i.e., the grid nodes, are considered. Each of these points, denoted here as r', is enclosed with the contour K in the manner

shown in Figure 1b. A new time at the point r' is then calculated by means of Statement 1. It is taken into account that the shortest path connecting the points k and r' may be either straight, or two-segmented (see Figure 1b). The point r' is assigned the new value only if it is less than the old one, otherwise the old value is kept.

2) All the points located at the boundaries of cells, but not coincident with the grid nodes, are considered. Each of these points, denoted here again as r', is enclosed with the contour K as indicated in Figure 1c. Then an attempt is made to lower the time at this point too.

Such sequence of operations is performed until no further decreasing of times is possible.

The following ray tracing is performed by means of the successive movement from the source to the receiver. A ray segment that connects the source to a boundary of the cell, the source lies inside, is found on the first step. For this purpose, the source is enclosed by the contour K, as shown in Figure 1d, and the associated shortest path $\hat{L}_{kr}$ is calculated (it is always straight here). The following sequence of operations is evident: the last found point of the raypath should be enclosed by the contour K as in Figure 1b if it coincides with a cell corner, or as in Figure 1c otherwise. The next subpath is then easily determined by means of the second part of Statement 1, and so on. Having reached the cell containing the source, one should turn to Note 2 to the mentioned statement. If it is found that the current subpath originates at the source point, then ray tracing ends.

It should be noted that, describing the time field computation and the following ray tracing, we haven't mentioned the case when a current point belongs to a boundary of the domain. This case may be easily taken into consideration by means of the first note to Statement 1.

We are coming now to justification of the routine described or, more exactly, to justification of the time field computation process, as correctness of the ray tracing step is undoubted due to Statement 1.

First of all, denote the path $\hat{L}_{kr'}$ connecting the current point r' to the contour K, which is constructed according to one of the rules described above, as an "elementary path". Then the essence of operations performed after the initialization is as follows. All the pairs of points located at different boundaries of a cell are considered, and an attempt is made to lower the time at one of them through linking it to the other by the corresponding elementary path. These operations are repeated until the following decrease of any time value is impossible.

During the time field computation, we can anytime determine the path that has provided the current time field value, by means of the ray tracing scheme described above. Therefore, no one point may be assigned the value less than the true one. Let us prove that, when the computation ends, no one point may be characterized by a value exceeding the true one either.

To begin with, consider those points disposed at boundaries of the cell containing the source, for which the true shortest path $L_{sr'_1}$ coincides with an elementary path linking the point with the source (hereafter these points are denoted as r'_1). It is evident that these points are assigned the times not exceeding the true ones just during the initialization. The following reasoning will be based on the mathematical induction method. Obviously, every true shortest path $L_{sr'}$ connecting the source to some point r' may be represented as a superposition of a limited number of elementary paths:

$$L_{sr} = L^1 \cup L^2 \cup \ldots L^n. \tag{4}$$

Suppose that all the points r'_{n-1}, for which the shortest paths $L_{sr'_{n-1}}$ consisting of $n-1$ elementary paths are finally characterized by the time values not exceeding the true ones. Let us show that the similar statement is valid for the points r'_n whose shortest paths consist of n elementary paths. Suppose it isn't the case for one of these points, where the finally computed time is equal to $T^{(n)}_{fin}$ ($T^{(n)}_{fin} > T^{(n)}$, where $T^{(n)}$ is the true time). The true time may be represented as $T^{(n)} = T^{(n+1)} + \Delta T$, where $T^{(n-1)}$ is the true time at one of the points r'_{n-1}; ΔT is the traveltime along an elementary path connecting that point with the point under consideration. According to the induction assumption, the true time $T^{(n-1)}$ is equal to that finally obtained. On the other hand, the time field computation ends only when neither value may be updated, whether elementary path would be considered. In other words, the inequation $T^{(n)}_{fin} \leq T^{(n-1)} + \Delta T$ is valid. The contradiction reached proves the statement.

Note that a set of the points after specification of rules for their pair consideration is an example of the oriented graph. The algorithm of the shortest path computation described above was introduced into the graph theory by Ford and Fulkerson (1962).

Needless to say that, using the described routine in practice, we have to deal with a limited number of points. To get the time field at other points located at the cell boundaries, one of interpolation schemes might be invoked. But the above justification becomes somewhat incorrect in that case. Strictly speaking, a computational procedure including the time field interpolation should be considered on the base of the hypergraph theory rather than on the base of the graph one, because relationships are set between several points, not between pairs. Unfortunately, hypergraphs are the objects introduced just recently and not investigated properly. Therefore, suppose, just as a hypothesis, that ray tracing is correct if the interpolation is "proper".

Consideration of the points order

The next problem arising when applying the time field computation routine to practice is a choice of the optimum consideration of the points order. There are various ways to do it. In particular, for the case where all the points located at a cell boundary are utilized for the interpolation, we propose the following algorithm.

To begin with, let us divide all the points, where times are explicitly computed, into two categories. Denote the points coincident with cell corners (i.e. the grid nodes) as "nodal points"; all the other points will be referred to as "intermediate points". Now suppose that each of the nodal points possesses its own status: it may be either "accounted" or "non-accounted". At the beginning the accounted points are absent. The essence of the next operations is as follows.

1) All the non-accounted points are considered and one of them with the minimum time is picked. The picked point becomes accounted.

2) An attempt is made to update (lower) times at all the neighbor points that may be linked to the picked one with an elementary path. If the updating at a neighbor point takes place, and that point is a nodal one too, then it becomes non-accounted. If the updating occurs at an intermediate point, one of the two adjacent nodal points (i.e. located at the ends of the boundary, the intermediate point belongs to) becomes non-accounted.

3) If the picked point is situated in such a way that after assigning it "accounted" status, some cell boundaries appear with accounted points at the both ends, then an attempt is made to update (lower) the times at all the neighbor points that may be linked to these boundaries with an elementary path. If the updating takes place, the

corresponding neighbor point becomes non-accounted according to the same rules as in the previous item.

Then proceeding to Item 1 follows, etc. The computation ends when running out of the non-accounted points.

Notice that in order to increase the performance, only those points should be considered at Item 1, where the times are finite (i.e. updated at least once). It is also possible to achieve the further growth of the computational efficacy, but two additional assumptions must be made for that. They are not strict, but we have all reasons to believe that they are true for the great majority of computations with models of real media.

Assumption 1. *Updating of the time at an intermediate point is always accompanied by that at at least one adjacent nodal point.*

Before proceeding to the second assumption, let us introduce the notion of a "preceding point" for nodal points. A nodal point s' is preceding to a given nodal point if an elementary path, used for the final determination of the time at the given point, originates at the point s'. Besides, two adjacent nodal points s'_1 and s'_2 are preceding to a given nodal point if an elementary path, used for the final determination of the time at the given point, originates at the cell boundary disposed between the points s'_1 and s'_2. Then the formulation of the second assumption may be as follows.

Assumption 2. *The final time at a nodal point always exceeds the time(s) at the point(s) preceding it.*

In the following, Assumption 2 will be referred to as the "causality principle". Prerequisites for fulfillment of the causality principle are:

- close to the linear changing of the time field at cell boundaries;
- close to the square shape of cells;
- close to a homogeneous medium.

It is easy to show that the causality principle holds, in particular, in the following cases:

- where the wave is plane;
- where the cells are square, and the medium is homogeneous.

If the causality principle fulfills, then the following statement is true.

Statement 2. *Once a nodal point becomes accounted, the time field at it (and consequently, its status) would not change anymore.*

Let us prove it. Evidently, the time at the nodal point becoming accounted first would not change if the causality principle is valid. The following reasoning is based on the mathematical induction method. Suppose the $n-1$ points that receive the "accounted" status at the first $n-1$ iterations would not change it anymore either. Let us show the validity of the similar statement for the point becoming accounted at the n-th iteration. Denote this point as q, and the time at it as T'_q. Suppose that in the process of following computations this value is updated and becomes finally equal to T_q ($T_q < T'_q$). If this is the case, at least one point preceding the point q is characterized at the n-th iteration by a time other than the final one. (Suppose all the points preceding the point q have already been characterized by final times at the n-th iteration. Therefore, all of them cannot be accounted at the n-th iteration, otherwise the time T'_q would be equal to the final one T_q. On the other hand, no one of them can be non-accounted, otherwise that point rather than the point p would become accounted at the n-th iteration.) Denote that point as p, and the final time at it as T_p . In accordance with the causality principle, $T_p < T_q$. Now consider the points preceding the point p. Similarly, at least one of them is also characterized at the n-th iteration by a time other than the final one. Denote that point as o, and the final time at it as T_o . In line with the causality principle, $T_o < T_p < T_q$. Pursuing such reasoning, we can infinitely extend the sequence of inequalities. But the number of points at our disposal is finite. Thus, we have reached the contradiction that proves the validity of Statement 2.

Therefore, if the causality principle holds, then during execution of Items 2 and 3, there is not a necessity to "turn back", i.e. to perform attempts to update times at accounted nodal points (as well as at intermediate points lying between accounted nodal ones, provided Assumption 1 is valid). After such a modification, the time field computation requires exactly N iterations (where N is a total number of the nodal points), because the number of non-accounted nodal points is decreased by one with each iteration.

In conclusion, it should be noted that the modified procedure for the time field computation described above may be considered as a generalization of the Dijkstra's (1959) method, traditionally used for computing the shortest paths in graphs.

Practical Realization of the Proposed Approach: Some Details

Basing on the described approach to ray tracing, we have developed a detailed computational algorithm and realized it as a computer code. Some peculiarities of our algorithm are discussed below.

Time field approximation at cell boundaries

The time field at a cell boundary is approximated, basing on known times at three points: two nodal and one intermediate (disposed in the middle of the boundary). Denote the times at the ends of the boundary as T_a and T_c, and the time at the middle point as T_b. Let the origin be at the middle point. Then the time field at other points belonging to the boundary is calculated by an interpolation according to the following rules.

1) Case $T_b \leq \frac{1}{2}(T_a + T_c)$, corresponding to a convex or a plane wave front. A parabolic interpolation is used here with regard for the following inequality evident from the eikonal equation:

$$\left|\frac{\partial T}{\partial x}\right| \leq V^{-1}, \tag{5}$$

where V is the greater of velocities in the two cells adjacent to the boundary. Namely,

$$T(x) = \begin{cases} -V^{-1}\left(x + \frac{d}{2}\right) + T_a, & x \leq \hat{x}; \\ Ax^2 + Bx + C, & \hat{x} < x < \tilde{x}; \\ V^{-1}\left(x - \frac{d}{2}\right) + T_c, & x \geq \tilde{x}; \end{cases} \tag{6}$$

where d is the boundary length; A, B, C, $\hat{x}$, $\tilde{x}$ are some unknowns. These unknowns are found from known function values at the three points and from the conditions of the first derivative continuity at the points $\hat{x}$ and $\tilde{x}$. In the case

$$\begin{cases} \hat{x} \leq -\frac{d}{2}, \\ \tilde{x} \geq \frac{d}{2}, \end{cases} \tag{7}$$

the time field varies at the boundary according to the ordinary parabolic law.

2) Case $T_b > \frac{1}{2}(T_a + T_c)$, corresponding to the presence of a time field gradient discontinuity. Since the information about the time field at three points is inadequate to locate the discontinuity point, just a combination of linear interpolations is implemented:

$$T(x) = \begin{cases} T_b + \frac{2x}{d}(T_b - T_a), & x \leq 0; \\ T_b + \frac{2x}{d}(T_c - T_b), & x > 0. \end{cases} \tag{8}$$

Computations in the vicinity of the source

When developing the algorithm, much attention was given to computations at the points near the source. The matter is that the isochrones in the vicinity of the source are strongly curved, so that neither linear, nor parabolic interpolation is sufficient to represent the time field with a proper accuracy. As a consequence, all the other times might be computed rather inexactly. To solve this problem, we could increase the number of intermediate points near the source. But we preferred to use another way: to extend the area where times are computed immediately at the initialization step. Denote the cell size along the X-axis as d_x, and that along the Y-axis as d_y . Then, at the initialization, times are computed at all the nodal points lying inside the area

$$\begin{cases} |x - x_s| < 2d_x, \\ |y - y_s| < 2d_y, \end{cases} \tag{9}$$

and at all the intermediate points lying between them. Correspondingly, when tracing a ray, attempts to link a given point to the source begin as soon as this point falls between nodal points lying inside the area described above (or happens to coincide with one of these nodal points).

Raypath deviations from the plane

Bearing in mind an application of our algorithm to interpretation of field experiments, we made provisions for correction of a raypath to take into account possible deviations of sources and receivers from the plane under consideration. The essence of such a correction is determination of values z_i for all straight segments of a raypath, with these values obeying the equality

$$\sum_{i=1}^{I} z_i = Z, \tag{10}$$

where Z is the difference of the source and the receiver Z-coordinates; I is the total number of the straight segments that the raypath consists of. This problem may be solved easily under the following assumptions:

1) The projection of a 3-D raypath onto the plane under consideration coincides with the 2-D raypath which is computed under the assumption that both the source and the receiver belong to the plane.

2) The angle between a raypath segment and the plane is small.

If the first assumption is valid, then this problem, in view of the Fermat's principle, may be stated as follows: to find the conditional minimum of the expression

$$T = \sum_{i=1}^{I} \frac{\sqrt{(l_i)^2 + (z_i)^2}}{V_i} \tag{11}$$

under condition (10), where l_i is the length of the i-th raypath segment projection onto our plane; V_i is the velocity in the cell crossed by the i-th segment. An easy sequence of manipulations with consideration for the second assumption leads to the following result:

$$z_i = Z\frac{V_i l_i}{\sum_{j=1}^{I} V_j l_j} \quad (i = 1, 2, \ldots, I). \tag{12}$$

Test Examples

Evaluation of the achievable accuracy with simplest velocity models

To evaluate the achievable accuracy of our algorithm, a number of tests with simplest velocity models was carried out. A square region of size 100x100 was considered (units are arbitrary), with the coordinate system origin at the lower left corner. The region was divided into square cells of unit size. The source was placed somewhere in the lower left cell. The time field at grid nodes as well as traveltimes obtained in process of the following ray tracing from the source to each of grid nodes were considered. As for velocity distributions, only homogeneous and two-layered media were considered so as to check easily the results. As a consequence, it was found that even for distant points, the accuracy of computations strongly depends on the source position with respect to the boundaries of the containing cell: the less the distance from the source to a boundary, the less the computational accuracy for all the points beyond this boundary. The accuracy depends also on the velocity model: the transition from a low velocity zone to a high velocity one is attended with an increase of the isochrone curvature and with a loss of the accuracy; the reverse effect is observed in the case of transition from a high velocity zone to a low velocity one. These effects are especially strong when the interface is near the source.

The obtained results were also compared with the results that would be achieved using algorithms of other authors. For this purpose, the computations were repeated under the assumption that the time field at the cell boundaries varies linearly, according to values at the grid nodes. Such a technique, which may be considered as a special case of our approach, had already been used in several publications (Asming and Buyanov, 1990; Podvin and Lecomte, 1991; Schneider et al., 1992).

Hereafter we present only one example of the considered test sequence, where the computational errors were close to maximum. In that example the source was located at the point (0.9999, 0.9999), whereas the velocity distribution was as follows:

$$V(x, y) = \begin{cases} 0.3333 & x < 2 \\ 1.0 & x \geq 2 \end{cases} \tag{13}$$

The distributions of the absolute computational error after some smoothing are demonstrated in Figure 2. It is easy to see that the absolute error of the time field computation is no more than 0.06 in the case of three-point approximation (Figure 2a). It corresponds to the relative error of about 0.05% at the farthest points. In the case of two-point approximation, the maximum error is equal to 0.4 (Figure 2b), i.e. it is about 7 times greater. The comparison of the traveltime errors received in the process of the following ray tracing leads to similar conclusions: the use of the three-point approximation (Figure 2c) results in the accuracy about 10 times as high as that in the case of the two-point approximation (Figure 2d).

Ray tracing in a complicated medium

The next example is devoted to ray tracing in a more complicated medium (see Figure 3a). It is a smoothed result of the tomographic inversion of physical modelling data (Gorev and Ditmar, 1992). The size of the investigated region was 1000x1000 mm, a grid of 41x41 cells was used. The task discussed below was to trace rays from the sources disposed at the lower edge of the region to the receivers located at the upper edge. The numbers of the sources and of the receivers were 21, i.e. both the sources and the receivers were situated 50 mm apart. Only pairs with the receiver disposed exactly above the source were considered.

The raypaths computed are demonstrated in Figure 3b, whereas the traveltimes as a function of the ray number are plotted in Figure 3c. It can be seen that the average traveltime is about 450 μs. The three high velocity zones manifest themselves as local traveltime minima. The average computation of the time field took 60 sec, whereas the following tracing of each ray required about 1 sec (computer IBM PC AT-286/287, 16 MHz).

The aim of the following operations was to evaluate the achieved accuracy. For that purpose, a more detailed division (121x121 cells) was considered. Firstly, two computational experiments were performed on its basis. The first experiment was similar to that described above; the second was carried out under the assumption that the positions of sources and receivers are interchanged. The traveltime differences between the results of the two experiments are very little (see the solid line in Figure 3d), so that both the computations may be treated as practically exact. In particular, we took the least traveltime of the two for each source-to-receiver pair as the reference one. The difference between the results of the experiment with 41x41 cells and the reference values is demonstrated in Figure 3d with the dashed line. It is easy to see that the maximum error of the performed ray tracing is about 0.6 μs, which corresponds to the relative error around 0.15%. As to the root-mean-square (rms) error, it is equal to 0.2 μs, i.e. less than the maximum one by a factor of 3. For comparison, ray tracing was repeated on the basis of the two-point time field approximation. The corresponding errors are shown in Figure 3d with the dotted line. The maximum error in this case exceeds 2 μs, and the rms error is about 1.2 μs, i.e. it is several times greater than in the case of three-point approximation.

Ray tracing in a "labyrinth"

In conclusion, consider an example that doesn't reflect a real geophysical situation but clearly demonstrates the uniqueness of the time-field-based ray tracing algorithms as compared to the conventional ones. A labyrinth-like velocity structure was used (Figure 4a). It was supposed that rays could only propagate across white cells characterized by the unit velocity. On the other hand, black cells were taken as "opaque", i.e. exhibiting an extremely low velocity (e.g., 0.0001). The problem was to trace a ray from the lower left corner to the upper right one. As shown in Figure 4b, our algorithm has met the challenge. Needless to say, that the conventional ray tracing methods, such as shooting or bending, would fail to solve this problem.

Conclusion

Our study was devoted to the recently developed method of ray tracing based on the preliminary time field computation. A new approach to ray tracing is described that allows us to achieve the high computational accuracy with the minimum computational cost. Some peculiarities of an algorithm developed in the context of the proposed approach are discussed. A characteristic feature of the algorithm is a three-point approximation of the time field at cell boundaries, two point coinciding with grid nodes and one being in the boundary middle. The accuracy achievable by this algorithm is compared with that by the algorithms described earlier, where a two-point approximation was used. The examples shown demonstrate in the first case an accuracy 5–10 times as higher as that in the second case. On the other hand, the question as to what accuracy level is sufficient was out of our consideration. Obviously, it depends on many factors: the accuracy of the field experiment; the number of cells that a ray crosses typically; the probable level of the discretization errors etc. Nevertheless, we believe that the three-point approximation meets all practical requirements.

The study is confined to the first arrival ray tracing. Meantime, there are papers whose authors showed that the ray tracing method based on the preliminary time field computation is applicable for an investigation of the following arrivals, for the reflected waves in particular (Moser, 1991; Matsuoka and Ezaka, 1992). Our approach may also be generalized to the case of reflected waves, though it needs the interfaces to coincide with cell boundaries.

Another possible direction of the following studies is a generalization of the described approach to the 3-D case. A conventional way to raise the computational accuracy, increasing the number of cells, may be too time-consuming in 3-D. That's why the proposed approach may be especially useful there.

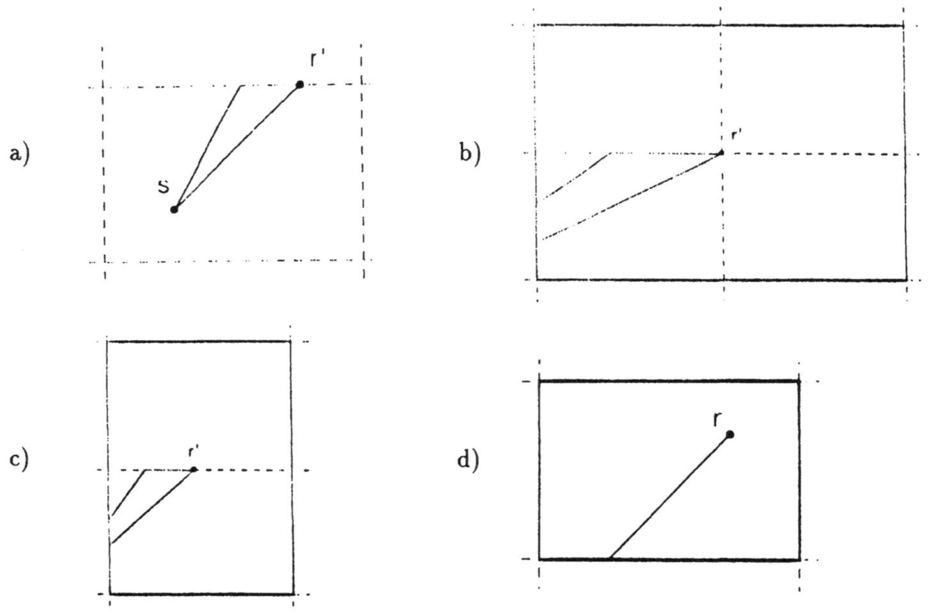

FIGURE 1. An illustration to procedures of time field computation and following ray tracing. The heavy line denotes the contour K, thin lines indicate possible shortest paths, dash lines represent cell boundaries. (a) Initialization. Computation of the time field at a point r' situated at a boundary of the cell containing the source. (b) Construction of the contour K round a point r' coincident with a grid node. (c) Construction of the contour K round a point r' not coincident with a grid node. (d) Construction of the contour K round the receiver.

a)

b)

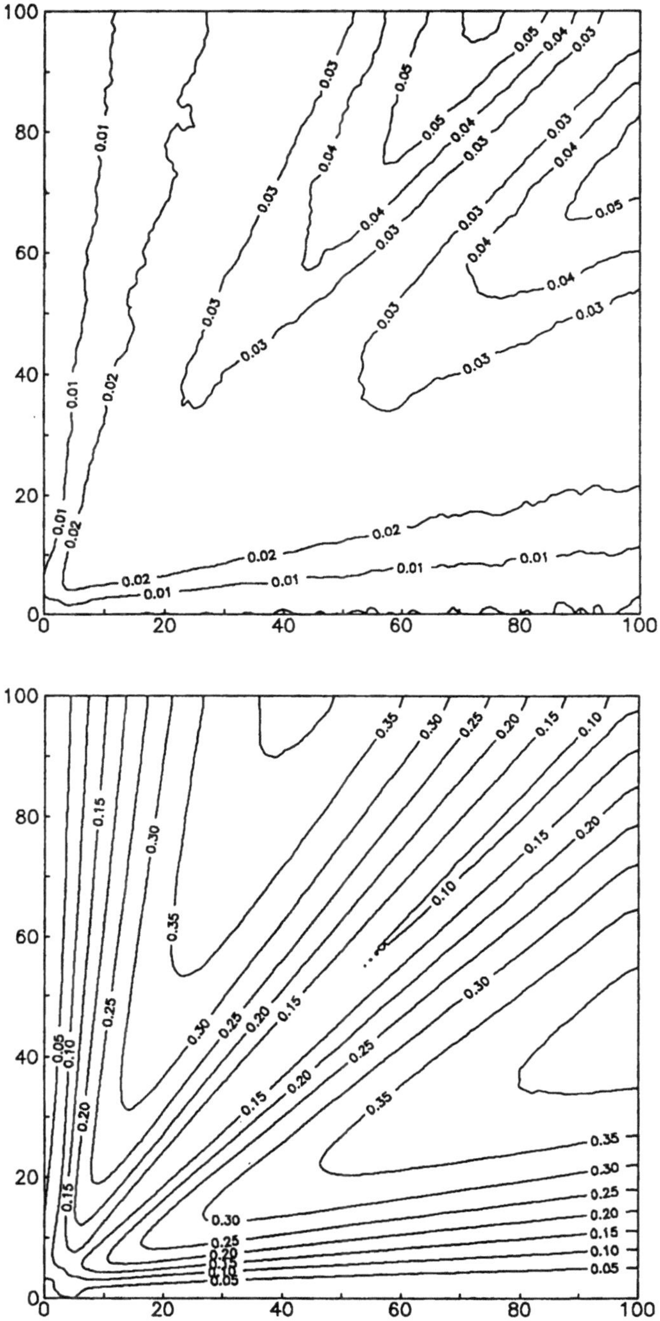

FIGURE 2 a), b).

c)

d)

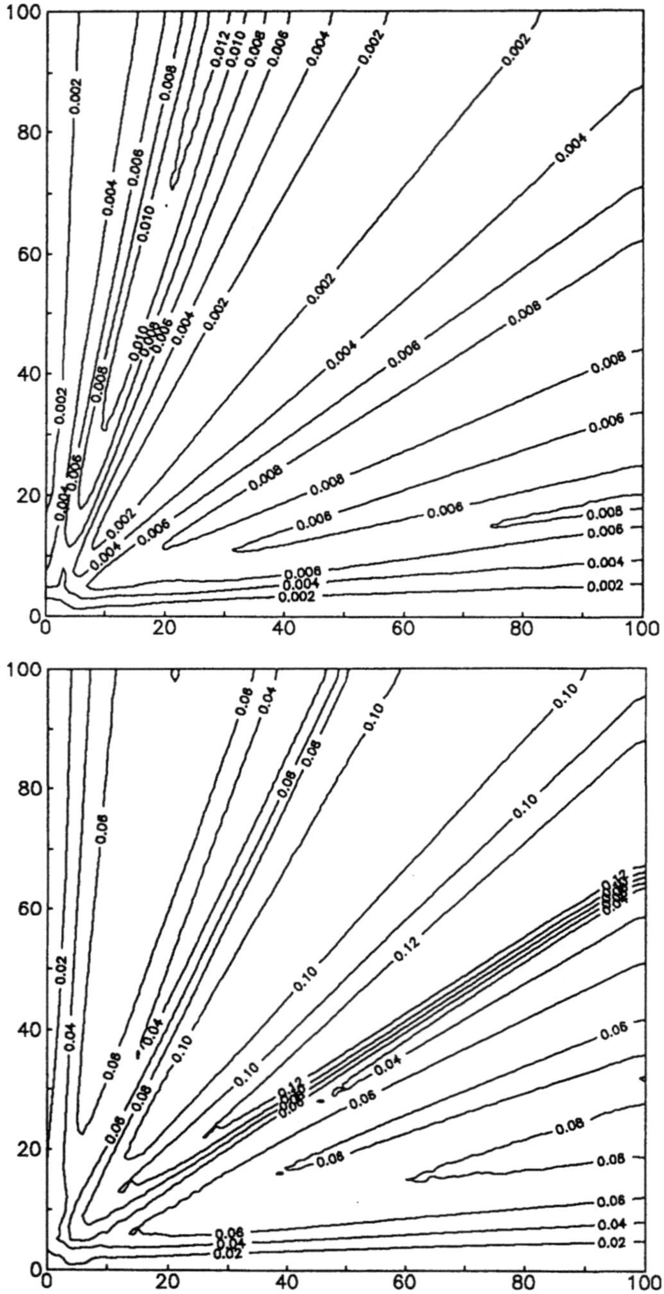

FIGURE 2. Absolute computational error in the medium with the velocity distribution $V(x, y) = 0.333$ $(x < 2)$; $V(x, y) = 1.0$ $(x \geq 2)$. (a) Error of the time field computation in the case of three-point approximation. (b) The same in the case of two-point approximation. (c) Error of the following ray tracing in the case of three-point approximation. (d) The same in the case of two-point approximation. Units are arbitrary.

a)

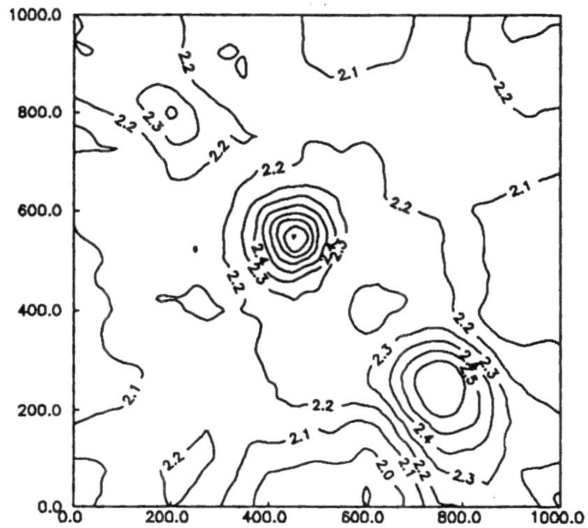

b)

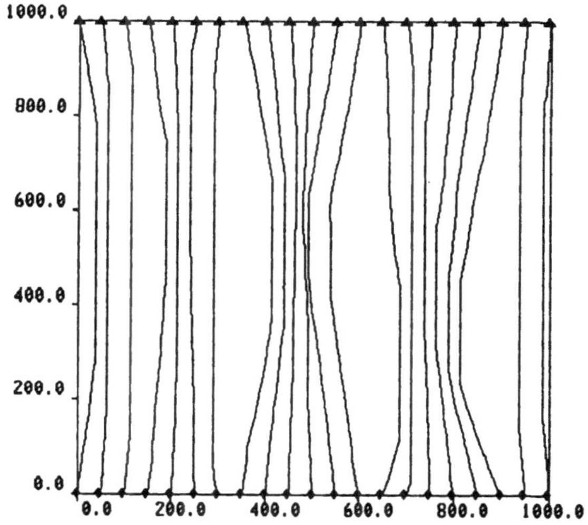

Figure 3 a), b).

c)

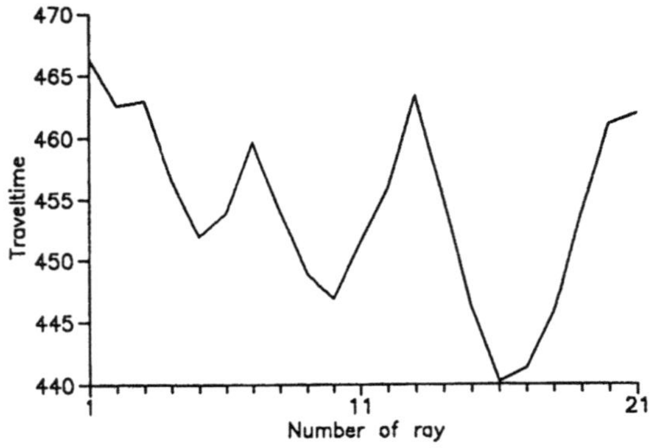

d)

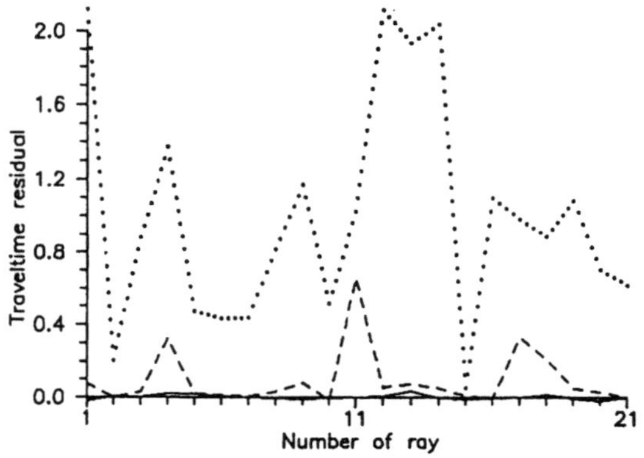

FIGURE 3. An example of ray tracing in a complicated medium. (a) Velocity model considered. (b) Raypaths. (c) Dependence of traveltimes (μs) on the ray number. (d) Residuals of "reference" ray tracing (solid line); the error of ray tracing in the case of three-point approximation (dashed line); the same in the case of two-point approximation (dotted line); units are μs.

a)

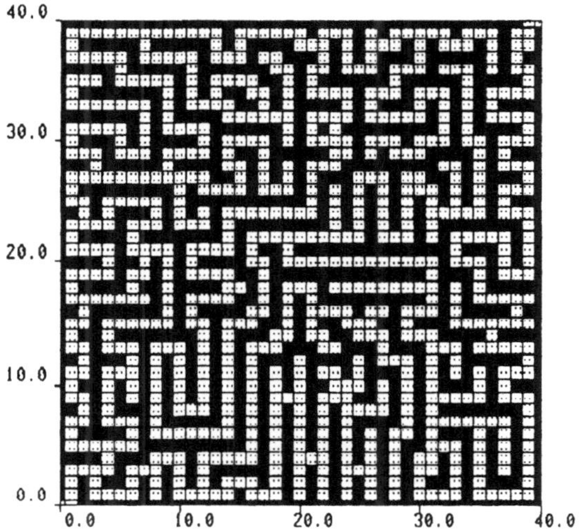

b)

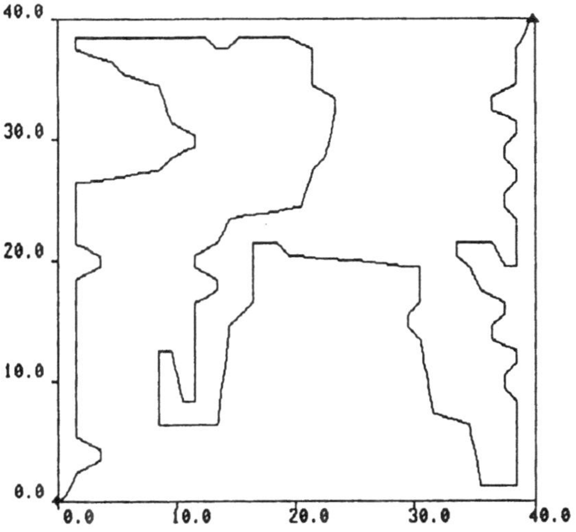

FIGURE 4. Ray tracing in a "labyrinth". (a) An appearance of the "labyrinth" (white cells are associated with the unit velocity, black cells demonstrate an extremely low velocity). (b) The raypath propagating from the lower left corner to the upper right one.

Acknowledgments

The author would like to thank prof. T. B. Yanovskaya and Dr. G. A. Ryzhikov for their valuable comments.

REFERENCES

Asming V. E., Buyanov A. F. (1990), *A wavefront algorithm and its application to location of seismic events.* GI KNC AN USSR (USSR Academy of Scienses Kola Scientific Center Institute of Geology), Apatity, Deposited in VINITI (Union's Institute of Scientific and Technical Information) on 05/21/90, no. 2763-B90 (in Russian)

Dijkstra E. W. (1959), A note on two problems in connection with graphs. *Numer. Math.*, vol. 1, 269–271.

Dines K. A., Lytle R. J. (1979), Computerized geophysical tomography. *Proc. IEEE*, vol. 67, 1065–1073.

Ditmar P. G. (1990), *Occam's tomographic inversion.* Ph.D. thesis, Leningrad University. (in Russian)

Ford L. R., Fulkerson D. R. (1962), *Flows in networks.* Princeton university, N.J.

Gorev V. A., Ditmar P. G. (1992), Non-linear tomographic inversion of 2-D physical modelling data. *International workshop on geotomography "Geotomography-92"*, Abstracts, 41–42, Apatity. (in Russian)

Matsuoka T., Ezaka T. (1992), Ray tracing using reciprocity. *Geophysics*, vol. 57, 326–333.

Moser T. J. (1991), Shortest path calculation of seismic rays. *Geophysics*, vol. 56, 59–67.

Nakanishi I., Yamaguchi K. (1986), A numerical experiment on nonlinear image reconstruction from first-arrival traveltimes for two-dimensional island arc structure. *J. Phys. Earth*, no. 34, 195–201.

Pilipenko V. N. (1979), Time field technique for imaging of seismic interfaces. *Inverse kinematic problems of explosion seismology*, Nauka, Moscow. (in Russian)

Podvin P., Lecomte I. (1991), Finite difference computation of traveltimes in very contrasted velocity models: a massively parallel approach its associated tools. *Geophys. J. Int.*, no. 105, 271–284.

Qin F., Luo Y., Olsen K. B., Cai W. and Schuster G. T. (1992), Finite-difference solution of the eikonal equation along expanding wavefronts. *Geophysics*, no. 57, 478–487.

Reshef M., Cosloff D. (1986), Migration of common shot gathers. *Geophysics*, no. 51, 324–331.

Schneider W. A., Jr. Ranzinger K. A., Balch A. H. and Kruse C. (1992), A dynamic programming approach to first arrival travel time computation in media with arbitrary distributed velocity. *Geophysics*, no. 57, 39–50.

Van Trier J., Symes W. W. (1991), Upwind finite-difference calculation of traveltimes, *Geophysics*, no. 56, 812–821.

Vidale J. (1988), Finite difference calculation of traveltimes, *Bull. Seism. Soc. Am.*, no. 78, 2062–2076.

Vidale J. (1990), Finite difference calculation of traveltimes in three dimensions, *Geophysics*, no. 55, 521–526.

Asymptotic Expansions of Continuous Solutions of System of Volterra Integral Equations of the First Kind

M. V. FALALEEV

Department of Mathematics Irkutsk University
Irkutsk, Russia

ABSTRACT

In the papers by Magnitskii (1983) and Falaleev (1986), a Volterra integral equation of the first kind was studied. We extend the results of these papers to systems of such equations.

We consider a system of integral equations

$$\int_0^t K(t,s)\overline{x}(s)ds = \overline{b}(t), \tag{1}$$

where $K(t,s)$ is a matrix function with elements $K_{pl}(t,s)$, $p,l = 1,\ldots,m$; $\overline{x}(t)$ and $\overline{b}(t)$ are m dimensional vector-functions.

We assume all the kernels $K_{pl}(t,s)$ to be analytic in a neighborhood of the point $(0;0)$. Suppose also that the functions $K_{pl}(t,s)$ and all their partial derivatives of the order not exceeding $(n-1)$ vanish at the point $(0;0)$, whereas their n-th derivatives are not zero at this point. We denote by K_{ij} the matrix consisting of the Maclaurins (i,j)-coefficients of the kernels $K_{pl}(t,s)$.

Introduce a matrix function

$$L_n(\lambda) = \sum_{i+j=n} \frac{1}{\lambda + j + 1} K_{ij}$$

and its derivatives with respect to λ: $L_n^{(k)}(\lambda)$, $k \in N$.

The equation

$$\det L_n(\lambda) = 0 \tag{2}$$

will be referred to as the *characteristic equation* of system (1). For *each natural* root λ_0 satisfying (2), the following condition is assumed to be fulfilled:

A) for a basis φ_i, $i = 1,\ldots,n_{\lambda_0}$, of the kernel $N(L_n(\lambda_0))$, there exists a complete

$$\left\{ \frac{(-1)^{(k+1)}}{k!} L_n^{(k)}(\lambda_0) \right\}\text{-set},$$

that is, *a linearly independent* system of vectors $\{\varphi_i^{(j)}\}$, $i = 1,\ldots,n_{\lambda_0}$, $j = 1,\ldots,p_i$, such that

$$\varphi_i^{(1)} = \varphi_i, \quad i = 1,\ldots,n_{\lambda_0},$$

$$L_n(\lambda_0)\varphi_i^{(1)} = 0,$$

$$L_n(\lambda_0)\varphi_i^{(2)} = L_n^{(1)}(\lambda_0)\varphi_i^{(1)},$$

$$L_n(\lambda_0)\varphi_i^{(3)} = L_n^{(1)}(\lambda_0)\varphi_i^{(2)} - \frac{1}{2!}L_n^{(2)}(\lambda_0)\varphi_i^{(1)},$$

$$\dots\dots\dots$$

$$L_n(\lambda_0)\varphi_i^{(p_i)} = L_n^{(1)}(\lambda_0)\varphi_i^{(p_i-1)} - \frac{1}{2!}L_n^{(2)}(\lambda_0)\varphi_i^{(p_i-2)} + \dots + \frac{(-1)^{p_i}}{(p_i-1)!}L_n^{(p_i-1)}(\lambda_0)\varphi_i^{(1)},$$

$$\det \| < L_n^{(1)}(\lambda_0)\varphi_i^{(p_i)} - \frac{1}{2!}L_n^{(2)}(\lambda_0)\varphi_i^{(p_i-1)} + \dots$$
$$+ \frac{(-1)^{p_i+1}}{p_i!}L_n^{(p_i)}(\lambda_0)\varphi_i^{(1)}, \psi_k >, i, k = 1, \dots, n_{\lambda_0} \| \neq 0,$$

where ψ_k, $k = 1, \dots, n_{\lambda_0}$, is a basis of the kernel $N(L_n^*(\lambda_0))$.

The number $p = \max(p_1, \dots, p_{n_{\lambda_0}})$ is called *the multiplicity* of the natural root λ_0.

Lemma 1. *If the characteristic equation (2) has a finite number of natural roots $\lambda_1, \dots, \lambda_l$ of multiplicities $m_1, \dots, m_l$, $0 \le \lambda_1 < \lambda_2 < \dots < \lambda_l$, then the homogeneous system of equations corresponding to (1) admits a formal $\sum_{k=1}^{l}\sum_{i=1}^{n_{\lambda_k}} p_i$-parametric solution in the form*

$$\overline{x}(s) = \sum_{\alpha=\lambda_1}^{\infty} s^\alpha\{\overline{x}_{\alpha 0} + \overline{x}_{\alpha 1}(\ln s) + \dots + \overline{x}_{\alpha r_\alpha}(\ln s)^{r_\alpha}\}, \tag{3}$$

where

$$r_\alpha = \begin{cases} m_1 - 1, & \lambda_1 \le \alpha \le \lambda_2 - 1, \\ m_1 + m_2 - 1, & \lambda_2 \le \alpha \le \lambda_3 - 1, \\ m_1 + m_2 + m_3 - 1, & \lambda_3 \le \alpha \le \lambda 4 - 1, \\ \dots & \dots \\ m_1 + m_2 + \dots + m_l - 1, & \lambda_l \le \alpha. \end{cases}$$

Lemma 2. *Assume that the characteristic equation (2) has a finite number of natural roots $\lambda_1, \dots, \lambda_l$ of multiplicities $m_1, \dots, m_l$, $0 \le \lambda_1 < \lambda_2 < \dots < \lambda_l$, the function $\overline{b}(t)$ is analytic in a neighborhood of 0, and $\overline{b}(0) = \overline{b}^{(1)}(0) = \dots = \overline{b}^{(n)}(0) = 0$. Then system (1) has a formal partial solution of the form (3), where*

$$r_\alpha = \begin{cases} 0, & 0 \le \alpha \le \lambda_1 - 1, \\ m_1, & \lambda_1 \le \alpha \le \lambda_2 - 1, \\ m_1 + m_2, & \lambda_2 \le \alpha \le \lambda_3 - 1, \\ \dots & \dots \\ m_1 + m_2 + \dots + m_l, & \lambda_l \le \alpha. \end{cases}$$

REMARK 1. To compute the vector-coefficients involved in Lemmas 1 and 2, we need to replace $K(t,s), \overline{b}(t)$ and $\overline{x}(s)$ in (1) by their expansions. We rearrange the resulting equation and equate expressions at the same degrees of t and $\ln t$. The obtained recurrence relations yield the vectors $\overline{x}_{\alpha i}$. The corresponding independent parameters in quantities of $(p_1 + \dots + p_{n_{\alpha_i}})$ appear when computing the vectors $(\overline{x}_{\lambda_i 0}, \dots, \overline{x}_{\lambda_i m_i}), i = 1, \dots, l$, in Lemma 1. The remaining vectors $\overline{x}_{\alpha i}$ are uniquely computed from those already found.

Note that condition A) *guarantees* the solvability of *all* the singular systems arising in the computation and uniquely determines the number of terms r_α (see Lemmas 1 and 2).

REMARK 2. If, for some λ_i, the system of vectors $\{\varphi_i^{(j)}\}$ in A) is linearly dependent, then the number of independent parameters reduces to the dimension of the linear span $\mathcal{L}(\{\varphi_i^{(j)}\})$.

Theorem. *Assume that the characteristic equation (2) has a finite number of natural roots* $\lambda_1, \ldots, \lambda_l$ *of multiplicities* $m_1, \ldots, m_l$, $0 \leq \lambda_1 < \lambda_2 < \ldots < \lambda_l$, $\det \| \sum_{i+j=n} K_{ij} \| \neq 0$, *the functions* $K(t,s)$ *and* $\overline{b}(t)$ *are smooth enough in a neighborhood of* 0, *and* $\overline{b}(0) = \overline{b}^{(1)}(0) = \cdots = \overline{b}^{(n)}(0) = 0$. *Then the general solution of (1) belonging to the class* $C[-R,R]$ *can be represented in the form*

$$\overline{x}(s) = \sum_{\alpha=0}^{m} \overline{x}_\alpha(s) s^\alpha + \overline{\eta}(s)$$

for a sufficiently large $m \in N$, *where* $\overline{x}_\alpha(s)$ *is the sum of the corresponding functional coefficients as in Lemmas 1 and 2,* $\overline{\eta}(s) = s^m \overline{\gamma}(s)$, $\overline{\gamma}(s) \in C[-R,R]$, $\overline{\gamma}(s) \longrightarrow 0$ *as* $s \longrightarrow 0$, $\overline{\eta}(s) = o(s^m)$ *as* $s \longrightarrow 0$.

REMARK 3. Assume the condition $\overline{b}(0) = \overline{b}^{(1)}(0) = \cdots = \overline{b}^{(n)}(0) = 0$ to be not fulfilled. In this case, if $\det \| K_{ij} \| \neq 0$, $i + j = n$, we may construct the generalized solution of system (1) in the form

$$\overline{c}_0 \delta(t) + \overline{c}_1 \delta^{(1)}(t) + \cdots + \overline{c}_n \delta^{(n)}(t) + u(t)\theta(t),$$

where $\delta(t)$ is the Dirac delta-function and $\theta(t)$ is the Heaviside function. The function $u(t)$ is constructed according to Lemmas 1,2 and the above theorem. The vectors $\overline{c}_i$ are uniquely determined.

REFERENCES

Falaleev F. M. (1986), *Asymptotic representations of the continuous and generalized solutions of the Volterra integral equation of the first kind*, Dep. VINITI, N1553-87. (in Russian)

Magnitskii N. A. (1983), Asymptotics of the solutions of Volterra integral equations of the first kind, *Doklady Akad. Nauk SSSR*, **296**, 29–32. (in Russian)

On New Methods for Determining Static Time Corrections from Phase Velocities of Surface Waves [1]

M. V. FOKIN, V. V. SKAZKA

Institute of Mathematics, Siberian Branch of Russian Acad. Sci.
Novosibirsk, Russia

ABSTRACT

A problem of determining the traveltime of seismic waves to a certain depth using the phase characteristics of surface waves is considered. For Love waves, an approximate algorithm with a priori estimate of its accuracy is presented. We describe also some approximate algorithms for Rayleigh waves which were tested for different types of nonhomogeneous media. The results of numerical experiments are given.

1. On an Algorithm for Determining the Static Time Corrections from Phase Characteristics of Love Waves

Of importance in seismic prospecting is the problem of determining the traveltime of seismic waves from surface points to a certain (as a rule, comparatively small) depth L or the so-called static corrections. Their further use in actual data processing allows us to improve appreciably the reliability of conclusions as to the structure of the deeper layers of a geological profile. In the model case of a vertically nonhomogeneous medium, the problem can be formulated for the surface Love waves.

Suppose that the signal velocity $v(z)$ is dependent only on the depth z and constant in the horizontal directions x, y, while the Lamé parameter μ is constant throughout the half-space $z \geq 0$. Then, for a given frequency ω, the amplitude $u(z)$ of the surface plane Love wave $u(z) \exp i(\omega t - \xi x)$ satisfies the equation (Fokin, 1993)

$$u'' + \frac{\omega^2}{v^2(z)} u = \xi^2 u. \tag{1}$$

The following conditions are stated for the surface $z = 0$ and for $z \to \infty$

$$u'(0) = 0, \qquad \lim_{z \to \infty} u(z) = 0. \tag{2}$$

If for a given ω there are such values of ξ that problem (1)–(2) has a non-trivial solution, then the largest possible value $\xi_0(\omega)$ corresponds to the first (principal) mode of the Love waves, and the ratio $\omega/\xi_0(\omega)$ characterizes its phase velocity. In seismogram processing this quantity may be determined in a certain range of ω values.

[1]This work was supported in part by Russian Foundation of Fundamental Research under grant No 93-05-8638.

Denote by $V(L, v_\infty)$ the class of functions $v(z)$ satisfying the following conditions: they are constant for $z \geq L$ ($v(z) = v_\infty$ for these z values); they are strictly positive and continuous at all the points of the segment $[0, L]$ except, possibly, for a finite set of points $z_1 < z_2 < \ldots < z_n \leq L$, where they have discontinuities of the first kind, this set being peculiar to each function. At the points of discontinuities of $v(z)$ we require, in addition, that the following matching conditions should be satisfied:

$$u(z_k - 0) = u(z_k + 0), \quad u'(z_k - 0) = u'(z_k + 0), \quad k = 1, \ldots, n. \tag{3}$$

We deal with the problem of determining the traveltime

$$T(v) = \int_0^L \frac{dz}{v(z)}$$

to the depth L for a vertically incident wave. Here the values of L, v_∞, ω, $\xi_0(\omega)$ are assumed to be given, and it is required to determine the approximate value $T_0(v)$ for $T(v)$ and to estimate the relative error

$$\Delta(v) = \frac{T_0(v) - T(v)}{T_0(v)}.$$

In order to calculate $T_0(v)$, the following natural algorithm is proposed. Instead of the unknown function $v(z)$ we consider the piecewise-constant function $\tilde{v}(z)$:

$$\tilde{v}(z) = \begin{cases} v_0, & 0 \leq z \leq L \\ v_\infty, & z > L. \end{cases}$$

Let us choose the value v_0 (an approximate value of the signal mean velocity in the layer of thickness L) so that problem (1)–(3) will have a non-trivial solution for the initial values of ω, $\xi_0(\omega)$, with $\xi_0(\omega)$ being as large as possible for the given ω. We regard the ratio L/v_0 as an approximate value $T_0(v)$ of the sought-for quantity $T(v)$. The relative error $\Delta(v)$ will be estimated in the set $V(\alpha, \beta, L, v_\infty)$ of functions $v(z) \in V(L, v_\infty)$ which satisfy the inequalities

$$\alpha \leq \frac{v_\infty}{v(z)} \leq \beta \tag{4}$$

for all $z \in [0, L]$ (α and β are positive constants, $\alpha \leq \beta$). Note that the constraints similar to (4) appear to be natural from the viewpoint of applications and allow us to take into account *a priori* information, if any, concerning the admissible values of $v(z)$.

We introduce a new variable $\tau = \omega z / v_\infty$ and assume

$$q(\tau) = \frac{v_\infty^2}{v^2(v_\infty \tau/\omega)}, \qquad H = \frac{\omega L}{v_\infty}, \qquad \lambda = \frac{\xi^2 v_\infty^2}{\omega^2}.$$

For the function $w(\tau) = u(v_\infty \tau/\omega)$ equation (1) reduces to

$$w'' + q(\tau) w = \lambda w, \tag{5}$$

while conditions (2)–(3) are readily replaced by

$$w'(0) = 0, \qquad \lim_{\tau \to \infty} w(\tau) = 0. \tag{6}$$

$$w(\tau_k - 0) = w(\tau_k + 0), \quad w'(\tau_k - 0) = w'(\tau_k + 0), \quad k = 1, \ldots, n, \tag{7}$$

where τ_k are the discontinuity points, if any, of the function $q(\tau)$. The class of functions $V(L, v_\infty)$ is thus transformed into the set $Q(H)$ of piecewise-continuous functions $q(\tau)$ for which $q(\tau) \equiv 1$ when $\tau > H$. If $v(z) \in V(\alpha, \beta, L, v_\infty)$, then $q(\tau)$ satisfies, for $\tau \in [0, H]$, the inequalities

$$\alpha^2 \leq q(\tau) \leq \beta^2. \tag{8}$$

The set of functions $q(\tau) \in Q(H)$ which satisfy (8) will be denoted by $Q(\alpha^2, \beta^2, H)$.

Each function $q(\tau) \in Q(H)$ is placed into correspondence with an $L_2(0,\infty)$-self-adjoint Sturm-Liouville operator $\mathcal{L}[q(\tau)]w \equiv w'' + q(\tau)w$ defined by the boundary condition $w'(0) = 0$. Thus, the largest value $\xi_0(\omega)$ for which there exists a non-trivial solution of problem (1)–(3) corresponds to the first (largest) eigenvalue $\lambda_0 = \xi_0^2(\omega)v_\infty^2/\omega^2$ of the operator $\mathcal{L}$. The sought-for quantity $T(v)$ is expressed in terms of $q(\tau)$ as

$$T(v) = \tilde{T}(q) = \frac{1}{\omega}\int_0^H \sqrt{q(\tau)}\,\mathrm{d}\tau.$$

The algorithm for calculating the approximate value $T_0(v)$ is to choose such a value q_0, that for the function

$$\tilde{q}(q_0, \tau) = \begin{cases} q_0, & 0 \leq \tau \leq H \\ 1, & \tau > H \end{cases}$$

the first eigenvalue λ_0 of the operator $\mathcal{L}[\tilde{q}]$ is also the first eigenvalue of the original operator $\mathcal{L}[q]$. The determination of q_0 reduces to solving an elementary equation. The approximate value $\tilde{T}_0(q)$ of the quantity $\tilde{T}(q)$ is calculated using the formula $\tilde{T}_0(q) = H\sqrt{q_0}/\omega$, and the relative error is

$$\tilde{\Delta}(q) = \frac{\tilde{T}_0(q) - \tilde{T}(q)}{\tilde{T}_0(q)} = \frac{1}{H\sqrt{q_0}}\int_0^H \left(\sqrt{q_0} - \sqrt{q(\tau)}\right)\mathrm{d}\tau.$$

Let us define the function $F(\alpha, \beta)$ for positive α, β as

$$F(\alpha, \beta) = \begin{cases} 1 - 2\sqrt{\alpha\beta}/(\alpha + \beta), & \alpha\beta \geq 1 \\ 1 - (1 + \alpha\beta)/(\alpha + \beta), & \alpha\beta < 1. \end{cases}$$

The following theorem is valid.

Theorem 1. *Suppose that $\beta > 1$, $\alpha < \beta$, and, for a function $v(z) \in V(\alpha, \beta, L, v_\infty)$, the operator $\mathcal{L}[q(\tau)]$ constructed by the corresponding function $q(\tau) \in Q(\alpha^2, \beta^2, H)$ has at least one eigenvalue. Then, for $H \leq H(\alpha, \beta)$, the relative error $\Delta(v)$ in the determination of the approximate value $T_0(v)$ for the quantity $T(v)$ using the above algorithm and the largest eigenvalue λ_0 of the operator $\mathcal{L}[q]$, is estimated as*

$$-C(\alpha, \beta)H^2 \leq \Delta(v) \leq F(\alpha, \beta) + C(\alpha, \beta)H^2, \tag{9}$$

with the constants $C(\alpha, \beta)$ and $H(\alpha, \beta)$ depending only on α and β.

REMARK. Using Theorem 1 as an example, we shall consider the case when the relation $v_\infty/2 \leq v(z) \leq v_\infty$ is known *a priori* for all $z \in [0, L]$. This corresponds to $\alpha = 1$ and $\beta = 2$. Irrespective of L and v_∞, the parameter $H = \omega L/v_\infty$ vanishes when $\omega \to 0$. Theorem 1 allows us to conclude that, for sufficiently small ω, the relative error $\Delta(v)$ does not exceed $F(1, 2) \approx 0.057$. The left-hand part of estimate (9) shows that the approximate result is usually greater than the sought-for value $T(v)$.

See the complete proof of Theorem 1 in Fokin (1993). The fact that estimate (9) cannot be improved results from the Theorem 2 in Fokin (1993).

2. Algorithms for the Construction of Static Corrections from the Phase and Group Velocities of Rayleigh Waves

The following model was considered in the determination of static corrections from the phase velocities of the surface Rayleigh waves: the medium was considered to be locally layered at each measurement point of the incoming signal, i.e. it was assumed that here waves propagate in the same way as in the medium, whose characteristic parameters (V_p — propagation velocity of longitudinal waves, V_s — propagation velocity of transverse waves, ρ — density) depend only on the vertical variable z and are piecewise-constant in z (there is a finite number of discontinuity points in z , and, beginning from the depth L , the values of V_p, V_s and ρ stabilize and the underlying layer begins).

The boundary value problem describing the dispersive properties of Rayleigh waves for such media is as follows (see Levshin, 1973):

$$\begin{cases} U''_{zz} + \xi(1 - \frac{V_p^2}{V_s^2})W'_z - \xi^2 \frac{V_p^2}{V_s^2} U = -\frac{\omega^2}{V_s^2} U, \\ W''_{zz} + \xi(1 - \frac{V_s^2}{V_p^2})U'_z - \xi^2 \frac{V_s^2}{V_p^2} W = -\frac{\omega^2}{V_p^2} W, \end{cases} \tag{10}$$

$$U'_z = \xi W \quad , \quad z = 0, \tag{11}$$

$$W'_z = -\xi(1 - \frac{2V_s^2}{V_p^2})U \quad , \quad z = 0, \tag{12}$$

$$U(z) \to 0, \quad W(z) \to 0 \quad \text{when} \quad z \to \infty. \tag{13}$$

Here $U(z)$ and $W(z)$ are the amplitude characteristics of horizontal and vertical oscillations, correspondingly, having the form $\exp i(\omega t - \xi x)$ for the Rayleigh wave propagating along the x -axis. It is considered that, for a layered medium, the velocities V_p, V_s and the density ρ are constant on the intervals separated by the points z_m ($m = 1, \ldots, n$; $0 < z_1 < \ldots < z_n \leq L$, where L is a depth at which the underlying layer begins). Let $V_p(m)$, $V_s(m)$, $\rho(m)$ be the values of the velocities V_p, V_s and the density ρ on the constancy interval $z_m < z < z_{m+1}$. The matching conditions for the solutions $U(z)$, $W(z)$ at the discontinuity points $z = z_m$ are given in the following way:

$$U(z_m - 0) = U(z_m + 0), \quad W(z_m - 0) = W(z_m + 0), \tag{14}$$

$$\begin{aligned} &\rho(m-1)V_s^2(m-1)(U'_z(z_m - 0) - \xi W(z_m - 0)) \\ &\quad = \rho(m)V_s^2(m)(U'_z(z_m + 0) - \xi W(z_m + 0)), \end{aligned} \tag{15}$$

$$\begin{aligned} &\rho(m-1)V_p^2(m-1)(W'_z(z_m - 0) + \xi(1 - \tfrac{2V_s^2(m-1)}{V_p^2(m-1)})U(z_m - 0)) \\ &\quad = \rho(m)V_p^2(m)(W'_z(z_m + 0) + \xi(1 - \tfrac{2V_s^2(m)}{V_p^2(m)})U(z_m + 0)). \end{aligned} \tag{16}$$

If problem (10)–(16) has a non-trivial solution for suitable values of ω, ξ, then the largest possible value of ξ at the given ω determines, as previously, the first (and the basic) mode of Rayleigh waves. Then the relation ξ/ω gives the phase velocity of the wave, and the derivative $d\xi/d\omega$ yields the group velocity. Several variants of the algorithm and the programs were developed that make it possible to determine, using the given seismograms, the parameters of the wave field characterizing the phase and group velocities of Rayleigh waves for all values of the signal frequency.

Essentially, the algorithm proposed for the determination of static corrections is as follows: pairs of the corresponding values (ω — signal frequency, ξ/ω — phase velocity of the surface wave) or (ω — signal frequency, $d\xi/d\omega$ — group velocity of surface waves) at each observation point obtained in the processing of data of seismograms, were considered to characterize the vertical distribution of the medium parameters under the observation points. In order to determine the traveltime of the vertically incident Rayleigh wave to the given depth L, the medium was locally replaced by a two-layered medium where the velocities V_p and V_p in the underlying layer (i.e. at $z > L$), the ratio of the rocks' densities in the upper layer and in the underlying layer, and also the ratio V_p/V_s in the upper layer were assumed to be given. Thereafter the value of V_p in the upper layer was determined so that the pair of parameters that was the same as for the studied medium, i.e. $(\omega, \xi/\omega)$ or $(\omega, d\xi/d\omega)$, could correspond to the basic mode of the surface Rayleigh waves in the model medium. And, finally, the static correction for a point at the surface was obtained as the ratio of L to the found value V_p in the upper layer.

The algorithm developed turned out to be highly stable relative to the different perturbations: the choice of the values of V_p and V_s in the underlying layer, variations in the ratio of densities in the upper layer and in the underlying layer, introduction of random deviations for ω, ξ/ω, $d\xi/d\omega$.

When the ratio V_p/V_s was changed in the test calculations, it was found that the mean travel velocity of Love waves to the given depth, i.e. V_s, can be determined in the best way. Using either $(\omega, \xi/\omega)$ or $(\omega, d\xi/d\omega)$ gives similar results. Calculation of one static correction value takes 0.5 - 1 sec. for $(\omega, \xi/\omega)$ and 3 - 5 sec. for the pair $(\omega, d\xi/d\omega)$ on the computer IBM PC 386/387. However, it is better to choose the pair $(\omega, d\xi/d\omega)$ as an input value: these parameter values can be determined more effectively and reliably using given seismograms. An algorithm for a more exact determination of the structure of deeper layers of the medium using the phase characteristics of surface waves was also studied. It was found that information on the phase velocity of the first (basic) mode is not sufficient in order to get more detailed information on the structure of these layers and the parameters of the higher modes must be determined using the data of seismic prospecting. However, the determination of these parameters needs the signal to be measured very accurately. Unfortunately, it is not real to achieve such an accuracy in the present-day field measurements of seismic prospecting.

The programs developed were tested using a large number of calculations both for the synthetic seismograms obtained by means of numerical modelling for different types of media and in the data processing of field sounding. It was shown that the performance of the developed algorithms and programs is rather good. It should be noted that the presence of low frequencies in the recorded signals makes it possible to improve the accuracy in the determination of static corrections and to increase the depth L of the investigated layer.

In conclusion we would like to give two examples of calculation of static corrections. In the course of numerical experiments, calculations of seismograms were carried out for media with the parameters described below. These calculations were based on the programs developed by the group headed by B.G. Mikhailenko at the Computing Center of Siberian Branch of the Russian Academy of Sciences. The basic frequency in the sources was 6 Hz. Then the full cycle of processing was carried out for the obtained synthetic seismograms: the phase velocities of Rayleigh waves at all points of observation were determined, and then the values of static corrections were computed. Values of the relative error of the algorithm for all observation points are also given.

Example 1.

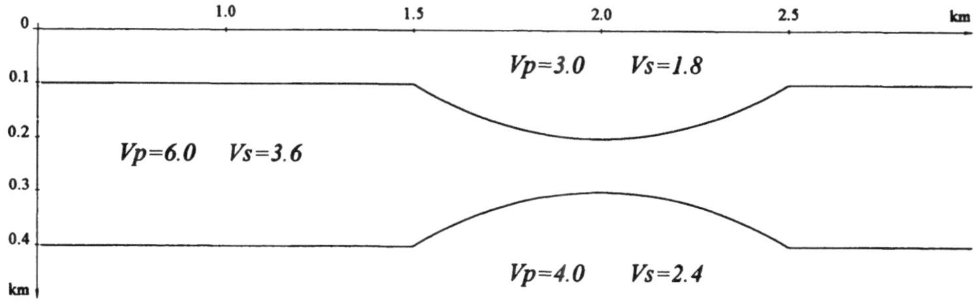

FIGURE 1. Sample media. Source at 0 km, receivers at every 10 m.

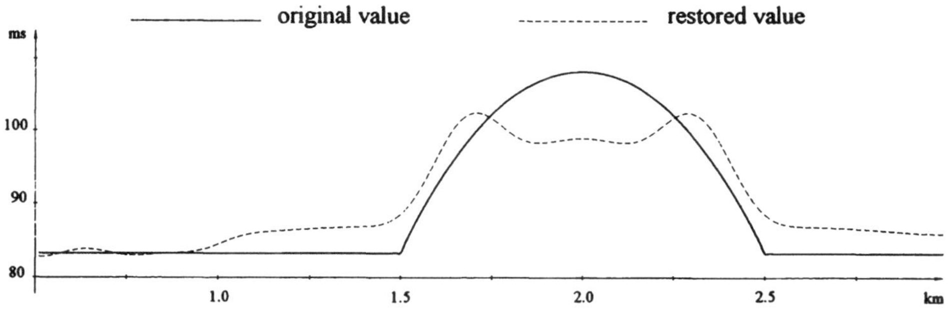

FIGURE 2. The result of calculation of average vertical propagation times of P-waves to depth 0.4 km.

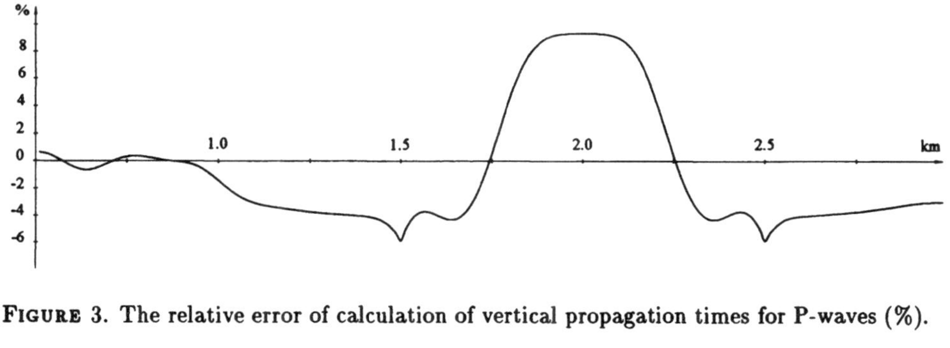

FIGURE 3. The relative error of calculation of vertical propagation times for P-waves (%).

Example 2.

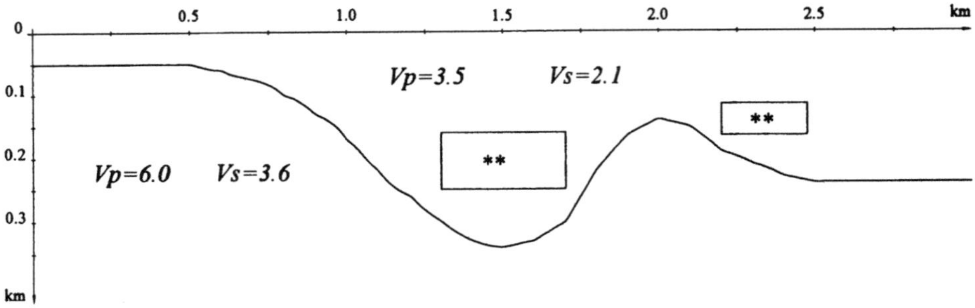

FIGURE 4. Sample media. Source at 0 km, receivers at every 25 m (** — $V_p = 6.0$, $V_s = 3.6$).

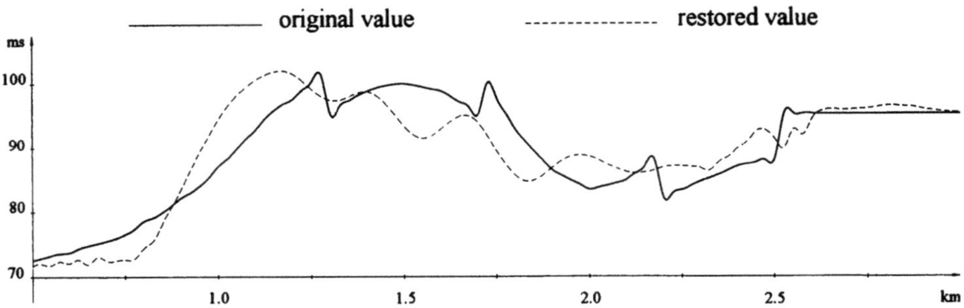

FIGURE 5. The result of calculation of average vertical propagation times of P-waves to depth 0.4 km.

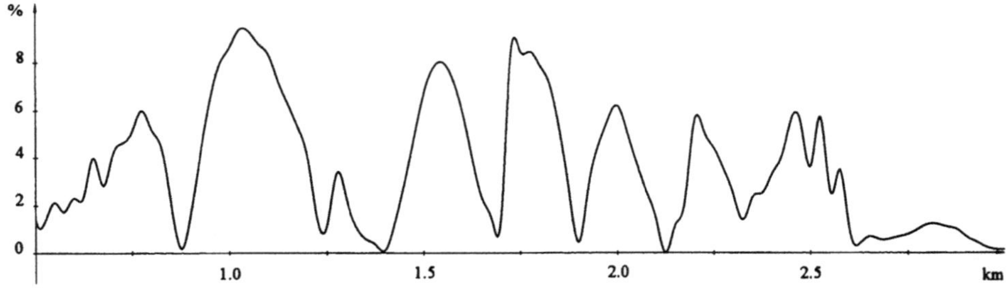

FIGURE 6. The relative error of calculation of vertical propagation times for P-waves (%).

REFERENCES

Fokin M. V. (1993), On the estimation of seismic wave traveltime to a certain depth. *Journ. of Inverse and Ill-Posed Problems.* vol. 1, no. 2, 101–114.

Levshin A. L. (1973), *Surface and Channel Seismic Waves.* Nauka, Moscow. (in Russian)

Soft X–Ray Tomography on Tokamaks Using Flux Coordinates

G. FUCHS[1], Y. MIURA and M. MORI

Department of Fusion Plasma Research
Naka Fusion Research Establishment
Naka–machi, Naka–gun, Ibaraki–ken, Japan
[1]Institut für Plasmaphysik Ass. EURATOM–KFA D–52425 Jülich, Germany

ABSTRACT

Methods to obtain tomograms of Hamiltonean flows, are presented. They are exemplified for measurements of soft x-rays on tokamaks of widely arbitrary poloidal cross–section. Boundary conditions can be met and cause less "ghosts" for most of the present day machines. The soft x-ray measurements are then used to improve the flux function Ψ as obtained from codes based on different input, e.g. magnetic measurements. Results are given for the JFT–2M tokamak.

1. Introduction

Plasma confinement machines, which are based on the concept of toroidal confinement, use magnetic configurations where the fieldlines are the stream lines of a Hamiltonean flow. The Hamiltonean of the system being the poloidal magnetic flux Ψ and action angle variables are given by the toroidal flux Φ and a poloidal angle $\vartheta = \iota\varphi$, where ι and φ are the rotational transform and the toroidal angle, they are analogous to ω and t for the anharmonic oscillator. In Figure 1 the case is demonstrated for the tokamak. Due to bremsstrahlung and line radiation the plasma emits photons whereby their intensity and energy depends on the plasma density and temperature. The particle mobility and the heat conduction are many orders of magnitude larger along field lines as compared to cross field transport and for that reason the density and temperature are expected to be constant on the flux surfaces $\Psi = \text{const}$.

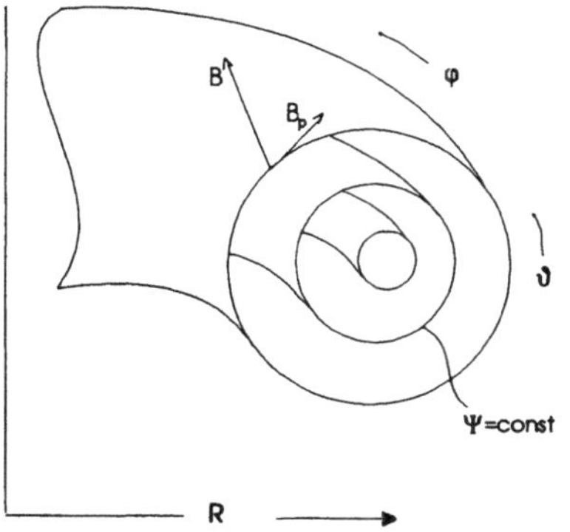

Figure 1: Nested flux surfaces on a circular tokamak

As the plasma density and the electron temperature is absolutely measured better by other diagnostics, however with low spatial resolution, the photon emission in the range of several (1..20) keV can be used to locate the flux surfaces to higher accuracy and to detect their degradation e.g. due to MHD activity, which produces magnetic islands and

regions of chaotic field lines and destroys the confinement. The typical frequencies of these oscillations are in the kHz range.
The number of diagnostic beams used for soft x–ray tomography on fusion machines (see e.g. Granetz and Smeulders, 1988) is usually very small – below 100 – this is because the number of ports is limited and has to be shared with other diagnostics. Furthermore, these beams belong to a very small number of cameras, i.e. viewing directions. Therefore it is advantageous to use a priori information provided either by theory or other diagnostics.

2. Experimental Setup

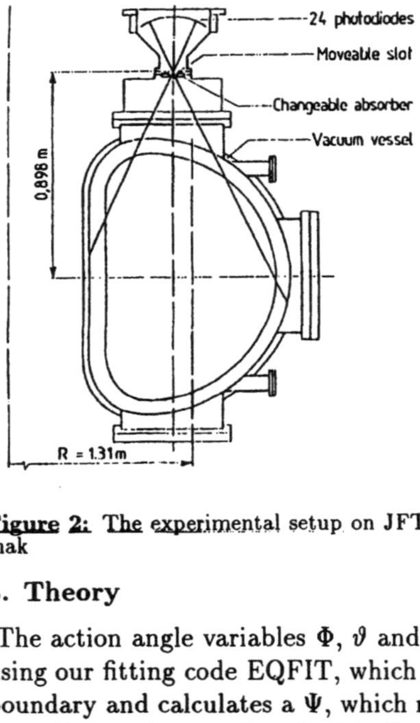

Figure 2: The experimental setup on JFT-2M tokamak

The experimental setup of the soft x–ray diagnostic on JFT–2M is shown in Figure 2. It consists of 1 pin hole camera with 24 beams viewing the whole poloidal cross–section from top. This camera is equipped with silicon detectors, slits of different size and with metal (Be and Ti) foil windows of different thickness. Slits and windows can be changed between subsequent discharges.
Camac data loggers are used to digitize and store the data acquired during the shot. The sampling frequency can be up to 10 kHz and it can be changed during the discharge, which lasts about 1 s. The storage depth per channel is up to 4 k.

3. Theory

The action angle variables Φ, ϑ and the Hamiltonean Ψ of the magnetic flow are found using our fitting code EQFIT, which takes the magnetic flux measured by 24 coils at the boundary and calculates a Ψ, which maintains the plasma inside in equilibrium; the code is similar to the one described in Lao et al. (1987). Given Ψ, Φ is the toroidal magnetic flux enclosed in the isocontours of Ψ and $\iota = \partial\Psi/\partial\Phi$.
For the luminosity we make an ansatz using set of functions, which are zero on the boundary. For convenience we take Bessel functions along the "radial" coordinate $\sqrt{\Phi}$ and a Fourier series in angular direction. We have

$$\begin{aligned} g(\Phi,\vartheta) &= \sum_{k=0}^{\infty}\sum_{m=-k}^{k} a_{km} J_m(\lambda_{km}\sqrt{\Phi}) \cdot e^{i\,m\vartheta}, \qquad (1) \\ a_{k,-m} &= a^*_{km}, \end{aligned}$$

whereby the $\lambda_{k,m}$ are the roots of the Bessel functions, $\vartheta = \iota\varphi$ and Ψ and Φ are normalized to give $\Phi = 1$ at the plasma boundary. Applying the Radon transform $\mathcal{R}$ we get in projection space p, β (similar to Nagayama, 1987)

$$f_{km}(p,\beta) = \mathcal{R}\left[J_m(\lambda_{km}\sqrt{\Phi})e^{i\,m\vartheta}\right], \qquad (2)$$

$$f(p,\beta) = \sum_{k,m} a_{km} f_{km}(p,\beta). \tag{3}$$

The $a_{k,m}$ were calculated from a measured set of line integrated data $f_i = f(p_i, \beta_i)$ using a linear least square fit and the procedures described in Granetz and Smeulders (1988). Postulating that the emission depends on Φ only we expect $a_{km} = 0$ for $m \neq 0$. If this is not the case, an incorrect flux function and incorrect action angle variables must have been used. The task is then to find the coordinate transform to new variables Φ' and ϑ' such that g becomes independent from ϑ':

$$g(\Phi') = \sum_k a'_{k0} J_0(\lambda_{k0}\sqrt{\Phi'}). \tag{4}$$

To find the canonical transform and in order to get its generating function S we should make the ansatz

$$\Phi' = \Phi + \sum b_m(\Phi) e^{i m \vartheta'} = \frac{\partial S(\Phi, \vartheta')}{\partial \vartheta'}, \tag{5}$$

$$\vartheta' = \vartheta + \sum c_m(\Phi') e^{i m \vartheta} = \frac{\partial S(\Phi', \vartheta)}{\partial \Phi'}. \tag{6}$$

Note that S contains the variables in mixed form, and it would be more convenient to use

$$\Phi' = \Phi + \sum b_m(\Phi) e^{i m \vartheta}. \tag{7}$$

Then we get simply

$$a'_{k0} = a_{k0}, \tag{8}$$

$$b_m = -\frac{\sum_k a_{km} J_m(\lambda_{km}\sqrt{\Phi})}{\sum_k a_{k0}\lambda_{k0} J_1(\lambda_{k0}\sqrt{\Phi})}. \tag{9}$$

The Φ' found in this way is not quite correct. However, the procedure may be repeated and will converge to the correct value in almost all the cases of interest.
To calculate a corrected Ψ function we follow standard procedures known from perturbation theory in classical mechanics (see, e.g., Lichtenberg and Liebermann, 1983):

$$\Psi' = \bar{\Psi}(\Phi') + \tilde{\Psi}(\Phi', \vartheta') + \iota \frac{\partial \tilde{S}(\Phi', \vartheta')}{\partial \vartheta'}, \tag{10}$$

whereby $\bar{\Psi}$ and $\tilde{\Psi}$ are the average and oscillating parts of the old Ψ function along the isocontours of the new Φ'. In order to make Ψ' independent of ϑ', the two last terms have to cancel each other, meaning that the new Ψ' is the average of the old Ψ on the new isocontours. This does not mean that erroneous Φ dependencies are corrected. The information gained from the soft x-ray measurements alone is not sufficient for this.
In a first approximation we may replace equation (6) by

$$\vartheta = \vartheta' - \sum c_m(\Phi') e^{i m \vartheta'} \tag{11}$$

which together with $\tilde{S}$ from equation (10) yield the c_m.

4. Data processing and results

The data retrieved from the data acquisition system either can be inverted for each subsequent snapshot or we can use the information acquired in time for filtering. The filtering technique we used is described in Nardone (1992); we decompose the matrix $A = f(p_i, \beta_i, t_j)$ into orthogonal matrices using singular value decomposition i.e. $A = U \cdot \Lambda \cdot V$ and cut the matrix λ containing the eigenvalues at the estimated noise level. With this method we can distinguish different contributions to U, which are related to different physical processes. During the flat top phase of the discharge the largest contribution is given by the average (in time) profile. Depending on discharge parameters, the next on the scale is the contribution from sawtoothing or MHD activity. Sawtoothing manifests by a periodic steepening and flattening of the profiles, and MHD activity shows up by rather localized structures rotating in most of the cases in paramagnetic direction; they are most likely magnetic islands.

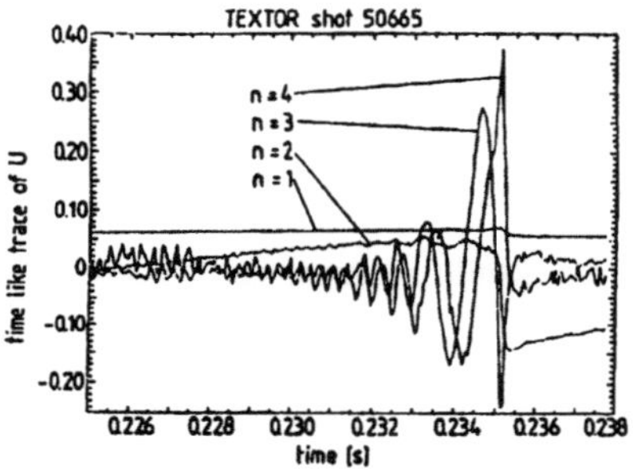

Figure 3: A typical tomogram for the x-ray emission as obtained from the measurements on JFT-2M e pin diodes. On the plasma boundary the calculated emission is zero.

An example for a tomogram as obtained from JFT-2M is given in Figure 3. The date for this profile have been taken with a rather thin foil of 30 μm explaining the rather broad shape. If we assume that the emission is a function of Ψ only, we must conclude that the results from the modeling of Ψ with EQFIT have to be corrected in such a way that the magnetic axis is shifted by 1..3 cm to the low field side. This is, because we find systematically values of $a_{11}/a_{00} = 0.1 \ldots 0.25$. This kind of result is not unexpected, because EQFIT gives better results at the boundary, whereas the soft x-rays do so near the center. We are not sure whether the nodge in the center is real or not. As we have seen that the shape of the tomogram near the magnetic axis is sensitive to noise, we suspect that it is an artifact.

Acknowledgement

G. Fuchs thanks JAERI for the kind invitation to Tokai.

REFERENCES

Granetz R. S. and Smeulders P. (1988), *Nucl. Fus.*, **28**, 457.

Lao L. L., John H. S., Stambaugh R. S., Kellman A. G. (1987), *J.Appl. Phys.*, **62 2702.**

Lichtenberg A. J. and Liebermann M. A. (1983), *Regular and Stochastic Motion*, Springer New York.

Nagayama Y. (1987), *J.Appl. Phys.*, **62 2702.**

Nardone C. (1992), *Plasma Physics and Controlled Nuclear Fusion* **34**, 1447.

Ray Reflection Tomography: Review and Comments

S. GOLDIN

Instutute of Geophysics RAS

Universitetsky pr. 3, Instutute of Geophysics, 630090 Novosibirsk, Russia

ABSTRACT

There are a lot of papers devoted to ray reflection tomography. In the most of them linear algebraic problems connected with the solution of big systems of linear equations are under considerations. Only in a few of them authors are discussing the connection between tomographic approach and classical travel-time inversion. this paper is an attempt to include the reflection ray tomography into the framework of general development of the kinematical interpretation during last two decades and to trace how peculiarities of the continual inverse problems influence tomographic solution. A special stress is laid on works of Russian geophysicists.

Introduction

To begin with our review in the best way we'll determine the subject itself and thus outline a range of questions under considerations. The most mysterious word in the title of the paper is "tomography". If to proceed from the etymology of the word (tomos is section) then to the seismic tomography there should be referred any works which propose interpretation in the frames of the two-dimensional model of a medium (such was suggestion, for example, by Barryman (1991) as opposite to the three-dimensional inversion) or, on the contrary, only three-dimensional interpretation of the essentially three-dimensional media with the aim of reconstructing crossections. In the first case the term is getting too broad including nearly 80% of the everything that is being done in the field of inverse kinematical problems. The second explanation seems to be more reasonable but if to accept this then it would embrace the three-dimensional migration, three-dimensional kinematical interpretation in the limits of the Dix approach (see below) which are not referred to the tomography traditionally. If to speak about proper tomographic works then there will be counted two-three of such (for example, the paper by Finn and Backus (1992)).

The appeal to the determination of the term of "tomography" in mathematics does not help either. According to Natterer (1986) tomography is defined as a reconstruction of a function from its integrals along straight lines or planes. The same source indicates of the possible widening of the term in a case when straight lines and planes are replaced by smooth arbitrary manifolds. The first definition I should consider too narrow even in respect to the practice available in the technical tomography. In respect to seismics it reduces the tomography to a linearized inverse kinematical problem with the homogeneous referent medium (without the subsequent iterative improvement). Most of researchers who attach themselves to the tomographic direction work with a more general model. At the same time the above mentioned second definition proves to be too wide, since any inverse kinematical problem is a reconstruction of the velocity function $v(x,y,z)$ on the integrals:

$$T(\gamma) = \int_{\gamma} v^{-1}(x,y,x)\mathrm{d}s, \tag{1}$$

which are given along a family of rays (Fermat extremals) identified by their edge points (i.e. by the points of shot–reception of a wave).

I don't reckon myself to rigorists and suppose that such a term as "seismic tomography" which came to geophysics on the wave of the fashion can be used relatively freely. However, there is a point in the definition of the subject of our discussion. For this purpose it is worth while of clearing up what common things are shared by the papers which contain the word "tomography" in their titles. This common element is a linearization of relation (1) in respect to parameters of a model. Just in this case there appears the common character of algorithmic problems with other fields of science and technique which make use of the term "tomography". The inversion itself usually is nonlinear: linearization is being made at each stage of the iterative process (for example, in the frame–work of optimization according to the Gauss–Newton method).

An important feature of the works concerning the tomographic direction is the depiction of a medium with the help of small cells as well as the application of algorithms of the linear–algebraic reconstruction (ART) which are designated for the solution of big systems of linear equations with "rare" matrices. The mentioned peculiarities are of importance but I don't think that they are essential. Recently a set of publications appeared in which the use is made of a smoother (though connected with grids) depiction of a medium. As for the techniques of the reconstruction of parameters (at each step of the iteration process) then, proceeding from the natural analogy, the methods connected with the inverse Radon's transformation or with its generalizations will be competing with ART (Beylkin, 1984 and Goldin, 1988, 1989).

In this paper we are going to pay the attention rather "geophysical" than computing aspects of the ray seismic tomography on reflected waves.

Retrospective Review

The history of the ray seismic tomography is not a topic of our paper. However, some comments should be done especially as in western publications any references used to be begun either from the work by Aki et al. (1977) or from that by Bois et al. (1972). Meanwhile, the method of linearization of the inverse kinematic problem has been theoretically developed in the works Lavrent'ev and Romanov (1966) as early as in the middle of the 60-es. Namely these authors have shown as follows. If the slowness function $v^{-1}(x,y,z)$ can be represented as a sum $v_0^{-1}(x,y,z) + w(x,y,z)$, where the function v_0^{-1} depicts the so–called referent medium and w is a small smooth addition:

$$|w(x,y,z)| < \varepsilon \tag{2}$$

then

$$\delta \equiv T - \int_{\gamma_0} v_0^{-1}(x,y,z)\mathrm{d}s = \int_{\gamma_0} w(x,y,z)\mathrm{d}s + \mathrm{O}\left(\|w\|\right), \tag{3}$$

where γ_0 is a ray (Fermat's extremal) in the referent medium, that connects the same points as a ray g does. The sense of the statement is the following. If to throw off a small value $O(\|w\|)$ that possesses a higher order of the smallness than w, then δT proves to be connected linearly with the addition w (since γ_0 does not depend on w). That is why the inverse problem of determining the function w from a set of values δT proves to be a linear one.

The first applied work in which the above mentioned linearized inverse problem was used and the medium was investigated by diving waves was that by Alekseev et al. (1969). The function under estimation was approximated by smooth functions (polynomials) and this made difference between the publication and classical tomographic works which used the cell approximation of the medium. The next one probably was Bois' et al. paper (1972) and this publication can be considered as the first one concerning cross well tomography.

In 1977 K. Aki published his work in which he developed a solution of the three-dimensional refraction kinematical problem on the basis of the cell approximation of $w(x, y, z)$. The referent medium was described by the function $v_0(z)$ (it was represented by a so-called normal velocity section). When using a cell approximation for the function $w(x, y, z)$ the linearized inverse kinematical problem is reduced to the solution of a linear equation system typical of the classical tomography

$$Aw = \delta T \tag{4}$$

where w is a vector witch components $w_i(x_i, y_i, z_i)$ (i is a number of the cell), δT is a vector with components δT_j (j a number of the ray) and A is a matrix,the component of with l_{ij} is equal to the segments length of the j-th ray in the limits of i-th cell.

Aki's paper was followed by a whole set of publications made by him and his collaborators with the explanations how to use the mentioned scheme for studying particular parts of the Earth's crust and Upper mantle. In all these publications system (4) was solved by the generalized inversion technique (or, that is the same, with the help of the singular decomposition of matrix A). At the same time Dziewonski et al. (1977) published the work in which they used the analogous scheme.

Beginning from 1979 a lot of works concerning the subject under discussion has been published both in our country and abroad. We should note here the works by Yanovskaya and her students in which they used successfully Backus-Gilbert technique, the works by S.R.Mishenkina and I.Yu.Sheludko, by P. Fibras and M.Novotny which differ both by techniques of the medium parametrization and by techniques of determining unknown parameters. It is natural that given a different (from sell's) parametrization of the medium, let us say, when using expression

$$w(x, y, z) = \sum w_j Q_j(x, y, z) \tag{5}$$

(where Q_j is a chosen system of basic functions), elements of the matrix A are determined in a different manner. In case (5)

$$a_{ij} = \int_{\gamma_i} Q_j(x, y, z) \mathrm{d}s \tag{6}$$

Dziewonski (1987) presentation where for the first time there was given a tomographic model for the Earth as a whole, can be rightly considered as the top of the development of this idea. In this work the Earth was described by the expression of the type

$$1/v_1 = 1/v_0(r) + w(r, \phi, \vartheta)$$

and the sought function $w(r, \phi, \vartheta)$ was described with the help of decomposition in a series of the spherical functions. Coefficients of a given expansion made up a vector of the sought parameters.

In a number of the cited works it was pointed, that a correction $w = w^{(1)}$ found as a result of solving equation (4) can be used for the subsequent elucidation of the velocity model of a medium if to take the function

$$1/v_1 = 1/v_0 + w^{(1)}$$

as a new initial approximation. The iteration process is originated in this manner

$$1/v_k = 1/v_{k-1}(r) + w^{(k)}$$

at each step of which the system of linear equations is being solved

$$A^{k-1} w^{(k)} = \delta T^{(k-1)}, \quad k = 1, 2, \ldots; \quad A^0 = A, \quad \delta T^0 = \delta T \tag{7}$$

with respect to the vector $w^{(k)}$.

The dependence of the matrix $A^{(k)}$ and of the right hand side on a number of the iteration is connected with that rays are traced for the medium found at the previous iteration. Because the system (7) is usually overdetermined, instead of it the system

$$\left(A^{(k-1)}\right)^T A^{(k-1)} w^{(k)} = \left(A^{(k-1)}\right)^T \delta T^{(k-1)}, \quad k = 1, 2, \ldots . \tag{7a}$$

is of use with accordance to least square method. Let us call iterative scheme (7) (or scheme (7a)) by the nonlinear tomography on transient waves or transmission tomography (Tr–tomography). The algorithm realizing the first iteration of the given scheme (or any one equivalent to it mathematically) will be called linear Tr–tomography.

The described iterative scheme (7a) is equivalent to the application of the Gauss-Newton technique for solving the following optimization task:

$$\sum \left|\hat{T}_i - \hat{T}_i(v^{-1})\right|^2 \to \min_{(v^{-1})} \tag{8}$$

where T_i is a functional that according to formula (1) puts a value of the travel time of a wave T_i for the fixed (i-th) pair of source–receiver in the accordance to the slowness field $1/v(x, y, z)$, $\hat{T}_i$ is an observed time. The first term in the right part of relationship (3) is noting but a linear part (Frechet derivative) of the functional $T_i(v^{-1})$ that will be designated as $L_i w$. The Gauss-Newton method is as follows. Making k-th iteration we approximate the left side of relationship (8) by the quadratic form replacing $\hat{T}_i - \hat{T}_i(v_k^{-1})$ by the linear expression $\delta T_i^{k-1} - L_i(w^k)$ where $\delta T_i^{k-1} = \hat{T}_i - T_i(v_{k-1}^{-1})$. Minimization of the obtained quadratic form we'll make according to the classical Newton method (keeping in mind that the slowness function w is a approximated by expression (5) and taking into account formula (6)). This will provide for $w^{(k)}$ equation (7). As any process of the Newton type this iteration process is not obliged to converge. A convergence is guaranteed with the use of rather good initial approximation. At the same time a speed of the Newton process convergence (if it takes place) is very high. The possibility of realization of the described iterative process was rather declared until the mid of the 80-es than realized factually, that doubtlessly was connected with a numerical difficulties of forward kinematical problems in the media which differ from the initial approximation.

It is worth while of noting that posing reverse kinematical problems as optimization ones (8) has appeared in the end of the 60-es. For the first time it might have been made for reflected waves (F.M.Goltsman, V.S.Kiseliov, 1969). It was exactly the Gauss-Newton algorithm that was applied. But it hardly can be said that the linearization of a functional $T(v^{(-1)})$ was used in the given presentation, since the calculation of times was made in the simplified model of average velocities with locally flat reflectors. A more advanced optimization algorithm for solving inverse kinematical problem for the reflected waves was developed by R.M.Bembel' (1974): a layered-homogeneous model of the medium with curvilinear reflecting and refracting interfaces was used by him. The optimization problem was solved by the techniques of nonlinear programming (without application of the linearization).

In the class of turning waves the optimization scheme was proposed a little later (Wesson, 1971).

Optimization algorithm for reconstructing velocity models of the medium on reflected and head waves are still in use,but in the 70-es some approach that can be conventionally called as that by Dix, proved to be the most popular. There are two principles in its basis: 1° The use is made of a layered model of the medium in which each layer is characterized by the interval velocity that depends in its turn on the horizontal coordinate (such models in the monograph by Goldin S.V.(1979) are called locally layered-homogeneous); 2° The inversion is made by means of functionals of the time field which coincide with first two coefficients of the Taylor's series of a quadratic of the common middle point travel-time function

$$t_X^2(l) = t_0^2(X) + l^2/v_e^2 \cdots \tag{9}$$

(X is a coordinate of middle point, l is a source-receiver distance). The inversion based on the system of data $t_0(X)$ and $v_e(X)$ is nonlinear, explicit and includes the use of the Krey-Gritsenko-Chernyak formula, the continuation of the front curvatures of a fictitious wave along the central ray and so on (Goldin, 1986). In the development of this approach decompositions of the middle point travel time curve or of its quadratic in the Teylor's set played an important role (P.Shakh, P.Hubral, A.Levin, S.Goldin, B.Ursin et al.). For a horizontally-layered medium this technique is reduced to the classic formulas by Dix. In the further description the layered medium models with curvilinear reflecting boundaries and homogeneous or locally-homogeneous layers will be designated by the symbol $\mathcal{M}_{\mathcal{D}}$.

The important problem of the Dix's approach was that estimations of the parameters t_0 and v_e obtained as a result of the standard processing are biased because of the hyperbolic approximation of travel time curves. This problem was successfully solved with the help of the iterative algorithm of the unbiased estimation (a so-called R-algorithm, Goldin, 1979). The developed scheme of kinematical interpretation in which the optimization algorithm was combined with the Dix's type inversion and R-algorithm for locally layered-homogeneous media was presented in the paper by Goldin et al. (1978).

It should be emphasized, that the approach by Dix is essentially non-linear and by no means connected with the method of perturbations. That is interesting that in theory of inverse kinematical problems nonlinear methods precede linearization. First of all it concerns vertically-inhomogeneous media: $v = v(z)$. In the class of refracted waves the theory put forward by the classic results by Gerglotz and Wiehert as early as in the beginning of the century has been accomplished in its main aspects in the end of the 60-es by Moscow mathematicians M.L.Gerver and V.M.Markushevich. In the class of reflected waves main results were get for the horizontal reflector by 1975 (Goldin S.V., Suvorov V.D., 1975, Geyko V.S.,1974) and a little bit later for the arbitrary curvilinear reflector (S.Goldin, 1977). In these investigations of extremely great importance was the analytical continuation of a travel time function into the complex plane. In the further development "exact" techniques were extended on the more complicated classes of the function $v(x,z)$ (Anikonov I., G.Beylkin, Goldin S.V. and others).

Finishing the review of the "Pre-tomographic" development of the theory and practice of inverse kinematical problems on reflected waves we'll note three schemes of the solution of general optimization problem (8) which were worked out on the 70-es and 80-es border and which were realized in the sub-sequent works directly connected with tomography on reflected waves. The first two were proposed by me in the book (Goldin, 1979) and were called there as schemes of incomplete optimization. In these schemes a vector of all the medium parameters is divided into two subvectors Θ_v and Θ_g, the first of the former describing a velocity distribution and the latter a location and geometry of reflectors. The first scheme is based on the following well-known statement: if the velocity distribution and location of reflectors's which are higher than k-th reflector, are known, then there is a method of constructing the k-th reflector (kinematical migration) from data $t_0^{(k)}(x)$ (travel time of the reflected wave on the source). Making this procedure with $k = 1, 2, \ldots$, we'll define $\Theta_g = F_1(\Theta_v)$. Now let $G(\Theta) \equiv G(\Theta_v, \Theta_g)$ designate the left part of relation (8) in the parametrization of the medium by the vector Θ. Then V-algorithm is the solution of the following optimization problem

$$G_1(\Theta_v) \Rightarrow \min_{(\Theta_v)}, \quad \text{where } G_1(\Theta_v) \equiv G(\Theta_v, F_1(\Theta_v)). \tag{10}$$

And thus, V-algorithm is the combination of non-linear optimization (on vector Ω_v) and kinematical migration.

If, on opposite, to make the vector Θ_g fixed, then the velocity distribution in layers can be defined using non-linear tomography according to iteration process (7). In this manner we'll define the mapping $\Theta_v = F_z(\Theta_g)$. H-algorithm is the solution of the next optimization problem:

$$G_2(\Theta_g) \Rightarrow \min_{(\Theta_g)}, \quad \text{where } G_2(\Theta_g) \equiv G(\Theta_g, F_2(\Theta_g)). \tag{11}$$

The operator $F_2(\Theta_g)$ although being constructed on reflected wave rays coincides in its essence with the non-linear Tr-tomography operator. This is why H-algorithm is made of the combination of nonlinear optimization (with respect to the vector Θ_g) with nonlinear Tr- tomography.

It should be noted here that operators $F_1(\Theta_v)$ and $F_2(\Theta_g)$ can follow each other (in any order). For this reason if we know an initial approximation Θ_v^0 then it is possible to define the following iterative subsequence:

$$\Theta_g^{(k)} = F_1(\Theta_v^{(k)}), \quad \Theta_v^{(k+1)} = F_2(\Theta_g^{(k)}), \quad k = 0, 1, \ldots \tag{12}$$

The similar scheme representing a combination of kinematical migration with the linear or non-linear Tr-tomography for the first time was published by Bjorn Ursin (1981) for the model of the layered medium with curvilinear interfaces. However, velocity distribution in this work was described (in the same manner as in the Dix's approach) by interval velocities and for this reason it can't be completely assigned to the tomographic direction.

The numerical realization of the linear approximation of the operator $F_2(\Theta_g)$ (i.e. of its first iteration $\hat{F}_2(\Theta_g)$) was made for the first time in 1980 (although the presentation that I know, appeared only three years later: Kjartsson, 1983). A reflecting boundary was proposed as horizontal and known, a referent medium as homogeneous and the function $w(x,z)$ was parametrized on the basis of the cell approximation typical of the tomography. We'll agree to designate this model by symbol $\mathcal{M}$. Let H be a depth to the reflector, $\zeta = H - z$ is taken at that. Let us introduce coordinates of a source s and a receiver r, designating by $X = (s+r)/2$ the coordinate of the middle point and by $h = (r-s)/2)$ the half-distance between the source and the receiver. Then simple geometrical considerations will yield:

$$\delta T(X,h) = \frac{\sqrt{H^2+h^2}}{H}\int_0^H [w(\zeta, X - hz/H) + w(\zeta, X + hz/H)]\,\mathrm{d}\zeta.$$

Further Kjartsson used the Fourier transformation with respect to the variable X :

$$\delta\tilde{T}(k_x,h) = 2\frac{\sqrt{H^2+h^2}}{H}\int_0^H \cos(k_x h\zeta/H)\tilde{w}(k_x,h)\mathrm{d}\zeta. \tag{13}$$

By a discretization of the last relationship the problem is reduced to the solution of linear equations system. It's easy to see, that the described construction "works" only in the linear approximation of the operator F_2 (i.a., at the first iteration). The next iteration would be much more complicated.

In 1981 Gertrude Neumann published the paper in which the same model of the medium as that used by Kjartsson was considered, but a system of equations was made for δT (but not for $\delta\tilde{T}$). We are not going into details as for the techniques used for solving linear equations systems: the excellent analysis of them is given in the presentation by Phillips and Fehler (1991) where our reader can be referred to.

The work by S.N.Medvedev (1979) that appeared even earlier than the two papers which we just mentioned, was devoted to reconstruction of a finite inhomogenuity $w(x,z)$ in a more complicated model μ_1 :

$$1/v = 1/v_0(z) + w(x,z), \quad 0 \le z \le H. \tag{14}$$

In the constructive aspect the problem was reduced to the task of moments (that is rather far from the tomographic tradition) but for a set of particular cases of functions $v_0(z)$ Medvedev had obtained Radon's type transformations.

The first works which can be doubtlessly assigned to the ray tomography on reflected waves (Rf-tomography) have probably appeared not earlier of 1984. So, it happened much later as compared which other types of the ray tomography. The reason might have been in some psychological barrier in the face of the linearization concerning reflector's geometry.

For the first time the method of perturbations as for interfaces geometry was realized in he limits of the Dix's approach by E.A.Blyas (1982). Taking $z = H_k\varepsilon \cdot d_k(x)$ (where ε is a small parameter) for each of n interfaces, he constructed series (on powers of ε) for coefficients of series (9), representing a middle point travel time function. In 1984 A. F. Glebov made his student research in which he proposed formulae for the derivatives $\partial t/\partial\Theta_v$ and $\partial t/\partial\Theta_g$ for the layered homogeneous media with curvilinear interfaces but his results remained practically unknown. Later Glebov (1988) deduced formulae not only for the first derivatives but for the second ones too for a very wide class of media including anisotropic. This made possible to use the classical Newton method in the KING's system

(Glebov, 1988). As for works of a pure tomographic direction it started to develop in two ways.

In the works of the first way there was proposed one system of equations for determining both velocity and structural parameters, in other words, perturbations concern both slowness function values and these of depths to each of reflectors.

The first work made in this direction was probably that by D.A.Spindler and P.I.Love (1984). However, the most comprehending description was presented in the paper by Bishop et al. (1985) (with the participation of the same authors) which is usually referred to in all works on ray reflection–tomography. As an initial model as a rule use is made of $\mathcal{M}_0$ for which the linearization with reference to depths is made easily.

The linearization with respect to $d(x)$ in the cited above paper was done on the ground of very simple geometrical considerations. Below we are giving a very strict but simple derivation for an arbitrary referent medium. We note that perturbations $w(x,z)$ and $d(x)$ disturb the operator K of a forward problem which puts each pair of functions $1/v(x,z)$ and $z = H(x)$ in accordance with the time field $t(x,h)$. This is why we are speaking here about a functional derivative (that is a Frechet's derivative) of the operator K calculated in the "point" $w \equiv 0, d \equiv 0$. Let us put $t = T_1 + T_2$, where $T_1 = T(S;x,H(x))$, $T_2 = T(R;x,H(x))$, $T(P;x,z)$ is travel–time along the ray from the point P to the point (x,z), S – the source, R - the receiver. In accordance with the Fermat's principle, abscissa of the reflection point x satisfies the equation

$$\frac{\partial t}{\partial x} \equiv T_{1x} + T_{2x} + (T_{1z} + T_{2z})\frac{\partial H}{\partial x} = 0. \tag{15}$$

It is clear that the perturbation of the travel–time (for a given pair of S and R) depends on perturbation of the reflector in a small vicinity of the reflection point x_0, which corresponds to the referent model, where the function $H(x)$ can be represented (with any accuracy) by it's Taylor's series:

$$H(x) = \theta_0 + \theta_1(x - x_0) + \theta_2(x - x_0)^2 + \ldots \equiv H(x;\theta_0,\theta_1,\ldots)$$

(it is presumed that $H(x)$ is sufficiently smooth). Therefore the reflector's disturbance can be expressed by the disturbance of a set of coefficients θ_i. It means that the functional derivative can be expressed with help of "usual" derivatives: $\delta_{d(x)}t = \sum(\partial t/\partial\theta_i)\cdot\delta\theta_i$, where

$$\frac{\partial t}{\partial\theta_i} = (T_{1z} + T_{2z})\frac{\partial H}{\partial\theta_i} + \frac{\partial t}{\partial x}\frac{\partial x}{\partial\theta_i} = (T_{1z} + T_{2z})\frac{\partial H}{\partial\theta_i}$$

(we have applied equation(15)). It is easily to notice that for the referent ray ($x = x_0$) all derivatives $\frac{\partial t}{\partial\theta_i}$ but $\frac{\partial t}{\partial\theta_0} = 1$ equal to 0. And since $d\theta_0 = d(x_0)$ then

$$\delta_{d(x)}t(S,R) = (T_{1z} + T_{2z})d(x_0) \tag{16}$$

where the dependence $x_0 = x_0(S,R)$ is known (for the referent medium).

For the simplest model $\mathcal{M}_0$ we have $T_{1z} = T_{2z} = H/\sqrt{H^2 + h^2}$ and $x_0 = X$, therefore

$$\delta_{d(x)}t(X,h) = 2\frac{\sqrt{H^2 + h^2}}{H}dX \tag{17}$$

This coincides with the formula from the paper by Bishop et al. (1985).

Frechet's derivative $\delta_{w(x,t)}(t)$ is determined accordingly to general equation (2) which for the referent model $\mathcal{M}_0$ is reduced to (13).

Although simultaneous linearization with respect to perturbations of $w(x,t)$ and $d(x)$ is correct, application of Gauss–Newton procedure needs additional justification because the role of nonlinear terms may be essential.

In the second direction the difficulty connected with a linearization with respect to reflectors's geometry was overcome be using scheme (12), in other words by the iterative subsequence consisting of the migration (that is not obligatory kinematical) and Tr-tomography (Cristoff Stork and Robert Clayton, 1985, 1986; R. Philip Bording et al., 1987). The work by S.V. Krylov and I.F. Sheludko (1987) stands by itself. First, it is oriented for the use of rays with big angles of reflection (that corresponds to specificity of Deep Seismic Soundings) and, second, kinematical migration is made in the one–layered μ_d-model, in which the interval (efficient) velocity was determined. For this aim they used the connection of first Taylor's coefficients of the time field $t(X,h)$ (in the vicinity of points belonging to the line $h = \text{const} \neq 0$) with parameters of a homogeneous layer.

In the works by Loinger (1983), Toldi (1989) and W.Harlan (1989) the attempt was made to carry out the tomography not from travel times of reflected waves but (as in Dix's approach) from results of the standard processing, in other words from values of t_0 and v_e. In this case the double linearization is conducted: times are linearized both in the relation of the medium perturbations and in the relation of the effective parameters' perturbations (according to the hyperbolic approximation of the CMP travel time curves:

$$t = \sqrt{t_0^2 + h^2/v_e^2}. \tag{18}$$

By this way a linear connection of the medium's perturbations with δt_0 and δv_e is established. After this procedure there can be solved systems of linear equation typical of tomography. As it was pointed at pages 247-248 (Goldin, 1986) a realization of this scheme with accordance to the method of the maximum likelihood is equivalent to the solution of initial problem (8) in the case, when $\hat{t}$ is calculated according to hyperbolic approximation. For this reason the linearization in the mention ed works in respect to the effective parameters was not necessary in fact since the tomography on parameters t_0 and v_e can be carried out in the same manner as it is done with the usual tomography defining "a measured" time according to hyperbolic approximation (18).

Algorithms of linearized inverse kinematical problems on the ground of effective parameters for horizontally inhomogeneous layered model $w = w_k(x)$ were developed by Blyas (1979). On opposite to cited work by Loinger Blyas have used direct linearization of connection between effective parameters and parameters of media.

A majority of tomographic presentations was made for the two dimensional case. The already mentioned work by C. Finn and M. Backus (1992) belongs to rare exceptions. In this connection it is worth of noting that all the above said refers to the same degree both to two and three–dimensional cases. In the latter case it is enough to say that x, X, s, r are two–dimensional vectors (in particular $x = (x_1, x_2)$ and so on), and $h = |s - r|/2$. (But in formulae (13) $h = (s - r)/2$ should be put assuming h as a vector).

If put the formal aspect aside, then there is a very strong difference between 2D and 3D tomographies. First, in 3D case it is possible to put the question about nontrivial sufficient system of observations. Second, systems of linear equations in 3D tomography have better conditional number then in 2D case. Third, there is possibility to study local regions of nonhomogeneous media, when we use overdetermined 3D system of observations (we are going to touch this problem further).

A further analysis of the publications which appeared after 1985 will be made in the thematic rather than in the chronological order in correspondence to the problems which were considered in them.

Finishing the retrospective review we can't but note a great number of works on the computerized tomography which were summed up in the well-known monograph by F. Natterer (1986) and influenced very much the development of the ray Rf-tomography.

Theoretical Problems of Reflection Tomography

Uniqueness of solution. In the non-linear Rf-tomography the problem of uniqueness (and in general of well-posedness) does not differ at all from the corresponding problem of a general theory of the reflection seismic kinematical problems. However in works of the tomographic direction the tradition of studying the uniqueness of the solution of a linearized in- verse problem has been developed. In other words, the subject under investigation is the operator's equation

$$\mathbf{K}'\delta\mathbf{m} = \delta t \tag{19}$$

where

$$\delta\mathbf{m} = \begin{pmatrix} w(x,z) \\ d(x) \end{pmatrix}$$

is a perturbation of the medium, $\mathbf{K}'$ is a Frechet's derivative of the operator $\mathbf{K}$ of forward problem. In this connection a question arises how to relate the uniqueness of nonlinear and linearized problems. Due to the fact that in practice both observations t and estimated functions $d(x)$ and $w(x,z)$ are elements of the finite-dimensional spaces we can apply the theorem about a implicit function according to which the operator $\mathbf{K}(\mathbf{m}) = t$ is reversed, if for all $\mathbf{m}$ from some set $\mathcal{M}$ the operator $\mathbf{K}'$ is not degenerated. (It is natural that the operator $\mathbf{K}^{-1}$ is considered on the set $\mathcal{J}$ which is an image of the set $\mathcal{M}$ with the mapping $\mathbf{K}$). In other words, uniqueness of the solution of a linearized problem leads to uniqueness of the solution of a non-linear one. The inverse statement is generally speaking wrong.

First, we shall bring a number of well-known results for the non-linear problem, then for the linearized one and after it we'll see how these problems are related.

T1. (Goldin S.V. and Suvorov V.D., 1975). Let $v = v(z)$ is unknown measured function and a horizontal reflector is located at unknown depth H. Then the function $\sigma(u) = \text{mes}\,\{z :\ v(z) \leq u\}$ and in particular $H = \text{var}\ \sigma(u)$ are uniquely determined. The problem is a very ill-posed one.

T2. (Goldin S.V., 1975). Let $v = v(z)$ at $0 \leq z \leq h(x)$. Both functions $v(z)$ and $h(x)$ are unknown. A function $h(x)$ as well as $v(z)$ under $z \geq H = \min h(x)$ assumed to be smooth. If the function $\sigma(u)$ up to the value H is known (it is sufficient that a point $\arg\min h(x)$ is the point of local minimum), then $h(x)$ and $v(z)$ (the latter on the interval $\min h(x) \leq z \leq \max h(x)$) is uniquely determined. The estimation of $v(z)$ and $h'(x)$ is well-posed in the Tikhonov's sense under variation of the time field in the C^2.

A unique solution is available also in the problem of determining parameters of the layered–homogeneous medium with curvilinear reflectors (as well as for a more general $\mathcal{M}_0$-model), in the problem for the horizontally–inhomogeneous layers $v = v(x)$ with a horizontal reflector. Also, it exists for the media models, which can be obtained (constructed) as a result of the turn by some angle ϕ of the models from T1 and T2. In all cases x can be both a scalar (in a two–dimensional problem) and a two-dimensional vector (in a three–dimensional problem).

While studying linearized operator's equation (19) of importance is the autonomous analysis of the equations

$$\mathbf{K}_1'w = \delta t, \tag{20}$$

$$\mathbf{K}_2'd = \delta t, \tag{21}$$

The study of the former empties the problem of uniqueness in the scheme "$Tr - tomography + migration$". According to formula (16), the latter can have the only solution (if any, and it should be said since a complete set of data $\delta t(s,r)$ is abundant). It means that the study of the first equation in the complete Rf-tomography empties the problem as well if there s a possibility to show that the intersection of sub-spaces

$$\mathcal{T}_w = \{\delta t = \mathbf{K}_1'w \mid w \in W\}, \quad \mathcal{T}_d = \{\delta t = \mathbf{K}_2'd \mid d(x) \in D\},$$

(where W and D are a priori given linear spaces of functions) contains zero only:

$$\mathcal{T}_w \bigcap \mathcal{T}_d = \{0\}. \tag{22}$$

The investigation of the operator's equation (20) as well as of general equation (19) was made for simple referent models only.

The review of results for equation (20) we'll start from the simplest ones.

T3. (Sven Ivansson (1986)). Let the referent medium be an homogeneous layers with a dip planar reflector $z = H_0 + x \tan\phi$ and let $w(x,z)$ have a finite support. Then the equation (20) has an unique solution.

To prove the theorem the problem is reduced to the classic "Radon" tomography under the location of receivers on the line $z = 0$ and sources on the line of imaginary sources $z = H_0 + x\tan 2\phi$ with the reconstructed function $w(x,z) = w(x,z) + w\,(x\cos^2\phi - 2H_0 \sin 2\phi + z\tan\phi(1+\cos^2\phi),\ (2H_0\cos\phi + x\sin\phi - z\cos\phi)\cos\phi)$ (the second term is a mirror image of the first one in respect to the line $z = H_0 + \tan\phi$). Strictly speaking, the problem is reduced to the "Radon tomography" only in the case, when dimension $n = 2$. With $n = 3$ we deal with the so-called "ray transformation", but the uniqueness takes place in this case too. (Natterer, 1986).

In the situation under consideration there is no necessity to measure $\delta t(s,r)$ with all s and r. Let $s \in (a,b)$ and $a'(s) \le r \le b'(s)$ with each s. From theorem 3.3 (Natterer, 1986) it can be concluded, that a solution's uniqueness (in the illuminated domain) is provided if $w(x,z) \in C^\alpha$ and has a compact support Ω, but there are rays from $s = a$ in $r = c(a)$ as well from $s = b$ in $r = c(b)$ have a common point neither with Ω nor which its mirror image. A sufficient system of data in the three-dimensional case will be defined considering a reflector as horizontal. Each observation is given by the vector $s = (s_1, s_2)$, the azimuth α and the distance l. It is assumed that α belongs to some fan of direction $\mathcal{A}$, and l to some interval of distances $\mathcal{L}$. A solution (in the illuminated domain) is unambiguous, if sources s belong to a certain smooth curve γ and if for each $a \in \mathcal{A}$ and $l \in \mathcal{L}$ there will be found such a point $s \in \gamma$, that the ray corresponding to observation (s,α,l) does not have common points with Ω.

It is reasonable to differ the problem of the reconstruction of the local inhomogeneity (when it is known a priori that beyond some Ω $w = 0$) from the general problem of the recreation $w(x,z)$ in the illuminated limited domain. The latter one in a two-dimensional case does not have an unambiguous solution. In the three-dimensional case can be received only with a abundant system of data available. (An observational system is a complete one with reference to Ω if it contains all the possible rays passing through Ω). In this case data on the rays lying in one plane can be summed up and after this procedure the problem will be reduced to the Radon transformation i.o. to stacking on planes. In this case the unique solution takes place.

T4. (Bube, 1989). Let the referent medium be $\mathcal{M}_0$. . A set of solutions of the operator's equation $\mathbf{K}_1'w(x,z) = 0$ (- the kernel of the operator $\mathbf{K}_1'$) is made up of the functions like

$$w(x,\zeta) = \sum_{k=0}^{N} \alpha_k(\zeta)x^k, \tag{23}$$

where $\zeta = H - z$ and for all the smooth $j \in (0,k)$

$$\int_0^H \alpha_k(\zeta)\zeta^j d\zeta = 0 \tag{24}$$

(for proving it is sufficient to substitute into (13)). A class of the functions satisfying (23) and (24) is designated by $\mathcal{W}_0$.

A comment. It's easy to see, that $\mathcal{W}_0$ does not contain the functions which would meet condition (2) i.e. on the infinite interval $-\infty \leq x \leq \infty$ a solution proves to be the only one.

T5. (Medvedev, 1979). Let the referent medium be a vertically inhomogeneous layer with a horizontal reflector and $w(x,z)$ has a finite support. Then the problem of reconstructing $w(x,z)$ has the unique solution.

Theorems 3-5 don't empty the problem of the solution uniqueness. There is one more thing to be done: to prove relationship (22).

T6.(Bube et al. 1989). Let the referent medium belongs to the class $\mathcal{M}_0$ and let $\delta t(x,h) \equiv \mathbf{K}_1'\delta\mathbf{m} \equiv 0$ for all $h \in (h_1,h_2)$ and $-\infty \leq h_1,h_2 \leq \infty$. Then $d(x) \equiv 0$ and $w(x,z)$ belongs to the class $\mathcal{W}_0$.

Authors suppose that functions d and w belong to the class of the slow growth functions but due to the assumed smallness of both functions this condition looks extra. Their proof is quite elegant. Making the Fourier transformation of both parts of the equation $\mathbf{K}_1'w + \mathbf{K}_2'd = 0$ and using formulae (13) and (17) after simple procedures we'll get the equation in which the left side is an integer function from h and the right one does not depend on h (i.e. both parts are equal to zero).

Since functions w with a finite support don't belong to the class $\mathcal{W}_0$, then operator's equation (19) has the unique solution in the situation when w is finite and the referent medium is $\mathcal{M}_0$.

T7. Relationship (22) fits any referent medium if w is finite and if a system of observations is complete enough.

PROOF. It is possible to proceed from coordinates (X,h) to coordinates (x_0,α) where x_0 is an abscissa of the reflection point and α is a reflection angle (it is sufficient to construct a family of referent rays). If $w = 0$ (but $d \neq 0$) then, according to formula (16), a residual travel time function of the fixed reflection point ($x_0 = const$) will be defined in this way:

$$\delta t_1(\alpha) = v_0^{-1}[\cos(\alpha+\phi) + \cos(\alpha-\phi)]d(x_0) \equiv \frac{2}{v_0}\cos\alpha\cos\phi\ d(x_0),$$

where ϕ is a reflector's slope angle in the reflection point and v_0 is a velocity in the same point (all the mentioned is for a referent medium). Since $|\alpha| < \pi/2$ then $\delta t_1(\alpha)$ will turn into zero at neither of interval values a by $d(x_0) \neq 0$. At the same time on condition that

$d = 0$ and $w \neq 0$ the travel time function of a fixed reflection point due to the finiteness w must have the intervals on which it is equal to zero identically. Thus, there is no such a travel time function $\delta t_1(\alpha)$ that under $d(x) \neq 0$ would coincide with a relative travel time curve for the situation $d = 0$ and $w \neq 0$. And this means that relationship (22) is true. In turns it evidences that $d(x)$ is defined unambiguously; as for the uniqueness of the definition w it remains unestablished yet.

From all that was said above on the connection between a uniqueness and non–uniqueness of the solution of non–linear and linearized problems it follows, that a class of non–uniqueness in the linearized case can turn to be much wider. It can be illustrated by the study of vertically–inhomogeneous media. According to Theorem 1, the non–uniqueness class is a set of functions $v(z)$ possessing by the same system of moments

$$\int_0^H v^k(z)\mathrm{d}z = m_k, \quad k = 1, 2, \ldots \tag{25}$$

meanwhile the non-uniqueness class in the relative linearized problem - as it follows from conditions (23)-(24) (where n should be taken equal to zero) - is a set of functions $1/v_0(z) + w(z)$ under condition

$$|w(z)| < \varepsilon, \quad \int w(z)\mathrm{d}z = 0 \tag{26}$$

(for the correct comparison of (25) and (26) it should be considered that $\left|v_0^{-1}(z) - v^{-1}(z)\right| < \varepsilon$). Is it possible to get the unique solution of a non–linear problem calculating the iteration subsequence of linearized problems, the latters don't have an ambiguous solution? Generally speaking, the answer is negative, but the non-uniqueness class can change in the iterative process. Let, for example, the initial referent medium have a horizontal reflecting boundary. In addition, there will be assumed a velocity dependence on z only. Even the first iteration will lead to the change of the referent medium: a reflector will be getting non–horizontal. In this case condition (26) can't be applied to the interval of the reflector's depths changes. A hypothesis can be proposed here: in the situation under question the iterative subsequence of linearized problems (the Newton–Gauss method) produces a solution that below a minimal depth coincides with the unique solution of a non–linear problem (both with the reference to $d(x)$ and to $w(x, z)$) and above it meets condition (26).

In conclusion we'll note the result by E. Blyas (1987) according to which in the horizontal-layered (with homogeneous layers) referent medium functions $d_k(x)$ and $w_k(x)$ can be reconstructed unambiguously. (Unfortunately, this result has not been formulated exactly). It can be concluded from it that all the non–uniqueness is mostly in the vertical change of slowness (and in those depths intervals only which don't have reflection points).

Stability. We have very little knowledge as for the stability of solution in linear and non–linear Rf-tomography. The next factors will be marked below.

1°. In the conditions of theorem 1 a strong ill–posedness (of the analytical continuation type) takes place.

2°. In the conditions of Theorem 2 definition of both $h'(x)$ and $v(z)$ (when $z \geq \min h(x)$) is stable under variation of the time field in C^2.

3°. Let the medium between the observation surface $z = 0$ and the reflector $z = h(x)$ be homogeneous. The inverse kinematical problem for the described model of the medium is well posed in the Tikhonov sense if it is possible to indicate such constants V_0, V_1, H_0,

H_1, $M_1 < 1$ and M_2 that on some interval (a, b)

$$V_0 \leq v \leq V_1, \quad H_0 \leq h(x) \leq H_1, \quad |h'(x)| \leq M_1, \tag{27}$$

$$\text{vrai max}\,|h''(x)| \leq M_2, \tag{28}$$

at that the function $h(x)$ is considered as an element of the space C^1 with the metric

$$p = c_1 \max_{a \leq x \leq b} |h_1(x) - h_2(x)| + c_2 \max_{a \leq x \leq b} |h_1'(x) - h_2'(x)| \tag{29}$$

and for the time–field $t(s, r)$ another metric is used

$$\text{vrai} \max_{(s,r) \in \Theta} |t_1(s, r) - t_2(s, r)| \tag{30}$$

where Θ is a set of observations illuminated some interval $(a', b') \subset (a, b)$ (Goldin, 1986).

For the problem of a simultaneous estimation of the finite function $w(x, z)$ and $d(x)$ we can bring the following considerations: the stability of the estimate $d(x)$ is not better than that in just cited theorem, and the stability of the definition $w(x, z)$ is not better than that in corresponding kinematical problems for transient waves. Principal results in the latter field have been obtained by R.G.Mukhometov, I.N.Bernstein and M.L.Gerver, Yu.E.Anikonov and V.G Romanov (one can get knowledge of them in the monograph by V.G.Romanov – 1987). Here we shall limit ourselves only by the stability estimation for a linearized two-dimensional problem.

4°. Let $1/v_0 \in C^2(\overline{D})$, a curve $\Gamma:\ x = x(l),\ z = z(l),\ l \in (0, l)$ which bounds the area D is smooth and concave in the respect to the ray family $\gamma(l, \lambda)$ which connect each two points l and λ of the curve Γ; the mentioned ray family is regular and $w \in C^2(\overline{D})$ then

$$\|w\|_{L_2(\overline{D})} \leq (2\pi)^{-1/2} \left\| \frac{\partial}{\partial l} \delta t(l, \lambda) \right\|_{L_2[(0,L)\times(0,L)]}. \tag{31}$$

Convergence. Here we'll talk about the convergence of iteration algorithms of the non–linear Rf-tomography. In the most part of the works devoted to Rf-tomography the convergence problem is considered only with reference to that finite-dimensional problem to which an initial continual problem is reduced as a result of the finite-dimensional parametrization of a medium and operator $\mathbf{K}'$, as well of the approximation of the target function used in the Gauss– Newton approach:

$$\|\mathbf{t} - \mathbf{K}(\mathbf{m})\|^2 \approx ((\delta\mathbf{t} - \mathbf{K}'\delta\mathbf{m}), (\delta\mathbf{t} - \mathbf{K}'\delta\mathbf{m})). \tag{32}$$

For the same reason the applied methods of regularization are routine techniques for the finite-dimensional problems which are not connected with the initial continual problem. As it follows from the review by W.Phillips and M.Fehler (1991) stabilizers based on the norms of the first and second differences of estimated functions are used only in a small number of works.

There are two sub-problems in the general convergence problem. The second is a consideration of the initial continual task and the former is establishing a possibility to use the Gauss–Newton technique.

A reasonable approach to the solution of the first subproblem is in that the conditions of smoothness under which the uniqueness and stability of tackling a continual problem are provided, are used in the construction of a finite-dimensional algorithm on the basis of simple heuristic considerations.

In the theorems, dealing with uniqueness and stability of a solution, the smoothness conditions enter as: 1) assumptions of the type $w \in C^2$, $d \in C^1$ and so on, 2) limitations of the kind $|d'(x)| \leq A$, 3) estimations of stability like (31). Conditions of the type 1) determine a method of parametrization of both a referent medium and perturbations as well. According to ideas which were proposed in this section the description would be smooth enough. In particular, the cell description as a rule does not fit. In the book by Natterer (1986) there are given estimations of stability in the Sobolev's spaces, assuming jumps. The results of my monograph (S.V.Goldin, 1988) suggest that the reconstruction of such media cause the use of the quasi Radon type formulae (when a rather big number of rays is present) and this is possible only in the scheme "*Migration+Tr−Tomography*".

The limitations of type 2) put conditions to the use of the stabilizers (demphers) based on the norms of first or second derivatives (to say it more exactly, of their finite-differential approximation).

Let's write the condition of type 3) in the common form:

$$\|\mathbf{m} - \tilde{\mathbf{m}}\|_{\mathcal{M}} \leq c\,\|\tau - \tilde{\tau}\|_{\mathcal{T}}, \tag{33}$$

where $\|\|_{\mathcal{M}}$ is a norm in the model's space and $\|\|_{\mathcal{T}}$ is a norm in the data space.

The general rule is as follows. The norm in the data space determines a target function (or affects it) mean-while $\|\|_{\mathcal{M}}$ affect a metric in the estimated parameters space. This rule can be illustrated. Norm (30) means that it is necessary to exclude (to throw off) deviated observations from the aim functions. The right part of expression (31) means that derivatives of the time field must contribute to the target function structure. Let's note here that given a fixed velocity distribution a connection between the common shot point travel time curves and the reflector's geometry is generated by mutually one-to-one continuous correspondence of two tangential layerings. With this point in view it is very desirable that time field derivatives would participate very "actively" in the synthesis of the target function.

A more effective (though more difficult mathematically) way for solving the first sub-problem is the construction of the target functional providing a continual inverse problem solution, the finite-dimensional approximation of this functional and, then, the study of the convergence of a finite-dimensional solution to a continuous one. The only work that was done in the close direction was that by F.Delpraite and P.Laily (1991). They showed that the operator $\mathbf{K}'$ and therefore the scalar product $(\mathbf{K}'\mathbf{m}, \mathbf{K}'\mathbf{m})$ are not continuous in the norm $L_2(\mathcal{M})$, but are continuous in the norm that corresponds to the scalar product

$$(\mathbf{m}_1, \mathbf{m}_2)_{D_\alpha} = (1-\alpha)\int \mathbf{m}_1\mathbf{m}_2 + \alpha\int \frac{\partial \mathbf{m}_1}{\partial x_i}\frac{\partial \mathbf{m}_2}{\partial x_i}, \quad (\alpha > 0).$$

The equation $(\mathbf{m}, \mathbf{m})_{D_\alpha} = (\mathbf{m}, D_\alpha \mathbf{m})_{Euc.}$ takes place here. If at the ends of the interval where the function $d(x)$ is determined $d(x)/dx = 0$, then as applied to the function $d(x)D_\alpha = (1-\alpha)I + \alpha d^2/dx^2$. If the function $w(x,z)$ has a normal derivative equal to zero on the border of the domain where w is determined then $D_\alpha = (1-\alpha)I + \alpha\nabla^2$.

On the basis of the mentioned facts authors introduce a posterior covariance operator $\mathbf{C_m}$ after the formula $\mathbf{C}_{\mathbf{m}}^{-1} = D_\alpha^{-1}(\mathbf{K}'^T\mathbf{K}')$ as well its regularization $\tilde{\mathbf{C}}_{\mathbf{m}} = (\mathbf{K}'^T\mathbf{K}' + \mu D_\alpha)^{-1}D_\alpha$. (It should be noted here that when $\alpha = \mu = 0$ the finite-dimensional parametrization of this operator is $(A^TA)^{-1}$). The result obtained is that a spectrum of the operator $\tilde{\mathbf{C}}_{\mathbf{m}}$ of a discrete problem is reduced to a spectrum of the relative operator of the continuous problem. In the other work (1992) authors make grounds for the correctness of the optimization algorithm in which use is made of a norm of the second derivatives of estimated functions in L_2 as a stabilizer.

The second subproblem is as follows. The Gauss-Newton approach used in practice possess a smaller (as compared with the Newton method) convergence domain and while using interference data the convergence does not occur at all if the non-linearity which is thrown away in the linearization, is not small. This circumstance can prove to be of importance just for the complete scheme of Rf-tomography because in the linear part of the operator $\mathbf{K}$ the influence of a reflector's slope angle is not taken into account but it is essential from the point of view of physics. As far as I know, this aspect of the convergence problem was not discussed in the literature at all.

Informativity. In a rather big number of works the topic of investigating is what structural peculiarities of velocity inhomogeneities are well mapped in the results of tomographic experiments and what are not (K.Bube, J.R.Resnick, 1984, K.Bube, D..Jovanovich et al, 1985, V.Farra, R.Madariaga, 1988, Cristof Stork, 1991 and others). As a rule, the use is made either of the singular values decomposition of the matrix A or eigenvalues of matrix $A^T A$. In the work by Delpraite and Laily that was cited already the operator $\mathbf{C}_{\mathbf{m}}$ and its finite-dimensional approximation were utilized. It should be stressed, that singular values (as well as eigenvalues) are concentrated close to maximum and zero values. Relative eigenvectors characterize mostly inhomogeneities with very small wave numbers. In particular, poorly-defined features are: compensated slope of a reflector in the combination with linear change of the slowness, linear change of the slowness in each layer with the zero average change between reflectors, vertical change of the slowness with the zero middle and so on. In all events a slowness variation along the vertical is reconstructed worse than along the horizontal. The opposite picture can be observed in the inter-well tomography. This difference follows directly form the ray geometry that is peculiar of the reflection and cross bore hole tomography (a connection between ray geometry and information on geometry of discontinuities can be revealed in the monograph by Goldin, 1988)

Conclusion

In spite of the intensive study of the ray tomography on reflected waves many cardinal problems of the method remain unsolved yet. First of all it concerns the solution uniqueness for rather complicated referent media. Besides, we still need the theory which would study the convergence of discretized solution to a continual one. In the most part of works there is proposed a linear-algebraic approach without any connection with the initial inverse kinematical problem.

Though the term "Ray seismic tomography" has come in wide use in the seismic literature, all barriers between pure tomographic algorithms and traditional methods of solving inverse kinematical problems including non-linear optimization, the use of sufficient systems of data and so on will disappear in the nearest future.

The application of combined observational systems which would make it possible to carry out the ray tomography simultaneously reflection and cross-well can be also foreseen to happen in the nearest future. In this case only we'll succeed to create a rather complete pattern of velocity inhomogeneities both lateral and vertical.

REFERENCES

Aki K., Cristofferson A., Husebye E. S. (1976), Determination of the three-dimensional seismic structure of the litosphere, *J. Geophys. Res.* vol. 82, 277-296.

Alekseev A. S., Lavrent'ev M. M., Mukhometov R. G., Romanov V. G. (1969), Numerical method of solution of three-dimensional inverse kinematical problem of sesmics. *In: Mathematical Problem of Geophysics.* vol. 1, part 2, Computer center of Siberian Branch of Soviet Academy of Sciences, 137-159, (in Russian)

Alekseev A. S., Lavrent'ev M. M., Mukhometov R. G., Nersesov I. L. (1971), Numerical reconstruction of Earth's mantle structure. *In: Mathematical Problems of Geophysics.* vol. 2, part 2, Computer center of Siberian Branch of Soviet Academy of Sciences, 143-165, (in Russian)

Barryman J. G. (1991), Lecture notes on nonlinear inversion and tomography. *In: Borehole Seismic Tomography,* Lawrence Livermore national laboratory. University of California. UCRL-LR-105358-Rev. 1.

Bembel R. M. (1974), Statistical processing of reflection travel-times with taking into account apriory information. *In: Seismic Processing with computers. Proceedings of Western siberian Petroleum Institute.* vol. 36, 136-153. (in Russian)

Beylkin G. (1984), The inverse problem and applications of the generalized Radon Transform. *Communication on Pure and Applied Mathematics,* vol. 37, 579–599

Bishop T., Bube K., Cutler R., Langan R., Love P., Resnick J., Shuey R., Spindler D., Wyld H. (1985), Tomography determination of velocity and depth in laterally varying media. *Geophysics.* no. 50, 903-923.

Blyas E. A. (1982), Estimation of internal velocities in horizontally layered media at hyperbolic approximation of travel-times. *Soviet Geology and Geophysics.*

Blyas E. A. (1987), Reflection travel-times in layered media with nonlinear variation of intervel velocities. *Problems of Dynamic Theory of Seismic Waves Propagation.,* vol. 27, Nauka, Leningrad, 149-157, (in Russian)

Blyas E. A. (1989), Doctor thesys. *Application of the Theory of Disturbances to Problems of Kinematical Interpretation of Reflection Seismic Data.* no. 32, Leningrad State University, (in Russian)

Bois M., Laporte M., Lavergne M., Thomas G. (1972), Well-to-well seismic measurements. *Geophysics.* vol. 37, 471-480.

Bording R. Ph., Gersztenkorn A. et al. (1987), Application of seismic travel-time tomography. *Journal of Royal Astr. Society.* no. 90, 285-303.

Bube K. and Resnick J. (1984), Well-determined and poorly determined features in seismic reflection tomography. *Technical Program, 54th Annual SEG Meeting.* Atlanta, D.C., 717-719.

Bube K., Jovanovich D., Langan R., Resnick J., Shuey R., Splinder D. (1985), Well-determined and poorly determined features in seismic reflection tomography, part II. *Technical Program, 55th Annual SEG Meeting.* Washington, D.C., 608-610.

Bube K., Langan R., Resnick J. R. (1989), Unique determination of reflector depth in seismic reflection tomography. *Technical Program, 59th Annual SEG Meeting.* Dallas, D.C., 918-921.

Carrion P. (1989), Robust constrained tomography. *Bollettino di Geofisica teorica ed appllicata.* vol. XXXI, no. 123-124, 167-200.

Chiu S. K. L., Stewart R. (1987), Tomographic determination of three-dimensioned velocity structure using well-logs, vertical seismic profiles and surface seismic date. *Geophysics.* vol. 52, 1085-1098.

Delprat-Jannaud F. and Lailly P. (1991), Ill-posed and well-posed formulations of the reflection tomography problem. *Technical Program, 61st Annual SEG Meeting.* Houston, 977-980.

Delprat-Jannaud F. and Lailly P. (1991), What is the information contained in reflection tomography? part I. *Theory. technical Program, 60th Annual SEG Meeting.* San-Francisco, 1290-1293.

Dziewonski A., Hager B., O'Connell R. J. (1977), Large-scale heterogeneities in the lower mantle. *J. Geophysics. Res.*, vol. 82, 239.

Dziewonski A. M. and Woodhous J. H. (1987), Global images of the earth's interior. *Sciences,* vol. 236, 37-40.

Farra V. and Madariaga R. (1988), Non-linear reflection tomography. *Geoph. Journal,* no. 95, 135-147.

Finn C. J. and Backus M. M. (1992), Maximum-likelihood travel-time inversion for a 3-D velocity-depth model. *Technical Program, 62st Annual SEG Meeting.* New-Orlean, 897-900.

Geyko V. S. (1974), Features and inversion of reflection travel-times. *Report of Soviet Academy of Sciences,* vol. 218, no. 1, 88-90.

Glebov A .F. (1988), A class of optimizational algorithms for seismic travel-time inversion. *In: Mathematical Problems of Interpretation of Seismic Data.* Nauka, Novosibirsk, 62-82, (in Russian)

Goldin S. V. (1979), *Interpretation of Seismic Reflection Data.* Nedra, Moscow, 1-350, (in Russian)

Goldin S. V. and Suvorov V. D. (1975), Analytical continuation of reflection travel-times. Report of Soviet Academy of Sciences, vol. 222, no. 4, 825-828.

Goldin S. V. (1977), An inverse kinematical problem. *Reports of Soviet Academy of Sciences,* vol. 233, no. 1, 64-67.

Goldin S. V., Sudvarg D. I., Chernyak V. S. (1978), Reconstruction of the velocity model of media using multiply coverage of seismic waves. *Geology and Geophysics,* no. 6, 103-115, (in Russian)

Goldin S. V. (1986), Seismic travel-time inversion. *Investigation in Geophysics,* vol. 1, SEG, Tulsa, 1-363.

Goldin S. V. (1988), Transformation and reconstruction of discontinuities in tomographic-type problems. *Soviet Academy of Sciences,* Institute of Geology and Geophysics. Novosibirsk, Preprint, 1-100, (in Russian)

Goltsman F. M. and Kiseliov V. S. (1969), Optimal algorithm for join interpretation of reflection travel-time systems connected with several different horizons. *Applied Geophysics (Pricladnaya Geofisica),* vol. 55,Nedra, Moscow, 7-16, (in Russian)

Harlan W. S. (1989), Tomographic estimation of seismic velocities from reflection raypath. *Technical Program, 59th Annual SEG Meeting,* Dallas, 922-924.

Ivansson S. (1986), Some Remarks Concerning Seismic reflection Tomography and velocity Analysis. *Geophys. J. R. Astr. Soc.,* no. 87, 539-557.

Krylov S. V. and Shelud'ko I. F. (1987), Principles of a method of linearized interpretation of reflection time–fields. *Soviet Geology and Geophysics*, vol. 28, no. 7, 110-120.

Kjartsson E. (1983), Analysis of amplitude and travel–time variations versus offset and middle point. *Numerical methods in seismic investigations*. Nauka, Novosibirsk, 221-233, (in Russian)

Lavrent'ev M. M. and Romanov V. G. (1966), About three linearized inverse problems for hyperbolic equations. *Report of Soviet Academy of Sciences*, vol. 171, no. 6, 1179-1281, (in Russian)

Loinger E. (1983), A linear model for velocities anomalies. *Geophysical Prospecting*, vol. 31, 98-118.

Medvedev S. N. (1979), Several solutions of linearized inverse kinematical problem. *In: Far East Acoustic Collection. (Dalnevostochny Acusticheski Sbornic)*, vol. 4, Vladivostok, (in Russian)

Mishenkina Z. R., Shelud'ko I. F., Krylov S. V. (1983), Usage of linearized inverse kinematical problem for two–dimensional time–fields of reflected waves. *In: Numerical methods in seismic investigations*, Nauka, Novosibirsk, 140-151, (in Russian)

Natterer F. (1986), *Mathematics of computerized tomography*. Stuggart, John Wiley & Sons.

Neumann G. (1981), Determination seismics by inversion of travel–time residuals. *Geophysical Prospecting*, no. 29, 161-177.

Phillips W. S. and Fehler M. C. (1991), Travel–time tomography: a comparison of popular methods. *Geophysics*, no. 56, 1639–1649.

Romanov V. G. (1984), *Inverse Problems of Mathematical Physics*. Nauka, Moscow, 1–264. (in Russian)

Spindler D. A. and Love P. A. (1984), Depth determination with laterally varying velocity using seismic tomography. *Technical Program, 54th Annual SEG Meeting*, Atlanta.

Stock C. and Clayton R. W. (1985), Iterative Tomographic and migration reconstruction of seismic images. *Technical Program, 55th Annual SEG Meeting*, Washington, D.C.

Stock C. and Clayton R. W. (1986), Analysis of the resolution between ambiguous velocity and reflector position for travel–time tomography. *Technical Program, 55th Annual SEG Meeting*, Houston, 545–550.

Stock C. (1991), High–resolution SVD analysis of the coupled velocity determination and reflector imaging problem. *Technical Program, 61th Annual SEG Meeting*, Houston, 981–985.

Toldi (1989), Velocity analysis without picking. *Geophysics*, vol. 54, 191-199.

Ursin B. (1981), Inversion of reflection times in three dimensions. *Geophysics*, no. 46, 972-983.

Wesson R. L. (1971), Travel–time inversion for laterally inhomogeneous crustal velocity model. *Bull. Seism. Soc. Am.*, no. 729.

Yanovskaya T. B. (1983), Methods for solution of inverse kinematical problems of seismology in case of two– and three–dimensional inhomogeneous media. *Numerical methods in seismic investigations*. Nauka, Novosibirsk, 79-89, (in Russian)

Tomographic Methods for Information Processing in Seismology. Nonlinear Version

V. I. GOLOV

Institute of Applied Mathematics
Shevchenko st., 9, Khabarovsk, 680000, Russia

ABSTRACT

A nonlinear tomography problem is considered which reduces to a problem of finding a conformal metric in a Riemannian space. An algorithm of finding such a metric is suggested.

1. Introduction

Tomographic methods for information processing (TMIP) in a ray approach are extensively used in theoretical and applied investigations. They have found wide use in seismology also. The direct and inverse Radon transform (Radon, 1917) are the basic part of mathematical tools used in TMIP. Their applications are restricted by the errors which we have after reconstruction of an investigated object. However, there is another factor which influences on the possibility of the Radon transform application. The point is that the "classical" Radon transform can be used only for the processes in which the wave propagates along a straight line. In practice, such problems arise in the case when the investigated object is examined with hard radiation such as X-rays, electrons, positrons, visible part of spectrum, etc.

A remarkable application of TMIP is given by medicine computing tomograph. But many important physical processes in which the wave signal propagates not along a straight line stay outside of consideration. Such an example can be found in seismology. Let us consider a general scheme of TMIP application in a ray approach. In this approach seismic waves propagate along curved trajectories determined by a velocity distribution within an investigated object. Some parameters which characterize properties of a media are changed along a trajectory. If we denote a trajectory which connects points P and Q as γ, then in practice the values of the below functional are accessible for measurements

$$R(P,Q) = \int_{\gamma} f(c)\mathrm{d}l, \tag{1}$$

where $f(c)$ is a function of a parameter c, $\mathrm{d}l$ is an element of the trajectory length.

In TMIP a value of the functional $R(P,Q)$ is called the projection of $f(c)$ on the trajectory γ. Most commonly, that $f(c) = c$. The problem of evaluating $c(r)$ in this case will be called the P-problem, coming from 'perception'. If $f(c) \neq c$, we shall call this problem as the GP-problem.

If the trajectory is a straight line, then we have a classical Radon transform. Mathematical tools used in this case are remarkably presented in the monograph by Natterer (1988) with the wide bibliography.

In the case of γ belonging to a certain family of curves, the P-problem was solved by Cormack (1981), Cormack and Quinto (1980), Helgason (1980). In the case of γ being a great circle on a sphere, the P-problem has been solved by Funk (1913) earlier than Radon tackled such a problem for the case of γ being a straight line. The formulation and solution of the P-problem in seismology in the case when γ is a great circle on a sphere were represented by Xu Guo-ming (1988). All the above problems are linear problems since a kind of the trajectory is known beforehand. If the kind of the trajectory γ is unknown, then the problem becomes essentially nonlinear. The P-problem is formulated in seismic tomography in terms of time delays in such a way. On the Earth surface time arrivals are known. Assume that the traveltimes of body waves are known. It is required to find a 3-D velocity distribution. In this case equation (1) looks as follows

$$T = \int_{\gamma} v(r)^{-1} \mathrm{d}l, \tag{2}$$

where T is a traveltime, r is a radius-vector, and v is a velocity modulus. As noted above, equation (2) is nonlinear. In practice, this equation is usually linearized. Some initial model is taken, for example, the Jeffreys-Bullen's model. Then the direct problem for equation (2) is solved and the trajectories of signal propagation and traveltimes are determined:

$$T_0 = \int_{\gamma_0} v_0(r)^{-1} \mathrm{d}l. \tag{3}$$

Then time delay are found from

$$\delta T = T - T_0 \approx -\int_{\gamma_0} \frac{\delta v(r)}{v_0(r)^2} \mathrm{d}l. \tag{4}$$

Equality (4) is an approximate one because the unknown trajectory γ is assumed to slightly differ from the trajectory γ_0 built on the initial model. Now the P-problem represented by equation (4) becomes linear and it can be solved by the well-known methods, such as Backus-Gilbert's method (Backus and Gilbert, 1967), the method of disturbances in separated blocks (Aki and Lee, 1976), the method of decomposition in terms of spherical harmonics (Dziewonski et al., 1977), and others.

2. Nonlinear Method

We demonstrate our approach in the case when a signal's propagates in the plane R^2. Let us return to equation (2) to do some transformations. Let us introduce the Cartesian coordinates xOy in the plane R^2. Now we assume that $1/v = g^{1/2}$. Then equation (2) is rewritten in the following way:

$$L = \int_{\gamma} g^{1/2} \mathrm{d}l. \tag{5}$$

One can easily recognize in this equation the formula expressing the length of the curve γ in the 2-D Riemannian space with a conform metric

$$\mathrm{d}l^2 = g(x, y)(\mathrm{d}x^2 + \mathrm{d}y^2). \tag{6}$$

The curve γ in (5) is a geodesic for metric (6) because a seismic signal propagates along the trajectory, for which the Fermat principle holds.

So the P-problem for (2) is formulated now in the following way. Let D be a restricted simple connected domain on the plane with a boundary ∂D. For any couple of points (P, Q) lying on ∂D, the geodesic length in the Riemann metric (6) is known. It's required to find a function g(x,y).

It is clear that under these conditions this problem does not have a unique solution because if we slightly change the function value at a certain point then we shall have the same geodesic lengths. Let us try to restore the uniqueness. We shall assume that for any couple of points (P, Q), belonging to ∂D not only the geodesic length is known but also the direction of the tangent vectors at these points. In such a formulation the problem was considered by Mukhometov (1975). It was proved that under certain natural restrictions on a geodesic family and some restrictions on the boundary's smoothness this problem has a unique solution.

To describe our nonlinear algorithm it is more convenient to change to into tensor denotations.

Further we use the following notation: let $f(x, y)$ be a function of two variables $x = x(t)$, $y = y(t)$, then $\mathrm{d}f/\mathrm{d}t$ is the derivative of f with respect to t, $\partial f/\partial x$, $\partial f/\partial y$ are its partial derivatives.

So, let $x^1 = x$, $x^2 = y$, then the following formula set up the Riemann conform metric in the domain D

$$\mathrm{d}l^2 = g_{ij}\mathrm{d}x^1\mathrm{d}x^2, \quad i,j = 1,2,$$

where the metric tensor g_{ij} has the followings components in our case:

$$g_{11} = g_{22} = g(x^1, x^2), \quad g_{12} = g_{21} = 0.$$

Let γ be a geodesic line connecting P and Q, $\mathbf{p}$ be a unit tangent vector at P, and α be an angle between $\mathbf{p}$ and x^1. Let Q be a fixed point, whereas the point P is allowed to move on the plane x^1Ox^2. Then the length of the geodesic connecting P and Q and expressed by (3) is a function of the point P. Let us denote this function by $L_Q(x^1, x^2)$ where x^1, x^2 are the coordinates of the point P. Assume p_i to be a covariant component of the unit tangent vector $\mathbf{p}$ to the geodesic at the point P. Then the following proposition is valid.

Theorem 1. *It holds that*

$$\frac{\partial L_Q(x^1, x^2)}{\partial x^i} = -p_i, \quad i = 1,2.$$

Let us remember the relations between covariant and contravariant components of any vector $\mathbf{p}$:

$$p_i = g_{ij}p^j, \quad p^i = g^{ij}p_j.$$

They look very simple in our case:

$$p_1 = g(x^1, x^2)p^1, \quad p_2 = g(x^1, x^2)p^2. \tag{7}$$

Now assume that on ∂D the following equation holds:

$$F(x^1, x^2) = 0. \tag{8}$$

Let the point P move not in the whole plane but only on ∂D and let equation (8) have the form

$$x^2 = f(x^1)$$

in some neighborhood of the point P.

Then

$$L_Q(x^1,x^2) = L_Q(x^1, f(x^1)) = L^*(x^1). \tag{9}$$

Taking into account that

$$\frac{\mathrm{d}L^*}{\mathrm{d}x^1} = \frac{\partial L_Q}{\partial x^1} + \frac{\partial L_Q}{\partial x^2} \cdot \frac{\mathrm{d}f}{\mathrm{d}x^1}, \tag{10}$$

we rewrite (10) in the following form using Theorem 1:

$$\frac{\mathrm{d}L^*}{\mathrm{d}x^1} = -(p_1 + p_2 \frac{\mathrm{d}f}{\mathrm{d}x^1}). \tag{11}$$

We recall that the direction of the tangent vector, i.e. the angle α, is known:

$$\frac{p^1}{p^2} = \mathrm{tg}\ \alpha. \tag{12}$$

Because $\mathbf{p}$ is a unit vector, we have

$$g(x^1,x^2)\left[(p^1)^2 + (p^2)^2\right] = 1. \tag{13}$$

The components p^1 and p^2 are easily obtained from (10), (11), (12), and Theorem 1. Then equation (13) combined with (7) yields

$$p_1^2 + p_2^2 = g(x^1,x^2). \tag{14}$$

Thus, given $\mathrm{d}L^*/\mathrm{d}x^1$ at the point P, we can calculate a value of $g(x^1,x^2)$ at this point. For the description of the algorithm we assume that we know

1. the discrete values of $L_Q(P_K)$, $k = -n,\dots,0,\dots,n$, where Q is a fixed point on ∂D;
2. the directions of the tangent vectors $\mathbf{p}_K$ to the geodesic lines γ_K connecting Q and P_K;
3. the boundary equation.

Using these data, we can calculate the values of $\mathrm{d}L^*/\mathrm{d}x^1$ or $\mathrm{d}L^*/\mathrm{d}x^2$ at the point P_0. Then we find p_1 and p_2 after solving system (11), (12). Finally, from (14) we evaluate $g(x^1,x^2)$ at the point P_0. Analogously we may estimate $g(x^1,x^2)$ for each $\mathbf{p}_K$, $k = -n,\dots,0,\dots,n$.

From the geodesic equation we may easily obtain the geodesic curvature radius

$$\rho_0 = \frac{2g^{1/2}}{|p^1 g_y - p^2 g_x|}, \tag{15}$$

where $g_x = \mathrm{d}g(x,y)/\mathrm{d}x$, $g_y = \mathrm{d}g(x,y)/\mathrm{d}y$.

Now, as the values of $g(x,y)$, the directions $\mathbf{p}_K$ and the curvature radius ρ_K are known at P_K, we can "penetrate" into the domain D in linear approximation to the "depth" Δs, where Δs is a small deviation of a canonical parameters of a geodesic γ. To this purpose, knowing $\mathbf{p}_K$ for each point P_K, we plot a point P'_K by the linear shift of P_K in the direction $\mathbf{p}_K$ for a distance of Δs and thus we obtain the domain D' within D. Now we can evaluate

$$\Delta l(P_K, P'_K) \approx g(P_K) \cdot \Delta s, \tag{16}$$

where Δs is a Cartesian distance between P_K and P'_K. Then $L(P'_K, Q)$ is evaluated for each P'_K by

$$L(P'_K, Q) \approx L(P_K, Q) - \Delta l(P_K, P'_K). \tag{17}$$

So, we have now two domains D and D', two discrete sets of points P_K and P'_K, belonging to D and D' respectively, and estimates of $g(x, y)$ in these sets. Then we can evaluate g_x and g_y at the points P_K and estimate ρ_K by (15). Now we have a possibility to "penetrate" into the domain D in quadratic approximation to the "depth" Δs. To this purpose, we plot a circle C_K of the radius ρ_K, which passes through P_K, for each P_K, and shift each P_K along the arc of the circle C_K for the Cartesian distance Δs. Thus we obtain the points P''_K and the domain D''.

Now let us evaluate $\Delta l(P_K, P''_K)$ in the metric g:

$$\Delta l(P_K, P''_K) \approx g(P_K) \cdot \Delta s. \tag{18}$$

We evaluate $L(P''_K, Q)$ by

$$L(P''_K, Q) \approx L(P_K, Q) - \Delta l(P_K, P''_K). \tag{19}$$

As a result of these steps we get the domain D'', the points P''_K on the boundary $\partial D''$, the tangent vectors $\mathbf{p}''_K$ at these points and the estimates of $L(P''_K, Q)$. Thus, we get enough information to proceed to the next step of the algorithm.

3. Discussion and Conclusions

It is evident that the algorithm accuracy depends on the number of points P_K, $k = -n, \ldots, 0, \ldots, n$. The greater is the difference between the conform metric g_{ij} and the Cartesian metric δ_{ij}, the greater should be n. It is seen from the description of the algorithm that we need in data about two points P_{-1} and P_1 to "penetrate" into D in one step Δs (if one estimates the derivative for (15) as the convolution with the δ-function, then the number of points may be more than two). We need the data about four points to "penetrate" into D in two steps, etc. Thus, the "depth of penetration" depends on n.

Let us take into consideration that the point Q is a fixed point in the suggested algorithm. Thus, on the first step the algorithm processes information that is contained on the surface of the geodesic cone with the center at Q and the axis line γ connecting Q and P_0. Hence one has a possibility to use additional information by considering the discrete sets Q_i, $i = 1, \ldots, m$, and applying the algorithm to $L_{Q_i}(Q_i, P_0)$. So, we obtain a series of estimates $g_i(P_0)$. Now we can sum them with some weights to get a more accurate estimate of $g(P_0)$.

Finally, it should be marked that the algorithm was tested on a model problem and produces a good approximation. The author thanks N. S. Ovsjannikov for the organization and realization of numerical tests and E. V. Gajdukov for the valuable discussion on some theoretical questions as to the suggested algorithm.

REFERENCES

Aki K. and Lee W. H. K. (1976), Determination of three-dimensional velocity anomalies under a seismic array using first P arrivals from local earthquakes. 1. A homogeneous initial model. *J. Geoph. Res.*, **81**, 4381–4399.

Backus G. and Gilbert J. F. (1967), Numerical applications of a formalism for geophysical inverse problems. *Geophys. J. R. Astr. Soc.*, **13**, 247–276.

Cormack A. M. (1981), The Radon transform on a family of curves on the plane, *Proceedings of the Amer. Math. Soc.*, **83**, 325–330.

Cormack A. M. and Quinto E. T. (1980), A Radon transform in spheres through the origin in R^2 and applications to Darboux equation, *Trans. Amer. Math. Soc.*, **260**, 575–581.

Dziewonski A. M., Hager B. H., and O'Connel R. J. (1977), Large-scale heterogenetics in the lower mantle, *J. Geophys. Res.*, **82**, 239–255.

Funk P. (1913), Uber Flachen mit lauter gescheossenen geodatischen Linien, *Math. Ann.*, **74**, 278–300.

Helgason S. (1980), *The Radon Transform*, Birkhauser.

Mukhometov R. G. (1975), An inverse kinematical problem of seismology on a plane, In:*Mathematical problems of Geophysics* no. 6, Computer Center, Siberian Branch of the Russian Academy of Sciences, Novosibirsk, 243–254. (in Russian)

Natterer F. (1988), *The Mathematics of Computerized Tomography*, John Wiley & Sons.

Radon J. (1917), Uber die Bestimmung von Funktionen durch ihre Integralwerte langs gewisser Mannigfaltigkeiten, *Math-Phys. k I.*, **69**, 262–267.

Xu Guo-ming. (1988), Spherical Radon transform. *Geophysical Journal*, **95**, 361–370.

Penalty Method with Internal Prox-Regularization for an Inverse Source Problem [1]

G. Yu. GRYAZINA and S. I. KABANIKHIN

Institute of Mathematics, Novosibirsk, Russia

ABSTRACT

We consider a problem of determining the external current density from measurements of the electric field generated by the current, on the medium surface. An iterative algorithm is proposed which is based on a combination of the penalty approach with the prox-regularization.

Unlike the usual regularized algorithms, the proposed method uses the proximal point algorithm for penalty functions at each iteration. The number of iterations is not fixed and depends on the requirements on the solution accuracy.

1. Introduction. Formulation of the Problem.

The paper deals with a problem of recovering the external current density from measurements of the electric field, generated by the current, on the surface of a medium. Assume the external current vector j^{ex} to be modelled as

$$j^{ex} = \begin{bmatrix} 1 \\ 0 \\ 0 \end{bmatrix} = a(x)f(x,t), \tag{1.1}$$

where $a(x)$ is the sought-for function of the current density, $f(x,t)$ is a given positive function.

Let a function $u(x,t,a)$ be a solution of the following boundary value problem

$$\varepsilon u_{tt} + \sigma u_t = \frac{1}{\mu}(u_{xx} - \lambda^2 u) + a(x)f(x,t), \tag{1.2}$$

$$x \in [0,T], \quad t \in [0,T],$$

$$u(x,0) = 0, \quad u(x,0) = 0, \quad x \in [0,T], \tag{1.3}$$

$$u(T,t) = 0, \quad u(0,t) = 0, \quad t \in [0,T], \tag{1.4}$$

where $\varepsilon(x)$, $\sigma(x)$ are known positive functions, μ is a positive constant. Then the problem is to minimize the functional

$$J(a) = \int_0^T [u(0,t,a) - h(t)]^2 dt \tag{1.5}$$

[1]The work was supported by the Russian Foundation of Fundamental Research under grant No. 93-01-01739

subject to

$$\int_0^T a(x)^2 w(x)\mathrm{d}x \le C, \tag{1.6}$$

where $w(x)$ is a prescribed function, C is a constant.

Condition (1.6) provides us with a priori additional information on the desired function which has to be taken into account when solving the problem.

Similar problems occur in electromagnetic prospecting when it is required to determine the source power $a(x)$ provided that the electric field (more precisely, the horizontal component of the electric field vector) is measured on the medium surface and a priori restrictions on the desired function, such as (1.6), are given (Romanov and Kabanikhin, 1991). It should be noted that problem (1.1)–(1.5), as well as its different modifications, has been studied properly enough (Lavrent'ev and Reznitskaya, 1979; Sacks ,1990; Sun, 1992).

In this paper we consider problem (1.1)–(1.6) and propose a numerical iterative algorithm for its solution which is based on the penalty method and uses the internal prox-regularization.

2. A Numerical Method for the Direct Problem.

To solve the direct problem (1.2)–(1.4) we use an explicit finite-difference scheme of the second order approximation. Let $h_t = T/N$, $h_x = T/N_x$, $\varepsilon_i = \varepsilon(h_i)$, $\sigma_i = \sigma(h_i)$. Then the finite-difference scheme can be written as

$$\varepsilon_i \frac{u_i^{k+1} - 2u_i^k + u_i^{k-1}}{h_t^2} + \sigma_i \frac{u_i^{k+1} - u_i^{k-1}}{2h_t} = \tag{2.1}$$

$$\frac{1}{\mu}\left(\frac{u_{i+1}^k - 2u_i^k + u_{i-1}^k}{h_x^2} - \lambda^2 u_i^k\right) + a_i f_i^k,$$

$$u_i^0 = 0, \quad u_i^1 = 0, \quad i = 0, 1, \dots, N_x, \tag{2.2}$$

$$u_0^k = u_1^k, \quad u_0^{N_t} = 0, \quad k = 0, 1, \dots, N_t. \tag{2.3}$$

It is commonplace that scheme (2.1) is stable under the Courant condition.

3. Formula for the Objective Functional Gradient.

To obtain the formula for the gradient $\nabla J(a)$, we use the well-known method presented in Romanov and Kabanikhin (1991).

Note that

$$J(a + \Delta a) - J(a) \approx 2\int_0^T [u(0,t,a) - h(t)]\Delta u(0,t)\mathrm{d}t, \tag{3.1}$$

where $\Delta u(x,t) = u(x,t,a + \Delta a) - u(x,t,a)$.

Consider the conjugate problem

$$\varepsilon\psi_{tt} - \sigma\psi_t = \frac{1}{\mu}(\psi_{xx} - \lambda^2\psi), \tag{3.2}$$

$$x \in [0,T], \quad t \in [0,T],$$

$$\psi(x,T)=0, \quad \psi_t(x,T)=0, \quad x\in[0,T], \tag{3.3}$$

$$\psi(T,t)=0, \quad \psi_x(0,t)=-2(u(0,t,a)-h(t)), \quad t\in[0,T]. \tag{3.4}$$

Using the integration by parts, one can easily verify that

$$\int_0^T\int_0^T \Delta u(x,t)\psi_{xx}(x,t)\mathrm{d}x\mathrm{d}t =$$

$$\int_0^T [\Delta u(T,t)\psi_x(T,t)-\Delta u(0,t)\psi_x(0,t)-\Delta u_x(T,t)\psi(T,t)+\Delta u_x(0,t)\psi(0,t)]\,\mathrm{d}t+$$

$$\int_0^T\int_0^T \Delta u_{xx}(x,t)\psi(x,t)\mathrm{d}x\mathrm{d}t.$$

Then the right-hand side of equality (3.1) can be rewritten, in view of conditions (1.6), (3.4), as

$$2\int_0^T[u(0,t,a)-h(t)]\Delta u(0,t)dt = -\int_0^T \Delta u(0,t)\psi_x(0,t)\mathrm{d}t =$$

$$\int_0^T\int_0^T \Delta u(x,t)\psi_{xx}(x,t)\mathrm{d}x\mathrm{d}t - \int_0^T\int_0^T \Delta u_{xx}(x,t)\psi(x,t)\mathrm{d}x\mathrm{d}t.$$

We express $\Delta u_{xx}(x,t)$ from (1.2), substitute this expression into the second integral and integrate it by parts taking into account conditions (1.5), (3.3).

As a result, we obtain

$$2\int_0^T[u(0,t,a)-h(t)]\Delta u(0,t)\mathrm{d}t = \mu\int_0^T\int_0^T \Delta a(x)f(x,t)\psi(x,t)\mathrm{d}x\mathrm{d}t-$$

$$\mu\int_0^T\int_0^T\left[\varepsilon\psi_{tt}(x,t)-\sigma\psi_t(x,t)-\frac{\psi_{xx}(x,t)-\lambda^2\psi(x,t)}{\mu}\right]\Delta u(x,t)\mathrm{d}x\mathrm{d}t.$$

Finally, in view of (3.2), we get the desired formula:

$$\nabla J(a)=\mu\int_0^T f(x,t)\psi(x,t)\mathrm{d}t. \tag{3.5}$$

4. Regularized Penalty Method.

As pointed out above, the source problem is reduced to the following optimization problem:

$$J(a)\longrightarrow \min, \quad a\in G, \tag{4.1}$$

$$G=\{a\in R^n:\ g^j(a)\le 0,\ j=1,\dots,m\}, \tag{4.2}$$

where $J,g\in C^1(R^n)$, g^j are convex functions.

To solve problem (4.1), (4.2), we use the penalty approach combined with internal prox-regularization. The penalty method is based on successive unconditional minimization of functions $J + V_k$, $k = 1, 2, \ldots$, where $V_k : R^n \to R^1 \bigcap \{\infty\}$ are convex functions such that

$$\lim_{k\to\infty} V_k(a) = \begin{cases} 0, & a \in G, \\ +\infty, & a \notin G, \end{cases}$$

almost everywhere in R^n. The V_k are called the penalty functions.

The advantages of the penalty method are its applicability to a wide range of the problems, the simplicity of the calculating scheme, and the possibility to use various unconditional minimization algorithms.

The most essential limitation of this method is an extent of level lines of the function $J + V_k$ with increasing k, which causes the efficiency of unconditional minimization algorithms to lower.

To overcome this drawback, we embed an iterative prox-regularization into a penalty method scheme. As a rule, each function $J + V_k$ is regularized only once (Grossman and Kaplan, 1981; Kaplan, 1988), which leads to auxiliary problems

$$J(a) + V_k(a) + \|a - a_{k-1}\|^2 \longrightarrow \min, \quad a \in R^n, \tag{4.3}$$

where a^{k-1} is the result of the preceding iteration. Obviously, the conditionality of the auxiliary functions in (4.3) is improved as compared to the classical penalty method, but it is inevitably impaired as the number k increases. For the degradation to slow down, we propose to repeat the proximal point algorithm for every penalty function until the difference between two last approximate solutions becomes small enough.

Suppose that $V_k(a) = V(r_k, a)$, where $V_k \in C^1(R^n)$ and r_k are some penalty parameters, are convex penalty functions satisfying the conditions

1)

$$\lim_{k\to\infty} V_k(a) = \begin{cases} 0, & a \in G, \\ +\infty, & a \notin G; \end{cases}$$

2)

$$V_k(a) \geq 0 \quad \text{for} \quad a \in R^n, \quad k = 1, 2, \ldots$$

For instance, the following functions widely used as penalty ones meet the above requirement:

$$V_k(a) = r_k \sum_{j=1}^{m} \max^2[0, g^j(a)], \quad k = 1, 2, \ldots \tag{4.4}$$

Suppose problem (4.1),(4.2) to obey the following conditions: first, there exists a Slayter's point, that is, such a point a_0 that $g(a) < 0$, $j = 1, \ldots, m$; second, the set of its solutions $G^* = \{a^* \in G : J(a^*) = \inf_{a\in G} J(a)\}$ is nonempty and bounded.

Denote $\hat{G} = a \in G : J(a) \leq D$, where D is an arbitrary number greater than $J_{\min} \equiv \inf_{a\in G} J(a)$. Since J is convex and G^* is nonempty and bounded, the set $\hat{G}$ is also bounded. For prescribed values $\delta_k > 0$ we can find penalty parameters $r_k > 0$ so that

$$V_k(a) < \delta_k \quad \text{for} \quad a \in \hat{G}, \quad k = 1, 2, \ldots \tag{4.5}$$

For example, (4.5) is always fulfilled for the penalty functions of type (4.4) since $V_k(a) \equiv 0$ for $a \in G$ in this case.

We proceed now to the description of the penalty method with the embedded prox-regularization.

Step 1. Let $\{\varepsilon_k\}$ and $\{\delta_k\}$ be sequences of positive numbers. We choose a system of convex penalty functions $V_k \in C(R^n)$ satisfying properties 1), 2), so that (4.5) holds. Take an arbitrary $a^{1,0} \in R^n$ as the starting point. Let $l = 0$, $k = 1$.

Step 2. For fixed l, k, a point $a^{k,l+1}$ is determined from the condition

$$\left\| \nabla \eta_{k,l}(a^{k,l+1}) \right\| \le \varepsilon_k, \tag{4.6}$$

where

$$\eta_{k,l}(a) = J(a) + V_k(a) + \left\| a - a^{k,l} \right\|^2. \tag{4.7}$$

Step 3. Unless a the stopping criterion holds, specify a controlling parameter $\Phi_{k,l+1}$ according to the requirements given below. If $\left\| a^{k,l+1} - a^{k,l} \right\|^2 \ge \Phi_{k,l+1}$, then we increase l by 1 and return to Step 2. If $\left\| a^{k,l+1} - a^{k,l} \right\|^2 < \Phi_{k,l+1}$ then set $l(k) = l + 1$, $a^{k+1,l} = a^{l,l(k)}$, assign $k = k + 1$, $l = 0$ and go to Step 2.

Hereafter the sequences $\{\varepsilon_k\}$ and $\{\delta_k\}$ are assumed to satisfy the following conditions:

$$\sum_{k=1}^{\infty} \varepsilon_k < \infty, \quad \sum_{k=1}^{\infty} \delta_k^{1/2} < \infty. \tag{4.8}$$

For $1 \le l \le l(k)$, $k \in \{1, 2, \ldots\}$, we denote

$$I_{k,l} = \{j = 1, \ldots, m : g^j(a^{k,l}) > 0\};$$

$$\mu_{k,l} = \begin{cases} \max\limits_{j \in I_{k,l}} \dfrac{g^j(a^{k,l})}{g^j(a^{k,l}) - g^j(a_0)}, & \text{if} \quad I_{k,l} \ne \emptyset; \\ 0, & \text{if} \quad I_{k,l} = \emptyset; \end{cases}$$

$$\rho_{k,l} = \|\nabla J(a_0)\| \, \mu_{k,l} \left\| a^{k,l} - a_0 \right\|.$$

Let values of $d_{k,l}$ be such that

$$\max_{a^* \in G^*} \left\| a^{k,l} - a^* \right\| \le d_{k,l}, \quad l = 0, 1, \ldots, l(k) - 1; \quad k = 1, 2, \ldots$$

Then $\Phi_{k,l}$ should be chosen so that

$$\Phi_{k,l} \ge \frac{1}{2}\varepsilon_k^2 + \delta_k + \rho_{k,l} + \varepsilon_k d_{k,l-1} + \Theta_k - V_k(a^{k,l}), \tag{4.9}$$

$$l = 1, 2, \ldots, l(k); \quad k = 1, 2, \ldots; \quad \text{where} \quad \Theta > 0, \quad \lim_{k \to \infty} \Theta_k = 0$$

In the paper by Gryazina (1993) conditions (4.8), (4.9) were shown as sufficient for the following assertions to hold:

- $l(k) < \infty$ for each k;
- the sequence $\{a^{k,l}\}$ is bounded;
- for arbitrary values of $\sigma_k > 0$, $k = 1, 2, \ldots$, the penalty parameters $r_k > 0$, $k = 1, 2, \ldots$, can be chosen so that

$$\max_{j=1,2,\ldots,m} g^j(a^{k,l(k)}) \le \sigma_k, \quad k = 1, 2, \ldots \tag{4.10}$$

It was proved by Gryazina (1993), that under the above assumptions inequality (4.9) is non-contradictory to

$$\lim_{k \to \infty} \Phi_{k,l(k)} = 0. \tag{4.11}$$

If, in addition, conditions (4.10), (4.11) and $\sum_{k=1}^{\infty} \delta_k^{1/2} < \infty$ are fulfilled, then the sequence $\{a^{k,l}\}$ constructed using the penalty method with the internal prox-regularization converges to the solution of problem (4.1), (4.2).

5. Results of Numerical Simulations.

The presented method was tested on problem (1.1)–(1.6), with $\varepsilon = 1$, $\mu = 1$, $\sigma = 1/2$, $\lambda = 1$, $T = 1$, $w(x) = 1 + 5x$, $C = 0.4$, $f(x,t) = 4 + \cos(x)\sin(t)$, $h(t) = \bar{u}(0,t)$, where $\bar{u}(x,t)$ is the solution of the direct problem (1.2)–(1.4) for $a(x) = \cos(5x) + 1$.

The calculations were carried out for two meshes with $N_x = N_t = 10$ and $N_x = N_t = 20$, respectively.

The regularized penalty method is realized through unconditional minimization of functions (4.7), where the function $J(a)$ is a discrete analogue of (1.5). We take for penalty functions ones of type (4.4), with $m = 1$ and $g^1(a)$ being the expression $\int_0^1 a(x)^2(1+5x)dx - 0.4$ related to the meshes. The parameters are chosen so that $\varepsilon = 1/(1.5)^{k-1}$, $r_k = 2^{k-1}$. The unconditional minimization of the functions $\eta_{k,l}$ is carried out using the conjugate gradient method. The gradient of the function $J(a)$ is calculated on the meshes by (3.5).

Figures 1 and 2 present the curves of the function $a(x)$ for the model problems mentioned. We used the following notation: "—" denotes the initial approximation of $a(x)$; "x", "□", and "*" indicate the approximate solution of the source problem after one, two and nine iterations of the regularized penalty method, respectively.

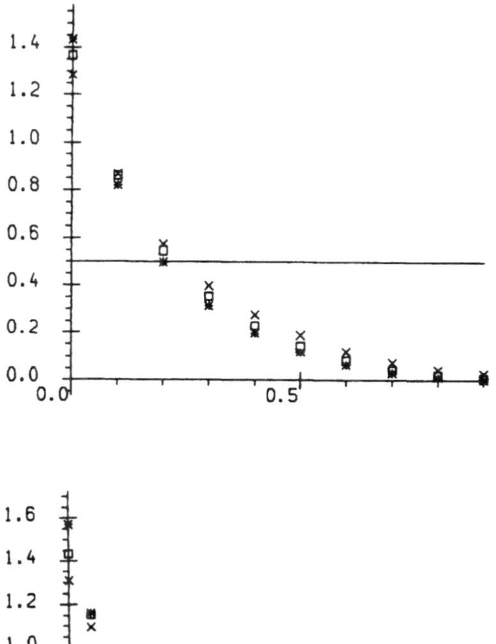

FIGURE 1.

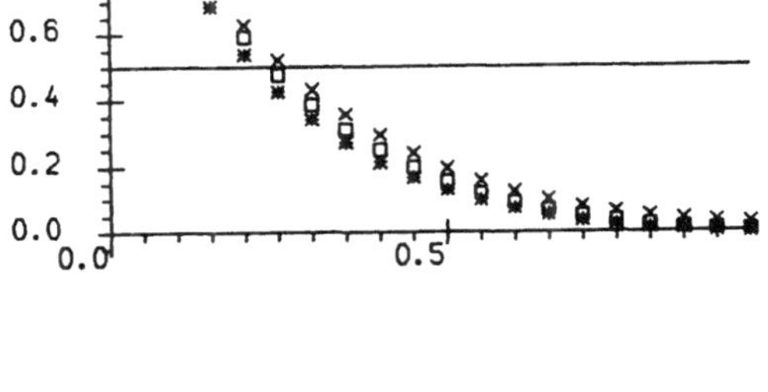

FIGURE 2.

REFERENCES

Grossman K. and Kaplan A. A. (1981), *Nonlinear programming based on unconditional minimization*, Nauka, Novosibirsk. (in Russian)

Gryazina G. Yu. (1983), *Algorithms with Internal Prox-Regularization Based on the Penalty Method*, Preprint no. 6, Institute of Mathematics, RAS, Novosibirsk. (in Russian)

Kaplan A. A. (1988), On stability of methods for solving convex programming problems and variational inequalities. In: *Models and Methods of Optimization*, no. 10, Institute of Mathematics, RAS, Novosibirsk. (in Russian)

Lavrent'ev M. M. and Reznitskaya K. G. (1979), An inverse problem with an unknown source, *Soviet Math. Dokl.*, **20**, 1416–1418. (in Russian)

Romanov V. G¿ and Kabanikhin S. I. (1991), *Inverse problems in geoelectrics*, Nauka, Moscow. (in Russian)

Sacks P. (1990), A velocity inversion problem involving an unknown source, *SIAM J. Appl. Math.*, **50**, 931–941.

Sun Z. (1992), Direct scattering and reconstruction of internal source, *Wave Motion*, **16**, 249–263.

Inverse Problems in Single Photon Emission Computed Tomography

Grant T. GULLBERG, Gengsheng L. ZENG and Rolf CLACK

Medical Imaging Research Laboratory, University of Utah
Salt Lake City, Utah, USA

ABSTRACT

Single photon emission computed tomography (SPECT) presents some of the most difficult and yet interesting inverse problems in computerized tomography. In recent years, the introduction of new collimator geometries and detector orbits, the need for efficient attenuation correction, the need to improve image quality degraded by physical factors (variable attenuation, spatially varying geometric response, scatter, and noise), and the new application of dynamic acquisitions have provided challenging inverse problems for which in many cases the mathematics is still not fully worked out. Several of the inverse problems do not have efficient analytical solutions, but through the development of new computer technology the solutions of several of these problems are possible using numerical methods. These advancements have improved sensitivity and specificity of diagnostic nuclear medicine. Present work is continuing to improve SPECT through mathematical algorithms which better model the physics of the imaging detection process.

1. Introduction

Single photon emission computed tomography (SPECT) has become a routine procedure in clinical nuclear medicine. New camera technology has increased geometric efficiency and improved image resolution which has enhanced diagnostic results. The development of new technology has required the derivation of new mathematical algorithms to solve non-standard tomographic inverse problems. In many cases, solutions to these inverse problems are best accomplished through solutions of systems of linear equations. This is true for the problem of improving the image degradation caused by photon attenuation, collimator geometric response, and scatter. Advanced camera and computer technology is making progress in solving these difficult inverse problems that have challenged SPECT for several years.

In recent years there have been several new developments in collimators and detector orbits for rotating camera SPECT systems (Gullberg et al., 1992b). New fan-beam, astigmatic, cone-beam, variable focal length collimators and various noncircular and nonplanar orbit configurations have been introduced to improve image resolution and geometric efficiency. These tomographic geometries have brought forth new and interesting inverse problems that are difficult to solve using analytical methods and for which traditional filtered backprojection and backprojection filtering algorithm have not yet been derived.

In routine clinical practice correction for attenuation is accomplished using approximate analytical methods that assume uniform attenuation. These methods give good approximations for imaging the brain but are inaccurate for imaging the heart. A filtered backprojection algorithm (Tretiak and Metz, 1980; Gullberg and Budinger, 1981) which inverts the exponential Radon transform has been investigated but it amplifies noise. Work continues in an attempt to suppress the significant noise amplification of the algorithm. Presently, it remains an open question as how to invert the exponential Radon transform for cone-beam tomography, especially since it does not fit into the frame-work of standard Radon inversion techniques that are typically used to solve cone-beam inverse problems.

More accurate methods for correcting for spatially-varying photon attenuation and other physical processes of spatially-varying geometric point response, and scatter have required that the solution to the SPECT inverse problem be worked out by solving large systems of linear equations. Typically, solutions are determined using iterative expectation-maximization algorithm (Lang and Carson, 1984) which maximizes a likelihood function that incorporates the underlying probability distribution for photon detection and the model of the imaging physics which is programmed into projection and backprojection operators. Present day computing capability has allowed researchers to demonstrate the improvement of image quality by appropriately modeling the imaging physics, but the computation times are not feasible for clinical application where analytical reconstruction algorithms are still the present method of choice.

With the development of multi-detector SPECT systems, dynamic SPECT is able to extract *in vivo* physiological information. Dynamic cardiac SPECT has already demonstrated the capability to estimate the three-dimensional distribution of tissue extraction of teboroxime in the heart (Smith, Gullberg et al. , 1992, 1993). However, dynamic imaging places considerable demands on computer memory, disk space, computer speed, and algorithms as kinetic parameters are determined using nonlinear estimation techniques. Not only is the solution to large system of linear equations required to compensate for image degradation, but the solutions are required for as many as 160 three-dimensional reconstructions over time for one dynamic study.

In this paper we review the recent developments in SPECT which have brought about interesting and challenging inverse problems. We first review the development of new collimator geometries and detector orbits. Then we present an analytical and a singular value decomposition approach to the solution of the constant attenuation problem. Then we show how the more general attenuation problem along with geometric response, and scatter can be solved using systems of linear equations. We finish our review with some of the recent applications of dynamic SPECT to cardiac imaging.

2. Tomographic Geometry

Today, the rotating scintillation camera is used more widely than fixed ring systems because of its ability to perform both body and brain SPECT as well as conventional planar imaging. Most of the systems are single detector systems but recently systems with three (Figure 1) and four rotating detectors have been commercially made available. Several manufacturers are also selling large field-of-view dual camera systems (Figure 2) that offer whole body imaging and computed tomography capability. With convergent beam collimators, these systems offer an advantage of a significant increase in sensitivity for imaging small organs because of the large field-of-view of up to 60 cm compared with the standard 40 cm detectors.

There is considerable variability in collimator geometries that offer significant potential for improved diagnostic utility. This is an area that still requires significant research both

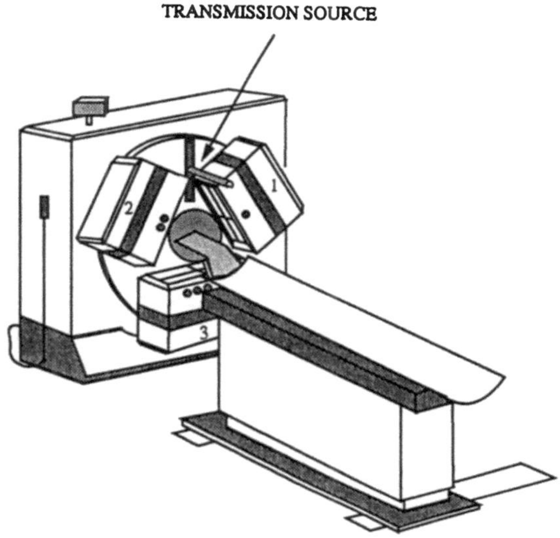

Figure 1. Schematic diagram of the Picker Prism®3000 three-detector SPECT system with a transmission line source.

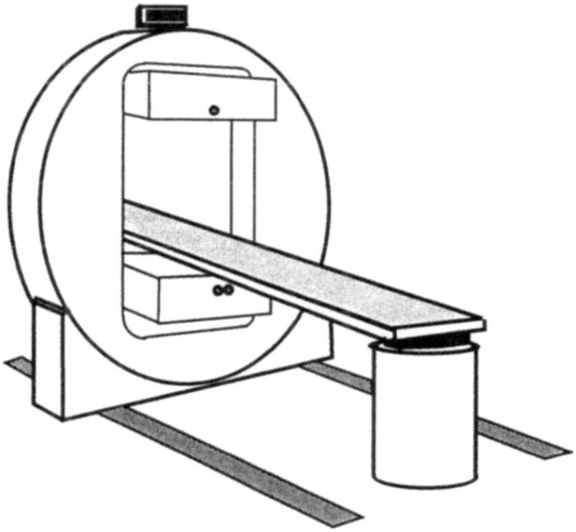

Figure 2. Schematic diagram of the Picker Prism®2000 two-detector SPECT system.

in hardware and mathematical algorithm development. The encouraging results of preliminary work from several investigators has stimulated the investigation of several new detector concepts using converging beam geometries (Gullberg et al., 1992b).

In a very general sense, we consider a convergent beam tomographic system to be specified by the shape of the envelope of rays, the orientation of the rays emanating from the focal point (or focal points) relative to the axis of rotation, and the scanning trajectory of the focal point (or focal points). The envelope of rays may converge to a single point or to several points or could be made up of rays that converge to straight or curved lines. The standard cone-beam collimator which consists of holes that converge to a single point behind the object, and the pin-hole collimator which has a focal point at the aperture, are two types of examples where the envelope of rays converges to a single point forming a symmetrically oriented cone beam. In brain imaging an asymmetric cone-beam geometry has been proposed where the focal point is at the bottom edge of the camera, and the image is magnified upward in a way that uses the entire crystal area, thus forming a half-cone beam geometry. Thus the patient's head can be placed at the lower edge of the crystal so that the camera can rotate with a small radius of rotation and still clear the patient's shoulders. Other geometries include those where the rays converge to a series of focal points. These include a fan-beam geometry (Jaszczak et al., 1979; Tsui et al., 1986) which has a series of parallel transaxial fan-beam slices with rays that focus at a series of points on a line parallel to the axis of rotation. A variation of this is an astigmatic geometry (Hawman and Hsieh, 1986) which has rays that converge to two lines that are orthogonal but at different focal lengths. The envelope of rays for this geometry becomes more cone shape in that the range of the distribution of focal points along each line is shorten to less than that of a fan-beam.

A second geometrical consideration is the specification of the orientation of the axis of the convergent beam relative to the axis of rotation. In cardiac SPECT, the orientation of the rays of the central plane is always perpendicular to the axis of rotation. This is in contrast to that which has been proposed for the brain imaging where the central plane is oblique to the axis of rotation. This allows the camera to be tilted so that the camera can clear the patient's shoulders.

A third geometrical consideration concerns the scanning trajectory of the focal point and the angles it subtends as it travels along the trajectory. The focal point itself may follow various type of paths that may include helical arc, dual orthogonal (Clack et al., 1991) or single-plane scanning geometries, or remain in a single plane throughout the rotation. Within the plane the focal point could follow a circular or noncircular orbit. In the noncircular orbit, the focal length remains constant but the distance between the focal point and the center of rotation varies depending upon the study. As for example in heart imaging, tomographic images are reconstructed from "short scan" data acquisitions in order to minimize image degradation due to reconstructing attenuated posterior projections which register very few photons from the heart.

2.1 *Collimator Geometries*

Several different collimator geometries have been used in SPECT with rotating gamma cameras. These include parallel, fan-beam (Jaszczak et al., 1979), cone-beam (Jaszczak et al., 1986), pin-hole (Palmer and Wollmer, 1990), and astigmatic geometries (Hawman and Hsieh, 1986). The astigmatic geometry is a generalization of all of these geometries.

2.1.1 **Astigmatic.** The astigmatic collimator (Figure 3) has two orthogonal lines of focus at different focal lengths. One focal line is aligned parallel and the other is aligned perpendicular to the axis of rotation. A fan-beam collimator is equivalent to placing the perpendicular line of focus at a distance far from the face of the collimator; whereas, a cone-beam collimator is an astigmatic collimator with both lines of focus at the same focal length. As both focal lengths increase, the astigmatic collimator approaches that of a parallel collimator.

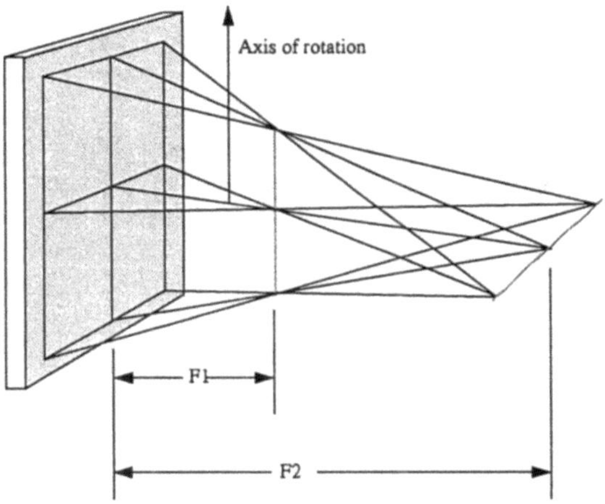

FIGURE 3. The astigmatic geometry has two lines of focus. One has a focal length of F1 and the other has a focal length of F2.

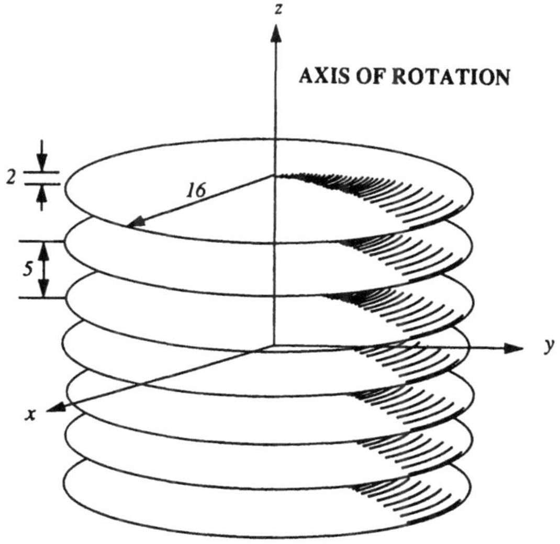

FIGURE 4. Defrise phantom consists of seven parallel ellipsoidal disks of identical uniform spaced at 5 voxels apart parallel to the axis of rotation.

Simulations using the Defrise phantom (Figure 4) have been performed comparing cone-beam, fan-beam, parallel and astigmatic geometries (Gullberg et al., 1992b). A generalized astigmatic reconstruction algorithm was applied to simulated projection data generated over one circular planar orbit. The astigmatic reconstruction algorithm uses the same ramp filter in the transaxial direction as the Feldkamp cone-beam reconstruction algorithm (Feldkamp et al., 1984) but backprojects along the astigmatic geometry. The simulations in (Gullberg et al., 1992b) illustrate that the cone-beam reconstruction has more slice-to-slice cross talk in the sagittal reconstructions of the Defrise phantom than fan-beam and parallel reconstructions (see also Zeng and Gullberg, 1990a). The parallel and fan-beam geometries give the best results and are expected to be equivalent since sufficient data are acquired in one circular planar orbit for these geometries. Comparisons of the astigmatic and cone-beam reconstructions indicate that an astigmatic geometry with the focal line F1 equal to the cone beam focal length and the focal line F2 less than the cone-beam focal length will give worse results. Whereas, an astigmatic geometry with F2 longer than the cone-beam focal length gives better results because it more closely approximates a fan-beam geometry.

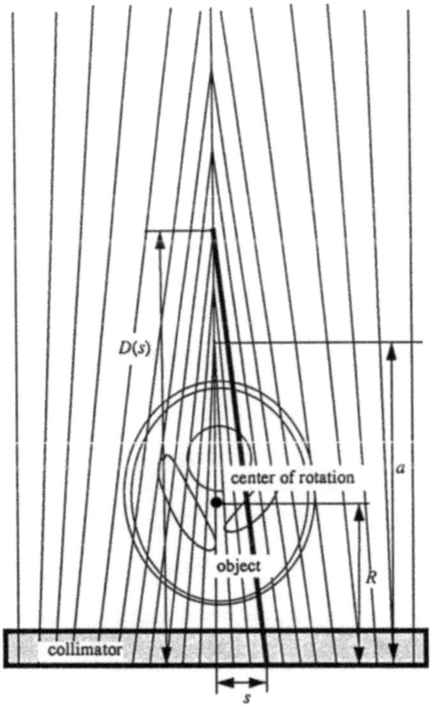

Figure 5. A spatially varying focal length fan beam collimator. The focal length $D(s)$ increases as $|s|$ increases, and a is the shortest focal length.

2.1.2 **Variable Focal Length.** The use of a fan-beam collimator with a variable focal length (Figure 5) has previously been proposed as a means of reducing or even eliminating the problem of truncation which is a major drawback of conventional fan-beam collimation. The collimator is constructed such that the focal length increases from the shortest length at the central holes to the longest length at the edge of the collimator and provides higher sensitivity in the central field of view than a parallel collimator of the similar resolution. This provides a projection geometry for which a filtered backprojection reconstruction algorithm does not exist. However, several other reconstruction algorithms have been developed. These include: iterative, fix-point numerical integration, rebinning, and series approximations of filtered backprojection (Zeng et al., 1993c) approaches, all of which are less efficient than a straight-forward filtered backprojection algorithm.

Another approach is to reconstruct variable focal length collimated projections using a backprojection filtering algorithm (Zeng et al., 1993a). For a circular orbit, the point spread function of the projection operation followed by the backprojection operation is spatially invariant and is equal to $1/r$. Images are reconstructed in two steps. First, the projection data is backprojected into the image array, and then a two-dimensional ramp filter is applied to the backprojected image using discrete Fast Fourier Transforms. In computer simulations, this algorithm is shown to be efficient and to give artifact-free reconstructions.

2.2 *Orbits*

Cone-beam collimation has the best sensitivity for large crystals but does not produce artifact-free reconstructions when using a scanning trajectory whose focal point remains in a plane. Presently, all SPECT systems acquire data by scanning only in single planar orbits where the focal point traverses either circular or elliptical curves lying within a single plane. These data acquisitions have been shown to give insufficient projection data (Tuy, 1983; Smith, 1985). The reconstruction, using the Feldkamp algorithm (Feldkamp et al., 1984), is an approximation except for the central plane which is reconstructed accurately. Results of cone-beam tomography of the heart (Gullberg et al., 1991a, 1991b) look promising when the heart is positioned near the central plane of the cone-beam geometry and ROC studies (Tsui et al., 1993) demonstrate improved lesion detection.

Data sufficiency conditions exist for cone-beam geometry (Tuy, 1983; Smith, 1985) but have not been generalized to astigmatic geometries.

2.2.1 **CONE-BEAM DATA SUFFICIENCY ORBITS.** For cone beam tomography, we know that a detector system will acquire sufficient data (Tuy, 1983; Smith, 1985) if the scanning trajectory of the focal point has at least one point of intersection for any plane passing through the reconstructed region of interest. A planar circular orbit does not satisfy this data sufficiency condition since it is easy to find many planes that pass through the object but does not intersect the orbit. Whereas, nonplanar orbits as a circle-and-line orbit (Zeng and Gullberg, 1992a), a dual orthogonal orbit (Clack et al., 1991), and a helical orbit (Weng et al., 1993) do satisfy the data sufficiency requirement.

These orbits could be implemented with present SPECT systems at little additional cost. Using the SPECT system illustrated in Figure 2, the implementation of the circle-and-line orbit first rotates through a planar orbit while the gantry remains stationary and then the gantry performs one linear translation while the detector remains fixed with respect to the gantry. A helical orbit can easily be implemented by translating the gantry while simultaneously rotating the detector. The dual orthogonal orbit could be applied to brain imaging and could be implemented using one transaxial orbit and one half orbit that scans over the top of the head. This is best accomplished using a single detector SPECT system.

2.2.2 **NONCIRCULAR.** During a patient scan the camera rotates in a noncircular orbit, with the center of the camera always pointing to the center of rotation, and closely follows the patient's body contour to minimize the fall off in resolution with distance from the collimator. New fan- (Weinstein, 1980) and cone-beam (Gullberg et al., 1991a, Gullberg and Zeng, 1992a) reconstruction algorithms were developed to reconstruct projections from noncircular detector orbits. A cone-beam reconstruction algorithm (Gullberg et al.,

1991a) was derived using the noncircular fan-beam algorithm developed by Weinstein (1980) and extending it to cone-beam geometry using the method of Feldkamp et al. (1984) to derive the cone-beam reconstruction algorithm for noncircular planar orbits.

2.2.3 **Short Scan.** In the application of SPECT to cardiac imaging, it was found that the reconstruction of the posterior projections of emissions from a radiopharmaceutical localized in the heart resulted in reconstruction artifacts. Therefore, an angular sampling range is usually chosen so as not to acquire the highly attenuated posterior projections. For fan- and cone-beam geometries new "short scan" reconstruction algorithms (Zeng and Gullberg, 1990b, 1991, Gullberg and Zeng, 1992) were developed for less than 360 degrees, noncircular detector orbit acquisitions. The algorithms require that data are sampled over an angular range of 180 degrees plus the fan or cone angle plus an additional small angle. For these conditions the algorithms accurately reconstruct fan-beam geometry but can only accurately reconstruct the central plane of a cone-beam projection acquisition since only the reconstruction of the midplane is equivalent to a fan-beam reconstruction. Therefore, "short scan" cone-beam reconstructions of the midplane should be equivalent to "full scan" cone-beam and parallel reconstructions if for the "short scan" acquisitions the central slice is sampled with a sufficient set of fan-beam projections.

The cone-beam reconstructions of transaxial slices off the central slice are approximations due to the nature of the incomplete data sampling. Differences become more apparent for slices further from the midplane. From simulations (Gullberg and Zeng, 1992a), it appears that the combination of the approximation of cone-beam reconstruction algorithm for non central slice voxels and the "short scan" data set, together result in artifacts that are seen in "short scan" cone-beam reconstructions but are not seen in "full scan" reconstructions.

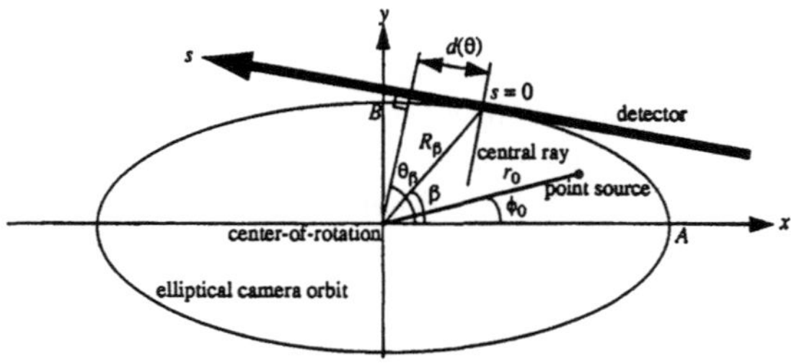

Figure 6. The scanning geometry. The detector rotates in an elliptical orbit with its face is always tangent to the ellipse. The detector is mounted with a parallel hole collimator.

2.2.4 **Orbit Modulated Angular Sampling.** An orbit geometry (Figure 6) where the central projection ray of the detector is not constrained to pass through the center-of-rotation for all projection angles can be useful to reduce uniformity artifacts and to avoid truncation when imaging non centrally located organs with converging collimators. For cardiac imaging, this would allow the central projection ray of the detector to always pass through the center of the heart. If the projections are sampled such that the line from the center of rotation to the center of the detector rotates about the center of rotation in equal angular increments, the point response of the projection followed by the backprojection operation is shift invariant and is not the common $1/r$ but is a function of the orbit geometry. Simulations show that if the common ramp filter is applied to the backprojected image, reconstruction artifacts appear. These artifacts are removed using a newly derived inverse filter (Gullberg and Zeng, 1993).

3. Constant Attenuation

Attenuation correction in clinical practice uses approximate analytical methods that assume uniform attenuation. These methods give good approximations for imaging the brain but are inaccurate for imaging the heart. An analytical method which was proposed uses a filtered backprojection algorithm to invert the exponential Radon transform (Tretiak and Metz, 1980; Gullberg and Budinger, 1981). The use of singular value decomposition has been investigated to suppress the significant noise amplification of the filtered backprojection algorithm (Gullberg and Zeng, 1992b).

3.1 *Filtered Backprojection*

A convolution reconstruction algorithm was derived that is applicable to SPECT in the presence of a constant attenuating medium. A high resolution digital convolver was obtained from digital samples of a continuous analytical convolution function:

$$c(x) = 0.5[\sin(\pi x)/(\pi x)] - 0.25[\sin(\pi x/2)/(\pi x/2)]^2$$
$$-\mu^2/(2\pi^2)[\sin(\mu x)/(\mu x)] + \mu^2/(4\pi^2)[\sin(\mu x/2)/(\mu x/2)]^2$$

obtained by evaluating the Fourier inverse of a truncated ramp function that is zero for frequencies less than $\mu/(2\pi)$ (where μ is the constant attenuation coefficient) and zero for frequencies greater than 0.5 cycles per projection bin. Other digital convolvers which suppress high frequency noise are obtained by digitally sampling the Fourier inverse of the same truncated ramp function multiplied by window functions. It is shown that the digital convolvers reconstruct simulated data accurately. Corresponding digital filters that are obtained by taking the discrete Fourier transform of the digital convolver give equivalent results. These digital filters differ from taking discrete samples of the truncated ramp function in that the DC component is a nonzero complex number and the filter is not zero for frequencies less than $\mu/(2\pi)$. These digital filters give more accurate reconstructions than a digitally sampled truncated ramp function.

3.2 *Singular Value Decomposition*

In the previous section, a convolution reconstruction algorithm was presented which mathematically can be shown to give an accurate reconstruction of projections of emission sources within a constant attenuating media. The major problem with the algorithm is the severe noise amplification. The algorithm was derived based upon being able to represent the projection data as projections of the exponential Radon transform and uses an exponential backprojector after applying a reconstruction filter to the modified projection data. Due to the exponential backprojector, the point spread function of the backprojection has a hyperbolic cosine factor, which makes the point spread function non local and noise amplifying. Even if window functions are applied to the filters to try to remedy the noise amplification, the window functions do not help very much to improve the image quality in a noisy reconstruction. Another approach is to derive an algorithm which uses the singular value decomposition (SVD) of the exponential Radon transform (Gullberg and Zeng, 1990). Using SVD to reconstruct an image, the projection data is first backprojected using the standard tomographic backprojection without the exponential term. The point spread function of the backprojection is local and easy to regularize. Using the SVD approach, regularization is accomplished by truncating the terms with small singular values. Computer simulations show an improvement in the SVD method over the convolution backprojection method when the projection data is corrupted with noise (Gullberg and Zeng, 1992).

4. Variable Physical Factors

Correcting for the physical processes of spatially-varying photon attenuation, geometric point response of the collimator, and scatter has required that the solution to the SPECT inverse problem be worked out by solving large systems of linear equations. Typically, solutions are determined using an iterative expectation-maximization algorithm (Lang and Carson, 1984) which maximize a likelihood function that incorporates the underlying probability distribution for photon detection and the model of the imaging physics. The model for the imaging physics is programmed into projection and backprojection operations. A non-overlapping square pixel basis is usually used to represent the image. The algorithm updates the new solution estimated by projecting and backprojecting data between the image and projection domains. Since non-overlapping square pixels is a poor representation of a continuous image, the approximation errors increase with each iteration. To eliminate these errors, one can use instead natural pixels which are the intersection of measured ray paths. The iterative algorithm assumes that a continuous image can be obtained by backprojecting the discrete array q, which is the least squares solution to Mq=p where p is the array of discrete measurements. The matrix M represents the operator obtained from the backprojection operator F^T followed by the projection operator F using the natural pixel basis. Since both p and q are projection arrays, the iterative algorithm acts only in the projection domain. In our implementation (Zeng and Gullberg, 1993b), the conjugate gradient algorithm is used to solve the equation Mq=p, and the final image is obtained by sampling the backprojection of the solution q at a discrete array of points. The challenge is to form the matrix M which is precalculated and stored. Each entry of M represents a natural pixel formed as the overlapped area of two projection rays on a finite support. When the natural pixel is at the edge of the finite support, its shape is irregular and the area is approximated by dividing each ray into 20 sub-rays. Non-boundary pixels are calculated exactly. The rate of convergence using natural pixels is slower than iterative algorithms with non-overlapping square pixels but reconstruction artifacts are reduced.

4.1 *Attenuation*

The use of a three-detector SPECT system (Figure 1) has been proposed as a way to simultaneously acquire transmission and emission data in a reasonable patient scanning time so that the reconstructed transmission map can be used to correct for photon attenuation in the emission reconstruction (Tung et al., 1992a). The three detectors are collimated with fan beam collimators and are arranged in a triad with a line source at the focal line of one of the detectors so that transmission and emission data is acquired in one detector while the other two detectors simultaneously acquire emission data. The triangular three- detector arrangement presents a geometrical configuration which requires particular attention to the data acquisition that will yield optimal emission photon statistics and obtain adequate data necessary to correct for cross talk between transmission and emission data.

In cardiac studies, we use a transmission line source of Tc-99m which is selected to be at a different energy than the emission source of Tl-201. A scanning procedure which requires each detector to scan 240 degrees has an initial start angle which gives the highest photon yield (anteriorly scanned angles) for the detectors adjacent to the line source and gives sufficient projections for these detectors which is useful for the implementation of a scatter correction technique. A simple algorithm uses the complimentary data to estimate the back scatter of the transmission source photons from the patient into the emission energy windows for these detectors and subtracts it from the data. A separate experiment estimates the scatter fraction for the detector simultaneously acquiring transmission and emission data. An iterative reconstruction algorithm with a finite support provided by the gantry position information is used to reconstruct the truncated transmission data. The cardiac patient results demonstrate a dramatic improvement in the quantitative accuracy of *in vivo* radionuclide concentration with attenuation correction.

The three detector SPECT system offers a unique detector arrangement that lends itself to easily acquiring simultaneous transmission and emission computed tomography studies and makes it possible to obtain a measure of the distribution of attenuation coefficients in an acceptable clinical imaging time. Fan beam collimation improves the sensitivity for both emission and transmission studies and significantly reduces the source strengths required for the transmission scans. The combination of multiple detectors with converging collimators offers for the first time a reasonable approach to correcting for the variable attenuation in the thorax which has constantly plagued cardiac SPECT imaging.

4.2 *Resolution*

Photon attenuation is a major factor contributing to the quantitative inaccuracy in using SPECT to measure the *in vivo* cardiac distribution of radioactivity. Another factor is the spatially-varying geometric point response of the collimator which causes reconstructed shape distortions, density non-uniformities, and poor resolution recovery of smaller regions of interest. Projection and backprojection operators have been programmed to model the non-uniform attenuation and the three-dimensional spatially-varying geometric point response. Originally, ray-driven projection and backprojection operators were proposed (Zeng et al., 1991b) but more recently pixel driven operators (Zeng and Gullberg, 1992b) have been programmed which are computationally much more efficient. The point response is modelled as a convolution which is efficiently calculated using Fast Fourier transforms.

The model has been used with iterative reconstruction algorithms to process cardiac patient emission data acquired using a three-detector SPECT system (Picker PRISM 3000) with fan-beam collimators and acquired simultaneously with transmission data used to determine the attenuation model. Tl-201 is usually used as the emission source and Tc-99m line source is the transmission source. The non-uniform attenuation map is reconstructed from the transmission data. The results of using an iterative reconstruction algorithm combined with a measured attenuation distribution and a known collimator point spread function demonstrate a dramatic improvement in the quantitative accuracy of *in vivo* radionuclide concentrations for cardiac patient studies. It is also observed that the point response correction improves the image resolution and reduces image noise.

4.3 *Scatter*

Another factor which degrades quantitation in SPECT is scatter. A significant number of photons undergo one or more Compton scatters and are detected within the primary photopeak window. Iterative reconstruction methods allow correction to be made for the effects of these scattered events by including their distribution in the model of the imaging system calculated by the projection and backprojection operations during the reconstruction process. However, predicting the distribution of scattered photons is complicated by the fact that it is inherently object dependant. Information about the structure of the object is collected as part of the attenuation correction scheme as discussed in 4.1. This information is in the form of a map of attenuation coefficients, produced from transmission data acquired simultaneously with the emission data. It was recognized that the number of photons which are scattered into a particular projection bin, via a single Compton interaction, is related to the integral of the linear attenuation coefficients over an area of the object defined by: the geometry of the imaging system; and the initial photon enginery (Welch et al., 1993). The form of this relationship was determined using data generated via the Monte Carlo technique. A simple method of estimating the required area integrals from the set of line integrals calculated as part of the attenuation correction scheme was developed. This method of simulating the distribution of scattered events was incorporated into the model of the imaging system, used by the iterative reconstruction algorithm, for the reconstruction of cardiac images from both phantom and patient data. The resulting images demonstrated both the qualitative and quantitative improvements consequent on using this scatter correction technique. These were most noticeable in regions of low activity near the myocardium.

5. Dynamic Imaging

Fitting compartment models to dynamically acquired data is possible with the development of multi-detector SPECT systems. Research is being performed to determine if regional kinetic parameters of blood volume fraction and exchange between blood and tissue compartments will give a more sensitive indication of myocardial perfusion defects than visual evaluations of static images. Studies use canine models injecting 15 mCi of the heart agent teboroxime (CardioTec, SQUIBB) and a three detector SPECT system (Picker PRISM 3000). Complete sets of sixty projections (each 64 x 64 matrix) are obtained every 5 seconds for 15-30 minutes, and blood samples were drawn from the femoral artery every 10 sec for the first 5 min, then every minute for the last 10-25 min. The projections are reconstructed and tracer concentration versus time curves are formed from regions of interest (ROIs) in the left intraventricular cavity (blood) and the left ventricular wall (tissue). The blood and tissue ROI curves are fit to a two-compartment model using the ROI curve sampled from the intraventricular cavity as the input function for the blood. Typical results can give a blood volume fraction in the heart of approximately 0.20 and 1.54 min^{-1} and 0.926 min^{-1} for blood to tissue and tissue to blood exchange, respectively. The modeling of teboroxime is possible with multi-detector rapid SPECT acquisition and may lead to more accurate evaluation of perfusion than visual analysis.

Dynamic imaging is capable of estimating the three-dimensional distribution of tissue kinetic parameters. However, dynamic imaging places considerable demands on computer memory, disk space, computer speed, and algorithms as kinetic parameters are determined using nonlinear estimation techniques.

6. Summary

The challenge for SPECT is to continue to improve computing hardware in order to provide, in a reasonable clinical time, images that are reconstructed using algorithms that solve inverse problems that better model dynamic biological processes, the physics of the imaging detection process, and complicated detector geometries and scanning orbits which are both utilized to improve image resolution and geometric efficiency.

REFERENCES

1. Bailey D. L., Hutton B. F., and Walker P. J., (1987), Improved SPECT using simultaneous emission and transmission tomography, *J. Nucl. Med.* **28**, 844–851.
2. Cao Z. J., and Tsui B. M. W., (1993 in press), Nonstationary filtering technique for improved cone-beam SPECT, *IEEE Trans. Nucl. Sci.*
3. Clack R., Zeng G. L., Weng Y., Christian P. E., and Gullberg G. T. (1991). *Cone beam single photon emission computed tomography using two orbits*, Information Processing in Medical Imaging, XIIth IPMI International Conference, Kent England, July 7–12, 1991 ed Colchester A. and Hawkes D., Springer–Verlag, New York, Lecture Notes in Computer Science **511**, 45–54.
4. Crawford C. R, Gullberg G .T.,and Tsui B. M. W. (1988). Reconstruction for fan beam with an angular-dependent displaced center-of-rotation, *Med. Phys.* **15**, 67–71.
5. Feldkamp L. A., Davis L. C., and Kress J. W. (1984). Practical cone-beam algorithm, *J. Opt. Soc. Am. A.* **1**, 612–619.
6. Floyd C. D. Jr., Jaszczak R. J., Greer K. L., and Coleman R. E. (1986). Cone beam collimation for SPECT: simulation and reconstruction. *IEEE Trans. Nucl. Sci.* **33**, 511–514.
7. Floyd C. D. Jr., Jaszczak R. J., Greer K. L., and Coleman R. E. (1986). Inverse Monte Carlo as a unified reconstruction algorithm for ECT, *J. Nucl. Med.* **27**, 1577–1585.
8. Formiconi A. R., Pupi A., and Passeri A. (1989). Compensation of spatial system response in SPECT with conjugate gradient reconstruction technique, *Phy. Med. Biol.* **34**, 69–84.
9. Frey E. C, Tsui B. M. W, and Perry J. R. (1992). Simultaneous acquisition of emission and transmission data for improved Tl-201 cardiac SPECT imaging using a Tc-99m transmission source, *J. Nucl. Med.* **33**, 2238–2245.
10. Grangeat P. (1987). *Analysis d'un Systeme D'Imagerie 3D par reconstruction a partir de radiographies X en geometrie conique*, Ph. D. Thesis, l'Ecole Nationale Superieure Des Telecommunications.
11. Gullberg G. T. (1979). The reconstruction of fan-beam data by filtering the backprojection, *Computer Graphics and Image Processing* **10**, 30–47.
12. Gullberg G. T. and Budinger T. F. (1981). The use of filtering methods to compensate for constant attenuation in single-photon emission computed tomography, *IEEE Trans. Biomedical Engineering* **28**, 142–157.
13. Gullberg G. T., Huesman R. H., Malko J. A., Pelc N. J, and Budinger T. F. (1985). An attenuated projector-backprojector for iterative SPECT reconstruction, *Phy. Med. Biol.* **30**, 799–816.

14. Gullberg G. T., Crawford C. R., and Tsui B. M. W. (1986). Reconstruction algorithm for fan beam with a displaced center-of-rotation, *IEEE Trans. on Med. Imag.* **5**, 23–29.

15. Gullberg G. T. (1987). An analytical approach to quantify uniformity artifacts for circular and noncircular detector motion in single photon emission computed tomography imaging, *Med. Phys.* **14**, 105–114.

16. Gullberg G. T., Tsui B. M. W., Crawford C. R., and Edgerton E. R. (1987). Estimation of geometrical parameters for fan beam tomography, *Phy. Med. Biol.* **32**, 1581–94.

17. Gullberg G'. T., Zeng G. L., Tsui B. M. W and Hagius J. T. (1989). An iterative reconstruction algorithm for single photon emission computed tomography with cone beam geometry, *Int. J. of Imag. Sys. and Techn.* **1** , 169–86.

18. Gullberg G. T., Tsui B. M. W., Crawford C. R., Ballard J. G, and Hagius J. T. (1990). Estimation of geometrical parameters and collimator evaluation for cone beam tomography, *Med. Phys.* **17**, 264–72.

19. Gullberg G. T. and Zeng G. L. (1990). *Spectral decomposition of the exponential Radon transform*, In Proceedings of SPIE - The International Society for Optical Engineering, July 9-13, San Diego, CA, SPIE 1351, pp. 288–299.

20. Gullberg G. T, Zeng G. L, Christian P. E., Tsui B. M. W, and Morgan H. T. (1991). *Single photon emission computed tomography of the heart using cone beam geometry and noncircular detector rotation.* Information Processing in Medical Imaging, XIth IPMI International Conference, Berkeley, California June 19–23, 1989 ed Ortendahl D. A and Llacer J., Wiley-Liss, New York, 123–138.

21. Gullberg G. T., Zeng G. L., Christian P. E., Datz F. L., and Morgan H. T. (1991). Cone beam tomography of the heart using SPECT, *Invest. Radiol.* **26**, 681–8.

22. Gullberg G. T. and Zeng G. L. (1992). A cone beam filtered backprojection reconstruction algorithm for cardiac single photon emission computed tomography, *IEEE Trans. on Med. Imag.* **11**, 91–101.

23. Gullberg G. T., Zeng G. L., Datz F. L., Christian P. E., Tung C-H, and Morgan H. T. (1992). Review of convergent beam tomography in single photon emission computed tomography, *Phys. Med. Biol.* **37**, 507–534.

24. Gullberg G. T., Zeng G. L. (1992). A reconstruction algorithm using singular value decomposition to compensate for constant attenuation in single photon emission computed tomography, *J. Nucl. Med.* **33**, 831.

25. Gullberg G. T. and Zeng G. L. (1993, in press) An elliptical orbit backprojection filtering algorithm for SPECT, *IEEE Nucl. Sci.*

26. Hahn L. J., Jaszczak R. J., Gullberg G. T., Floyd C. E., Manglos S. H., Greer K. L., and Coleman R. E. (1988). Noise characteristics for cone beam collimators: a comparison with parallel hole collimator, *Phy. Med. Biol.* **33**, 541–55.

27. Hawman E. G. and Hsieh J. (1986). An astigmatic collimator for high sensitivity SPECT of the brain. *J. Nucl. Med.* **27**, 930.

28. Hsieh Y-L, Zeng G. L, Gullberg G. T, and Morgan H. T. (1993). A method for estimating the parameters of a fan-beam and cone-beam SPECT system using five point sources, *J. Nucl. Med.* **34**, 191.

29. Hu H., Gullberg G. T, and Kruger R. A. (1988). Convolutional reconstruction algorithm for fan beam with concave and convex circular detectors, *IEEE Trans. on Med. Imag.* **7**, 233–238.

30. Jaszczak R. J., Chang L. T., and Murphy P. H. (1979). Single photon emission computed tomography using multislice fan beam collimator, *IEEE Trans. Nucl. Sci.* **26**, 610–18.

31. Jaszczak R. J., Floyd C. E., Manglos S. H., Greer K. L., and Coleman R. E. (1986). Cone beam collimation for single photon emission computed tomography: Analysis, simulation, and image reconstruction using filtered backprojection, *Med. Phys.* **13**, 484–9.

32. Jaszczak R. J., Greer K. L., and Coleman R. E. (1988). SPECT using a specially designed cone-beam collimator, *J. Nucl. Med.* **29**, 1398–1405.

33. Jaszczak R. J., Greer K. L., Floyd C. E., Manglos S. H., and Coleman R. E. (1988). Imaging characteristics of a high resolution cone-beam collimator, *IEEE Trans. Nucl. Sci.* **35**, 644–648.

34. Lang K. and Carson R. (1984). EM reconstruction algorithms for emission and transmission tomography, *J. Comp. Assist. Tomogr.* **8**, 306–316.

35. Li J., Jaszczak R. J., Wang H., Gullberg G. T., Greer K. L., Coleman R. E. (1993 in press) A cone beam SPECT reconstruction algorithm with a displaced center-of-rotation, *Med. Phys.*

36. Liang Z. R., Jaszczak R. J., Turkington T. G., Gilland D. R., and Coleman R. E. (1992). Simultaneous compensation for attenuation, scatter, and detector response of SPECT reconstruction in three dimensions, *Phy. Med. Biol.* **37**.

37. Malko J. A., Van Heertum R. L., Gullberg G. T., and Kowalsky W. P. (1986). SPECT liver imaging using an iterative attenuation correction algorithm and an external flood source, *J. Nucl. Med.* **27**, 701–705.

38. Manglos S. H. (1989). Maximum likelihood reconstruction for cone beam SPECT with camera tilt, *IEEE Trans. Nucl. Sci.* **36**, 1117–21.

39. Manglos S. H., Jaszczak R. J., and Greer K. L. (1989). Cone beam SPECT reconstruction with camera tilt, *Phys. Med. Biol.* **34**, 625–631.

40. Manglos S. H., Jaszczak R. J., and Floyd C. E. (1989). Maximum likelihood reconstruction for cone beam SPECT: Development and initial tests, *Phys. Med. Biol.* **34**, 1947–1957.

41. Manglos S. H., Bassano D. A., Duxbury C. E., and Capone R. B. (1990). Attenuation maps for SPECT determined using cone beam transmission computed tomography, *IEEE Trans. Nucl. Sci.* **37**, 600–608.

42. Manglos S. H., Thomas F. D., and Capone R. B. (1991). Attenuation compensation of cone beam SPECT images using maximum likelihood reconstruction, *IEEE Trans. Med. Imag.* **10**, 66–73.

43. Palmer J. and Wollmer P. (1990). Pinhole emission computed tomography: Method and experimental evaluation, *Phys. Med. Biol.* **35**, 339–350.

44. Smith A. M., Gullberg G. T., Datz F. L., and Christian P. E. (1992). Kinetic modeling of teboroxime using dynamic SPECT imaging, *J. Nucl. Med.* **33**, 878.

45. Smith A. M., Gullberg G. T., Christian P. E., and Datz F. L. (1993). Using teboroxime kinetics as an indicator of myocardial blood flow, *J. Nucl. Med.* **34**, 91P.

46. Smith B. D. (1985). Image reconstruction from cone-beam projections: Necessary and sufficient conditions and reconstruction methods, *IEEE Trans. Med. Imag.* **4**, 14–25.

47. Smith B. D. (1990). Cone-beam tomography: recent advances and a tutorial review, *Opt. Eng.* **29**, 524–34.

48. Tretiak O. J. and Metz C. (1980). The exponential Radon transform, *SIAM J. Appl. Math.* **39**, 341–354.

49. Tsui B. M. W., Gullberg G. T., Edgerton E. R., Gilland D. R., Perry J. R., and McCartney W. H. (1986). Design and clinical utility of a fan beam collimator for SPECT imaging of the head, *J. Nucl. Med.* **27**, 810–19.

50. Tsui B. M. W., Hu H. B., Gilland D. R., and Gullberg G. T. (1988). Implementation of simultaneous attenuation and detector response correction in SPECT, *IEEE Trans. Nucl. Sci.* **35**, 778–783.

51. Tsui B. M. W., Hu H. B., Gilland D. R., and Gullberg G. T. (1988). Implementation of simultaneous attenuation and detector response correction in SPECT, *IEEE Trans. on Nucl. Sci.* **35**, 778–783.

52. Tsui B. M. W., Gullberg G. T., Edgerton E. R., Ballard J. G., Perry J. R., McCartney W. H., and Berg J. (1989). Correction of nonuniform attenuation in cardiac SPECT imaging, *J. Nucl. Med.* **30**, 497–507.

53. Tsui B. M. W and Gullberg G. T. (1990). The geometric transfer function for cone and fan-beam collimators, *Phys. Med. Biol.* **35**, 81–93.

54. Tsui B. M. W., Terry J. A., and Gullberg G. T. (1993 in press). An observer evaluation study of defect detection in cardiac SPECT using cone beam collimation, *Inv. Rad.*

55. Tung C. H., Gullberg G. T., Zeng G. L., Christian P. E., Datz F. L., and Morgan H. T. (1992). Non-uniform attenuation correction using simultaneous transmission and emission converging tomography, *IEEE Trans on Nucl. Sci.* **39**, 1134–1143.

56. Tung C-H, Gullberg G. T. (1992). Emission and transmission noise propagation in cardiac SPECT imaging with non-uniform attenuation correction, *J. Nucl. Med.* **33**, 902.

57. Tuy H. K. (1983). An inversion formula for cone-beam reconstruction, *SIAM J. Math.* **43**, 546–52.

58. Weinstein F. S. (1980). Formation of images using fan beam scanning and noncircular source motion, *J. Opt. Soc. Amer.* **70**, 931–5.

59. Welch A. E., Gullberg G. T., and Christian P. E. (1993). A simple non-uniform scatter correction for SPECT, *J. Nucl. Med.* **34**, 187.

60. Welch A. E., Gullberg G. T., Christian P. E., and Morgan H. T. (1993). Simultaneous transmission-emission Tl-201 cardiac imaging, *J. Nucl. Med.* **34**, 71.

61. Weng Y., Zeng G. L., and Gullberg G. T. (1993 in press) A reconstruction algorithm for helical cone-beam SPECT, *IEEE Nucl. Sci.*

62. Zeng G. L. and Gullberg G. T. (1990). A study of reconstruction artifacts in cone beam tomography using filtered backprojection and iterative EM algorithms, *IEEE Trans. Nucl. Sci.* **37**, 759–67.

63. Zeng G. L. and Gullberg G. T. (1990). *Short-scan cone beam algorithm for circular and noncircular detector orbits*, SPIE Medical Imaging IV Conference, Newport Beach, CA, February 4–9, SPIE 1233, 453–63.

64. Zeng G. L., Tung C. H., and Gullberg G. T. (1990). New approaches to reconstructing truncated projections in cardiac fan beam and cone beam tomography, *J. Nucl. Med.* **31**, 867.

65. Zeng G. L. and Gullberg G. T. (1991). *Short-scan fan beam algorithm for non-circular detector orbits*, SPIE Medical Imaging V Conference, San Jose, CA, February 27—March 1, SPIE 1445, 332–340.

66. Zeng G. L., Gullberg G. T., Tsui B. M. W., and Terry J. A. (1991). Three-dimensional iterative reconstruction algorithms with attenuation and geometric point response correction, *IEEE Trans. Nucl. Sci.* **38**, 693–702.

67. Zeng G. L. and Gullberg G. T. (1992). A cone-beam tomography algorithm for orthogonal circle-and-line orbit, *Phys. Med. Biol.* **37**, 563–577.

68. Zeng G. L. and Gullberg G. T. (1992). Frequency domain implementation of the three-dimensional geometric point response correction in SPECT imaging, *IEEE Trans. on Nucl. Sci.* **39**, 1444–1453.

69. Zeng G. L. and Gullberg G. T. (1993). A backprojection filtering algorithm for a variable focal length collimator, *J. Nucl. Med.* **34**, 81.

70. Zeng G. L. and Gullberg G. T. (1993). An iterative reconstruction algorithm with natural pixel basis, *J. Nucl. Med.* **34**, 19.

71. Zeng G. L., Gullberg G. T., Jaszczak R. J., and Li J. (1993 in press). Fan-beam convolution reconstruction algorithm for a spatially varying focal length collimator, *IEEE Trans. on Med. Imag.*

Application of GCV-Aided Phillips-Tikhonov Regularization to Direction Finding of Magnetospheric VLF/ELF Radio Waves

K. HATTORI[1], M. YAMAGUCHI[2], N. IWAMA[1] and M. HAYAKAWA[2]

[1] Department of Electronics and Informatics, Faculty of Engineering
Toyama Kenritsu University, Kosugi, Toyama 939-03, Japan

[2] The University of Electro-Communications, 1-5-1 Chofugaoka, Chofu
Tokyo 182, Japan

ABSTRACT

A new ground-based direction finding method based on the wave distribution function(WDF) with use of the linear regularization is proposed with emphasis on the optimization by means of the generalized cross validation(GCV). The simulations are examined on this problem and their results indicate the effectiveness of the proposed method.

1. Introduction

The direction finding(DF) of magnetospheric VLF/ELF radio waves is known to be a very important technique for the study of not only their generation and propagation mechanisms, but also of the dynamics of the magnetospheric plasma (e.g. Hayakawa et al., 1990; Hayakawa, 1991). Several methods for DF by using simultaneous recording of multiple field components have been developed and the present paper deals with the wave distribution function(WDF) method (Storey and Lefeuvre, 1979; Lefeuvre et al., 1979; Shimakura and Hayakawa, 1987,1989; Shimakura et al., 1992) which indicates the spatial distribution of wave energy. This WDF method is based on a numerical inversion from a spectral matrix composed of the auto- and cross-power spectra among observed field components. The maximum inversion procedure has been developed and so far used to reconstruct WDF images (Lefeuvre et al., 1979; Shimakura et al., 1992). However, this method involves a few problems;(1)the estimation of the optimum WDF image is less objective, (2)the computation time is large due to the nonlinear regularization by maximum entropy method(MEM). A method to overcome these difficulties involved the maximum entropy inversion is strongly required and we have to develop an objective method to estimate the optimum WDF image with small computation time. For this aim, a method of the linear regularization together with the generalized cross validation(GCV) is examined on the WDF imaging for the ground-based DF in this paper.

2. Reconstruction Procedure

2.1 *Phillips-Tikhonov Regularization*

This study on the DF is reduced to a problem on the two-dimensional distribution image. A general formulation is given in the following. We partition a reconstruction region into K pixels and take a fully discrete formula,

$$\boldsymbol{S} = \boldsymbol{AF} + \boldsymbol{e} \tag{1}$$

where $\boldsymbol{S}$ and $\boldsymbol{F}$ are M-dimensional datum vector and K-dimensional image vector by M available observational values and K image intensities ordered lexicographically. The M $\times$ K matrix $\boldsymbol{A}$ is the projection matrix with the geometrical weight A_{mk} of the k-th pixel is producing the m-th projection. $\boldsymbol{e}$ represents the M-dimensional noise vector which includes the error associated with the measurement and/or the discretization. Therefore, this is an inversion problem to estimate $\boldsymbol{F}$ from given data $\boldsymbol{S}$ and the projection constant $\boldsymbol{A}$.

In the case of a small number of data M<K, the solution may not be determined. To determine the solution, the regularization is needed subject to

$$\frac{1}{M}|\boldsymbol{S} - \boldsymbol{AF}|^2 \leq \varepsilon^2 \tag{2}$$

where ε^2 is the error limit. The previous MEM adopted to DF is one possible nonlinear regularization. This paper takes a linear regularization of Phillips-Tikhonov (Phillips, 1962; Tikhonov, 1965; Tikhonov and Arsenin, 1986) such that an objective function G(F) as the double integral of the square of the spatial double differential of image, that is Laplacian, is made minimum with constraint of ε^2.

$$G(F) = \int\int |\nabla^2 F(x,y)|^2 dxdy \tag{3}$$

This minimization is corresponding to the minimization of the following quantity J.

$$J = \gamma|\boldsymbol{CF}|^2 + \frac{1}{M}|\boldsymbol{S} - \boldsymbol{AF}|^2 \tag{4}$$

where γ is the reciprocal of the Lagrangian multiplier, and $\boldsymbol{C}$ is a K $\times$ K matrix representing the Laplacian operator (Iwama et al., 1989; Yoshida, 1988). And $\boldsymbol{C}$ is described as follows.

$$\boldsymbol{C} = \begin{bmatrix} -4 & 1 & 0 & \cdots & 0 & 1 & 0 & 0 & 0 & \cdots & 0 \\ 1 & -4 & 1 & 0 & \cdots & 0 & 1 & 0 & 0 & \cdots & 0 \\ 0 & 1 & -4 & 1 & 0 & \cdots & 0 & 1 & 0 & \cdots & 0 \\ \vdots & \ddots & \ddots & \ddots & \ddots & \ddots & \ddots & \ddots & \ddots & \ddots & \vdots \\ \\ 0 & \cdots & 0 & 0 & 1 & 0 & \cdots & 0 & 1 & -4 & 1 \\ 0 & \cdots & 0 & 0 & 0 & 1 & 0 & \cdots & 0 & 1 & -4 \end{bmatrix}$$

By performing partial differential at F_i, and replacing it 0, we obtain the solution $\hat{\boldsymbol{F}}(\gamma)$ as a linear operation to $\boldsymbol{S}$.

$$\hat{\boldsymbol{F}} = (\boldsymbol{A}^t\boldsymbol{A} + M\gamma\boldsymbol{C}^t\boldsymbol{C})^{-1}\boldsymbol{A}^t\boldsymbol{S} \tag{5}$$

In this formula, the weight γ is a regularization parameter. With increase of γ, more importance is given to $|\boldsymbol{CF}|^2$ in minimizing J and as a result, the image estimate $\hat{\boldsymbol{F}}(\gamma)$ will get the smoothing of profile with a loss of spatial resolution and with an increased ε^2. With the decrease of γ, the reverse will occur.

2.2 Generalized Cross Validation(GCV)

It is important to choose an appropriate γ, which provides an optimum reconstructed image $\hat{\boldsymbol{F}}$. For this aim, the statistical criterion of PRESS(prediction sum of squares) and its generalized version named the generalized cross validation(GCV) (Golub et al., 1979) may be useful. The PRESS(γ) is defined as the mean squared error in "predicting" the i-th projection datum S_i from the estimating $\hat{\boldsymbol{F}}^{(i)}(\gamma)$ which we obtain from the data without S_i.

$$PRESS(\gamma) = \frac{1}{M}\sum_{m=1}^{M}(\boldsymbol{A}\hat{\boldsymbol{F}}^{(i)}(\gamma)_i - \boldsymbol{S}_i)^2, \tag{6}$$

This quantity indicates a measure of the goodness of the estimation operator $(\boldsymbol{A}^t\boldsymbol{A} + M\gamma\boldsymbol{C}^t\boldsymbol{C})^{-1}\boldsymbol{A}^t$ as an ensemble of the projection value $\boldsymbol{S}$. The value of γ which gives the minimum PRESS(γ) is expected optimum in this sense that the corresponding projection estimate given by $\hat{\boldsymbol{S}}(\gamma) = \boldsymbol{A}\hat{\boldsymbol{F}}(\gamma)$ is best approximation to true projection. However, we have to repeat the image reconstruction M-times in order to obtain PRESS(γ) directly, so that a long computation time may be required. Furthermore, in the extreme case, the matrix $\boldsymbol{B}(\gamma) = (\boldsymbol{A}^t\boldsymbol{A} + M\gamma\boldsymbol{C}^t\boldsymbol{C})^{-1}\boldsymbol{A}^t$ is diagonal in relating $\boldsymbol{S}$ to $\hat{\boldsymbol{S}}(\gamma) = \boldsymbol{B}(\gamma)\boldsymbol{S}$, this criterion breaks down because of the prediction is damaged by excepting S_i. In order to overcome this difficulty, the GCV is adopted by the singular value decomposition(SVD) (Iwama et al., 1989; Yoshida, 1988) of $\boldsymbol{A}\boldsymbol{C}^{-1} = \boldsymbol{U}\Sigma\boldsymbol{V}^t$, where Σ is the $M \times K$ diagonal matrix having singular values $\sigma_j (j = 1, 2, \ldots, M; \sigma_1 \geq \sigma_2 \geq \ldots \geq \sigma_M \geq 0)$ as its diagonal, and $\boldsymbol{U}$ and $\boldsymbol{V}$ are the related M $\times$ M and K $\times$ K orthogonal matrices, respectively. A unitary transformation of the spectral matrix and the kernel, which keeps the norm of $|\boldsymbol{S} - \boldsymbol{A}\boldsymbol{F}|^2$ lead a newly defined PRESS to the standard form of the Generalized Cross Validation(GCV) written as

$$GCV(\gamma) = M|(\boldsymbol{I} - \boldsymbol{B}(\gamma))\boldsymbol{S}|^2/\{Tr[\boldsymbol{I} - \boldsymbol{B}(\gamma)]\}^2, \tag{7}$$

and we finally obtain

$$GCV(\gamma) = M\sum_{j=1}^{M}(\rho_j z_j)^2/(\sum_{j=1}^{M}\rho_j)^2 \tag{8}$$

with the j-th components z_j of $\boldsymbol{U}^t\boldsymbol{S}$ and $\rho_j = \boldsymbol{M}\gamma/({\sigma_j}^2 + \boldsymbol{M}\gamma), \rho_j$ being assumed zero for $j = M+1, M+2, \ldots, K$; $\boldsymbol{I}$ denotes the $M \times M$ unit matrix. The SVD also facilitates the computations of $PRESS(\gamma)$ and $\hat{\boldsymbol{F}}(\gamma)$ since we have

$$PRESS(\gamma) = \frac{1}{M}\sum_{k=1}^{M}\sum_{j=1}^{M}\rho_j u_{kj} z_j / \sum_{j=1}^{M}\rho_j u_{kj}^2 \tag{9}$$

and

$$\hat{\boldsymbol{F}}_k(\gamma) = \sum_{j=1}^{M}\frac{\sigma_j}{\sigma_j^2 + M\gamma}v'_{kj}z_j \tag{10}$$

where u_{kj} and v'_{kj} denote the kj-th components of $\boldsymbol{U}$ and $\boldsymbol{V}' \equiv \boldsymbol{C}^{-1}\boldsymbol{V}$, respectively. The GCV is expected to indicate a minimum as a function of γ and give the optimally reconstructed image there. Also, the regularization yields that the final solution $\hat{\boldsymbol{F}}$ and GCV are obtained by simple linear calculations with saving a lot of computation time.

2.3 *Application to Direction Finding*

The basic formula for DF of magnetospheric VLF/ELF radio waves using the WDF is given with use of the polar coordinate system in the following.

$$S(\omega) = \frac{\pi}{2} \int_0^{2\pi} \int_0^1 A(\omega, \theta, \phi) F(\omega, \theta, \phi) d(\cos\theta) d\phi \tag{11}$$

where $F(\omega, \theta, \phi)$ is the WDF to be determined, $S(\omega)$ is the spectral matrix composed of auto- and cross-power spectra of observed multi-field components and $A(\omega, \theta, \phi)$ indicates the integral kernel matrix which shows the propagation constants in free space (Shimakura and Hayakawa, 1989). And this equation is given at an angular frequency ω. This paper is discussed on the ground-based DF which investigates the location of the exit region of magnetospheric VLF/ELF waves at the ionospheric base of altitude 80km. A sufficiently large rectangular region($300 \times 300 km^2$) region at the ionospheric base over the observing point is considered as the reconstruction region. We partition this region into K_x in the x direction and into K_y in the y direction, therefore there are $K = K_x \times K_y$ pixels.

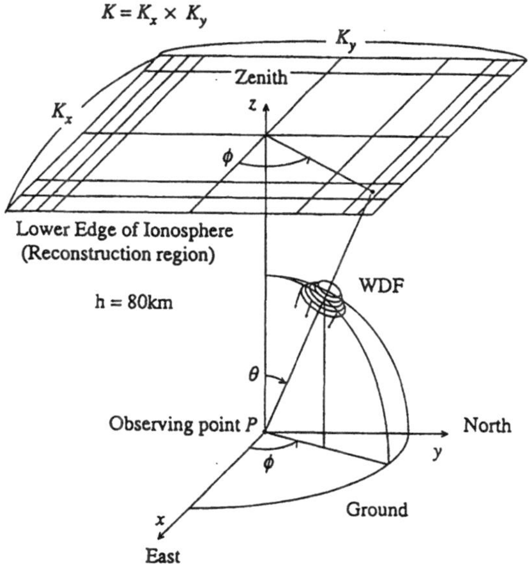

FIGURE 1. Reconstruction region at the lower ionospheric base and WDF on the spherical plane.

The coordinate system for the WDF reconstruction is described in Fig. 1 with the incident angle θ and the azimuthal angle ϕ. And we take a fully discrete formula, same as Eq. (1),

$$\boldsymbol{S} = \boldsymbol{A}\boldsymbol{F} + \boldsymbol{e} \tag{12}$$

where $\boldsymbol{S}$ is M-dimensional datum vector by M available observation values. Since elements of the spectral matrix are defined by the auto- and cross-power spectra of observed three field components(two horizontal magnetic and one vertical electric fields) for the ground-based observation based on the assumption that the ground is the perfect conductor for VLF/ELF radio waves, the component of the spectral matrix S_{ij} is expressed in complex except $i = j$. Therefore, there are 15 elements which are corresponding to the real part and the imaginary part of the individual spectral matrix. But we can omit conjugate components such as $S_{ij} = S^*_{ji}$. Therefore, the number of the elements of $\boldsymbol{S}$ is reduced to 9, that is M=9 for the ground-based DF. $\boldsymbol{e}$ represents M-dimensional residual vector which includes the error associated with the measurement and/or the discretization. $\boldsymbol{A}$ is M $\times$ K matrix corresponding to the integral kernel, in which A_{mk} ($m = 1, \ldots, M$) is the weight for the wave propagation relating the wave energy distribution of the k-th pixel to the m-th spectral data.

As mentioned above, M=9 for the ground-based DF and we take K=400. The Phillips-Tikhonov regularization with the GCV is examined on the reconstruction of WDF for the ground-based DF.

3. Numerical Simulations

We have carried out the application of this new ground-based DF method to simulated data which provide the Gaussian-shaped wave source.

3.1 *Generation of Simulated Data*

Numerical simulations have been performed in the following procedure.

1. The received signal is assumed to be a superposition of whistler-mode waves or right-handed circularly polarized waves and their energy distribution on unit sphere is assumed to have the Gaussian-shaped distribution around the wave source center, which is given by the incident angle θ relative to the zenith of $\theta = 0^o$ and azimuthal angle ϕ, as shown Fig. 1.

2. As for above WDF, we make a corresponding spectral matrix to be observed on the ground and add the statistical error/noise to each component in the following procedure.

$$\hat{S}_{ii}(\omega) = S_{ii}(\omega)(1 + \frac{N}{S}r) \tag{13}$$

$$\hat{S}_{ij}(\omega) = S_{ij}(\omega) + (S_{ii}(\omega) \times S_{jj}(\omega))^{\frac{1}{2}}\frac{N}{S}r) \tag{14}$$

   where r indicates the uniform random number between +1 and -1, and N/S is a given SN ratio.

3. The detailed data are processed by the proposed method of reconstruction to get the WDF at the ionospheric base.

3.2 *Reconstruction of the WDF for a Single Wave Source*

Fig. 2 illustrates that the WDF of the assumed single wave source at the ionospheric base with use of the contour map. Concentric circles indicate θ values; the innermost one corresponds to $\theta = 15^o$ and the circles are drawn by every 15°, therefore the outermost one represents $\theta = 60^o$. The outer square means the boundary of the reconstruction region and the corresponding length of a side is 300km. The location of the maximum WDF of the assumed source is $(\theta, \phi) = (30^o, 45^o)$ on the spherical surface, and it is shown by symbol * in Fig. 2. The values of the WDF are normalized by it's peak value and the contour curve nearest to the peak represents 90% of the maximum density. The outer contour lines are drawn at every 20% decrease of the density. The standard deviation σ_Θ

of the assumed Gaussian-shaped WDF corresponds to the width of wave source and it is 30° in this case. The SN ratio is assumed to be 10dB in this case. The reconstructed WDF images are compared to the assumed one from the view of the location of wave center and the shape of energy distribution and the optimally reconstructed image will be determined.

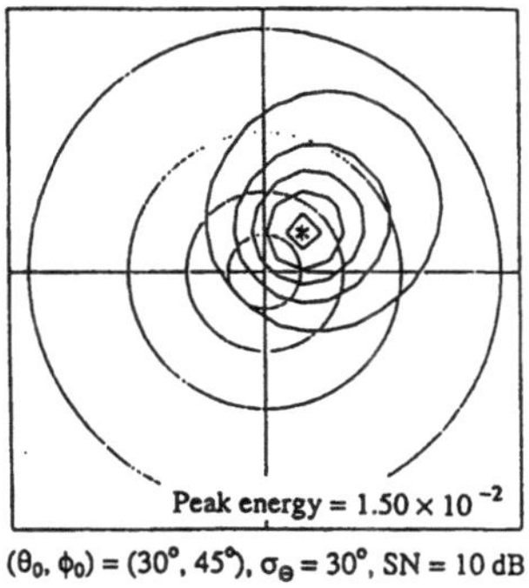

FIGURE 2. A simulated WDF image at the lower ionospheric base.

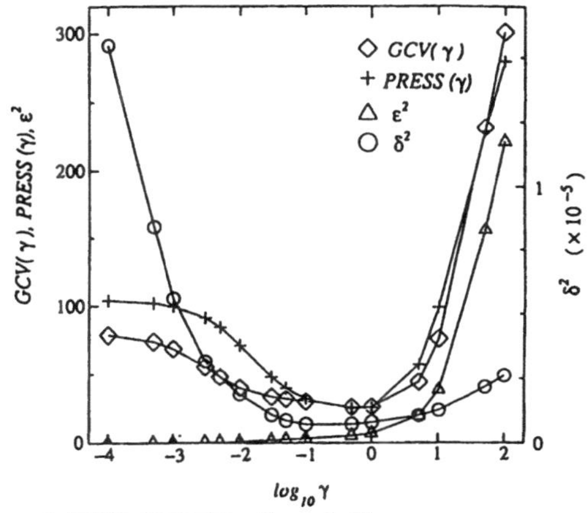

FIGURE 3. γ dependence of GCV, PRESS, ε^2, and δ^2.

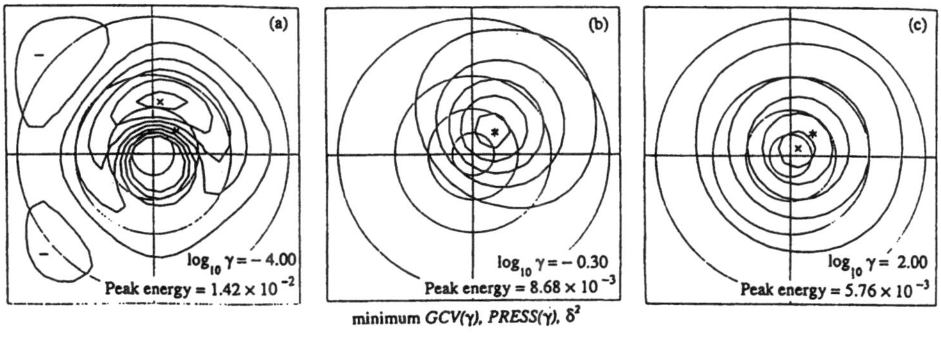

FIGURE 4. Profiles of the reconstruction WDF at the specific γ value.

(a)$\log_{10}\gamma = -4.0$,

(b)$\log_{10}\gamma = -0.3$, minimum GCV(γ), PRESS(γ), and $\delta^2(\gamma)$,

(c)$\log_{10}\gamma = 2.0$.

Figs.3 and 4 show typical results: the variations of PRESS(γ), GCV(γ), $\varepsilon^2(\gamma)$ and

$\delta^2(\gamma)$, which are indicated +, ◇, × and o, respectively, and the profiles of the reconstructed WDF images at the specific γ values. $\varepsilon^2(\gamma)$ and $\delta^2(\gamma)$ are the mean squared errors in approximating the spectral data and the assumed WDF at the ionospheric base, respectively and they are given in the following.

$$\varepsilon^2(\gamma) = \frac{1}{M}|\boldsymbol{S} - \boldsymbol{A}\hat{\boldsymbol{F}}(\gamma)|^2 \tag{15}$$

$$\delta^2(\gamma) = \frac{1}{K}|\hat{\boldsymbol{F}}(\gamma) - \boldsymbol{F}_0|^2 \tag{16}$$

$\hat{\boldsymbol{F}}(\gamma)$ and $\boldsymbol{F}_0$ mean the reconstructed and assumed WDF, respectively.

values of PRESS(γ) and GCV(γ) are asymptotic to the curve of $\varepsilon^2(\gamma)$ when γ becomes large. δ^2 is a powerful estimator for choosing the optimally reconstructed image when an input image is known in advance such as simulations.

When γ is sufficiently small at ($\log_{10}\gamma = -4.0$) as in Fig. 4(a), the value of ε^2 is also small and the reconstructed image is completely different from the assumed one shown in Fig. 3. It is the weakly regularized profile with negative values, which cannot be excluded in this linear algebraic analysis, and the wave center, which is marked ×, is shifted to the larger θ. With increase of γ, the reconstructed WDF is separated from the spectral data as indicated by the monotone increase of ε^2. Its shape is gradually improved with decrease of δ^2. δ^2 takes a minimum at $\log_{10}\gamma = 0.3$ and turns to increase. It is found that this best improvement in reconstruction is well indicated by the behaviors of GCV(γ) and PRESS(γ). Both evaluating quantities are minimized at the same γ value. The corresponding WDF image is Fig. 4(b). Negative values of WDF are well eliminated in this figure and the estimation error of the location of the wave center is found to be in a few degrees in the θ direction and there is no error in the azimuthal ϕ direction. And also the half-width of 50% contour indicates 27.8° in the θ direction. That of the assumed WDF is 25.1°. Therefore, it is found that the wave center and the shape of the wave source are well-reconstructed. Then, with further increase of γ, the reconstructed WDF gets the excessive smoothing and increasing δ^2 as we find increase of GCV(γ) and PRESS(γ). The curves of GCV(γ) and PRESS(γ) seem to be asymptotic to that of $\varepsilon^2(\gamma)$ in large γ region, which are theoretically expected. As shown in Fig. 4(c). the reconstructed WDF at $\log_{10}\gamma = 2.0$ is singly peaked near the zenith and it is far from the assumed one, again. Therefore, the best estimate of WDF is considered to be obtained at the minima of GCV(γ) and PRESS(γ).

Similar simulations have been carried out with changing the position of the wave center, standard deviation of the WDF and SN ratio in spectral data. The results of analysis show that the behaviors of GCV(γ) and PRESS(γ) are similar on the γ values for which they are minimized, and good enough in indicating the optimum estimate of WDF.

3.3 *Reconstruction of WDF for Two Wave Sources*

Similar simulations are performed for two wave sources, which is considered to be essential in studying the usefulness of our proposed method, and $300 \times 300 km^2$ scheme is also adapted. One source is fixed in space and the second source is widely varied. We show a typical result. Fig. 5 illustrates the assumed WDF; the first peak is located at the point $(\theta_1, \phi_1) = (40°, 300°)$ and the second peak with 70% amplitude of the first one, at $(\theta_2, \phi_2) = (40°, 120°)$. The peak positions are indicated by symbol $*$. We take 20° as the standard deviation for both sources and 10dB as the SN ratio. Fig. 6 indicates the variation of GCV(γ), PRESS(γ), $\varepsilon^2(\gamma)$, and $\delta^2(\gamma)$ with the regularization parameter. The variations of $\varepsilon^2(\gamma)$ and $\delta^2(\gamma)$ are similar to the case on the single wave source as shown in Fig. 3. Figs. 7 show the profiles of the reconstructed WDFs. When γ is very small as shown in Fig. 7(a)($log_{10}\gamma = -4.0$), the reconstructed WDF indicates two wave sources but it is completely different from the assumed one; it has a big cavity near the zenith and looks like a crescent. With increase of γ, these structures disappear gradually and the obtained WDF image becomes closer to the assumed one. When $\log_{10}\gamma = -1.0$ in Fig. 7(b), the optimal function GCV(γ) indicates minimum. It is found that there still exists the negative region of WDF. Fig. 7(c) is the reconstructed WDF at $log_{10}\gamma = -0.3$, which is corresponding to the minimum PRESS(γ) and $\delta^2(\gamma)$. This is most fitted to the assumed one. With further increase of γ, the reconstructed WDF exhibits a considerable difference. The reconstructed image represents a widely extended single wave source at $\log_{10}\gamma = 2.0$ as shown in Fig. 7(d).

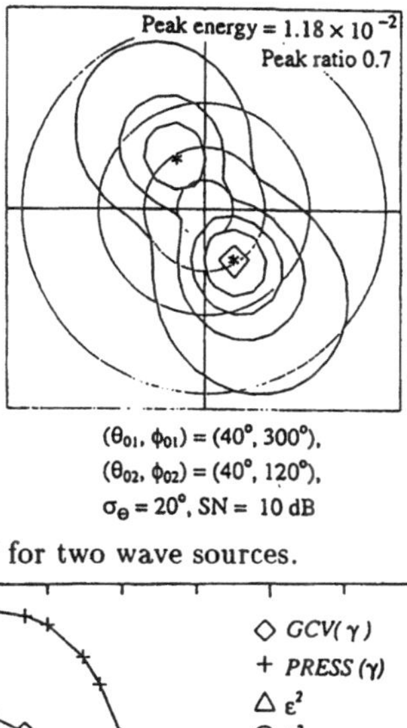

FIGURE 5. The assumed WDF for two wave sources.

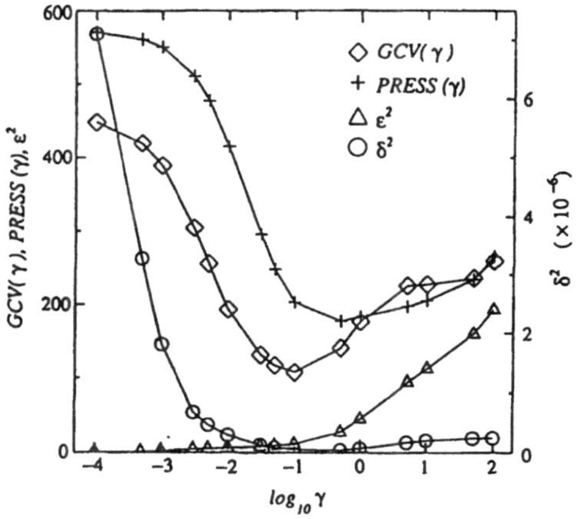

FIGURE 6. γ dependence of GCV, PRESS, ε^2, and δ^2.

The optimally reconstructed WDF is obtained at the γ value which indicates the minimum PRESS(γ) in this case. However, the optimal images are found to be obtained at the minimum GCV(γ) and/or PRESS(γ) through a lot of two wave simulations. It is confirmed that there is no statistical superiority of the criterion, which is the minimum

GCV(γ) or PRESS(γ), for the selection of the optimum WDF. The corresponding γ values giving the minimum GCV(γ) and PRESS(γ) are close in every case and they seem to predict well-reconstructed images. Therefore, a similar conclusion to the single wave source is obtained.

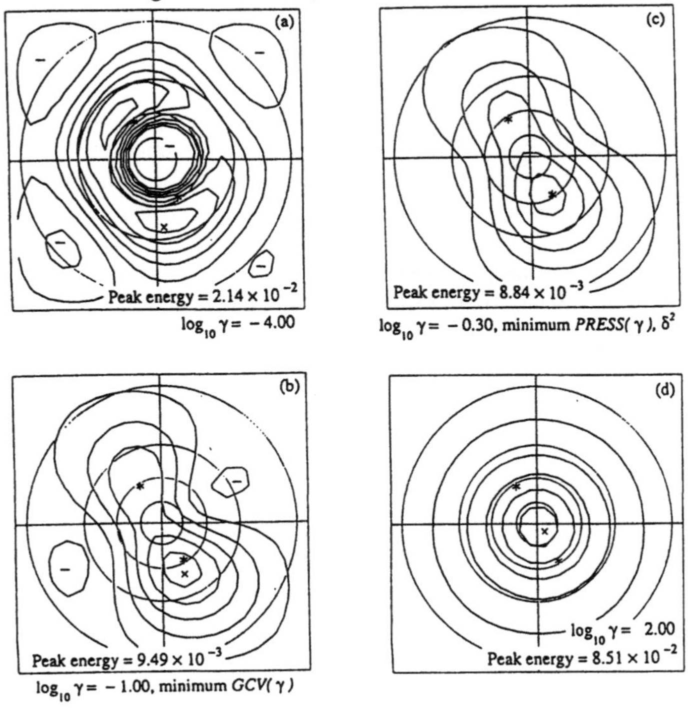

FIGURE 7. Profiles of the reconstructed WDF. (a) $\log_{10}\gamma = -4.0$, (b) $\log_{10}\gamma = -0.3$, the minimum GCV(γ), (c) $\log_{10}\gamma = -0.3$, the minimum PRESS(γ) and $\delta^2(\gamma)$, (d) $\log_{10}\gamma = 2.0$.

The results of various simulations are summarized in Table.1. This table indicates the critical values in θ or ϕ direction to be successful in the two wave separation in the points of the prediction of two wave center and the shape of WDF at the minimum GCV(γ). The upper columns refer to the case of various θ values, with $\Delta\phi = 180^o$, and relative ratio of peak energy of two wave source is 0.7. The same θ values are taken for two wave sources. The lower, corresponds to the case in changing the ϕ with $\Delta\theta = 40^o$, and relative energy ratio is assumed 1.0. The assumed standard deviation, the estimation error in distance, the width of reconstructed WDF normalized by the assumed one at the ionospheric base, and γ value at the minimum GCV(γ). We can understand that the standard deviation of the assumed sources is significant in distinguishing two wave sources.

4. Conclusion

The linear regularization of Phillips–Tikhonov type in least squares estimation was adopted to the WDF reconstruction for DF of VLF/ELF radio waves. With the objective function of Laplacian type, the SVD of $\boldsymbol{AC}^{-1}$ lead to the easily computed expressions of GCV(γ) and PRESS(γ), and also the WDF estimate. Numerical simulations with respect to the maximum entropy method in objectively determining the optimum value of the regularization parameter and in short computation time. The proposed method can be extended to the other types of spatial resolution and noise elimination, and can also be applied to the satellite observation by changing the integration kernel in Eq. (1).

σ_Θ width	$\log_{10}\gamma$ at min. GCV	$\Delta\theta$ critical distance	Δr error of peak position	$(\hat{L}-L_0)/L_0$ error of width
peak ratio = 0.7,	SN = 10dB			
15°	−1.00	70°	22.5*km*	0.52
20°	−1.00	80°	15.0*km*	0.37
30°	−0.30	90°	12.2*km*	0.36
peak ratio = 0.7,	SN = 20dB			
15°	−2.00	80°	9.4*km*	0.52
20°	−2.30	80°	~ 0.0*km*	0.73
30°	−3.30	80°	11.3*km*	0.19
peak ratio = 0.7,	SN = 30dB			
15°	−2.52	80°	7.5*km*	0.64
20°	−3.30	80°	~ 0.0*km*	0.43
30°	−2.00	100°	3.8*km*	0.23

σ_Θ width	$\log_{10}\gamma$ at min. GCV	$\Delta\phi$ critical distance	Δr error of peak position	$(\hat{L}-L_0)/L_0$ error of width
peak ratio = 1.0,	SN = 10dB			
15°	−1.00	120°	18.8*km*	0.52
20°	−1.00	135°	16.8*km*	0.37
30°	–	–	–	–
peak ratio = 1.0,	SN = 20dB			
15°	−2.00	135°	13.1*km*	0.68
20°	−2.30	135°	5.6*km*	0.42
30°	–	–	–	–
peak ratio = 1.0,	SN = 30dB			
15°	−2.52	135°	13.1*km*	0.73
20°	−3.30	135°	5.6*km*	0.39
30°	–	–	–	–

TABLE. 1. The summary of the reconstruction for two wave sources.

REFERENCES

Golub G. H., Heath H., and Wahba G. (1979), Generalized cross validation as a method for choosing a good ridge parameter, *Technometrics*, **21**, 215–223.

Hayakawa M., Ohta K., and Shimakura S. (1990), Spaced direction finding of whistlers at low and equatorial latitudes and their propagation mechanism, *J. Geophys. Res.*, **95**, 15091–15102.

Hayakawa M. (1991), *Direction Finding of Magnetospheric VLF/ELF Emissions and their Generation Mechanism*, In: "Environmental and Space Electromagnetics", Ed. by H.Kikuchi, Elsevier Pub. Comp., 155–167.

Iwama N., Yoshida H., Takimoto H., Shen Y., Takamura S., and Tsukishima T. (1989), Phillips-Tikhonov regularization of plasma image reconstruction with the generalized cross validation, *Appl. Phys. Lett.*, **54**, 502–504.

Lefeuvre F., Parrot M., and Delannoy C. (1979), Wave distribution functions estimation of VLF electromagnetic waves observed onboard GEOS-1, *J. Geophys. Res.*, **86**, 2359–2375.

Phillips D. L. (1962), A technique for the numerical solution of certain integral equations of the first kind, *J. Assoc. Comput. Mach.*, **9**, 84–97.

Shimakura S. and Hayakawa M. (1987), *On the Estimation of Wave Energy Distribution of Magnetospheric VLF Waves at the Ionospheric Base with Ground-Based Multiple Electromagnetic Field Components*, Proc. Chapman Conf. on Plasma Waves and Instabilities in Magnetospheres and at Comets, Sendai, 238–241.

Shimakura S. and Hayakawa M. (1989), *A New Direction Finding Technique for Magnetospheric VLF Waves Observed on the Ground*, Proc. Int'l Symp. on Antennas and Propagation, Tokyo, 751–754.

Shimakura S., Hayakawa M., Lefeuvre F., and Lagoutte D. (1992), On the estimation of wave energy distribution of magnetospheric VLF waves at the ionospheric base with ground-based multiple electromagnetic field components, *J.Geomag.Geoelectr.*, **44**, 573–590.

Storey L. R. O. and Lefeuvre F. (1979), The analysis of 6-component measurements of a random electromagnetic wave field in a magneto-plasma-I. The direct problem, *Geophys. J. Roy. Astron. Soc.*, **50**, 255–269.

Tikhonov A. N. (1965), Solution of incorrectly formulated problems and the regularization method, *Dokl. Akad. Nauk.*, **151**, 1035. (in Russian)

Tikhonov A. N. and Arsenin V. J. (1986), *Solution methods of ill-posed problems*, Nauka, Moscow. (in Russian)

Yoshida H. (1988), *Plasma Image Reconstruction Using the Phillips–Tikhonov Regularization*, Thesis, Nagoya Univ.

A Problem of Emission Tomography for a Domain of Small Diameter

R. Sh. ISANOV

Novosibirsk State University

ABSTRACT

We consider a system of emission tomography in $\mathbf{R}^n$ for the case when the absorption coefficient is a smooth function with a finite support. The main result is the theorem of existence, uniqueness and stability of the solution of the emission tomography problem.

In the present paper we consider a problem of emission tomography in $\mathbf{R}^n$ for the case when the absorption coefficient is a smooth function with a finite support. The main result of the paper is the theorem of existence, uniqueness and stability of the solution of the emission tomography problem. The works by Natterer (1980, 1986) deal with problems of transmission tomography, whereas in Natterer (1986), Yakhno and Klimenko (1991) some problems of emission tomography are considered which can be reformulated in a differential form (Yakhno and Isanov, 1991). In this paper we develop the results of Tikhonov et al. (1987), Tretiak and Metz (1980), Gryn (1991), Isanov and Yakhno (1991).

Let $x = (x_1, x_2, \cdots, x_n) \in \mathbf{R}^n$, $\theta \in S^{n-1} = \{\theta \in \mathbf{R}^n \mid |\theta| = 1\}$, $\Theta^\perp$ be a subspace orthogonal to θ, $T = \{(\theta, x) \mid \theta \in S^{n-1},\ x \in \Theta^\perp\}$ be a tangential stratification of S^{n-1}, and $\Omega \subseteq \mathbf{R}^n$ be a domain of a finite diameter d.

Let $H_0^\alpha(\Omega)$ be a space of functions $\psi(x)$ equipped with the norm

$$\|\psi(x)\|_\alpha = \left(\int_{\mathbf{R}^n} (1 + |\xi|)^\alpha \left|\hat{\phi}(\xi)\right|^2 d\xi \right)^{1/2} < \infty,$$

where $\hat{\psi}(\xi)$ is the Fourier transform of $\psi(x)$ and $\alpha \in \mathbf{R}$. Denote $H_0^\alpha(\Omega) = \{f(x) \subseteq H^\alpha(\mathbf{R}^n) \mid \text{supp } f(x) \subseteq \bar{\Omega}\}$. Introduce now a P - transformation by the formula

$$Pf(x, \theta) = \int_{-\infty}^{\infty} f(x, \theta t) dt,$$

where $x \in \mathbf{R}^n$, $\theta \in S^{n-1}$. The P - transformation is the integral of the function $f(x)$ along the line passing through the point x in the direction θ. Let η^α be the image of the space $H_0^\alpha(\Omega)$ under P-transformation, that is, $PH_0^\alpha(\Omega) = \eta^\alpha$. The set η^α is a linear subspace of the space $H^\beta(T)$, where $\beta = \alpha + \frac{1}{2}$. From Theorem 1-3 by Natterer (1980) it follows that $P^{-1} : \eta^\alpha \to H_0^\alpha(\Omega)$ is a linear and continuous, and the following estimate holds:

$$\frac{1}{\Gamma(\alpha)} \|g\|_{\beta,T} \leq \left\|P^{-1} g\right\|_\alpha \leq \frac{1}{\gamma(\alpha)} \|g\|_{\beta,T},$$

where $g \in \eta^\alpha$, $\Gamma(\alpha)$ and $\gamma(\alpha)$ are constant. Introduce now operators D and P_μ as:

$$D_\mu(x,\theta) = \int_{-\infty}^{\infty} \mu(x+\theta t)dt,$$

$$(P_\mu f)(x,\theta) \int_{-\infty}^{\infty} f(x+\theta t)\exp\left\{-\int_0^{\infty} \mu(x+\theta t+\theta t_1)dt_1\right\} dt, \tag{1}$$

for any function $\mu(x)$, $f(x)$ of the space $H_0^\alpha(\Omega)$.

We consider the equality

$$(P_\mu f)(\theta,x) = g(\theta,x), \quad x \in \Theta^\perp, \quad \theta \in S^{n-1}. \tag{2}$$

The main subject of the paper is the following problem.

PROBLEM. Let $\mu(x)$ be a known function in $H^l(\Omega)$. We need to determine the function $f(x)$ in $H_0^\alpha(\Omega)$ from a given $g(\theta,x)$, $x \in \Theta^\perp$, $\theta \in S^{n-1}$, where l and α are fixed real numbers.

REMARK. This problem is called a problem of emission tomography. From a physical point of view this problem is to find of the distribution density of radiation sources for a given coefficient of absorption and given values of the radiation intensity along any lines intersecting the cross domain Ω.

Let α be a fixed number in $(-\frac{1}{2},\infty)$, $\beta = \alpha + \frac{1}{2}$; $\mu(x)$ be a given function in $H^l(\mathbf{R}^n)$ such that supp $\mu(x) \in \bar{\Omega}$, where $l \geq (\alpha+\frac{1}{2})+1$. assume that K is a positive number such that $\left|\frac{\partial^j \mu(x)}{\partial x_i^j}\right| \leq K$ almost everywhere, $j = \overline{1,l}$, $i = \overline{1,n}$. For fixed $x \in \mathbf{R}^n$, $\theta \in S^{n-1}$ the function $\mu(x+\theta t)$ depends on $t \in \mathbf{R}$ and supp $\mu(x+\theta t) \in \mathbf{R}$.

We define

$$m = \sup_{x\in\mathbf{R}^n} \sup_{\theta\in S^{n-1}} \text{mes}\,[\text{supp}\ \mu(x+\theta t)].$$

The main result of this paper is the following theorem.

Theorem. *Let α, β, l be fixed positive numbers; $\mu(x)$ be a given function , $g(\theta,x)$ be a known function in the class $\eta^\alpha = PH_0^\alpha(\Omega)$. Then there exists $d^* > 0$, such that, if* diam$(\Omega) < d^*$, *then the considered problem of emission tomography has a unique solution in the space $H_0^\alpha(\Omega)$ and the following estimate holds :*

$$\|f\|_{H_0^\alpha(\Omega)} \leq C\,\|g\|_{H^\beta(T)},$$

where C is a positive number depending on α, diam(Ω) *and* $\|\mu\|_{H^l(\mathbf{R}^n)}$.

To prove the theorem, we write equality (2) in the following form:

$$(Pf)(x,\theta) + (T_\mu f)(x,\theta) = g(x,\theta).$$

Here,

$$T_\mu f(x,\theta) = \int_{-\infty}^{\infty} f(x+\theta t)E(x+\theta t)\{\exp(-D\mu(x+\theta t,\theta)) - 1\}dt,$$

where $E(x)$ is such a function that $E(x) \in C_0^l(\mathbf{R}^n)$, $E(x) \equiv 1$ for $x \in \Omega$, supp $E(x) \supset \bar{\Omega}$.

Introduce a consequence of functions $\{f_n(x)\}$, $n = 0, 1, \dots$ in the space $H_0^\alpha(\Omega)$ by the following relations:

$$\begin{cases} f_0(x) = (P^{-1}g)(x), \\ f_n(x) = -(P^{-1}T_\mu f_{n-1})(x), \quad n \geq 1. \end{cases} \tag{3}$$

One can show that there exists a number $d^* > 0$ such that if $\operatorname{diam}(\Omega) < d^*$, then the consequence (3) converges in $H_0^\alpha(\Omega)$ to a function $f(x) \in H_0^\alpha(\Omega)$. It is the function that solves uniquely the problem of emission tomography.

For the problem of emission tomography in the space $\mathbf{R}^2$, a numerical algorithm was worked out and tested. This algorithm is based on formulas (2), (3).

In the space $\mathbf{R}^2$ the operator P is, in fact, the Radon transform, which we define by the formula

$$Rf(\theta, s) = \int\limits_{x\cdot\theta=s} f(x)\mathrm{d}x = \int\limits_{\Theta^\perp} f(s\theta + y)\mathrm{d}y, \quad \theta \in S^{n-1}, \quad s \in \mathbf{R}^1.$$

For reconstructing the function $f(x)$, we used the filtered backprojection algorithm described in details by Natterer (1986). Then we construct successive approximations by formula (3).

REMARK. We considered a problem of finding simultaneously the absorption coefficient $\mu(x)$ and the function $f(x)$ of the distribution density of radiation sources. We propose to solve this problem successively, namely: to solve first the problem X-ray transform and then the problem of emission tomography.

REFERENCES

Gryn V. I. (1991), A joint reconstruction of the absorption coefficient and the sources function for a equation of stationary transport of a radiation in $\mathbf{R}^3$. *Dokl. Akad. Nauk. SSSR* vol. 319, no. 4, 849–853.

Isanov R. Sh. and Yakhno V. G. (1991), A problem of emission tomography. *In: Questions of Well-Posedness of Analysis Problems.* Novosibirsk., 67–78.

Natterer F. (1980), A Sobolev space analysis of picture reconstruction, *SIAM J. Appl. Math*, vol. 39, 402–411.

Natterer F. (1986), *The mathematics of computerized tomography.* Stuttgart, John Wiley & Sons Ltd.

Tikhonov A. N., Arsenin V. Ya., Timonov A. A. (1987), *Mathematical Problems of Computerized Tomography.* Nauka, 158 p.

Tretiak O., Metz. Ch. (1980), The exponential Radon transform. *SIAM J.Appl.Math.*, vol. 39, no. 2.

Yakhno V. G and Klimenko O. A. (1989), Problem of integral geometry. *In: Questions of Well-Posedness of Analysis problems* Novosibirsk, 173–180. (in Russian)

Direct and Inverse Problems of Thermoelasticity

R. Sh. ISANOV and V. G. YAKHNO
Novosibirsk State University, Institute of Mathematics of RAS

ABSTRACT

We consider two direct problems for a system of differential equations of the binding thermoelasticity with special boundary conditions. We study those properties of their solutions which are necessary to solve a related inverse problem.

The main results of our investigations are some theorems about the existence and uniqueness of the solutions of these direct and inverse problems.

In the present paper we consider the following system of differential equations of the binding thermoelasticity.

Let $x = (x_1, x_2, x_3) \in R^3$, $x_3 \in R^+$, $t \in R^+$, $R^+ = (0, \infty)$,

$$\begin{cases} \rho D_t^2 u_i = \sum\limits_{j=1}^{3} dx_j T_{ij}[u](x,t), & (1) \\ D_t\Theta - k\Delta\Theta - (3\lambda + 2\mu)\,\mathrm{div}\, D_t u \int\limits_0^{\Theta(x,t)} a(y)dy, & (2) \\ u_i \mid_{t=+0} = D_t u_i \mid_{t=+0} = 0, & (3) \\ \Theta \mid_{t=+0} = 0, & (4) \\ T_{3i}[u] \mid_{x_3=+0} = f_i(x_1, x_2, t), \qquad i = 1,2,3, & (5) \\ \Theta \mid_{x_3=+0} = f_4(x_1, x_2, t). & (6) \end{cases}$$

Here $\rho = \rho(x_3)$ denotes a density function; $\mu(x_3), \lambda(x_3)$ are the Lame functions, $u = (u_1(x,t), u_2(x,t), u_3(x,t))$ is a vector of displacement of an elastic body; $\Theta(x,t)$ is a temperature function of the body; γ, k are fixed positive numbers; $T_{ij}[u](x,t)$ is a stress of the displacement which is defined by

$$T_{ij}[u](x,t) = \mu\left[Dx_j u_i + Dx_i u_j\right] + \delta_{ij}\left[\lambda\,\mathrm{div}\; u - (3\lambda + 2\mu)\int\limits_0^{\Theta(x,t)} a(y)dy\right];$$

$a(y)$ is a thermal extension function; δ_{ij} are the Kronecker's symbols $(i, j = 1,2,3)$.

The papers by Kovalenko (1975) and Danilovskaya (1952) deal with direct problems of binding and unbinding thermoelasticity with special boundary conditions, whereas in the papers by Yakhno (1990) and Lomazov (1987) some inverse problems of unbinding thermoelasticity were investigated. Our research continues the investigations of these papers. In this paper we consider two direct problems with special boundary conditions and study those properties of their solutions which are necessary for solving a related inverse problem. Actually, the inverse problem is the main object of our investigation.

Assume $\lambda(x_3)$, $\mu(x_3)$, $\rho(x_3) \in \Lambda$, $a(y) \in C^1[0,\infty)$, where

$$\Lambda = \left\{c(x_3) \in C^2[0,\infty) \mid c(x_3) > 0,\ c'(+0) = 0\right\}.$$

If $\lambda, \mu, \rho \in \Lambda$, $a \in C^1$, $f_i \in C^1$ $(i = 1,2,3)$, $g(t) \in C^2[0,T]$ are given functions, the problem of defining the unknown vector-function $(u_1(x,t), u_2(x,t), u_3(x,t), \Theta(x,t))$ satisfying equations (1)-(6) is called the direct pfoblem of thermoelasticity.

We can prove the following theorems.

Theorem 1. *Let k, T be fixed positive numbers, let $\lambda(x_3)$, $\mu(x_3)$, $\rho(x_3) \in \Lambda$, $a(y) \in C^1[0,\infty)$ be given functions, let $f_i = f_i^s \equiv -\frac{1}{2}x_i^s\delta(t)$, $i = 1,2,3$; $s = 0,1$; $f_4 = 0$, where $\delta(t)$ is the Dirac delta-function. Then there exists a unique solution (u_1, u_2, u_3, Θ), $u_i = u_i(x,t)$, $\Theta = \Theta(x,t)$, of the direct problem (1)-(6) such that*

$$u_i(x,t) \in C\left([0,T]; C^\infty(R^2; W_2^1(R^+))\right) \bigcap C^1\left([0,T]; C^\infty(R^2; L_2(R^+))\right), \qquad i = 1,2,3;$$

$$\Theta(x,t) \in C\left([0,T]; C^\infty(R^2; W_2^2(R^+)\right) \bigcap C^1\left([0,T]; C^\infty(R^2; L_2(R^+))\right).$$

Theorem 2. *Let k, T be fixed positive numbers, let λ, μ, $\rho \in \Lambda$, $a \in C^1[0,\infty)$, $g(t) \in C^2[0,T]$ be given functions; $f_i = 0$, $i = 1,2,3$; $f_4 = g(t)$. Then there exists a unique solution $(u_1(x,t), u_2(x,t), u_3(x,t), \Theta(x,t))$ of the direct problem (1)-(6) satisfying*

$$u_i(x,t) \in C\left([0,T]; C^\infty(R^2; W_2^1(R^+))\right) \bigcap C^1\left([0,T]; C^\infty(R^2; L_2(R_+))\right), \qquad i = 1,2,3;$$

$$\Theta(x,t) \in C\left([0,T]; C^\infty(R^2; W_2^2(R^+))\right) \bigcap C^1\left([0,T]; C^\infty(R^2; L_2(R^+))\right).$$

We denote by $(u_1^s, u_2^s, u_3^s, \Theta^s)$ and $(\hat{u}_1, \hat{u}_2, \hat{u}_3, \hat{\Theta})$ the solutions of the direct problem (1)-(6) for $f_i = f_i^s \equiv -\frac{1}{2}x_i^s\delta(t)$, $i = 1,2,3$; $s = 0,1$; $f_4 = 0$, and for $f_i = 0$, $i = 1,2,3$; $f_4 = g(t)$, $g(t) \in C^2[0,T]$, respectively.

The main object of our investigations consists in the following inverse problem.

Inverse problem.

Find the unknown functions $\lambda, \mu, \rho \in \Lambda$, $a \in C^1[0,Y]$ if information on the solutions $(u_1^s, u_2^s, u_3^s, \Theta^s)$, $(\hat{u}_1, \hat{u}_2, \hat{u}_3, \hat{\Theta})$ of the direct problem (1)-(6) is given by

$$\begin{aligned} D_{x_1}^s u_1^s \,|_{x_3=+0,\ x_1=x_2=0} &= h_{s+1}(t), \qquad s = 0,1, \\ u_3^0 \,|_{x_3=0,\ x_1=x_2=0} &= h_3(t), \\ D_{x_3}\hat{u}_3 \,|_{x_3=+0,\ x_1=x_2=0} &= h_4(t), \end{aligned}$$

where $h_i(t)$, $i = 1,2,3$; $h_4(t) \in [0,T]$ are known functions.

The main results of this paper are the following theorems.

Theorem 3. *Let k, T be fixed positive numbers, and let $g(t) \in C^2[0,T]$ be a given function such that $g(0) = 0$, $g'(t) \neq 0$, $t \in [0,T]$. Then for solvability of the inverse problem it is necessary that the functions $h_i(t)$, $i = 1,2,3,4$, satisfy the following conditions*

$$h_i(t) \in C^2[0,T], \qquad i = 1,2,3,4; \qquad h_j(+0) = \frac{1}{2},$$

$$h'_j(+0) = 0, \qquad j = 1,3; \qquad h_4(+0) = 0. \tag{7}$$

The proof of Theorems 1,2,3 is based on the results of Yakhno (1990), Kovalenko (1975) and Danilovskaya (1952).

Theorem 4. *Let k, T be fixed positive numbers, let $g(t) \in C^2[0,T]$ be a given function such that $g(0) = 0$, $g'(t) > 0$; and let $h_i(t)$, $i = 1,2,3,4$, satisfy conditions (7). Then there exist three positive numbers T^*, X, Y and four functions $\lambda, \mu, \rho \in \Lambda$, $a \in C^1[0,Y]$ such that $\lambda(x_3), \mu(x_3), \rho(x_3), a(y)$, for $x_3 \in [0,X]$, $y \in [0,Y]$, are a unique solution of the inverse problem with the data $h_i(t)$, $i = 1,2,3,4$, $t \in [0,T^*]$.*

We prove this theorem by the following scheme.

Our inverse problem reduces to the following two inverse problems.

1. Find three functions $\lambda, \mu, \rho \in \Lambda$ when the information $h_i(t)$, $i = 1,2,3$, $t \in [0,T]$, concerning the solution $(u_1^s, u_2^s, u_3^s, \Theta^s)$ of the direct problem (1)-(6) is known.

2. Let $\lambda, \mu, \rho \in \Lambda$ be known functions. Find the function $a(y) \in C^1[0,\infty)$ when the information $h_4(t)$, $t \in [0,T]$, concerning the solution $(\hat{u}_1, \hat{u}_2, \hat{u}_3, \hat{\Theta})$ of problem (1)–(6) is known.

The problem of finding the functions $\lambda, \mu, \rho \in \Lambda$ was investigated in Yakhno (1990).

Let, further, $\lambda, \mu, \rho \in \Lambda$, $f_i \equiv 0$ $(i = 1,2,3)$, $f_4 = g(t)$ be known functions. From the boundary conditions of problem (1)-(6) we deduce

$$H_4(t) = \left(\frac{3\lambda(0) + 2\mu(0)}{\lambda(0) + 2\mu(0)}\right) \int\limits_0^{\Theta(0,t)} a(y)dy.$$

Using the operator D_t, we derive the following equations:

$$a(\Theta(0,t)) = \frac{\lambda(0) + 2\mu(0)h'_4(t)}{3\lambda(0) + 2\mu(0)} \cdot \frac{h'_4(t)}{\Theta_t(0,t)},$$

$$a'(\Theta(0,t)) = h'_5(t)\frac{\lambda(0) + 2\mu(0)h'_4(t)}{3\lambda(0) + 2\mu(0)},$$

where $h_5(t) = \frac{h'_4(t)}{\Theta_t(0,t)}$.

In the proof of Theorem 4 it was used mathematical apparatus of the monograph by Yakhno (1990).

REFERENCES

Danilovskaya V. I. (1952), On a Dynamic Problem in Thermoelasticity. *Appl. Math. and Mechanics.* **16**, no. 3, 341–344.

Kovalenko A. D. (1975), *Thermoelasticity.* Kiev, Vyshcha shkola. (in Russian)

Lomazov V. A. (1987), *Inverse Problems of Diagnostics for the Equations of Thermoelasticity.* Novosibirsk, Ph. D. thesis. (in Russian)

Yakhno V. G. (1990), *Inverse Problems for Differential Equations of Elasticity.* Novosibirsk, Nauka. 300 p. (in Russian)

GCV-Aided Regularizations in Sparse-Data Computed Tomography with an Application to Plasma Imaging

N. IWAMA, H. YOSIDA*, H. TAKIMOTO* and M. TERANISHI

Department of Electronics and Informatics, Toyama Kenritsu University
Kosugi, Toyama 939-03, Japan

*Department of Electrical Engineering, Nagoya University
Furo-cho, Chikusa-ku, Nagoya 464, Japan

ABSTRACT

For image reconstruction in sparse-data computed tomography, two linear algebra methods of Phillips-Tikhonov and of pseudoinverse are studied to regularize the ill-posed least squares solution in taking a special interest in optimization with the generalized cross validation (GCV). The methods are examined and compared by simulations and experiments on the plasma imaging for nuclear fusion research, which is currently a typical example of image reconstruction based on poor projection measurement. The Phillips-Tikhonov method with an objective function of Laplacian type is found effective with a useful assistance of the minimum GCV criterion.

1. Introduction

The image reconstruction based on sparse projection data is an interesting subject of tomography study. The purpose of this paper is to examine algebraic approaches in application to the plasma imaging with regard to the experimental difficulty of small data number and the expect ed smoothness of plasma, which allows a fairly rough partition of the image domain. On the least squares solutions easily ill-posed, interest is given to the linear methods of regularization for an advantage of short computing time, and also to the image optimization using the generalized cross-validation (GCV), that is, a neg-entropy in mathematical statistics (Golub et al., 1979).

In the fully discrete scheme of reconstruction, we divide the two-dimensional image domain into K pixels, and express the corresponding K values of image and the M projection data with K and M dimensional vectors $\boldsymbol{E}$ and $\boldsymbol{S}$, respectively. For relating the unknown $\boldsymbol{E}$ to the given $\boldsymbol{S}$, we employ a linear model $\boldsymbol{S} = L\boldsymbol{E} + \boldsymbol{e}$ with the so-called projection matrix L, whose (m, k) element L_{mk} is the geometrical weight of the k'th pixel in integrating the image along the m'th line of sight to produce the m'th projection; $\boldsymbol{e}$ denotes the residual vector. With an observation system L fixed, we take the least squares solution $\hat{\boldsymbol{E}}$ of minimizing the norm of the residual and find it ill-posed easily for small M and undetermined for $M < K$. To regularize the solution, we study two linear algebra methods and adopt the criterion of minimum GCV to get the best regularized solution. These numerical techniques of reconstruction are experimentally examined on an emission type tomography of toroidal plasma.

2. Numerical Procedures of Reconstruction

2.1 *Phillips-Tikhonov Method*

In strategy of minimizing the mean squares $\varepsilon^2 = |\boldsymbol{S} - L\boldsymbol{E}|^2/M$, the use of objective functions of Phillips-Tikhonov type may lead to solutions well-behaved and linear on data $\boldsymbol{S}$. Giving importance to meeting the sparceness of data, and assuming a spatially smooth image $E(x,y)$ having no sharp boundary, we take the functional $Q = \int[\nabla^2 E(x,y)]^2 dxdy$ to be minimized since it tends to give relatively strong regularization in simple mathematical scheme. Its minimization with constraint of constant ε^2 can be achieved by the Lagrangian procedure of minimizing the quantity $\Phi = \gamma|C\boldsymbol{E}|^2 + |\boldsymbol{S} - L\boldsymbol{E}|^2/M$, where γ is an unknown multiplier and C a square matrix approximating the Laplacian operator. The solution is given by $\hat{\boldsymbol{E}} = (L^T L + M\gamma C^T C)^{-1} L^T \boldsymbol{S}$, where γ works as a parameter for adjusting the strength of regularization.

2.2 *Pseudoinverse Method*

The pseudoinverse solution with the minimized norm $|\boldsymbol{E}|^2$ is also linear on $\boldsymbol{S}$, and an auxiliary regularization can be adopted by tapering the obtained series expansion with a window function $w_i(\gamma)$ as follows:

$$\hat{\boldsymbol{E}} = \sum_{i=1}^{N} w_i(\gamma) a_i \boldsymbol{v}_i, \quad a_i = \sigma_i^{-1} \boldsymbol{u}_i^T \boldsymbol{S}, \tag{1}$$

where $\boldsymbol{u}_i$ and $\boldsymbol{v}_i$ are the i-th column vectors of the unitary matrices U and V, respectively, in the singular value decomposition (SVD) of L given as $L = U\Sigma V^T$ with $\Sigma = \text{diag}\,(\sigma_1, \sigma_2, \ldots, \sigma_N)$ $(\sigma_1 \geq \sigma_2 \geq \ldots \geq \sigma_N;\ N = \min(M, K))$. The window function $w_i(\gamma)$ is a nonnegative-valued function decreasing with the singular value σ_i . In this paper, we choose a function of the form $w_i(\gamma) = (1 + \gamma\sigma_i^{-4})^{-1}\ \ (i \leq R), = 0\ \ (i > R)$ with regularization parameters γ and R .

2.3 *Expressions of GCV and PRESS*

In the Phillips-Tikhonov method, the SVD of LC^{-1} given as $LC^{-1} = XDY^T$ allows us to get the expressions of the GCV, $V(\gamma)$, and also of its primitive version PRESS (prediction sum of squares), $P(\gamma)$ (Iwama et al., 1989; Girard, 1987):

$$V(\gamma) = M\sum_{i=1}^{M}(\rho_i Z_i)^2 / (\sum_{i=1}^{M}\rho_j)^2, \quad P(\gamma) = M^{-1}\sum_{i=1}^{M}[(\sum_{k=1}^{M}\rho_k x_{ik} Z_k)/(\sum_{k=1}^{M}\rho_k x_{ik}^2)]^2, \tag{2}$$

where Z_i and x_{ik} are the components of $X^T\boldsymbol{S}$ and X, respectively, and we have $\rho_i = M\gamma(\sigma_i^2 + M\gamma)^{-1}$ with $\sigma_i = 0$ for $i = K+1, K+2, \ldots, M$ in the case of $M > K$. The quantities $V(\gamma)$ and $P(\gamma)$ are expected to take minimum at the optimum value of γ. Additionally, the SVD of LC^{-1} gives a useful reduction of $\hat{\boldsymbol{E}}$ to a computationally convenient form, for which any large scale matrix inversion is unnecessary:

$$\hat{E}_k = M^{-1}\sum_{i=1}^{M}[\sigma_i(\sigma_i^2 + M\gamma)^{-1}](C^{-1}Y)_{ki} Z_i \tag{3}$$

for the k'th component $\hat{E}_k$ of $\hat{\boldsymbol{E}}$ $(k = 1, 2, \ldots, K)$; $(C^{-1}Y)_{ki}$ is the ki element of $C^{-1}Y$ (Iwama et al., 1989).

Meanwhile, the GCV and PRESS expressions in the pseudoinverse method are obtained by the SVD of L ; that is, we have

$$V(\gamma) = M\sum_{i=1}^{M}(\beta_i\zeta_i)^2/(\sum_{i=1}^{M}\beta_i)^2, \quad P(\gamma) = M^{-1}\sum_{k=1}^{M}[b_{kk}(\gamma,R)(\sum_{i=1}^{M}\beta_i u_{ik}\zeta_i)], \tag{4}$$

where ζ_i and u_{ik} are the components of $U^T\boldsymbol{S}$ and U, respectively, and we have $\beta_i = \gamma\sigma_i^{-4}\alpha_i$, $b_{kk}(\gamma,R) = [1-(UFU^T)_{kk}]^{-1}$, $F = \mathrm{diag}\,(\alpha_1,\alpha_2,\ldots,\alpha_M)$ and $\alpha_i = (1+\gamma\sigma_i^{-4})^{-1}$ with $\sigma_i = 0$ $(R+1 \le i \le M)$; γ and R are regularization parameters.

The GCV and the PRESS in each method of regularization are expected to take minimum as functions of the regularization parameters, and thereby give the best regularized image $\hat{\boldsymbol{E}}$, for which the contradiction between the spatial resolution and the reliability is compromised in an appropriate manner.

3. Simulations and Experiments

The above numerical procedures were examined on a visible light emission CT, whose system layout is shown in Fig. 1. Numerical simulations were made on two cas es of emission intensity distribution: two Gaussian shaped peaks and point sources. Projection data were produced by corrupting their theoretical values with normally distributed white noises having zero mean and variance σ_o^2. Results are summarized in Figs. 2 and 3, where $\hat{\varepsilon}^2 = M^{-1}|\boldsymbol{S} - L\hat{\boldsymbol{E}}|^2$, $\delta_E^2 = K^{-1}|\boldsymbol{E}_o - \hat{\boldsymbol{E}}|^2$ and $\delta_S^2 = M^{-1}|\boldsymbol{S}_o - L\hat{\boldsymbol{E}}|^2$ are the attained least mean squares, the mean squared errors in approximating the true image $\boldsymbol{E}_o$ and its projection $\boldsymbol{S}_o$, respectively. For the pixel number up to $K = 900$ and for data number $M = 100 \sim 600$, both neg-entropies, except $P(\gamma)$ for point sources, favorably indicate good approximates of $\boldsymbol{E}_o$ by taking minima at γ values large in the Gaussian case and small in point source case. Precisely, the γ values minimizing $V(\gamma)$ and $P(\gamma)$ agree well with those minimizing rather δ_S^2 than δ_E^2 as expected from the fact that the GCV and PRESS are essentially the final prediction errors defined on projection data. Compared with the goodness of the Phillips-Tikhonov regularization, the pseudoinverse regularization is weak, giving undesirable distortions to the images. The superiority of the former is recognized also on the plasma experiment results in Fig. 4: the comparatively smooth plasma shape for a similar value of the attained $\hat{\varepsilon}^2$, and the lower minima of $V(\gamma)$ and $P(\gamma)$. The GCV is advantageous to the PRESS in easy computation. In Fig. 5 is displayed a time evolution of plasma obtained with the GCV-aided Phillips-Tikhonov method.

4. Conclusions

In two linear algebraic approaches to sparse-data CT, the neg-entropies GCV and PRESS behave well, attaining minima clearly in comparison to the AIC which has been used for series expansion approaches (Shen et al., 1988; Iwama et al., 1987). The Phillips-Tikhonov method is excellent in giving well-regularized image shapes as a result of optimization.

The authors would like to thank Prof. S. Takamura and Dr. Y. Shen of the Nagoya University for providing experimental data and also for useful discussions.

REFERENCES

G.H. Golub, M. Heath and G. Wahba (1979), *Technometrics*, **21**, 215.

N. Iwama, H. Yoshida and H. Takimoto et al. (1989), *Applied Phys. Lett.*, **54**, 502.
D.A. Girard (1987), *SIAM J. Sci. Stat. Comput.*, **8**, 934.
Y. Shen, S. Takamura, N. Iwama et al. (1988), *J. Plasma and Fusion Res.*, **59**, 30.
N. Iwama, H. Takami et al. (1987), *IEEE Trans. on Plasma Sci.*, **PS-15**, 609.

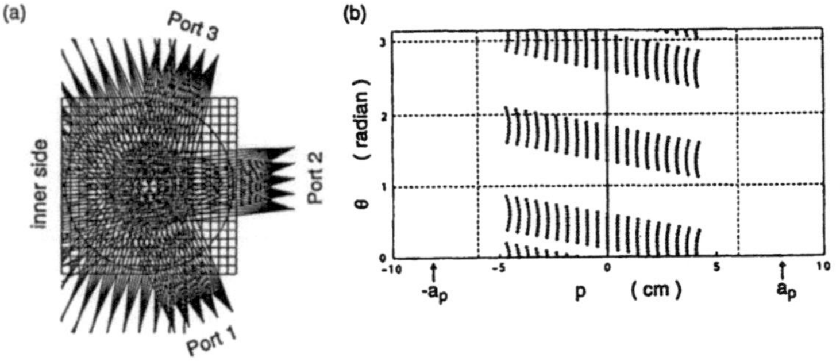

FIGURE 1. (a) Hα line emission CT layout of the tokamak CSNT-III (Nagoya Univ.) having three observation ports in a poloidal plane (plasma radius a_p = 8cm; optical fibers of M = 600). (b) (p, θ) plane locations of the lines of sight.

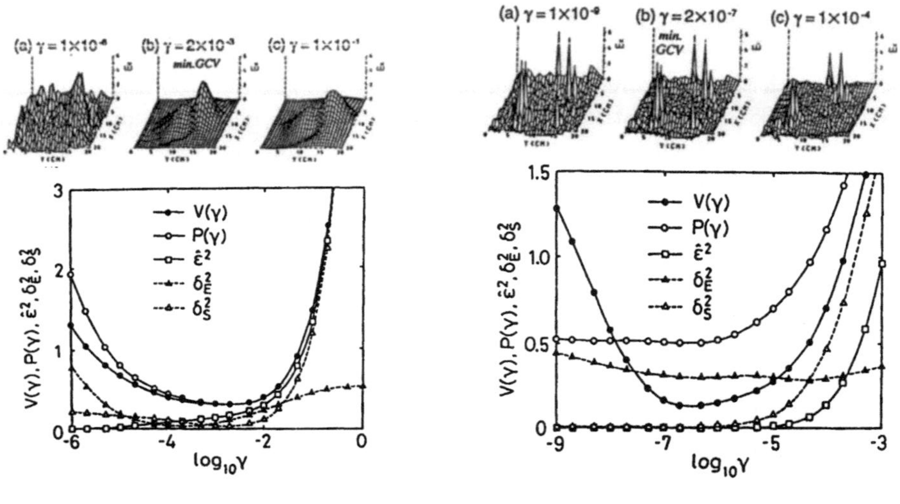

FIGURE 2. Simulation of the Phillips-Tikhonov reconstruction ($M = 600$, $K = 900$). Left: result on Gaussian shaped peaks ($\sigma_o^2 = 1.0$) ; γ-dependencies of $V(\gamma)$, $P(\gamma)$, $\hat{\varepsilon}^2$, δ_E^2, δ_S^2 and the obtained image $\hat{\boldsymbol{E}}$. Right: those on point sources ($\sigma_o^2 = 0.01$).

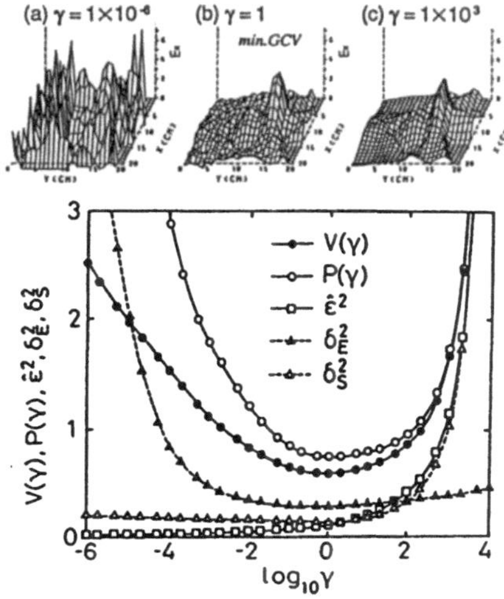

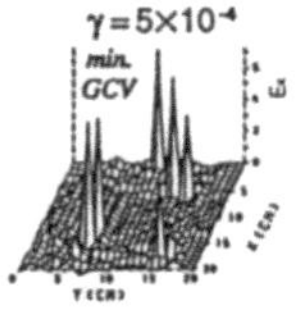

FIGURE 3. Simulation of the pseudoinverse reconstruction ($M = 600$, $K = 625$, $R = 567$). Left: result on Gaussian shaped peaks ($\sigma_o^2 = 0.25$). Right: result on point sources ($\sigma_o^2 = 0.01$); image at the minimum of GCV.

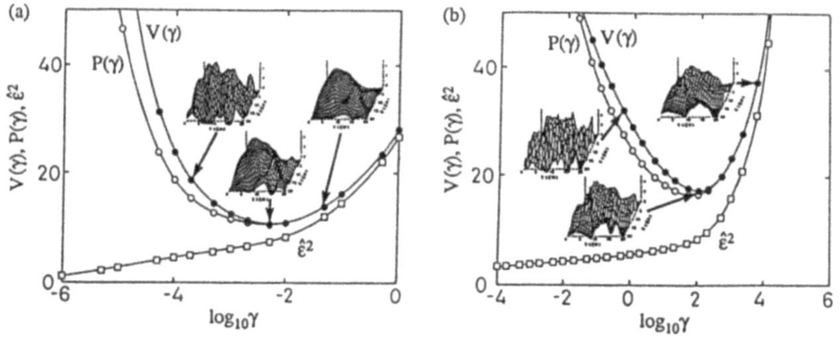

FIGURE 4. γ-dependencies of $V(\gamma)$, $P(\gamma)$ and $\hat{\varepsilon}^2$ in plasma experiment: (a) Phillips-Tikhonov method ($M = 600$, $K = 900$) and (b) pseudoinverse method ($M = 600$, $K = 625$; $R = 567$). Obtained images are illustrated by insets.

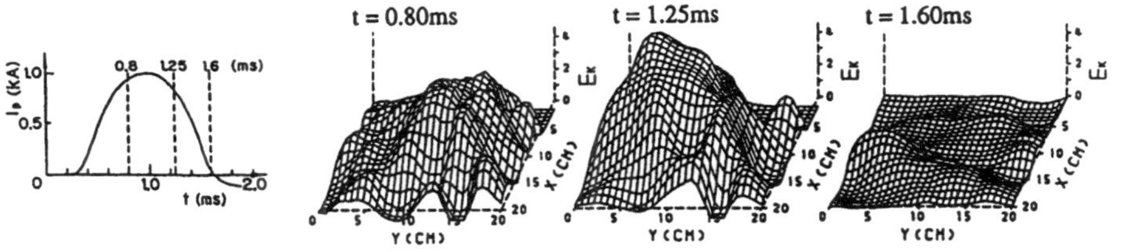

FIGURE 5. Time evolution of Hα elmission intensity distribution: time history of plasma current I_p, and Phillips-Tikhonov images obtained at the minimum GCV ($M = 600$, $K = 900$).

Frames on the Range of the Radon Transform[1]

Steven H. IZEN

Department of Mathematics, Case Western Reserve University

ABSTRACT

We show how to define frames on the range of the Radon transform. An operator is constructed which will induce from any frame on $L^2(\mathbf{R}^n)$ a frame on the Radon transform space $H^{(n-1)/2}(\mathbf{Z}) \cap$ Range(R). This operator will preserve the frame structure, in that it will also map the dual frame on $L^2(\mathbf{R}^n)$ to the dual frame on $H^{(n-1)/2}(\mathbf{Z}) \cap \text{Range}(R)$.

Wavelets have recently attracted much attention as a tool for the analysis of images. In most cases the wavelet transform (either discrete or continuous) is computed from an image to be examined. The wavelet transform is then subjected to analysis. If the data which are available are projection data for an image, and it is desired to perform such an analysis, one way to proceed is to reconstruct via the usual discrete Radon inversion and feed the resulting image to a wavelet analysis. The alternate approach which we are proposing is to perform a wavelet decomposition directly on the projection data. This will allow the computation of the wavelet transform of the original function without the intermediate approximation of the Radon inversion.

The first step in such a procedure is to construct frames on the Radon transform space. Ideally such frames should be related to frames on the image space, so any decomposition on the projection space can be pulled back to a decomposition on the image space. This report will document the construction of an isometry from the image space to the Radon space which preserves the entire frame structure.

In order to apply this to a wavelet reconstruction from projection data, much more will be required. Still to be done is the generalization of a multiresolution analysis to the projection space, along with the development of an fast algorithm to perform the multiresolution analysis on the projection data.

We note here in passing that in Holschneider (1991) it was shown how to interpret the Radon transform as a continuous wavelet transform with a singular mother wavelet. The relationship between a discretization of his construction and the frame construction described here is a topic for future investigation.

1. Frame Basics

We review briefly some fundamental facts about frames. For a more detailed exposition, see Daubchies (1992), Chapter 3.

Consider the set $\{\phi_j\}$, for all j in an index set J, and a Hilbert space K with the properties that

[1]This research was supported by NASA grant NAG 3-832 and the NASA/ASEE Summer Faculty Fellowship Program.

1. for each j, $\phi_j \in K$;

2. there exist constants A and B called frame bounds such that, for each $f \in K$,

$$A\|f\|_K^2 \leq \sum_{j\in J} |\langle \phi_j, f\rangle_K|^2 \leq B\|f\|_K^2. \tag{1}$$

Such a set is called a *frame*. The operator F from K to $\ell^2(J)$ defined by $(Ff)_j = \langle \phi_j, f\rangle_K$ is called the frame operator. It turns out that the family $\left\{\tilde{\phi}_j\right\}$, where $\tilde{\phi}_j = (F^*F)^{-1}\phi_j$, is also a frame, and it has bounds $1/B$ and $1/A$. This frame is called the dual frame, and the corresponding frame operator $\tilde{F}$ is given by $\tilde{F} = F(F^*F)^{-1}$.

One has the following identities (at least in a weak sense):

1. $f = \sum_{j\in J} \langle \phi_j, f\rangle_K \tilde{\phi}_j = \sum_{j\in J} \langle \tilde{\phi}_j, f\rangle_K \phi_j$,

2. $F^*Ff = \sum_{j\in J} \langle \phi_j, f\rangle_K \phi_j$.

If the frame bounds are equal, the frame is called *tight* and the recovery of f from Ff is numerically stable. For *snug frames*, where $B/A \approx 1$, the reconstructions are also numerically stable.

A wavelet frame is a one obtained by the action of a discrete subgroup of $IG(n)$ (Euclidean motions plus dilations in $\mathbf{R}^n$) on a mother wavelet $g \in L^2(\mathbf{R}^n)$. Conditions on the step sizes for such a discretization to be a frame have been given by Daubchies (1990) for one-dimensional wavelets, and by Murenzi (1990) for the multidimensional case. It is necessary that the discrete wavelets be a frame in order to perform a stable inversion from discrete wavelet transform data.

2. Frames in Radon Space

In this section we show how to construct frames on the range of the Radon transform from a frame given on $L^2(\mathbf{R}^n)$.

2.1. *Preliminaries*

The spaces used will be defined and some properties of the Radon transform will be listed.

Let $\mathbf{Z}$ be the cylinder $\mathbf{R} \times S^{n-1}$. The function space $H^s(\mathbf{Z})$ is a Hilbert space. The inner product on this space is given by

$$\langle f, g\rangle_{H^s(\mathbf{Z})} = \int_{\mathbf{R}} \int_{S^{n-1}} \overline{\hat{f}(p,\omega)}\hat{g}(p,\omega)(1+p^2)^s d\omega dp.$$

Let L be an abbreviation for $L^2(\mathbf{R}^n)$. The Radon transform R maps $f \in L$ into an even function on $\mathbf{Z}$. That is, $Rf(p,\omega) = Rf(-p,-\omega)$.

Theorem 2.1. *Let $f \in L$. Then $Rf \in H^{(n-1)/2}(\mathbf{Z})$, and*

$$c(n)\|f\|^2 \leq \|Rf\|^2_{H^{(n-1)/2}(\mathbf{Z})} \leq C(n)\|f\|^2,$$

where $C(n) = 2(2\pi)^{n-1}\frac{2^{(n+1)/2}\pi^{n/2}}{\Gamma(n/2)}$ *and* $c(n) = 2(2\pi)^{n-1}$.

Proof. This is Theorem II.5.1 in Natterer (1986) with $\alpha = 0$.

Use H to denote $H^{(n-1)/2}(\mathbf{Z}) \cap \mathrm{Range}(R)$. It is easy to see that H is a closed subspace of $H^{(n-1)/2}(\mathbf{Z})$ and hence is a Hilbert space with the same inner product as $H^{(n-1)/2}(\mathbf{Z})$.

The convention used for the Fourier transform $\hat{f}(k)$ of a function $f(x)$ is that

$$\hat{f}(k) = (2\pi)^{-n/2} \int_{\mathbf{R}^n} e^{-ix\cdot k} f(x)\, dx.$$

The Riesz potential I^α operates on a function f by

$$(I^\alpha f)\hat{}\,(k) = |k|^{-\alpha} \hat{f}(k).$$

If the Fourier transform or the Riesz potential is applied to a function on $\mathbf{Z}$, it is understood to be applied in the radial variable.

With $d_n = (2\pi)^{1-n}$, The Radon inversion formula becomes

$$f = \frac{1}{2} d_n R^\# I^{1-n} R f,$$

where $R^\#$ is the usual transpose of R.

2.2. *The operator* Λ^s

In order to construct a map of the frame structure, it is necessary to consider the integral operator Λ^s.

Let $\Lambda(t) = d_n \left(\frac{|t|^2}{1+|t|^2}\right)^{(n-1)/2}$. If $-\Box$ denotes differentiation twice in the radial variable, then we can define the operator $\Lambda_\Box{}^s = \Lambda^s(\Box)$ on functions on $\mathbf{Z}$ by

$$\widehat{\Lambda_\Box{}^s \eta}(k,\omega) = d_n^s |k|^{s(n-1)} (1+k^2)^{s(1-n)/2} \hat{\eta}(k,\omega).$$

Similarly, if Δ is the positive Laplacian operator on functions on $\mathbf{R}^n$, then the operator $\Lambda_\Delta{}^s = \Lambda^s(\Delta)$ can be defined by

$$\widehat{\Lambda_\Delta{}^s f}(k) = d_n^s |k|^{s(n-1)} (1+|k|^2)^{s(1-n)/2} \hat{f}(k).$$

Lemma 2.1. *Let $g \in L$. Then, for $s > -\frac{1}{2} - \frac{1}{2(n-1)}$, it holds that $\Lambda_\Delta{}^s g \in L$, with norm estimated by*

$$\|\Lambda_\Delta{}^s g\|^2 \le e_1(n,s)\, \|g\|^2. \tag{2}$$

Moreover, if $s < \frac{1}{2} + \frac{1}{2(n-1)}$ as well, then there is a lower bound for $\|\Lambda_\Delta{}^s g\|$,

$$\|\Lambda_\Delta{}^s g\|^2 \ge e_2(n,s)\, \|g\|^2.$$

The constants $e_1(n,s)$ and $e_2(n,s)$ can be explicitly given.

We omit the proof for brevity.

Lemma 2.2. *$\Lambda_\Box{}^s$ and $\Lambda_\Delta{}^s$ intertwine over R. That is, $R\Lambda_\Delta{}^s = \Lambda_\Box{}^s R$.*

Proof.

$$(\Lambda_\Box{}^s Rf)\hat{}\,(t,\omega) = \Lambda^s(t)\widehat{Rf}(t,\omega) = (2d_n)^{-1/2}\Lambda^s(t)\hat{f}(t\omega) = (2d_n)^{-1/2}(\Lambda_\Delta{}^s f)\hat{}\,(t\omega)$$

$$= (R\Lambda_\Delta{}^s f)\widehat{\ }(t,\omega).$$

Lemma 2.3. *Let $\eta \in H$ and $s > -\frac{1}{2} - \frac{1}{2(n-1)}$. Then $\Lambda_\Box{}^s\eta \in H$.*

PROOF. Use Theorem 2.1 and Lemmas 2.1 and 2.2.

Define $Q_s = R\Lambda_\Delta{}^s = \Lambda_\Box{}^s R$.

Lemma 2.4. *Let $f, g \in L$. Then*

$$\langle f,g\rangle = \langle Q_s f, Q_{1-s} g\rangle_H.$$

In particular, $Q_{1/2}\colon L \to H$ is an isometry.

PROOF.

$$\langle Q_s f, Q_{1-s} g\rangle_H = \int_{\mathbf{R}}\int_{S^{n-1}} \Lambda(p)\overline{\widehat{Rf}(p,\omega)}\widehat{Rg}(p,\omega)(1+p^2)^{(n-1)/2}\mathrm{d}\omega\mathrm{d}p$$

$$= \int_{\mathbf{R}^n} \overline{\hat{f}(k)}g(k)\mathrm{d}k = \langle f,g\rangle.$$

2.3. *The induced frame on the Radon space*

The operator mapping L to H and preserving the frame structure turns out to be $Q_{1/2}$. This follows as a special case of our main result, Theorem 2.2.

Theorem 2.2. *Let $\{\phi_j\}$ be a frame on $L^2(\mathbf{R}^n)$ with a frame operator F and frame bounds A and B. Define $(\psi_s)_j = Q_s\phi_j$. Then, for $\frac{1}{2} - \frac{1}{n-1} < s < \frac{3}{2} + \frac{1}{n-1}$, $\{(\psi_s)_j\}$ will be a frame on H, with the frame operator denoted by T_s. The dual frame is given by $\left\{Q_{1-s}\tilde{\phi}_j\right\}$.*

PROOF. To show that the ψ_j give a frame for H, we must find A' and B' so that, for all $g \in H$,

$$A'\|g\|_H^2 \le \sum_j |\langle(\psi_s)_j, g\rangle_H|^2 \le B'\|g\|_H^2.$$

Since H is the image of L under R, it suffices to show that, for each $f \in L$,

$$A'\|Rf\|_H^2 \le \sum_j |\langle(\psi_s)_j, Rf\rangle_H|^2 \le B'\|Rf\|_H^2. \tag{3}$$

In addition,

$$\sum_{j\in J} |\langle(\psi_s)_j, Rf\rangle_H|^2 = \sum_{j\in J} |\langle Q_s\phi_j, Q_{1-s}\Lambda_\Delta{}^{s-1}f\rangle_H|^2 = \sum_{j\in J} |\langle\phi_j, \Lambda_\Delta{}^{s-1}f\rangle|^2. \tag{4}$$

Since $\{\phi_j\}$ is a frame,

$$A\left\|\Lambda_\Delta{}^{s-1}f\right\|^2 \le \sum_{j\in J} |\langle\phi_j, \Lambda_\Delta{}^{s-1}f\rangle|^2 \le B\left\|\Lambda_\Delta{}^{s-1}f\right\|^2. \tag{5}$$

By Lemma 2.1 and Theorem 2.1,

$$\left\|\Lambda_\Delta{}^{s-1}f\right\|^2 \le e_1(n,s-1)\|f\|^2 \le \frac{e_1(n,s-1)}{c(n)}\|Rf\|^2, \tag{6}$$

$$\left\|\Lambda_\Delta{}^{s-1} f\right\|^2 \geq e_2(n, s-1)\,\|f\|^2 \geq \frac{e_2(n, s-1)}{C(n)}\,\|Rf\|^2. \tag{7}$$

Equations 4, 5, 6 and 7 together imply Equation 3, with frame bounds

$$A' = \frac{Ae_2(n, s-1)}{c(n)} \quad \text{and} \quad B' = \frac{Be_1(n, s-1)}{C(n)}.$$

For the dual frame,

$$T_s^* T_s\left(Q_{1-s}\tilde{\phi}_k\right) = \sum_{j\in J} \langle(\psi_s)_j, Q_{1-s}\tilde{\phi}_k\rangle_H (\psi_s)_j = Q_s\left(\sum_{j\in J} \langle(\psi_s)_j, Q_{1-s}\tilde{\phi}_k\rangle_H \phi_j\right)$$

$$= Q_s\left(\sum_{j\in J} \langle\phi_j, \tilde{\phi}_k\rangle\phi_j\right) = Q_s\left(F^*F\tilde{\phi}_k\right) = Q_s\phi_k = (\psi_s)_k.$$

Hence,

$$\left(\widetilde{\psi_s}\right)_k = \left(T_s^* T_s\right)^{-1}(\psi_s)_k = Q_{1-s}\tilde{\phi}_k.$$

Corollary 2.5. *$Q_{1/2}$ intertwines the operators F^*F and $T_{1/2}^* T_{1/2}$.*

Since $Q_{1/2}$ intertwines F^*F and $T_{1/2}^* T_{1/2}$, $Q_{1/2}\tilde{\phi}_j = \left(\widetilde{\psi_{1/2}}\right)_j$.

Thus the images of both the frame $\{\phi_j\}$ and its dual $\left\{\tilde{\phi}_j\right\}$ are a dual pair of frames on H. That is, $Q_{1/2}$ preserves the frame structure.

We close by showing that, for $\left\{(\psi_{1/2})_j\right\}$, the estimate given in Theorem 2.2 for the frame bounds can be improved considerably.

Proposition 2.6. *The frame bounds for $\left\{(\psi_{1/2})_j\right\}$ are the same as those for $\{\phi_j\}$.*

PROOF. Start with the frame bounds for $\{\phi_j\}$,

$$A\,\|f\|^2 \leq \sum_j |\langle\phi_j, f\rangle|^2 \leq B\,\|f\|^2.$$

Applying the isometry $Q_{1/2}$ yields

$$A\left\|Q_{1/2}f\right\|_H^2 \leq \sum_j |\langle\phi_j, f\rangle|^2 \leq B\left\|Q_{1/2}f\right\|_H^2,$$

$$A\left\|Q_{1/2}f\right\|_H^2 \leq \sum_j |\langle Q_{1/2}\phi_j, Q_{1/2}f\rangle_H|^2 \leq B\left\|Q_{1/2}f\right\|_H^2. \tag{8}$$

By Lemma 2.3, $\Lambda_\Box{}^{1/2}$ is invertible on H. Apply $\Lambda_\Box{}^{-1/2}$ to $Q_{1/2}f$ in Equation 8. Since $\Lambda_\Box{}^{-1/2} Q_{1/2} f = Rf$,

$$A\,\|Rf\|_H^2 \leq \sum_j |\langle(\psi_{1/2})_j, Rf\rangle_H|^2 \leq B\,\|Rf\|_H^2.$$

This is Equation 3, with $A' = A$, and $B' = B$.

REFERENCES

Daubchies I. (1990), The wavelet transform, time-frequency localization and signal analysis. *IEEE Transactions on Information Theory,* **36**, no. 5, 961–1005.

Daubchies I. (1992), *Ten Lectures on Wavelets.* CBMS-NSF Regional Conference Series in Applied Mathematics, vol. 61, S.I.A.M.

Holschneider M. (1991), Inverse Radon transforms through inverse wavelet transforms, *Inverse Problems*, **7**, 853–861.

Murenzi R. (1990), *Ondelettes Multidimensionelles et Applications à l'Analyse d'Images*, Ph.D. thesis, UCL (in French).

Natterer F. (1986), *The Mathematics of Computerized Tomography*, Wiley.

On the Stability Estimation of Finite–Difference and Differential–Difference Analogues of a Two–Dimensional Integral Geometry Problem[1]

S. I. KABANIKHIN and G. B. BAKANOV

Institute of Mathematics
Universitetskii Pr., 4, 630090, Novosibirsk, Russia

ABSTRACT

Finite–difference and differential-difference analogues of a two–dimensional integral geometry problem are studied. The stability estimates for the considered problems are obtained.

1. Introduction

A great deal of the work has been devoted to problems of integral geometry, that is, the problems of determining a function from its integral along a given family of curves. One of the stimuli for studying such problems is their connection with multidimensional inverse problems for differential equations (Lavrent'ev and Romanov, 1966). In Romanov (1972) some inverse problems for hyperbolic equations were shown to reduce to integral geometry problems and, in particular, a problem of integral geometry was considered in the case of shift–invariant curves. Mukhometov (1975) showed the uniqueness and estimated the stability of the solution of a two–dimensional integral geometry on the whole. His results were mainly based on the reduction of the two–dimensional integral geometry problem

$$V(\gamma, z) = \int\limits_{K(\gamma,z)} U(x,y) \mathrm{d}s, \quad \gamma \in [0,l], z \in [0,l], \tag{1}$$

where $U \in C^2(\overline{D})$, to the boundary value problem

$$LW = \frac{\partial}{\partial z}\left(\frac{\partial W}{\partial z}\cos\theta + \frac{\partial W}{\partial y}\sin\theta\right) = 0, \quad (x,y,z) \in \Omega_1; \tag{2}$$

$$W(\xi(\gamma), \eta(\gamma), z) = V(\gamma, z),$$
$$V(z,z) = 0, \quad \gamma, z \in [0,l]. \tag{3}$$

Here D is a plane bounded by a simply connected domain with a smooth boundary Γ :

$$x = \xi(z), \;\; y = \eta(z), \;\; z \in [0,l], \;\; \xi(0) = \xi(l), \;\; \eta(0) = \eta(l), \tag{4}$$

where a parameter z is the length of the curve Γ;

$$\Omega_1 = \Omega \backslash \{(\xi(\gamma), \eta(\gamma), z) : \; z \in [0,l]\}, \quad \Omega = \overline{D} \times [0,l];$$

[1]The work was supported by the Russian Foundation of Fundamental Research under the grant No. 93-01-01-739

$K(x,y,z)$ is the part of the curve $K(\gamma,z)$ included between the points $(x,y) \in \overline{D}$ and $(\xi(z),\eta(z))$, $z \in [0,l]$;

$$W(x,y,z) = \int\limits_{K(x,y,z)} U(x,y)ds,$$

$\theta(x,y,z)$ is an angle between the tangent to $K(x,y,z)$ at the point (x,y) and the x-axis; and the variable parameter s is the curve length.

A multidimensional problem of integral geometry was investigated by Romanov (1978).

In this paper we study a finite-difference analogue of problem (2), (3) (Section 2), a finite-difference analogue of problem (2), (3) with a weight function (Section 3), a differential-difference analogue of problem (2), (3) (Section 4), and a differential-difference analogue of problem (2), (3) with a weight function (Section 5). The main goal of the paper is to estimate the stability of the solution of finite-difference and differential-difference problems.

2. Discrete Problem of Integral Geometry

Suppose that the requirements on the family of curves $K(\gamma,z)$ and the domain D that are necessary for the problem (1) to reduce to problem (2), (3) are met (Mukhometov, 1975). Assume also that every line parallel to either the x- or the y-axis intersects the boundary of D at no more than two points. Let

$$a_1 = \inf_{(x,y)\in D}\{x\}, \quad b_1 = \sup_{(x,y)\in D}\{x\};$$

$$a_2 = \inf_{(x,y)\in D}\{y\}, \quad b_2 = \sup_{(x,y)\in D}\{y\};$$

$h_j = (b_j - a_j)/N_j$, $j = 1,2$; $h_3 = l/N_3$, and the N_j, $j = 1,2,3$, are natural numbers. Suppose that

$$0 < \varepsilon < \min\{(b_1 - a_1)/3,\ (b_2 - a_2)/3\},$$

$$D^\varepsilon = \left\{(x,y) \in D: \min_{(\alpha,\beta)\in\Gamma} \rho((x,y),(\alpha,\beta)) > \varepsilon\right\},$$

$$R_h = \{(x_i,y_j): \ x_i = a_1 + ih_1,\ y_j = a_2 + jh_2,\ i = 1,\dots,N_1,\ j = 1,\dots,N_2\}.$$

A neighborhood $B(ih_1,jh_2)$ of the point (a_1+ih_1, a_2+jh_2) is defined as the five-point set

$$\{(a_1+ih_1,a_2+jh_2),(a_1+(i\pm1)h_1,a_2+(j\pm1)h_2)\}.$$

A set D_h^ε consists of all points (a_1+ih_1,a_2+jh_2) such that their neighborhoods $B(ih_1,ih_2)$ are contained in $D^\varepsilon \cap \mathbf{R}_h$. A set Γ_h^ε is made of all points $(a_1+ih_1,a_2+jh_2) \in D_h^\varepsilon$ such that $B(ih_1,ih_2)\bigcap(D^\varepsilon \bigcap \mathbf{R}_h)\backslash D_h^\varepsilon \neq 0$. Finally,

$$\Delta_h^\varepsilon = \bigcup_{\Gamma_h^\varepsilon} B(ih_1,jh_2),$$

$$\Omega_h^\varepsilon = \{(a_1+ih_1,a_2+jh_2,kh_3): \ (a_1+ih_1,a_2+jh_2) \in D_h^\varepsilon,\ k = 0,1,\dots,N_3-1\}.$$

From here on we suppose that the coefficients and the solution of problem (2), (3) have the following properties:

$$W(x,y,z) \in C^3(\Omega^\varepsilon),\ \frac{\partial\theta}{\partial z} \geq 0,\ \theta(x,y,z) \in C^2(\Omega^\varepsilon),\ \Omega^\varepsilon = \overline{D^\varepsilon} \times [0,l]. \tag{5}$$

As in Samarskii (1983), we use the notation:

$$f_{\overset{\circ}{x}} = \left(f^k_{i+1,j} - f^k_{i-1,j}\right)/(2h_1), \quad f_{\overset{\circ}{y}} = \left(f^k_{i,j+1} - f^k_{i,j-1}\right)/(2h_1),$$

$$f_z = \left(f^{k+1}_{i,j} - f^k_{i,j}\right)/h_3.$$

Then the finite-difference problem can be formulated as

$$L_h\Phi_h = \left[\Phi_{\overset{\circ}{x}}A + \Phi_{\overset{\circ}{y}}B\right]_z = 0, \quad (a_1 + ih_1, a_2 + ih_2, kh_3) \in \Omega^\varepsilon_h, \tag{6}$$

with the boundary conditions

$$\Phi^k_{i,j} = F^k_{i,j}, \quad (a_1 + ih_1, a_2 + ih_2) \in \Delta^\varepsilon_h, \quad k = 1, \ldots, N_3 - 1; \tag{7}$$

$$\Phi^0_{i,j} = \Phi^{N_3}_{i,j}, \quad (a_1 + ih_1, a_2 + ih_2) \in D^\varepsilon_h. \tag{8}$$

Here $A = A^k_{i,j} = \cos\theta^k_{i,j}$, $B = B^k_{i,j} = \sin\theta^k_{i,j}$, $\theta^k_{i,j} = \theta(a_1 + ih_1, a_2 + ih_2, kh_3)$, and $\Phi^k_{i,j}$ is a mesh function on Ω^ε_h.

Note that in the above formulation the information on the solution is given not only on the boundary but also in some its neighborhood.

It is not difficult to verify the following assertion.

Lemma 1. *If $W(x,y,z) \in C^3(\Omega^\varepsilon)$ and $\theta(x,y,z) \in C^2(\Omega^\varepsilon)$, then the difference operator L_h approximates the differential operator L with an accuracy of the order of $O(h_1^2+h_2^2+h_3)$ in the domain Ω^ε.*

The following lemmas will also be needed below.

Lemma 2. *Suppose that conditions (5) are fulfilled and $\|\partial\theta/\partial z\|_{C(\Omega^\varepsilon)} = P$. Then, for $(a_1 + ih_1, a_2 + ih_2, kh_3) \in \Omega^\varepsilon_h$, it holds that*

$$M = M^k_{i,j} = (AB_z - A_zB) \geq 0, \quad 0 \leq h_3 \leq \pi/P, \tag{$\star$}$$

$$|AA_z + BB_z| \leq P^2h_3. \tag{$\star\star$}$$

PROOF. By the mean-value theorem,

$$\theta^{k+1}_{i,j} = \theta^k_{i,j} + \frac{\partial\theta}{\partial z}(x,y,\xi_1)h_3, \quad \xi_1 \in (kh_3, (k+1)h_3).$$

Hence, omitting subscripts for simplicity, we arrive at

$$\cos(\theta^{k+1}) = \cos(\theta^k + h_3\frac{\partial\theta}{\partial z}) = \cos(\theta^k)\cos(\frac{\partial\theta}{\partial z}h_3) - \sin(\theta^k)\sin(\frac{\partial\theta}{\partial z}h_3),$$

$$\sin(\theta^{k+1}) = \sin(\theta^k + h_3\frac{\partial\theta}{\partial z}) = \sin(\theta^k)\cos(\frac{\partial\theta}{\partial z}h_3) - \cos(\theta^k)\sin(\frac{\partial\theta}{\partial z}h_3).$$

Taking into account that $A = \cos(\theta^k)$, $B = \sin(\theta^k)$, and

$$A_z = \left[\cos(\theta^{k+1}) - \cos(\theta^k)\right]/h_3, \text{ and } B_z = \left[\sin(\theta^{k+1}) - \sin(\theta^k)\right]/h_3,$$

we obtain

$$AA_z + BB_z = \left[\cos\left(\frac{\partial\theta}{\partial z}h_3\right) - 1\right]/h_3 = -\sin(\xi_2)\frac{\partial\theta}{\partial z}(x,y,\xi_1),$$

where

$$\xi_2 \in (0, \frac{\partial \theta}{\partial z} h_3), \quad AB_z - A_z B = \sin\left(\frac{\partial \theta}{\partial z} h_3\right) / h_3.$$

It can easily be shown now that inequalities ($\star$) and ($\star\star$) hold.

Lemma 3. *If u and v are mesh functions, then*

$$(uv)_{\overset{\circ}{x}} = u_{\overset{\circ}{x}} v_i + u_i v_{\overset{\circ}{x}} + (1/2h_1)\left[(u_{i+1} - u_i)(v_{i+1} - 2v_i + v_{i-1}) \right.$$

$$\left. + (u_{i+1} - 2u_i + u_{i-1})(v_i - v_{i-1})\right]; \tag{9}$$

$$(uv)_z = u^k v_z + u_z v^k + h_3 u_z v_z. \tag{10}$$

PROOF. One can easily verify the lemma assertion taking into account that equalities (9) and (10) are equivalent to

$$(u_{i+1}v_{i+1} - u_{i-1}v_{i-1})/2h_1 = \left[(u_{i+1} - u_{i-1})v_i\right]/2h_1 + \left[u_i(v_{i+1} - v_{i-1})\right]/2h_1$$

$$+ \left[(u_{i+1} - u_i)(v_{i+1} - 2v_i + v_{i-1}) + (u_{i+1} - 2u_i + u_{i-1})(v_i - v_{i-1})\right]/2h_1;$$

$$(u^{k+1}v^{k+1} - u^k v^k)/h_3 = u^k(v^{k+1} - v^k)/h_3 + v^k(u^{k+1} - u^k)/h_3 + \left[(u^{k+1} - u^k)(v^{k+1} - v^k)\right]/h_3.$$

Lemma 4. *Let $\Phi_{i,j}^k$ be mesh functions. Then*

$$\Phi_{\overset{\circ}{x}} \Phi_{\overset{\circ}{x}z} + \Phi_{\overset{\circ}{y}} \Phi_{\overset{\circ}{y}z} = \frac{1}{2}\left[\Phi_{\overset{\circ}{x}}^2 + \Phi_{\overset{\circ}{y}}^2\right]_z - (h_3/2)\left[\Phi_{\overset{\circ}{x}z}^2 + \Phi_{\overset{\circ}{y}z}^2\right]. \tag{11}$$

PROOF. Note that

$$\frac{1}{2}\left[\Phi_{\overset{\circ}{x}}\right]_z = \frac{1}{2}\left[\left(\Phi_{\overset{\circ}{x}}^{k+1}\right)^2 - \left(\Phi_{\overset{\circ}{x}}^{k}\right)^2\right]/h_3 = \left(\Phi_{\overset{\circ}{x}}^{k+1} - \Phi_{\overset{\circ}{x}}^{k}\right)\left(\Phi_{\overset{\circ}{x}}^{k+1} + \Phi_{\overset{\circ}{x}}^{k}\right)/2h_3$$

$$= \Phi_{\overset{\circ}{x}z} \Phi_{\overset{\circ}{x}}^k + \Phi_{\overset{\circ}{x}z}\left(\Phi_{\overset{\circ}{x}}^{k+1} - \Phi_{\overset{\circ}{x}}^{k}\right)/2 = \Phi_{\overset{\circ}{x}} \Phi_{\overset{\circ}{x}z} + (h_3/2)\Phi_{\overset{\circ}{x}z}^2.$$

Similarly,

$$\frac{1}{2}\left[\Phi_{\overset{\circ}{y}}^2\right]_z = \Phi_{\overset{\circ}{y}} \Phi_{\overset{\circ}{y}z} + (h_3/2)\Phi_{\overset{\circ}{y}z}^2.$$

Combining two these equalities, we obtain (11).

Theorem 1. *Suppose that $\theta(x, y, z)$ obeys (5). Assume also the solution of problem (6)-(8) to exist and, moreover,*

$$\left|\Phi_{\overset{\circ}{x}z}\right| \le C_1, \quad \left|\Phi_{\overset{\circ}{y}z}\right| \le C_1, \text{ where } C_1 \text{ is constant.}$$

Then, for all $N_j > 9$, $j = 1, 2$, and $N_3 > Pl/\pi$, the following inequality holds true:

$$\sum_{\Omega_h^\epsilon}(M/2)\left[\Phi_{\overset{\circ}{x}}^2 + \Phi_{\overset{\circ}{y}}^2\right] h_1 h_2 h_3 \le C_2 \sum_{\Delta_h^\epsilon}\left[\Phi_{\overset{\circ}{x}}^2 h_1 h_3 + \Phi_{\overset{\circ}{y}}^2 h_2 h_3 + F_z^2(h_2 + h_1)h_3\right] + C_1 h_3^2, \tag{12}$$

with a constant C_2 independent of h_j, $j = 1, 2, 3$.

PROOF. From (6) it follows that

$$I_1 = (-B\Phi_{\overset{\circ}{x}} + A\Phi_{\overset{\circ}{y}})(A\Phi_{\overset{\circ}{x}} + B\Phi_{\overset{\circ}{y}})_z = 0.$$

Using (10), we get

$$I_1 = \left[(-B\Phi_{\overset{\circ}{x}} + A\Phi_{\overset{\circ}{y}})(A\Phi_{\overset{\circ}{x}} + B\Phi_{\overset{\circ}{y}})\right]_z - (-B\Phi_{\overset{\circ}{x}} + A\Phi_{\overset{\circ}{y}})_z(A\Phi_{\overset{\circ}{x}} + B\Phi_{\overset{\circ}{y}})$$
$$-h_3(-B\Phi_{\overset{\circ}{x}} + A\Phi_{\overset{\circ}{y}})_z(A\Phi_{\overset{\circ}{x}} + B\Phi_{\overset{\circ}{y}})_z. \tag{13}$$

We sum (13) with the first expression for I_1 and rearrange the resulting equation using (10). Thus, we obtain

$$\Phi^2_{\overset{\circ}{x}}(AB_z - A_zB) + \Phi^2_{\overset{\circ}{x}}(AB_z - A_zB) + h_3(AB_z - A_zB)\Phi_{\overset{\circ}{x}}\Phi_{\overset{\circ}{x}z} - \Phi_{\overset{\circ}{x}}\Phi_{\overset{\circ}{y}z}(A^2 + B^2)$$
$$-h_3(AA_z + BB_z)\Phi_{\overset{\circ}{x}}\Phi_{\overset{\circ}{y}z} + (A^2+B^2)\Phi_{\overset{\circ}{y}}\Phi_{\overset{\circ}{x}z} + h_3(AA_z+BB_z)\Phi_{\overset{\circ}{y}}\Phi_{\overset{\circ}{x}z} + h_3(AB_z - A_zB)\Phi_{\overset{\circ}{y}}\Phi_{\overset{\circ}{y}z}$$
$$+\left[(-B\Phi_{\overset{\circ}{x}} + A\Phi_{\overset{\circ}{y}})(A\Phi_{\overset{\circ}{x}} + B\Phi_{\overset{\circ}{y}})\right]_z - h_3(A\Phi_{\overset{\circ}{x}} + B\Phi_{\overset{\circ}{y}})_z(-B\Phi_{\overset{\circ}{x}} + A\Phi_{\overset{\circ}{y}})_z = 0.$$

Since $A^2 + B^2 = 1$, $M = (AB_z - A_zB)$ and (6) holds, we rewrite the last identity as follows:

$$M(\Phi^2_{\overset{\circ}{x}} + \Phi^2_{\overset{\circ}{y}}) + \Phi_{\overset{\circ}{y}}\Phi_{\overset{\circ}{x}z} - \Phi_{\overset{\circ}{x}}\Phi_{\overset{\circ}{y}z} + h_3(AA_z + BB_z)(\Phi_{\overset{\circ}{y}}\Phi_{\overset{\circ}{x}z} - \Phi_{\overset{\circ}{x}}\Phi_{\overset{\circ}{y}z})$$
$$+h_3M(\Phi_{\overset{\circ}{x}}\Phi_{\overset{\circ}{x}z} + \Phi_{\overset{\circ}{y}}\Phi_{\overset{\circ}{y}z}) + \left[(-B\Phi_{\overset{\circ}{x}} + A\Phi_{\overset{\circ}{y}})(A\Phi_{\overset{\circ}{x}} + B\Phi_{\overset{\circ}{y}})\right]_z = 0. \tag{14}$$

From (9) it follows that

$$\Phi_{\overset{\circ}{y}}\Phi_{\overset{\circ}{x}z} = \left(\Phi_{\overset{\circ}{y}}\Phi_z\right)_{\overset{\circ}{x}} - \Phi_{\overset{\circ\circ}{xy}}\Phi_z - (h_1^2/2)\left(\Phi_{\overset{\circ}{y}x}\Phi_{zx}\right)_{\overline{x}}, \tag{15}$$

$$-\Phi_{\overset{\circ}{x}}\Phi_{\overset{\circ}{y}z} = -\left(\Phi_{\overset{\circ}{x}}\Phi_z\right)_{\overset{\circ}{y}} + \Phi_{\overset{\circ\circ}{xy}}\Phi_z + (h_2^2/2)\left(\Phi_{\overset{\circ}{x}y}\Phi_{zy}\right)_{\overline{y}}. \tag{16}$$

Taking into account (11), we have

$$h_3M\left(\Phi_{\overset{\circ}{x}}\Phi_{\overset{\circ}{x}z} + \Phi_{\overset{\circ}{y}}\Phi_{\overset{\circ}{y}z}\right) = (h_3M)/2\left[\left(\Phi^2_{\overset{\circ}{x}} + \Phi^2_{\overset{\circ}{y}}\right)_z - h_3(\Phi^2_{\overset{\circ}{x}z} + \Phi^2_{\overset{\circ}{y}z})\right]$$
$$= (M/2)\left[(\Phi^{k+1}_{\overset{\circ}{x}})^2 + (\Phi^{k+1}_{\overset{\circ}{y}})^2 - (\Phi^k_{\overset{\circ}{x}})^2 - (\Phi^k_{\overset{\circ}{y}})^2\right] - (Mh_3^2/2)(\Phi^2_{\overset{\circ}{x}z} + \Phi^2_{\overset{\circ}{y}z}). \tag{17}$$

Substitution of (15), (16), (17) in (14), multiplication of the resulting equation by $h_1h_2h_3$ and summation over i, j, k yields

$$\sum_{\Omega_h^\epsilon}\Big\{(M/2)\left[(\Phi^{k+1}_{\overset{\circ}{x}})^2 + (\Phi^{k+1}_{\overset{\circ}{y}})^2 + (\Phi^k_{\overset{\circ}{x}})^2 + (\Phi_{\overset{\circ}{y}})^2\right] + h_3(AA_z + BB_z)(\Phi_{\overset{\circ}{y}}\Phi_{\overset{\circ}{x}z} - \Phi_{\overset{\circ}{x}}\Phi_{\overset{\circ}{y}z})$$
$$-(Mh_3^2/2)(\Phi^2_{\overset{\circ}{x}z} + \Phi^2_{\overset{\circ}{y}z})\Big\}h_1h_2h_3 = \sum_{\Omega_h^\epsilon} R_{i,j,k}h_1h_2h_3, \tag{18}$$

where

$$R_{i,j,k} = (\Phi_{\overset{\circ}{x}}\Phi_z)_{\overset{\circ}{y}} - (\Phi_{\overset{\circ}{y}}\Phi_z)_{\overset{\circ}{x}} - \left[(-B\Phi_{\overset{\circ}{x}} + A\Phi_{\overset{\circ}{y}})(A\Phi_{\overset{\circ}{x}} + B\Phi_{\overset{\circ}{y}})\right]_z$$
$$+(h_1^2/2)(\Phi_{\overset{\circ}{y}z}\Phi_{zx})_{\overline{x}} - (h_2^2/2)(\Phi_{\overset{\circ}{x}z}\Phi_{zy})_{\overline{y}}.$$

Using ($\star\star$) and inequality $|ab| \leq (a^2+b^2)/2$, we have

$$\left|h_3(AA_z+BB_z)(\Phi_{\overset{\circ}{y}}\Phi_{\overset{\circ}{x}z}-\Phi_{\overset{\circ}{x}}\Phi_{\overset{\circ}{y}z})\right| = \left|(AA_z+BB_z)(\Phi^k_{\overset{\circ}{y}}\Phi^{k+1}_{\overset{\circ}{x}}-\Phi^k_{\overset{\circ}{x}}\Phi^{k+1}_{\overset{\circ}{y}})\right|$$

$$\leq (P^2h_3/2)\left[(\Phi^{k+1}_{\overset{\circ}{x}})^2+(\Phi^{k+1}_{\overset{\circ}{y}})^2+(\Phi_{\overset{\circ}{x}})^2+(\Phi_{\overset{\circ}{y}})^2\right].$$

Since $\left|\Phi_{\overset{\circ}{x}z}\right| \leq C_1$, $\left|\Phi_{\overset{\circ}{y}z}\right| \leq C_1$, it follows from (18) that

$$\sum_{\Omega^\varepsilon_h}(M/2)(\Phi^2_{\overset{\circ}{x}}+\Phi^2_{\overset{\circ}{y}})h_1h_2h_3 \leq \left|\sum_{\Omega^\varepsilon_h} R_{i,j,k}h_1h_2h_3\right| + C_1h_3^2. \tag{19}$$

Due to conditions (7), (8), we can transform (19) so as to obtain the desired estimate (12).

Corollary. *Suppose that $W(x,y,z)$ belong to $C^3(\Omega^\varepsilon)$ and satisfies (2), and that $\Phi^k_{i,j}$ satisfies (6), (8) and the conditions*

$$\Phi^k_{i,j} = W^k_{i,j}, \quad (a_1+ih_1, a_2+ih_2) \in \Delta^\varepsilon_h, \quad k=0,1,\ldots,N_3-1;$$

$$\left|\Phi_{\overset{\circ}{y}z}\right| \leq C_1, \quad \left|\Phi_{\overset{\circ}{x}z}\right| \leq C_1,$$

where C_1 is constant and $W^k_{i,j}$ is the function W related to the Ω^ε_h.

Let $\theta(x,y,z)$ satisfy (5), and there exists a constant $M^\star > 0$ such that

$$M^k_{i,j} \geq M^\star, \quad (a_1+ih_1, a_2+ih_2, kh_3) \in \Omega^\varepsilon_h.$$

Then one can find positive constant $N^\star$ and C_3 such that, for all $N_j > N^\star$, $j=1,2,3$,

$$\left\{\sum_{\Omega^\varepsilon_h}(\Psi^2_{\overset{\circ}{x}}+\Psi^2_{\overset{\circ}{y}})h_1h_2h_3\right\}^{1/2} \leq C_3(h_1^2+h_2^2+h_3),$$

where

$$\Psi^k_{i,j} = W^k_{i,j} - \Phi^k_{i,j}, \quad (a_1+ih_1, a_2+ih_2, kh_3) \in \Omega^\varepsilon_h.$$

Note that the solution to problem (6)-(8) can be found using some well-known methods for solving overdetermined algebraic systems of equations (Voevodin, 1977).

3. Discrete Integral Geometry Problem with a Weight Function

Let us consider a finite-difference analogue of the two-dimensional problem of integral geometry with a weight function:

$$V(\gamma,z) = \int_{K(\gamma,z)} U(x,y)\rho(x,y,z)\mathrm{d}s, \quad \gamma \in [0,l], \quad z \in [0,l], \tag{1'}$$

where $\rho(x,y,z)$ is a known function. If we introduce function

$$W(x,y,z) = \int_{K(x,y,z)} U(x,y)\rho(x,y,z)\mathrm{d}s,$$

then problem (1′) is reduced to the following boundary value problem (Mukhometov, 1975):

$$\frac{\partial}{\partial z}\left(\frac{\partial W}{\partial x}\frac{\cos\theta}{\rho}+\frac{\partial W}{\partial y}\frac{\sin\theta}{\rho}\right)=0, \quad (x,y,z)\in\Omega_1; \tag{20}$$

$$W(\xi(\gamma),\eta(\gamma),z)=V(\gamma,z), \quad V(z,z)=0, \quad \gamma,z\in[0,l]. \tag{21}$$

From here on we shall supposed that the coefficients and the solution to problem (20), (21) have the following properties:

$$W(x,y,z)\in C^3(\Omega^\varepsilon), \quad \theta(x,y,z)\in C^2(\Omega^\varepsilon), \quad \Omega^\varepsilon=\overline{D^\varepsilon}\times[0,l] \tag{22}$$

$$\rho(x,y,z)\in C^2(\Omega), \quad \rho(x,y,z)>C^*>0, \quad \frac{\partial\theta}{\partial z}>\left|\frac{\partial\rho}{\partial z}/\rho\right|. \tag{23}$$

We formulate the finite–difference problem as follows:

$$\left[\Phi_{\overset{\circ}{x}}\frac{A}{C}+\Phi_{\overset{\circ}{y}}\frac{B}{C}\right]_z=0, \quad (a_1+ih_1,a_2+ih_2,kh_3)\in\Omega_h^\varepsilon, \tag{24}$$

$$\Phi_{i,j}^k=F_{i,j}^k, \quad (a_1+ih_1,a_2+ih_2)\in\Delta_h^\varepsilon, \quad k=0,1,\dots,N_3-1; \tag{25}$$

$$\Phi_{i,j}^0=\Phi_{i,j}^{N_3}, \quad (a_1+ih_1,a_2+ih_2)\in D_h^\varepsilon, \tag{26}$$

where

$$A=A_{i,j}^k=\cos\theta_{i,j}^k, \quad B=B_{i,j}^k=\sin\theta_{i,j}^k,$$

$$\theta_{i,j}^k=\theta(a_1+ih_1,a_2+ih_2,kh_3), \quad C=\rho_{i,j}^k=\rho(a_1+ih_1,a_2+ih_2,kh_3).$$

Theorem 2. *Suppose that the solution of problem (24)–(26) exists on Ω_h^ε and*

$$\left|\Phi_{\overset{\circ}{x}z}\right|\le C_1, \quad \left|\Phi_{\overset{\circ}{y}z}\right|\le C_1,$$

where C_1 is constant, and that

$$(AB_z-A_zB)-\left|\frac{c_z}{c}\right|\ge\alpha>0$$

for all N_j, $j=1,2,3$. Then there exists a positive constant N^{**} such that, for all $N_j>N^{**}$, $j=1,2,3$,

$$\sum_{\Omega_h^\varepsilon}(\Phi_{\overset{\circ}{x}}^2+\Phi_{\overset{\circ}{y}}^2)h_1h_2h_3$$

$$\le C_4\sum_{\Delta_h^\varepsilon}\left(F_{\overset{\circ}{x}}^2h_1h_3+F_{\overset{\circ}{y}}^2h_2h_3+F_z^2(h_2+h_1)h_3\right)+C_1h_3^2.$$

Here C_4 is a constant dependent on the function $\rho(x,y,z)$ and on the family of curves $K(\gamma,x)$.

Proof. The proof of the theorem is similar to that of Theorem 4. The only difference is that finite–difference analogues of differentiation formulas of product and quotient have several additional terms to be estimated.

4. Differential–Difference Analogue of a Two–Dimensional Problem of Integral Geometry

Suppose that the coefficients and the solution of problem (2)–(3) enjoy the properties (5).

The differential-difference problem we consider is to find the functions $\Phi_{i,j}(z)$, $U_{i,j}$ which satisfy the equation

$$\Phi_{\overset{\circ}{x}} A + \Phi_{\overset{\circ}{y}} B = U_{i,j}, \quad (a_1 + ih_1, a_2 + ih_2) \in D_h, \quad z \in [0, l] \tag{27}$$

and the boundary condition

$$\Phi_{i,j}(z) = F_{i,j}(z), \quad (a_1 + ih_1, a_2 + ih_2) \in \Delta_h^\varepsilon, \quad z \in [0, l]. \tag{28}$$

Here $A = \cos\theta_{i,j}(z)$, $B = \sin\theta_{i,j}(z)$, $\theta_{i,j} = \theta(a_1 + ih_1, a_2 + ih_2, z)$.

We note that in the differential–difference formulation information on the solution is given not only on the boundary Γ but also in its ε-neighborhood, because the partial derivatives Θ_z, W_{xz}, W_{yz}, W_{xy} have singularities of the type $[(x - \xi(z))^2 + (y - \eta(z))^2]^{-1/2}$ in a neighborhood of an arbitrary point $(\xi(z), \eta(z), z)$ (see Mukhometov, 1975).

Theorem 3. *Suppose that problem (27)-(28), with the function $\theta_{i,j}(z)$ such that $\theta_{i,j}(0) = \theta_{i,j}(l)$ and $\frac{\partial\theta}{\partial z} \geq 0$, has a solution $\Phi_{i,j}(z) \in C^1[0, l]$, $\Phi_{i,j}(0) = \Phi_{i,j}(l)$. Then for all $N_j > 9$, $j = 1, 2$, it holds that*

$$\sum_{D_h^\varepsilon} U_{i,j}^2 h_1 h_2 \leq C_5 \int_0^l \sum_{\Delta_h^\varepsilon} \left[F_{\overset{\circ}{x}}^2 h_1 + F_{\overset{\circ}{y}}^2 h_2 + \left(\frac{\partial F}{\partial z}\right)^2 (h_1 + h_2) \right] dz, \tag{29}$$

where C_5 is a positive constant.

PROOF. Multiplying both sides of (27) by

$$(-B\Phi_{\overset{\circ}{x}} + A\Phi_{\overset{\circ}{y}}) \frac{\partial}{\partial z},$$

we obtain

$$(-B\Phi_{\overset{\circ}{x}} + A\Phi_{\overset{\circ}{y}}) \frac{\partial}{\partial z} (A\Phi_{\overset{\circ}{x}} + B\Phi_{\overset{\circ}{y}}) = 0. \tag{30}$$

Using the differentiation formula for product, equation (30) can be written as

$$(-B\Phi_{\overset{\circ}{x}} + A\Phi_{\overset{\circ}{y}}) \frac{\partial}{\partial z} (A\Phi_{\overset{\circ}{x}} + B\Phi_{\overset{\circ}{y}}) = \frac{\partial}{\partial z} \left[(A\Phi_{\overset{\circ}{x}} + B\Phi_{\overset{\circ}{y}}) \times (-B\Phi_{\overset{\circ}{x}} + A\Phi_{\overset{\circ}{y}}) \right]$$

$$-(A\Phi_{\overset{\circ}{x}} + B\Phi_{\overset{\circ}{y}}) \frac{\partial}{\partial z} (-B\Phi_{\overset{\circ}{x}} + A\Phi_{\overset{\circ}{y}}) = 0.$$

Summing the latter equation with (30) and taking into account that $A = \cos\theta_{i,j}(z)$, $B = \sin\theta_{i,j}(z)$, we obtain

$$(-B\Phi_{\overset{\circ}{x}} + A\Phi_{\overset{\circ}{y}}) \left[-B\frac{\partial\theta}{\partial z}\Phi_{\overset{\circ}{x}} + A\frac{\partial}{\partial z}(\Phi_{\overset{\circ}{x}}) + A\frac{\partial\theta}{\partial z}\Phi_{\overset{\circ}{y}} + B\frac{\partial}{\partial z}(\Phi_{\overset{\circ}{y}}) \right]$$

$$+\frac{\partial}{\partial z} \left[(A\Phi_{\overset{\circ}{x}} + B\Phi_{\overset{\circ}{y}})(-B\Phi_{\overset{\circ}{x}} + A\Phi_{\overset{\circ}{y}}) \right]$$

$$-(A\Phi_{\mathring{x}} + B\Phi_{\mathring{y}}) \times \left[-A\frac{\partial\theta}{\partial z}\Phi_{\mathring{x}} - B\frac{\partial}{\partial z}(\Phi_{\mathring{x}}) - B\frac{\partial\theta}{\partial z}\Phi_{\mathring{y}} + A\frac{\partial}{\partial z}(\Phi_{\mathring{y}})\right] = 0.$$

Rearrange this equation using the identity $\sin^2\theta + \cos^2\theta = 1$, i.e., $A^2 + B^2 = 1$. As a result, we arrive at

$$\frac{\partial\theta}{\partial z}(\Phi^2_{\mathring{x}} + \Phi^2_{\mathring{y}}) - \Phi_{\mathring{x}}\frac{\partial}{\partial z}(\Phi_{\mathring{y}}) + \Phi_{\mathring{y}}\frac{\partial}{\partial z}(\Phi_{\mathring{y}}) + \frac{\partial}{\partial z}\left[(A\Phi_{\mathring{x}} + B\Phi_{\mathring{y}})(-B\Phi_{\mathring{x}} + A\Phi_{\mathring{y}})\right] = 0. \quad (31)$$

Using the formula

$$(uv)_{\mathring{x}} = u_{\mathring{x}}v + uv_{\mathring{x}} + (h_1^2/2)[u_x v_x]_{\bar{x}},$$

we have

$$\left[\Phi_{\mathring{y}}\frac{\partial\Phi}{\partial z}\right]_{\mathring{x}} = \Phi_{\mathring{y}\mathring{x}}\frac{\partial\Phi}{\partial z} + \Phi_{\mathring{y}}\left(\frac{\partial\Phi}{\partial z}\right)_{\mathring{x}} + (h_1^2/2)\left[\Phi_{\mathring{y}x}(\frac{\partial\Phi}{\partial z})_x\right]_{\bar{x}}. \quad (32)$$

Similarly, we can write

$$-\left[\Phi_{\mathring{x}}\frac{\partial\Phi}{\partial z}\right]_{\mathring{y}} = -\Phi_{\mathring{y}\mathring{x}}\frac{\partial\Phi}{\partial z} - \Phi_{\mathring{y}}\left(\frac{\partial\Phi}{\partial z}\right)_{\mathring{x}} - (h_2^2/2)\left[\Phi_{\mathring{x}y}(\frac{\partial\Phi}{\partial z})_y\right]_{\bar{y}}. \quad (33)$$

It follows from (32), (33) that

$$\Phi_{\mathring{y}}(\frac{\partial\Phi}{\partial z})_{\mathring{x}} - \Phi_{\mathring{x}}(\frac{\partial\Phi}{\partial z})_{\mathring{y}} = \left[\Phi_{\mathring{y}}\frac{\partial\Phi}{\partial z}\right]_{\mathring{x}} - \left[\Phi_{\mathring{x}}\frac{\partial\Phi}{\partial z}\right]_{\mathring{y}} - (h_1^2/2)\left[\Phi_{\mathring{y}x}(\frac{\partial\Phi}{\partial z})_x\right]_{\bar{x}} + (h_2^2/2)\left[\Phi_{\mathring{x}y}(\frac{\partial\Phi}{\partial z})_y\right]_{\bar{y}}. \quad (34)$$

Taking into account that

$$(\frac{\partial\Phi}{\partial z})_{\mathring{x}} = (\frac{\partial}{\partial z})(\Phi_{\mathring{x}}), \quad (\frac{\partial\Phi}{\partial z})_{\mathring{y}} = (\frac{\partial}{\partial z})(\Phi_{\mathring{y}}), \quad (35)$$

we can conclude from (31) and (34) that

$$\frac{\partial\theta}{\partial z}(\Phi^2_{\mathring{x}} + \Phi^2_{\mathring{y}}) = \left[\Phi_{\mathring{x}}\frac{\partial\Phi}{\partial z}\right]_{\mathring{y}} - \left[\Phi_{\mathring{y}}\frac{\partial\Phi}{\partial z}\right]_{\mathring{x}} + (h_1^2/2)\left[\Phi_{\mathring{y}x}(\frac{\partial\Phi}{\partial z})_x\right]_{\bar{x}} - (h_2^2/2)\left[\Phi_{\mathring{x}y}(\frac{\partial\Phi}{\partial z})_y\right]_{\bar{y}}$$

$$-\frac{\partial}{\partial z}\left[(A\Phi_{\mathring{x}} + B\Phi_{\mathring{y}})(-B\Phi_{\mathring{x}} + A\Phi_{\mathring{y}})\right]. \quad (36)$$

From (27) it follows that

$$A\Phi_{\mathring{x}} + B\Phi_{\mathring{y}} = U_{i,j}.$$

Squaring both sides of the last equality and taking into account that $A = \cos\theta, B = \sin\theta$, we obtain

$$\Phi^2_{\mathring{x}} + \Phi^2_{\mathring{y}} = U^2_{i,j} + \left(B\Phi_{\mathring{x}} - A\Phi_{\mathring{y}}\right)^2. \quad (37)$$

Substitute (37) in (36), sum the resulting equation over i, j using condition (28) and integrate it with respect to z taking into account that

$$\int_0^1 \frac{\partial\theta}{\partial z}\mathrm{d}z = 2\pi$$

and the functions $\Phi_{i,j}(z)$, $\theta_{i,j}(z)$ are periodic in z. As a result, we obtain

$$\sum_{D_h^\epsilon} U^2_{i,j}h_1h_2 \le C_5\int_0^1 \sum_{\Delta_h^\epsilon}\left[\Phi^2_{\mathring{x}}h_1 + \Phi^2_{\mathring{y}}h_2 + \left(\frac{\partial F}{\partial z}\right)^2(h_1 + h_2)\right]\mathrm{d}z,$$

where C_5 is a positive constant. Thus, Theorem 3 is proved.

5. Differential–Difference Analogue of the Two–Dimensional Problem of Integral Geometry with a Weight Function

Note that the integral geometry problem (1′) is reduced to the boundary value problem (20)–(21). We suppose hereafter that the coefficients and the solution of problem (20)–(21) obey (22)–(23).

We consider the differential-difference problem of finding the functions $\Phi_{i,j}(z)$, $U_{i,j}$ which satisfy the equation

$$\Phi_{\overset{\circ}{x}}\frac{A}{C}+\Phi_{\overset{\circ}{y}}\frac{B}{C}=U_{i,j},\quad (a_1+ih_1,a_2+jh_2)\in D_h,\quad z\in[0,l], \tag{38}$$

and the boundary condition

$$\Phi_{i,j}(z)=F_{i,j}(z),(a_1+ih_1,a_2+jh_2)\in\Delta_h^\epsilon,\quad z\in[0,l], \tag{39}$$

where $A=\cos\theta_{i,j}(z), B=\sin\theta_{i,j}(z), \theta_{i,j}(z)=\theta(a_1+ih_1,a_2+jh_2,z), C=\rho(a_1+ih_1,a_2+jh_2,z)$.

Theorem 4. *Suppose that problem (38)-(39), with the functions $\theta_{i,j}(z)$, $\rho(x,y,z)$ such that $\theta_{i,j}(0)=\theta_{i,j}(l)$, $\rho(x,y,z)\in C^1(\Omega)$, $\rho(x,y,z)>C^\star>0$, and $\frac{\partial\theta}{\partial z}>\left|\frac{\partial\rho}{\partial z}\frac{1}{\rho}\right|$, has a solution*

$$\Phi_{i,j}(z)\in C^1[0,l],\quad \Phi_{i,j}(0)=\Phi_{i,j}(l).$$

Then, for all $N_j>9$, $j=1,2$, the following inequality holds:

$$\sum_{\Delta_h^\epsilon}U_{i,j}^2h_ih_j\le C_6\int_0^1\sum_{\Delta_h^\epsilon}\left[F_{\overset{\circ}{x}}^2h_1+F_{\overset{\circ}{y}}^2h_2+(\frac{\partial F}{\partial z})^2(h_1+h_2)\right]\mathrm{d}z, \tag{40}$$

where C_6 is a constant dependent on the function $\rho(x,y,z)$ and the curves family $K(\gamma,z)$.

PROOF. Multiplying both sides of (38) by $C(-B\Phi_{\overset{\circ}{x}}+A\Phi_{\overset{\circ}{y}})\frac{\partial}{\partial z}$, we get

$$C(-B\Phi_{\overset{\circ}{x}}+A\Phi_{\overset{\circ}{y}})\frac{\partial}{\partial z}\left(\frac{A}{C}\Phi_{\overset{\circ}{x}}+\frac{B}{C}\Phi_{\overset{\circ}{y}}\right)=0. \tag{41}$$

Using the product differentiation formula, we rewrite (41) as follows:

$$0=C(-B\Phi_{\overset{\circ}{x}}+A\Phi_{\overset{\circ}{y}})\frac{\partial}{\partial z}\left(\frac{A}{C}\Phi_{\overset{\circ}{x}}+\frac{B}{C}\Phi_{\overset{\circ}{y}}\right)$$

$$=\frac{\partial}{\partial z}\left[C(-B\Phi_{\overset{\circ}{x}}+A\Phi_{\overset{\circ}{y}})(\frac{A}{C}\Phi_{\overset{\circ}{x}}+\frac{B}{C}\Phi_{\overset{\circ}{y}})\right]-\frac{\partial}{\partial z}\left[C(-B\Phi_{\overset{\circ}{x}}+A\Phi_{\overset{\circ}{y}})\right]\left(\frac{A}{C}\Phi_{\overset{\circ}{x}}+\frac{B}{C}\Phi_{\overset{\circ}{y}}\right)$$

$$=\frac{\partial}{\partial z}\left[(A\Phi_{\overset{\circ}{x}}+B\Phi_{\overset{\circ}{y}})(-B\Phi_{\overset{\circ}{x}}+A\Phi_{\overset{\circ}{y}})\right]-\frac{\partial C}{\partial z}\left(-B\Phi_{\overset{\circ}{x}}+A\Phi_{\overset{\circ}{y}})(\frac{A}{C}\Phi_{\overset{\circ}{x}}+\frac{B}{C}\Phi_{\overset{\circ}{y}}\right)$$

$$-\frac{\partial}{\partial z}\left(-B\Phi_{\overset{\circ}{x}}+A\Phi_{\overset{\circ}{y}})(A\Phi_{\overset{\circ}{x}}+B\Phi_{\overset{\circ}{y}})\right]=\frac{\partial}{\partial z}\left[(A\Phi_{\overset{\circ}{x}}+B\Phi_{\overset{\circ}{y}})(-B\Phi_{\overset{\circ}{x}}+A\Phi_{\overset{\circ}{y}})\right]$$

$$+AB(\frac{\partial C}{\partial z}/C)\Phi_{\overset{\circ}{x}}-(\frac{\partial C}{\partial z}/C)A^2\Phi_{\overset{\circ}{x}}\Phi_{\overset{\circ}{y}}+(\frac{\partial C}{\partial z}/C)B^2\Phi_{\overset{\circ}{x}}\Phi_{\overset{\circ}{y}}-(\frac{\partial C}{\partial z}/C)AB\Phi_{\overset{\circ}{y}}^2$$

$$+\frac{\partial\theta}{\partial z}A^2\Phi_{\overset{\circ}{x}}^2+AB\Phi_{\overset{\circ}{x}}\frac{\partial}{\partial z}(\Phi_{\overset{\circ}{x}})+\frac{\partial\theta}{\partial z}AB\Phi_{\overset{\circ}{x}}\Phi_{\overset{\circ}{y}}-A^2\Phi_{\overset{\circ}{x}}\frac{\partial}{\partial z}(\Phi_{\overset{\circ}{y}})$$

$$+\frac{\partial\theta}{\partial z}AB\Phi_{\mathring{x}}\Phi_{\mathring{y}} + B^2\Phi_{\mathring{y}}\frac{\partial}{\partial z}(\Phi_{\mathring{x}}) + \frac{\partial\theta}{\partial z}B^2\Phi^2_{\mathring{y}} - AB\Phi_{\mathring{y}}\frac{\partial}{\partial z}(\Phi_{\mathring{y}}). \tag{42}$$

At the same time, rearranging (41) yields

$$0 = C(-B\Phi_{\mathring{x}} + A\Phi_{\mathring{y}})\frac{\partial}{\partial z}\left(\frac{A}{C}\Phi_{\mathring{x}} + \frac{B}{C}\Phi_{\mathring{y}}\right)$$

$$= C(-B\Phi_{\mathring{x}} + A\Phi_{\mathring{y}}) \times \left[\frac{\partial}{\partial z}\left(\Phi_{\mathring{x}}\right)\frac{A}{C} + \Phi_{\mathring{x}}\left((-B\frac{\partial\theta}{\partial z}C - -A\frac{\partial C}{\partial z})/C^2\right)\right.$$

$$\left.+\frac{B}{C}\frac{\partial\theta}{\partial z}(\Phi_{\mathring{y}}) + \Phi_{\mathring{y}}\left((-A\frac{\partial\theta}{\partial z}C - B\frac{\partial C}{\partial z})/C^2\right)\right]$$

$$= -AB\Phi_{\mathring{x}}\frac{\partial}{\partial z}\left(\Phi_{\mathring{x}}\right) - \frac{\partial\theta}{\partial z}AB\Phi_{\mathring{x}}\Phi_{\mathring{y}} + \frac{\partial C}{\partial z}B^2\Phi_{\mathring{x}}\Phi_{\mathring{y}} - B^2\Phi_{\mathring{x}}\frac{\partial}{\partial z}B^2(\Phi_{\mathring{y}})$$

$$+\frac{\partial\theta}{\partial z}B^2\Phi^2_{\mathring{x}} + (\frac{\partial C}{\partial z}/C)AB\Phi^2_{\mathring{x}}$$

$$+A^2\Phi_{\mathring{y}}\frac{\partial}{\partial z}\left(\Phi_{\mathring{x}}\right) - \frac{\partial\theta}{\partial z}AB\Phi_{\mathring{x}}\Phi_{\mathring{y}} - (\frac{\partial C}{\partial z}/C)A^2\Phi_{\mathring{x}}\Phi_{\mathring{y}}$$

$$+AB\Phi_{\mathring{y}}\frac{\partial}{\partial z}\left(\Phi_{\mathring{y}}\right) + \frac{\partial\theta}{\partial z}A^2\Phi^2_{\mathring{y}} - (\frac{\partial C}{\partial z}/C)AB\Phi^2_{\mathring{y}}. \tag{43}$$

Sum (43) with (42) and denote

$$D = \sin 2\theta = 2\sin\theta\cos\theta = 2AB,\ \ E = \cos 2\theta = \cos^2\theta - \sin^2\theta = A^2 - B^2.$$

As a result, we obtain

$$\left(\frac{\partial\theta}{\partial z} + \frac{\partial C}{\partial z}D\right)\Phi^2_{\mathring{x}} - 2\Phi_{\mathring{x}}\Phi_{\mathring{y}}\frac{\partial C}{\partial z}E$$

$$+\left(\frac{\partial\theta}{\partial z} - \frac{\partial C}{\partial z}D\right)\Phi^2_{\mathring{y}} + \Phi_{\mathring{y}}\frac{\partial}{\partial z}(\Phi_{\mathring{x}}) - \Phi_{\mathring{x}}\frac{\partial}{\partial z}(\Phi_{\mathring{y}})$$

$$+\frac{\partial}{\partial z}\left[(A\Phi_{\mathring{x}} + B(\Phi_{\mathring{y}})(-B\Phi_{\mathring{x}} + A(\Phi_{\mathring{y}})\right] = 0. \tag{44}$$

Combined (44) with (35) and (34), we arrive at

$$I_z + \frac{\partial}{\partial z}\left[(A\Phi_{\mathring{x}} + B(\Phi_{\mathring{y}})(-B\Phi_{\mathring{x}} + A(\Phi_{\mathring{y}})\right]$$

$$+\left[\Phi_{\mathring{y}}\frac{\partial\Phi}{\partial z}\right]_{\mathring{x}} - \left[\Phi_{\mathring{x}}\frac{\partial\Phi}{\partial z}\right]_{\mathring{y}} - (h_1^2/2)\left[\Phi_{\mathring{y}x}\left(\frac{\partial\Phi}{\partial z}\right)_x\right]_{\overline{x}} + (h_2^2/2)\left[\Phi_{\mathring{x}y}\left(\frac{\partial\Phi}{\partial z}\right)_y\right]_{\overline{y}} = 0. \tag{45}$$

Here

$$I_2 = \left(\frac{\partial\theta}{\partial z} + \frac{\partial C}{\partial z}D\right)\Phi^2_{\mathring{x}} - 2\Phi_{\mathring{x}}\Phi_{\mathring{y}}\frac{\partial C}{\partial z}E + \left(\frac{\partial\theta}{\partial z} - \frac{\partial C}{\partial z}D\right)\Phi^2_{\mathring{y}}.$$

The expression for I_2 is a quadratic form with respect to $\Phi_{\mathring{x}}$ and $\Phi_{\mathring{y}}$, whose determinant is

$$\left(\frac{\partial\theta}{\partial z}\right)^2 - \left(\frac{\partial C}{\partial z}/C\right)^2.$$

Then, from the condition

$$\frac{\partial\theta}{\partial z} > \left|\frac{\partial C}{\partial z}/C\right|$$

the positive definiteness of the quadratic form I_2 follows. Using the inequality

$$ax^2 + 2bxy + cy^2 \geq \frac{2(ac-b^2)}{a+c+((a-c)^2+4b^2)}(x^2+y^2)$$

for a positively definite quadratic form $ax^2 + 2bxy + cy^2$, we have

$$I_2 \geq \left(\frac{\partial\theta}{\partial z} - \left|\frac{\partial C}{\partial z}/C\right|\right)(\Phi^2_{\overset{\circ}{x}} + (\Phi^2_{\overset{\circ}{y}}). \tag{46}$$

Due to (38), we obtain

$$(\Phi^2_{\overset{\circ}{x}} + (\Phi^2_{\overset{\circ}{y}}) = U^2_{i,j}C^2 + (B\Phi_{\overset{\circ}{x}} - A(\Phi_{\overset{\circ}{y}})^2. \tag{47}$$

Since

$$C = \rho(x,y,z), \quad \rho(x,y,z) > C^* > 0, \quad \frac{\partial\theta}{\partial z} - \left|\frac{\partial C}{\partial z}/C\right| > 0,$$

there exist a positive constant C_6 such that

$$\int_0^1 \left(\frac{\partial\theta}{\partial z} - \left|\frac{\partial C}{\partial z}/C\right|\right) C^2 \mathrm{d}z \geq (1/C_6^2) > 0. \tag{48}$$

Sum (47) over i,j with the use of (39) and integrate it with respect to z. Then using (46), (47), (48), the inequality $|ab| \leq (a^2+b^2)/2$ and the periodicity of the functions $\Phi_{i,j}(z)$, $\theta_{i,j}(z)$ in z, we can transform the resulting equation so as to obtain the desired estimate:

$$\sum_{D_h^\epsilon} U^2_{i,j}h_1h_2 \leq C_6 \int_0^1 \sum_{\Delta_h^\epsilon} \left(F_{\overset{\circ}{x}}h_1^2 + F_{\overset{\circ}{y}}h_2^2 + \left(\frac{\partial F}{\partial z}\right)^2 (h_1^2+h_2^2)\right) \mathrm{d}z.$$

Here C_6 is a constant dependent on the function $\rho(x,y,z)$ and on the family of curves $K(\gamma, z)$. Thus, the proof of Theorem 4 is completed.

Conclusions

In the paper we derive stability estimates for finite–difference and differential–difference analogues of a two–dimensional problem of integral geometry. The present results can be used in evaluating the convergence rate of numerical algorithms for the considered class of integral geometry problems.

REFERENCES

Lavrent'ev M. M. and Romanov V. G. (1966), Three linearized inverse problems for hyperbolic equations. *Soviet Math. Dokl.*, vol. 7, 1650–1652.

Mukhometov R. G. (1975), On a problem of integral geometry. *In: Mathematical Problems of Geophysics*, vol. 6, part 2, 212–242. Computer Center of Siberian Branch of the Russian Academy of Sciences. (in Russian)

Romanov V. G. (1974), *Integral Geometry and Inverse Problems for Hyperbolic Equations*, Springer–Verlag.

Romanov V. G. (1978), Integral geometry on the geodesics of an isotropic Riemannian metric. *Soviet Math. Dokl.*, vol. 19, 847–851.

Samarskii A. A. (1984), *Theory of Difference Schemes.* Akademische Verlag Geest and Portig, Leipzig.

Voevodin V. V. (1977), *Computational Foundations of Linear Algebra.* Nauka, Moscow. (in Russian)

Hyperfunctions and Radon Transform

A. KANEKO and T. TAKIGUCHI

Department of Mathematical Sciences, University of Tokyo
3-8-1, Komaba, Meguro-ku, Tokyo 153 Japan

ABSTRACT

We define the Radon transform for a class of hyperfunctions which are not necessarily of compact support, and give a variant of support theorem by Helgason-Boman.

1. Introduction. Hyperfunctions

In this section we first give a short review on (Fourier) hyperfunctions, and then discuss about the possibility of their Radon transform. What a catastrophe if we have a hyperfunctional singularity in our brains! We expect, however, there are some applications in some fields, such as astronomy. See Kaneko A., (1988) for more details on hyperfunctions.

A hyperfunction $f(x)$ on $\mathbf{R}^n$ is the equivalence class of formal expressions of the form

$$f(x) = \sum_{j=1}^{N} F_j(x + \mathrm{i}\Gamma_j 0), \tag{1.1}$$

where Γ_j denotes an open convex cone with vertex at the origin and $F_j(z)$ a function holomorphic in a wedge-like domain with asymptotic form $\mathbf{R}^n + \mathrm{i}\Gamma_j$. The equivalence means the natural rewriting among the defining functions F_j, and its precise expression is given by Martineau's edge of the wedge theorem which is a concrete expression of the zero cohomology class. There is a theory of localization based on the cohomology theory, and we can legally speak of the local vanishing, local analyticity, support or singular support of a hyperfunction. Distributions are included in hyperfunctions in the sense that they are realized as the limit to the real axis of $F_j(x + \mathrm{i}y_j 0)$ as $y_j \in \Gamma_j$ tends to 0 in respective sense instead of the formal one.

A Fourier hyperfunction $f(x)$ is a similar expression as above, but the defining functions are defined on a wedge conserving a fixed breadth up to infinity and there satisfy the infra-exponential growth estimate locally uniformly in Im y:

$$|F_j(z)| \le C_\varepsilon \mathrm{e}^{\varepsilon|\mathrm{Re}\, z|} \qquad \text{for any } \varepsilon > 0. \tag{1.2}$$

For Fourier hyperfunctions equivalence are understood under the same growth condition. In this sense, the space of Fourier hyperfunctions $\mathcal{Q}(\mathbf{D}^n)$ is not a subspace of the space of hyperfunctions $\mathcal{B}(\mathbf{R}^n)$, but is a kind of extension of the latter to the directional compactification $\mathbf{D}^n$ of $\mathbf{R}^n$ by adding the points at infinity. There is a natural restriction mapping defined by "forgetting the growth condition"

$$\mathcal{Q}(\mathbf{D}^n) \longrightarrow \mathcal{B}(\mathbf{R}^n),$$

which is surjective but not injective, contrary to the case of distributions $\mathcal{S}' \hookrightarrow \mathcal{D}'(\mathbf{R}^n)$. Fourier hyperfunctions are also characterized as the dual of the space $\mathcal{P}_*$ of exponentially decaying real analytic functions on a strip-neighborhood of the real axis. Since $\mathcal{P}_* \hookrightarrow \mathcal{S}$, we have $\mathcal{S}' \hookrightarrow \mathcal{Q}(\mathbf{D}^n)$. Hyperfunctions with compact supports $\mathcal{B}_*(\mathbf{R}^n)$, which are characterized as the dual of the space $\mathcal{A}(\mathbf{R}^n)$ of real analytic functions, are also contained in $\mathcal{Q}(\mathbf{D}^n)$. Exponentially decreasing hyperfunctions, simply defined as

$$\mathcal{Q}_*(\mathbf{D}^n) := \bigcup_{\varepsilon>0} e^{-\varepsilon\sqrt{1+x^2}}\mathcal{Q}(\mathbf{D}^n) \tag{1.3}$$

also form a subclass of $\mathcal{Q}(\mathbf{D}^n)$.

Fourier transform is easily seen to act isomorphically to the space $\mathcal{P}_*$, hence it can be defined on $\mathcal{Q}(\mathbf{D}^n)$ by the duality, in consistency with that for $\mathcal{S}'$. If the defining functions $F_j(z)$ are exponentially decaying on $(\mathbf{R}^n \backslash \Delta_k^\circ) + \mathrm{i}y_j$ locally uniformly in $y_j \in \Gamma_j$, then the Fourier transform can be directly calculated by

$$g(\xi) = G(\xi + \mathrm{i}\Delta_k 0); \qquad G(\zeta) = \sum_{j=1}^{N} \int_{x+\mathrm{i}y_j} F_j(x+\mathrm{i}y_j)\, e^{\mathrm{i}(x+\mathrm{i}y_j)\zeta} dx \quad \text{for } \zeta \in \mathbf{D}^n + \mathrm{i}\Delta_k.$$

The integral converges even on Δ_k° because of the exponential decay factor appearing by the shift $\xi \mapsto \zeta = \xi + \mathrm{i}\eta$ with $\eta \in \Delta_k$. For a general Fourier hyperfunction we can decompose it to such ones with various $\Delta_k' s$. Especially, the Fourier transform of an exponentially decaying hyperfunction can be directly calculated and becomes an analytic function which grows infra-exponentially on a strip-neighborhood of the real axis.

The Radon transform has no converging factor available as $e^{\mathrm{i}x\xi}$ in the case of Fourier transform. Thus we must restrict the growth order in some sense to calculate it. We can adopt several ways: The first is to use the formula

$$Rf(\omega, t) = \frac{1}{2\pi} \int_{-\infty}^{\infty} \tilde{f}(r\omega)\, e^{\mathrm{i}tr} dr.$$

In this case, the validity of the substitution $r\omega$ to $\tilde{f}$ requires some regularity of $\tilde{f}$ at the origin, and this is equivalent to some decay condition of f. The second is to define the Radon transform by duality. For this we have to examine the image of the adjoint Radon transform on the space of real analytic functions exponentially decaying to the radial direction. The third is to calculate the Radon transform directly by means of the defining functions

$$Rf(\xi) = \sum_{j=1}^{N} \int_{\xi} F_j(x+\mathrm{i}y_j) dx.$$

Under a suitable decay condition for $F_j(z)$, the result is well-defined independent of the choice of y_j.

In the next section we practice these ideas.

2. Radon Hyperfunctions and Its Radon Transform

We first collect Paley-Wiener type theorems for the Radon transform in relation with the function spaces concerned. As the fundamental space of test functions on the image side we adopt the space $\mathcal{P}_*(\mathbf{S}^{n-1} \times \mathbf{R})$ consisting of functions $g(\omega, t)$ satisfying the following

conditions: g is analytic and exponentially decaying in t when ω runs a complex neighborhood of $\mathbf{S}^{n-1}$ and t a strip neighborhood of $\mathbf{R}$. Now recall the homogeneity condition of Helgason S., (1980) for functions $g(\omega,t)$ on $\mathbf{S}^{n-1}\times\mathbf{R}$:
(2.1) for each $k=0,1,2,\ldots$

$$p_f^k(\omega)=\int_{-\infty}^{\infty} g(\omega,t)t^k\mathrm{d}t$$

becomes a homogeneous polynomial of order k.

Theorem 2.1. *Let $\mathcal{P}_{*H}(\mathbf{S}^{n-1}\times\mathbf{R})$ denote the subspace of $\mathcal{P}_*(\mathbf{S}^{n-1}\times\mathbf{R})$ consisting of even functions $g(\omega,t)$ satisfying the homogeneity condition* (2.1) *such that the arising polynomials as a whole satisfy the estimate*

$$\overline{\lim_{k\to\infty}}\sqrt[k]{\sup_{\omega\in\mathbf{S}^{n-1}}|p_f^k(\omega)|/k!}<\infty. \tag{2.2}$$

Then the Radon transform induces an isomorphism of $\mathcal{P}_*$ onto $\mathcal{P}_{*H}(\mathbf{S}^{n-1}\times\mathbf{R})$.

For elements of $\mathcal{P}_*$ the usual definition of Radon transform applies. The inverse formula is the same as given in Helgason S., (1980) or Natterer F., (1986). For a hyperfunction $f(x)$ with compact support, the Radon transform can be defined either by the hyperfunctional integration along fiber

$$Rf(\omega,t)=\int_{\mathbf{R}^n}\delta(t-x\omega)f(x)\mathrm{d}x,$$

or via the formula

$$Rf(\omega,t)=\frac{1}{2\pi}\int_{-\infty}^{\infty}\mathrm{e}^{itr}\tilde{f}(r\omega)\mathrm{d}r.$$

Here the right hand side is interpreted as a hyperfunction

$$g(\omega,t)=G(\omega,t+\mathrm{i}0)-G(\omega,t-\mathrm{i}0), \tag{2.3}$$

$$G(\omega,\tau)=\frac{1}{2\pi}\int_0^{\pm\infty}\mathrm{e}^{i\tau r}\tilde{f}(r\omega)\mathrm{d}r,\quad\text{for}\quad\pm\operatorname{Im}\tau>0. \tag{2.4}$$

Actually in this case $G(\omega,\tau)$ can be patched to one analytic function outside a compact interval on $\operatorname{Re}\tau$.

Theorem 2.2. *Let $\mathcal{B}_{*H}(\mathbf{S}^{n-1}\times\mathbf{R})$ denote the space of even hyperfunctions with compact support, containing ω as real analytic parameters and satisfying the homogeneity condition* (2.1) *such that the arising polynomials are subordinate to the estimate*

$$\overline{\lim_{k\to\infty}}\sqrt[k]{\sup_{|\zeta|\le 1}|p_f^k(\zeta)|}<\infty. \tag{2.5}$$

Then the Radon transform induces an isomorphism of $\mathcal{B}_*(\mathbf{R}^n)$ onto $\mathcal{B}_{*H}(\mathbf{S}^{n-1}\times\mathbf{R})$.

Corollary 2.3. *Let $\mathcal{E}'_H(\mathbf{S}^{n-1}\times\mathbf{R})$ denote the space of even distributions with compact support, containing ω as real analytic parameters and satisfying the homogeneity condition* (2.1) *such that the arising polynomials are subordinate to the estimate* (2.5). *Then the Radon transform induces an isomorphism of $\mathcal{E}'(\mathbf{R}^n)$ onto $\mathcal{E}'_H(\mathbf{S}^{n-1}\times\mathbf{R})$.*

In order to define the Radon transform for more general hyperfunctions we rather need the characterization of the image by the adjoint Radon transform of $\mathcal{P}_*(\mathbf{S}^{n-1} \times \mathbf{R})$:

$$R^\sharp\varphi(x) = \int_{\mathbf{S}^{n-1}} \varphi(\omega, x\omega)d\omega.$$

Theorem 2.4. *The adjoint Radon transform of a function $\varphi(\omega, t) \in \mathcal{P}_*(\mathbf{S}^{n-1} \times \mathbf{R})$ becomes analytic in a strip neighborhood of the real axis and there satisfies the decay condition:*
(2.6) *$|z|R^\sharp\varphi(z)$ is bounded.*

In view of this, we let $\mathcal{R}$ denote the space of functions $\varphi(x)$ such that each φ admits an analytic continuation to a strip neighborhood of $\mathbf{R}^n$ and there $|z|\varphi(z)$ is bounded. Equipping $\mathcal{R}$ with the natural topology, we shall call its dual the space of Radon hyperfunctions. For a Radon hyperfunction $f(x)$ we define its Radon transform as a Fourier hyperfunction on $\mathbf{S}^{n-1} \times \mathbf{R}$ via the duality:

$$\langle Rf(\omega,t), \varphi(\omega,t)\rangle = \langle f(x), R^\sharp\varphi(x)\rangle, \qquad \text{for} \quad \varphi(\omega,t) \in \mathcal{P}_*(\mathbf{S}^{n-1} \times \mathbf{R}).$$

We can also calculate Rf directly via the formula

$$Rf(\omega,t) = \int_{\mathbf{R}^n} \delta(t - x\omega)f(x)dx.$$

Here the integral is understood in the sense of hyperfunctions as follows: Decompose f into the sum of f_k, $k = 1, 2, \ldots, N$ such that f_k decreases exponentially outside a closed proper convex cone Δ_k°. Then

$$Rf(\omega,t) = \sum_{k=1}^{N} \left(G_k(\omega - i\Delta_k 0, t + i0) - G_k(\omega + i\Delta_k 0, t - i0)\right),$$

$$G_k(\zeta,\tau) = -\frac{1}{2\pi i}\int_{\mathbf{R}^n} \frac{1}{\tau - x\zeta} f_k(x)dx \quad \text{for} \quad \operatorname{Im}\zeta \in \Delta_k.$$

Hyperfunctions with compact supports are Radon hyperfunctions. So are the exponentially decaying Fourier hyperfunctions, or more generally, Fourier hyperfunctions which admit representations by defining functions of order $O(|x|^{-n})$. In such cases, the Radon transform contain ω as continuous parameters and specialization with respect to ω is legitimate.

3. Support Theorem

In this article we use the term "rapidly decreasing Fourier hyperfunctions" to indicate those hyperfunctions which admit representation by defining functions of order $O(|x|^{-N})$ for any $N > 0$. This subclass of Radon hyperfunctions has a special meaning in the support theorem for the Radon transform:

Theorem 3.1. *Let $f(x)$ be a rapidly decreasing Fourier hyperfunction. Assume that $Rf(\omega,t)$ vanishes for $|t| \geq A$. Then $f(x)$ vanishes on $|x| \geq A$.*

We can prove this either by Helgason's method (Helgason, 1980) or by Boman's method (Boman, 1991). For the former we cannot employ the regularization by convolution and have to prove the sphere theorem directly for hyperfunctions. For the latter, we should notice that near the hyperplane at infinity, we can consider $Rf(\omega, 1/t)$ as a hyperfunction containing t as real analytic parameter and with compact support in ω. The latter condition allows us to use the uniqueness theorem even for hyperfunctions. In the case of distributions, Boman (1991) gave a local version of this theorem. It holds also for non-quasi-analytic type ultradistributions without modification (Tanabe, 1993). As remarked by Boman, the argument does not apply to hyperfunctions in view of M. Sato's counter-example to the local uniqueness. Nevertheless we are pursuing a local type support theorem for hyperfunctions introducing the exponential decay. Recently, Boman, (1993) proved a local uniqueness theorem for quasi-analytic parameter, which can be applied to the support theorem for the Radon transform with quasi-analytic kernels.

Our discussion can be generalized to the x-ray transform or to the integral transform with respect to general linear subvarieties. We can treat as well the two point homogeneous spaces as in Helgason, (1980).

The details of proofs of these results and further extensions are to be given in our forthcoming paper (Takiguchi and Kaneko).

REFERENCES

Boman J. (1991), Helgason's Support Theorem for Radon Transforms - A New Proof and a Generalization, *Mathematical Methods in Tomography*, Lecture Notes in Math, Springer, vol. 1497, 1–5.

Boman J. (1992), A local vanishing theorem for distributions, *C. R. Acad. Sci. Paris*, **315**, 1231–1234.

Boman J. (1993), *A Vanishing Theorem for Microlocally Quasianalytic Distributions*, Preprint, Univ. of Stockholm.

Boman J. and Quinto T (1987), Support theorems for real-analytic Radon transforms, *Duke Math. J.*, **55**, 943–948.

Helgason S. (1980), *The Radon Transform*, Birkhäuser.

Kaneko A. (1988), *Introduction to Hyperfunctions*, Kluwer.

Natterer F. (1986), *The Mathematics of Computerized Tomography*, John-Wiley.

Takiguchi T. and Kaneko A., Support theorems for the Radon transforms of hyperfunctions, in preparation.

Tanabe S. and Takiguchi T. (1993), *A Local Vanishing Theorem for Ultradistributions with Analytic Parameters*, Preprint, Univ. of Tokyo.

Inversion of Some Projection Matrices

I .G .KAZANTSEV

Computing Center, 630090, Novosibirsk, Russia

ABSTRACT

Special cases of simple projection data acquisition geometries with integer rays are considered and their projection matrices inversions are obtained.

1. Introduction

Possibilities of additional measurements in few-views tomography and plasma diagnostics are very restricted and all data and a priori information must be embedded in image reconstruction procedures in the most useful way. Contribution of data into the reconstruction accuracy is different and it is interesting to develop algorithms that take into account informativity of single detectors and projection groups. The most appropriate method for the problem analysis is algebraic reconstruction techniques.

Iterative reconstruction techniques are applicable to a very general formulation of the reconstruction problem based on the series expansion approach especially in the case of incomplete data. In few-views tomography the information content of projection data, or informativity, has a great value for the image reconstruction quality. The greater information content of data, the greater closeness of reconstruction to the initial image in a chosen measure. There arises a problem of informativity calculation before reconstruction as such using only projection data, a priori constraints and the projection matrix structure. Preliminary account of data content for predicting the best data acquisition geometry could be useful when we have a priori knowledge about important features of the image to be reconstructed. This is the case in non-destructive testing, plasma physics and others.

Some efficient iterative algorithms converge to the minimal norm solution which could also be obtained by applying the generalized inverse projection matrix to the data vector. However, generalized inversion of an arbitrary projection matrix is complicated due to the matrix nonregularity. In this work we consider projection matrices for the so–called 'integer' projections. These matrices have a sufficiently regular structure of (0,1)-matrices. Inversion formulas are obtained for a few (horizontal, vertical and two arbitrary additional) projections in parallel geometry. It is shown that inversion of data from a few projections with an appropriate (0,1)-projection matrix is not more complicated than transposition of the projection matrix.

Using the formulas derived, a criterion of informativity for this simple case is suggested. Divergent beam geometry of the integer projections is also considered. The advantage of using these matrices is their invertibility, which allows us to compare the convergence of iterative algorithms and to analyze informativity effects.

2. An Informativity Problem in Algebraic Formulation

Let us consider a problem of computerized tomography in the algebraic form:

$$Ax = y, \tag{1}$$

where $x \in R^n$, $y \in R^m$, $A \in M_{m,n}$, $n = N \times N$, and the vector x presents a digital image defined on a grid of sizes $N \times N$. In general, the data vector y is a set of ray sums; A is a large sparse matrix of weights evaluating contribution of the image x pixels into the projection data y formation : a_{ij} is a length of that part of the i-th ray which passes through the j-th pixel. We shall refer to A as the projection matrix. Let us consider a subsystem of system (1)

$$Bx = u, \tag{2}$$

where $u \in R^k$, $B \in M_{k,n}$ for some $k < m$, which is obtained by eliminating some $m-k$ rows from system (1). We denote by x_A^+ and x_B^+ the minimal norm solutions of systems (1) and (2), correspondingly.

DEFINITION. Subsystem (2) is called the most essential (informative) one for system (1) if the minimal norm solution x_B^+ is the closest one to the solution x_A^+, i.e.,

$$\left\|x_A^+ - x_B^+\right\| = \min_{C \subset A, C \in M_{k,n}} \left\|x_A^+ - x_C^+\right\|,$$

where the inclusion $C \subset A$ denotes that C is a submatrix of A.

Assertion 1. *If A is a projection matrix of the full rank, the criterion for a choice of the most essential subsystem B consists in maximizing the quantity $Q(u,B) = \left(u, (BB^T)^{-1}u\right)$.*

PROOF. The matrix A rows are independent, therefore $x_A^+ = A^T(AA^T)^{-1}y$, $x_B^+ = B^T(BB^T)^{-1}u$. Taking into account that $BA^T(AA^T)^{-1}y = u$, we obtain

$$\begin{aligned}\left\|x_A^+ - x_B^+\right\|^2 &= \left\|x_A^+\right\|^2 - 2\left(BA^T(AA^T)^{-1}y\,,\,(BB^T)^{-1}u\right) + \left(u,(BB^T)^{-1}u\right)\\ &= \left\|x_A^+\right\|^2 - \left(u,(BB^T)^{-1}u\right).\end{aligned}$$

Assertion 2. *The most informative single ray with the number i in system (1) maximizes the ratio $\frac{y_i^2}{\|a_i\|^2}$, where a_i is the i-th row of the matrix A . The most essential group of two detectors (the rays i and j) is chosen so as to maximize the ratio*

$$\frac{y_i^2\left\|a_j\right\|^2 - 2y_iy_j(a_i,a_j) + y_j^2\left\|a_i\right\|^2}{\left\|a_i\right\|^2\left\|a_j\right\|^2 - (a_i,a_j)^2}$$

Assertion 3. *If rows of the matrix A are orthogonal, the informativity criterion is to maximize*

$$Q(u,B) = \frac{u_1^2}{\|b_1\|^2} + \ldots + \frac{u_k^2}{\|b_k\|^2}.$$

Generally, tomographic projection matrices are arbitrary and the Gram matrices BB^T are singular, as a rule, so investigation of informativity contribution needs commonly expensive generalized inversion computation . In this work we consider data acquisition systems (parallel and divergent beam geometries) for which every row of the projection matrix consists of equal quantities or zeroes. We call them "integer" rows, or rays. Multiplying the matrix A by an appropriate diagonal matrix $S \in M_{m,m}$, we obtain an equivalent system $Px = z$, where $P = SA$, $z = Sy$ and P is a $(0,1)$-matrix. Some "integer" ray geometries have analytical inverse matrices.

3. Inversion Formulas

We consider a partial case of parallel beam geometry with the projection matrix consisting of zeroes and unities. In comparison with the traditional tomographic parallel scheme, the projection matrix considered is constructed by summing two appropriate rows of the former projection matrix so that the sums of elements of the new projection matrix rows are the same. On the one hand, this summation extends the domain of initial system solutions; on the other hand, we can see that analytical generalized pseudoinversion of the projection matrix becomes possible. It was revealed occasionally that such projections are used also in the magic squares definition in combinatorics (Figure 1). Let us consider the new projection matrix $A \in M_{m,n}$ (after summing the rows), where $m = M \times N$, M is a number of directions, and nonzero elements from N rows of the k-th projection, for any $k, 1 \leq k \leq M$, have the form $a_{ij} = \sigma_k$, $(k-1)N + 1 \leq i \leq kN$, $1 \leq j \leq n$ for some σ_k. Introducing a diagonal matrix $S \in M_{m,m}$ such that $s_{ii} = \sigma_k^{-1}$ for $(k-1)N+1 \leq i \leq kN$, we obtain an equivalent system $Px = z$, where $P = SA$, $z = Sy$. The matrix P is a block $(0,1)$-matrix whose blocks are of the sizes $N \times N$ with N of unities and $\sum_{j=1}^{n} a_{ij} = N$, $1 \leq i \leq m$; $\sum_{i=1}^{m} a_{ij} = M$, $1 \leq j \leq n$.

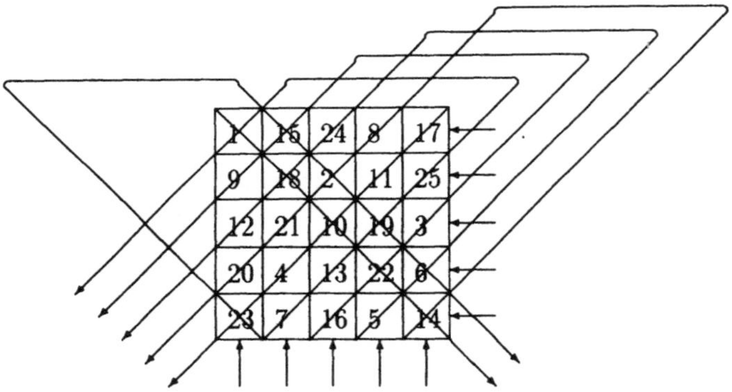

Figure 1. Parallel beam geometry of integer projections, $M = 4$, $N = 5$.

$$x = \begin{array}{|c|c|c|c|} \hline x_1 & x_5 & x_9 & x_{13} \\ \hline x_2 & x_6 & x_{10} & x_{14} \\ \hline x_3 & x_7 & x_{11} & x_{15} \\ \hline x_4 & x_8 & x_{12} & x_{16} \\ \hline \end{array}, \quad P =$$

1	1	1	1												
				1	1	1	1								
								1	1	1	1				
												1	1	1	1
1				1				1				1			
	1				1				1				1		
		1				1				1				1	
			1				1				1				1
			1		1					1		1			
1							1			1			1		
	1			1							1			1	
		1			1			1							1
1					1					1					1
			1	1					1					1	
		1					1	1					1		
	1					1					1	1			

.

Figure 2. Projection matrix in the case $M = 4$, $N = 4$. Empty squares denote zeroes.

Assertion 4. *For $M = 1,2,3$ and even N, the generalized pseudoinverse matrix of the matrix P has the form*

$$P^+ = \frac{1}{N}P^T + \frac{M-1}{MN^2}W, \tag{3}$$

where $P^+, P^T, W \in M_{n,m}$ and W is the matrix consisting of unities.

PROOF. The cases $M = 1,2,3$ are similar. Let us $M = 3$. We use straightforward calculation by the formula

$$P^+ = \lim_{\varepsilon \to 0} P^T(PP^T + \varepsilon I)^{-1},$$

where I is the identity matrix. We note that PP^T is of the form

$$PP^T = \begin{pmatrix} V & W_N & W_N \\ W_N & V & W_N \\ W_N & W_N & V \end{pmatrix},$$

where $V, W_N \in M_{N,N}$, $V = N \cdot I$, and W_N is a matrix of unities. After some transformations, we obtain

$$(PP^T + \varepsilon I)^{-1} = \begin{pmatrix} B & C & C \\ C & B & C \\ C & C & B \end{pmatrix},$$

where

$$B = \begin{pmatrix} a & b & \dots & b \\ b & a & \dots & b \\ \vdots & \vdots & & \vdots \\ b & b & \dots & a \end{pmatrix}, \quad C = \begin{pmatrix} c & \dots & c \\ \vdots & & \vdots \\ c & \dots & c \end{pmatrix}$$

and

$$a = \frac{(M-1)N}{\varepsilon(N+\varepsilon)(MN+\varepsilon)} + \frac{1}{N+\varepsilon}, \quad b = \frac{(M-1)N}{\varepsilon(N+\varepsilon)(MN+\varepsilon)}, \quad c = -\frac{1}{\varepsilon(MN+\varepsilon)}.$$

The matrix $(PP^T + \varepsilon I)^{-1}$ is decomposed into a sum of the matrices

$$\frac{1}{N+\varepsilon}I + c \cdot \left| \begin{matrix} \frac{(M-1)N}{N+\varepsilon} & \dots & \frac{(M-1)N}{N+\varepsilon} & -1 & \dots & -1 & -1 & \dots & -1 \\ \vdots & \dots & \vdots & \vdots & & \vdots & \vdots & & \vdots \\ \frac{(M-1)N}{N+\varepsilon} & \dots & \frac{(M-1)N}{N+\varepsilon} & -1 & \dots & -1 & -1 & \dots & -1 \\ -1 & \dots & -1 & \frac{(M-1)N}{N+\varepsilon} & \dots & \frac{(M-1)N}{N+\varepsilon} & -1 & \dots & -1 \\ \vdots & \dots & \vdots & \vdots & & \vdots & \vdots & & \vdots \\ -1 & \dots & -1 & \frac{(M-1)N}{N+\varepsilon} & \dots & \frac{(M-1)N}{N+\varepsilon} & -1 & \dots & -1 \\ -1 & \dots & -1 & -1 & \dots & -1 & \frac{(M-1)N}{N+\varepsilon} & \dots & \frac{(M-1)N}{N+\varepsilon} \\ \vdots & \dots & \vdots & \vdots & & \vdots & \vdots & & \vdots \\ -1 & \dots & -1 & -1 & \dots & -1 & \frac{(M-1)N}{N+\varepsilon} & \dots & \frac{(M-1)N}{N+\varepsilon} \end{matrix} \right|.$$

From the properties of the integer projections and the matrix P^T structure it follows that

$$P^T(PP^T + \varepsilon I)^{-1} = \frac{1}{N+\varepsilon}P^T + \frac{\frac{(M-1)N}{N+\varepsilon} - (M-1)}{\varepsilon(MN+\varepsilon)}W = \frac{1}{N+\varepsilon}P^T - \frac{M-1}{(MN+\varepsilon)(N+\varepsilon)}W,$$

which proves formula (3).

Assertion 5. *For $M = 4$ and even N, the generalized inversion is as follows:*

$$P^{+} = \frac{1}{N}P^{T} + \frac{M-1}{MN^2}W + \frac{1}{MN^2}U,$$

where the matrix U has the block structure , its blocks have zero mean rows and columns; these blocks are similar to the subdiagonal blocks of the matrix PP^T.

Let us consider a data acquisition system which has features of divergent beam geometry. Figure 3 presents an example of such a geometry for $N = 4$, and the corresponding projection matrix is presented in Figure 4.

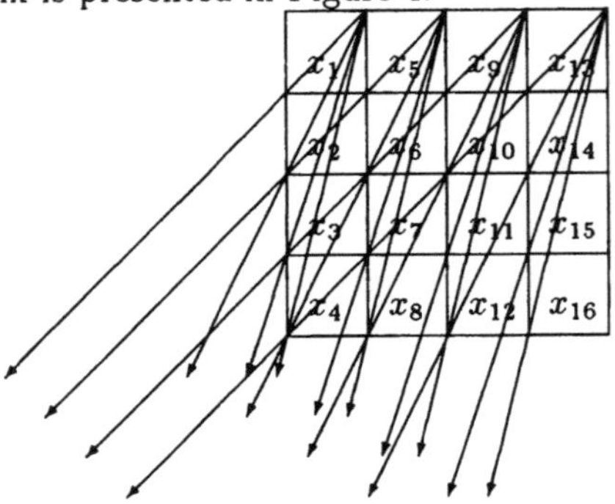

Figure 3. Divergent beam geometry of integer projections, $N = 4$.

$P =$

1															
1	1														
1	1	1													
1	1	1	1												
	1			1											
		1	1	1	1										
			1	1	1	1									
				1	1	1	1								
		1			1			1							
						1	1	1	1						
							1	1	1	1					
								1	1	1	1				
			1			1			1			1			
										1	1	1	1		
											1	1	1	1	
												1	1	1	1

Figure 4. Projection matrix of divergent beam geometry, $N = 4$.

Assertion 6. *The matrix $P \in M_{n,n}$ is triangular and nonsingular. Its inverse matrix has the form $P^{-1} = U + W$, where*

$$U = \begin{pmatrix} D & O & \dots & O \\ D & D & \dots & O \\ \vdots & \vdots & & \vdots \\ D & D & \dots & D \end{pmatrix}, \quad D = \begin{pmatrix} 1 & 0 & \dots & 0 & 0 \\ -1 & 1 & \dots & 0 & 0 \\ \vdots & \vdots & & & \vdots \\ 0 & 0 & \dots & -1 & 1 \end{pmatrix},$$

$D, O \in M_{N,N}$, O is the null matrix and W is a block-diagonal sparse matrix that can be recursively restored.

We can regroup rays into N projections with N parallel rays in each projection and then note that almost all projections (except for a vertical one) do not cover the grid entirely. It means that we use only parts of the projections to achieve the projection rows indepedence. We conclude that there are situations when some rays add no information to data. It depends on the way in which we expand the series, on the grid type, on the manner of the image domain decomposition and so on.

4. Conclusions

In the cases $M = 1, 2, 3$ of Assertion 3 we can calculate the closeness of the reconstructed image $x^+ = P^+ z$ to the initial image vector x. Without loss of generality, let us assume that x is a zero mean vector. Then

$$\left\| x - x^+ \right\|^2 = (x, x) - \left(\frac{2}{N}(z, z) - \frac{1}{N^2}(z, PP^T z) \right).$$

Since z is also a zero mean vector, we have $(z, PP^T z) = N(z, z)$ and, therefore,

$$\left\| x - x^+ \right\|^2 = \|x\|^2 - \frac{1}{N} \|z\|^2 .$$

Hence, we conclude that in the case of few-views integer projections the best ones could be chosen using their $\|\cdot\|_2$ -norms. The simplicity of the projection informativity criterion in this case is due to identity the $PP^T z = Nz$. In the general case, evaluating informativity of the data y in terms of the $\|\cdot\|_2$-norm of the initial image space could not be suggested. Taking into account only system (1) right-hand side variance, influence of the projection matrix is lost. In general, it is impossible to avoid completely the reconstruction process when investigating the informativity. In transform-based techniques we must either perform the ρ-filtration of projections and then calculate the ρ-weighted scalar product and the norms $(y, \rho * y)$ (Medoff, 1987) or apply the backprojection and evaluate the summed image in norm. Some practical aspects of the problem and computational experiment results are discussed by Kazantsev (1991). Generalizing the above consideration, we choose a measure of integer projection informativity so as to avoid inverting the Gram matrix, namely:

$$Q(z, P) = \left(z, (2N \cdot I - PP^T) z \right).$$

We conclude that the problem of evaluating the informativity of projection data groups in tomography is algebraically formulated as a search for the most essential rows of a given linear algebraic system. Special cases of simple geometries with integer rays are considered and some their inversions are obtained.

Acknowledgments

This work was partially supported under grant No. 713-224 of NWO (The Netherlands Scientific Society). The author would like to thank V.Pickalov for useful discussions and help for preparing this work.

REFERENCES

Kazantsev I. G. (1991), Projection information content. *Inverse Problems*, vol7, 887–898.

Medoff B. P. (1987), Image reconstruction from limited data. *In: Image Recovery: Theory and Application*, Stark H. (Ed.). Academic Press Inc., Chapter 9.

Interference Tomography of the Ocean Bottom

A. Yu. KAZAROVA, L. Ya. LYUBAVIN, A. G. NECHAEV

Institute of Applied Physics, Russian Academy of Sciences
Nizhny Novgorod, Russia

ABSTRACT

We consider a problem of interference tomography of the ocean bottom. The problem is to restore sound speed, density and absorption coefficient in the bottom by measuring sound field intensity (interference structure). For its solution we used the perturbation theory and the mode approach in adiabatic approximation. Numerical simulation results are presented.

In low-frequency acoustics the kinematic methods based on measuring a godograph or a difference godograph are usually employed to find the vertical structure of sea bottom parameters. Other approaches use such data as characteristics of the scattering field, the angle dependence of the sound reflection coefficient obtained by the Doppler method or the amplitude and phase of the pressure field measured by vector receivers. Meanwhile, it is possible to measure the space interference structure (IS) of the sound field so as to get information about acoustic characteristics of the sea bottom. In this paper we consider a method to determine the vertical profile of the sound speed, density and absorption from the measurements of IS and a priori estimates of bottom parameters.

Let the ocean waveguide be modelled as having a jump in the sound speed and density across the water ($0 \leq z \leq H$) - bottom ($H \leq z$) interface. The z-axis is directed downward. Let the sound speed profile be of the form $\tilde{c} = c(z) + \delta c$, and the density profile be as $\tilde{\rho} = \rho(z) + \delta\rho(z)$. We also take into account the absorption α ($\alpha = \xi + 4\eta/3$, where ξ,η are viscosity coefficients). Here $c(z)$ and $\rho(z)$ are previously known estimates of the sound speed and density (for example, averaged over the considered area), δc and $\delta\rho$ are their disturbances.

The sound propagation will be described by the equation for the complex amplitude of the sound pressure $p \times \exp(i\omega t)$:

$$\tilde{\rho}\nabla\left(\frac{\nabla p}{\tilde{\rho}}\right) + \frac{\omega^2}{\tilde{c}^2}(1 + 2i\gamma)p = p_s, \tag{1}$$

$$\gamma = \frac{\omega\alpha}{2\tilde{c}^2\tilde{\rho}}, \quad p_s = \frac{2i\delta(z - z_0)\delta(r - r_0)}{\pi a},$$

where ω denotes the frequency; $a = |\vec{r} - \vec{r}_0|$; $\vec{r}_0$, $\vec{r}$ are the source and the receiver coordinates, respectively.

The solution of equation (1) in the adiabatic approximation can be represented as

$$p = \sum_n \tilde{A}_n \tilde{\varphi}_n \tilde{R}_n, \tag{2}$$

$$\tilde{R}_n = \sqrt{\frac{2}{a\tilde{h}_n\pi}} \exp\left[\mathrm{i}\left(\int\limits_{r_0}^{r} d\vec{r}'\tilde{h}_n(\vec{r}') - \frac{\pi}{4}\right)\right], \tag{3}$$

where $\tilde{A}_n$ are mode excitement coefficients, $\tilde{h}_n$ are horizontal wave numbers, $\tilde{\varphi}_n$ are eigenfunctions. All these mode characteristics differ from those for the basic ocean model which are denoted by A_n, h_n, φ_n, respectively.

To find the dependence of mode characteristics on disturbances of the bottom parameters, we apply the perturbation theory to the Sturm-Liouville problem

$$(\hat{H} + \hat{V}_n)\tilde{\varphi}_n = \frac{\tilde{h}_n^2\tilde{\varphi}_n}{\rho}, \tag{4}$$

$$\tilde{\varphi}_n = 0 \quad \text{for} \quad z = 0, \quad \text{and the radiation condition for} \quad z \to \infty, \tag{5}$$

$$\hat{H} = \frac{\partial}{\partial z}\left(\frac{1}{\rho}\frac{\partial}{\partial z}\right) + \frac{\omega^2}{\rho c^2}, \tag{6}$$

$$\hat{V}_n = -\frac{2\omega^2\delta c}{\rho c^3} - \frac{\partial}{\partial z}\left(\frac{\delta\rho}{\rho^2}\frac{\partial}{\partial z}\right) - \frac{\omega^2\delta\rho}{\rho^2 c^2} - \frac{h_n^2\delta\rho}{\rho^2} - \frac{2\mathrm{i}\omega^2\gamma}{\rho c^2}. \tag{7}$$

Equation (4) was found by separating the variables in equation (1). From the theory of first-order perturbations it follows that

$$\delta h_n = \frac{V_{nn}}{2h_n}, \quad \delta\varphi_n = \sum_n a_{nm}\varphi_m, \quad a_{nk}(r) = \frac{V_{kn}}{h_n^2 - h_k^2}, \quad V_{kn} = \int \mathrm{d}z\varphi_k\hat{V}_n\varphi_n, \tag{8}$$

where $\delta\varphi_n = \tilde{\varphi}_n - \varphi$, $\delta h_n = \tilde{h}_n - h_n$.

To obtain $\tilde{A}_n$, we substitute (2) into (1) and then expand the terms of the resulting equation into series according to the theory of first-order perturbations. As a result, for $\delta A_n = \tilde{A}_n - A_n$ we obtain

$$\delta A_n = -\sum_m a_{mn}(r_0)\varphi(z_0). \tag{9}$$

Substituting (8) – (9) into (2), we can write an expression for the pressure field in the form

$$p = \sum_n \sqrt{\frac{2}{\pi h_n a}}(\alpha_n + \mathrm{i}\beta_n)\exp\left[\mathrm{i}(h_n a - \frac{\pi}{4} + \varepsilon_n) - H_n\right], \tag{10}$$

$$\varepsilon_n + \mathrm{i}H_n = \int\limits_{r_0}^{r} d\vec{r}'\delta h_n, \quad \alpha_n = (\varphi_n(z) + J_n)(\varphi_n(z_0) - \tilde{J}_n) - \Psi_n\tilde{\Psi}_n,$$

$$\beta_n = \varphi_n(z_0)\Psi_n - \varphi_n(z)\tilde{\Psi}_n, \quad J_n + \mathrm{i}\Psi_n = \delta\varphi(z), \quad \tilde{J}_n + \mathrm{i}\tilde{\Psi}_n = \delta A_n. \tag{11}$$

From (10)-(11) we obtain

$$I = \frac{2}{\pi a}\sum_n\sum_m\sqrt{\frac{1}{h_n h_m}}\{(\alpha_n\alpha_m + \beta_n\beta_m)\cos\left[(h_n - h_m)a + (\varepsilon_n - \varepsilon_m)\right]$$

$$+(\alpha_n\beta_m - \alpha_m\beta_n)\sin\left[(h_n - h_m)a + (\varepsilon_n - \varepsilon_m)\right]\}\exp(-H_n - H_m). \tag{12}$$

Formula (12) can be used to determine the coefficients $\alpha_n$and β_n from which we can reconstruct the desired bottom parameters.

Formula (12) can be linearized for traces short enough. Then the disturbance $\delta I = \tilde{I} - I$ of the sound field intensity is represented as

$$\delta I = \frac{4}{\pi a}\sum_n\sum_m\sqrt{\frac{1}{h_n h_m}}\{[\varphi_m(z)\varphi_m(z_0)\delta J_n - \varphi_n(z)\varphi_n(z_0)\varphi_m(z)\varphi_m(z_0)H_n]$$
$$\times\cos[(h_n - h_m)a] - [\varphi_m(z)\varphi_m(z_0)\delta\Psi_n + \varphi_n(z)\varphi_n(z_0)\varphi_m(z)\varphi_m(z_0)\varepsilon_n]\sin[(h_n - h_m)a]\},\tag{13}$$

$$H_n = \int_{r_0}^{r}\int d\vec{r}\,'dz'\frac{\omega^2}{h_n\rho c^2}\gamma\varphi_n^2(z'),\tag{14}$$

$$\delta\Psi_n = \sum_m\frac{2}{h_n^2 - h_m^2}\int dz'\varphi_n(z')\varphi_m(z')\left[\frac{\omega^2\gamma(\vec{r},z')}{\rho(\vec{r},z')c^2(\vec{r},z')}\varphi_n(z_0)\varphi_m(z)\right.$$
$$\left.+\frac{\omega^2\gamma(\vec{r}_0,z')}{\rho(\vec{r}_0,z')c^2(\vec{r}_0,z')}\varphi_n(z)\varphi_m(z_0)\right],\tag{15}$$

$$\varepsilon_n = \int_{r_0}^{r}\int d\vec{r}\,'dz'\left\{\left[(h_n^2 - k^2)\varphi_n^2(z') + (\varphi'(z'))^2\right]\frac{\delta\rho}{2h_n\rho c^2} - \frac{\omega^2\delta c\varphi_n^2(z')}{h_n\rho c^3}\right\},\tag{16}$$

$$\delta J_n = \sum_m\frac{1}{h_n^2 - h_m^2}\int dz'$$
$$\times\left\{\left\{\frac{\varphi_n(z_0)\varphi_m(z)\delta\rho(\vec{r},z')}{\rho^2(\vec{r},z')}[(h_n^2 - k^2)\varphi_n(z')\varphi_m(z') + \varphi_n'(z')\varphi_m'(z')]\right.\right.$$
$$\left.+\frac{\varphi_m(z_0)\varphi_n(z)\delta\rho(\vec{r}_0,z')}{\rho^2(\vec{r}_0,z')}\left[(h_m^2 - k^2)\varphi_n(z')\varphi_m(z') + \varphi_n'(z')\varphi_m'(z')\right]\right\}$$
$$\left.+\left[\frac{\omega^2\delta c(\vec{r},z')\varphi_n(z_0)\varphi_m(z)}{\rho(\vec{r},z')c^3(\vec{r},z')} + \frac{\omega^2\delta c(\vec{r}_0,z')\varphi_m(z_0)\varphi_n(z)}{\rho(\vec{r},z')c^3(\vec{r}_0,z')}\right]\varphi_n(z')\varphi_m(z')\right\}.\tag{17}$$

Formulae (13)–(17) is a linear integral equation, and its solution yields the bottom parameters.

In the general case, reconstruction can be realized from measurements of the acoustic field on a horizontal, a vertical or a mixed net. Depending on the quality of the measurements and a priori information, we can reconstruct either the bottom parameters or combinations of them. The coefficient $\delta\gamma$ can always be singled out, whereas separating δc from $\delta\rho$ needs the sound intensity to be measured on a vertical net. Besides, in reconstructing the bottom parameters, we can use the smoothing of some interference structure scales, in particular, by employing the spatial Fourier transform (Lyubavin and Nechaev, 1989; Kazarova and Nechaev, 1993).

Consider some results of numerical reconstruction of the sound absorption coefficient in a homogeneous bottom from the measurements of the sound intensity on a horizontal net in the case of a random noise signal.

If the sound speed and the density do not vary, the formulae become

$$\delta I(\vec{r},z) = \int dz'\gamma\left[K_1(z';\vec{r},z;\vec{r}_0,z_0) + K_2(z';\vec{r},z;\vec{r}_0,z_0)\right] + I_{\text{ns}},\tag{18}$$

$$K_1 = -\frac{4}{\pi a}\sum_n\sum_m\sqrt{\frac{1}{h_n h_m}}\int_{r_0}^{r} d\vec{r}\,'\varphi_n(z)\varphi_n(z_0)\varphi_m(z)\varphi_m(z_0)\cos[(h_n - h_m)a]\frac{\omega^2}{h_n\rho c^2}\gamma\varphi_n^2(z'),\tag{19}$$

$$K_2 = -\frac{8}{\pi a}\sum_n\sum_m\sqrt{\frac{1}{h_n h_m}}\frac{1}{h_n^2 - h_m^2}\varphi_m(z)\varphi_m(z_0)\sin[(h_N - h_m)a]\varphi_n(z')\varphi_m(z')$$
$$\left[\frac{\omega^2\gamma(\vec{r}, z')}{\rho(\vec{r}, z')c^2(\vec{r}, z')}\varphi_n(z_0)\varphi_m(z) + \frac{\omega^2\gamma(\vec{r}_0, z')}{\rho(\vec{r}_0, z')c^2(\vec{r}_0, z')}\varphi_n(z)\varphi_m(z_0)\right]. \quad (20)$$

We considered the following model: the sea depth was 295m, the sound speed in water was 1,483km/s, the bottom sound speed was 1,750km/s, the bottom density was $1,13 \times 10^3 \text{kg/m}^3$, the source was located at a depth of 75m, and the receiver was at a depth of 55m. We took for a noise intensity a random variable I_{ns} with the uniform distribution in the interval $[0, 2\langle I_{\text{ns}}\rangle]$ and the dispersion $\langle I_{\text{ns}}\rangle/6$. Values of the random variable were added to values of δI in the points of measurements. In the numerical reconstruction we solved the integral equation using programs based on the Tikhonov regularization technique (Tikhonov et al., 1983). The sound intensity was measured on a horizontal net situated at a distance of 0,5km–8km from the source.

The dependence of the reconstruction error ($\eta = |\gamma - \gamma_r|/\ \gamma$, where γ_r is the value obtained in the reconstruction) on the absorption coefficient (Figure 1) and the noise intensity (Figure 2) are shown in Figures 1, 2 (Curves 1 and 2 describe the reconstruction from K_1 and K_2, respectively) in the case when the density and the sound speed are invariable. The noise value is defined as $\langle I_{\text{ns}}\rangle/I_m$, where I_m is the maximum of the noise intensity.

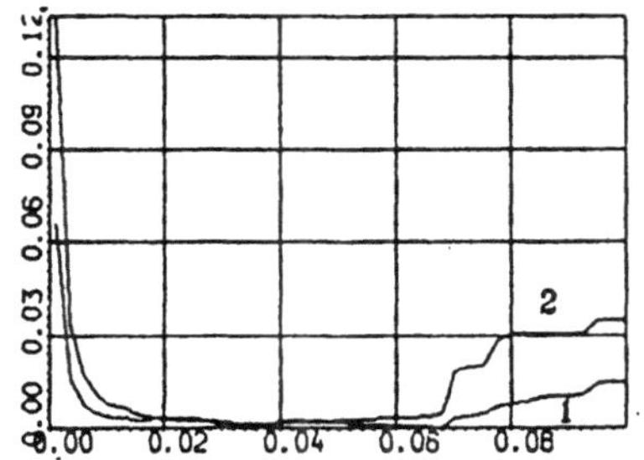

FIGURE 1.

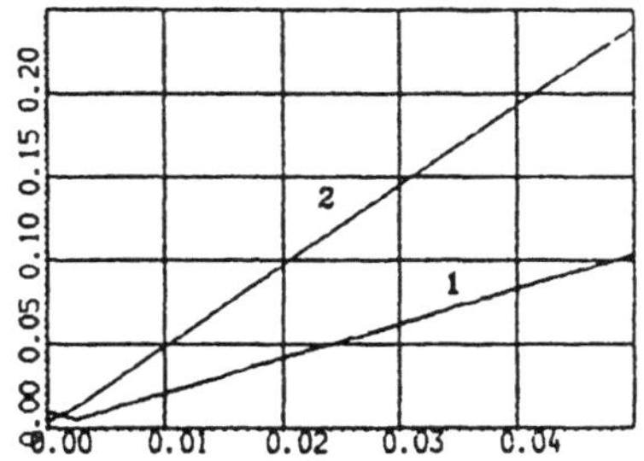

FIGURE 2.

The obtained results show that the reconstruction quality lowers with a growth of the noise intensity. The reconstruction quality also lowers for small (the absorption influences weakly on the sound interference structure) and large (the conditions for applicability of the perturbation theory are not satisfied) values of the absorption coefficient. Graphical displays of the density and the sound speed reconstruction are analogous.

We can conclude from the results of the numerical experiment that there is a range of the absorption coefficient where this coefficient may be suitably reconstructed for the values of the signal-noise ratio achievable in the ocean.

REFERENCES

Lyubavin L. Ya., Nechaev A. G. (1989), Acoustic interference ocean tomography. *Akust. Zh.*, vol. 35, no. 4, 703–710. (in Russian)

Kazarova A. Yu., Nechaev A. G. (1993), The determination of the intermode phase difference of the ocean waveguide normal waves by the acoustic field interference structure. *Akust. Zh.*, vol. 39, no. 5. (in Russian)

Tikhonov A. N., Goncharsky A. V., Stepanov V. V., Yagola A. G. (1983), *Regularization Algorithms and a Priory Information.* Moscow, Nauka. Fizmatgiz. 200 p. (in Russian)

Determination of Sea Bottom Characteristics Using Acoustic Field Spatial Dependence

B .V. KERZHAKOV, V. N. FOKIN, M. S. FOKINA

Institute of Applied Physics Russian Academy of Sciences
Ulyanov Str. 46, Nizhny Novgorod, Russia

ABSTRACT

In this paper we present the results of investigating the influence of sea bottom geoacoustic parameters on the reflection coefficient and on spatial dependence of the acoustic field. We consider also the inverse problem of determining geoacoustic parameters of a layered sea bottom from measurements of spatial dependence of acoustic fields. In solving this problem numerically, we used the method of matrix propagator.

In the present paper we give the results of investigations of sea bottom geoacoustic parameters influence on spatial dependence of the acoustic field and consider the possibility to determine these parameters using the regularities found in the course of these investigations. A system consisting of liquid or elastic layers placed in an elastic (liquid) half-space was taken as a model. To determine the modulus and the phase for coefficients of the plane wave reflection at the bottom, the technique of matrix propagator (suggested by Thomson (1950), Haskell (1953) and developed by Brekhovskikh (1989)) was numerically realized. Using the method of matrix propagator, we get the following system equations for finding the reflection coefficient V and the penetration factors $W_{\ell,t}$ of longitudinal and transverse waves for an arbitrary number of elastic (liquid) layers:

$$W_t = -W_l q_7/q_8, \tag{1}$$

$$V = 1 + W_l q_3/id_0 - W_l q_7 q_4/q_8 i\alpha_0, \tag{2}$$

$$W_l = 2/(-q_5/\omega^2\rho_0 + q_7 q_6/q_8\omega^2\rho_0 - q_3/i\alpha_0 + q_7 q_4/q_8 i\alpha_0), \tag{3}$$

where q_j, $(j = 1, 2, \ldots, 8)$ can be found in an explicit form after awkward but not complex transformation.

The mathematical analysis of equation (2) being rather complex, let us numerically estimate $V(\theta)$ and consider the changes in this dependence which are caused by variations of one of the layered bottom parameters. Such a consideration allows us to determine some regularities in $V(\theta)$ behavior and to qualitatively estimate the expected field level at a finite distance from the source.

The calculations were made using a waveguide with an even layered bottom which was homogeneous along the trace. The effect of the bottom in this model is rather strong. In Figures 1a,b,c the results of calculations of the modulus of the reflection coefficient at the bottom (mod V), the phase of the reflection coefficient at the bottom (arg V), and losses

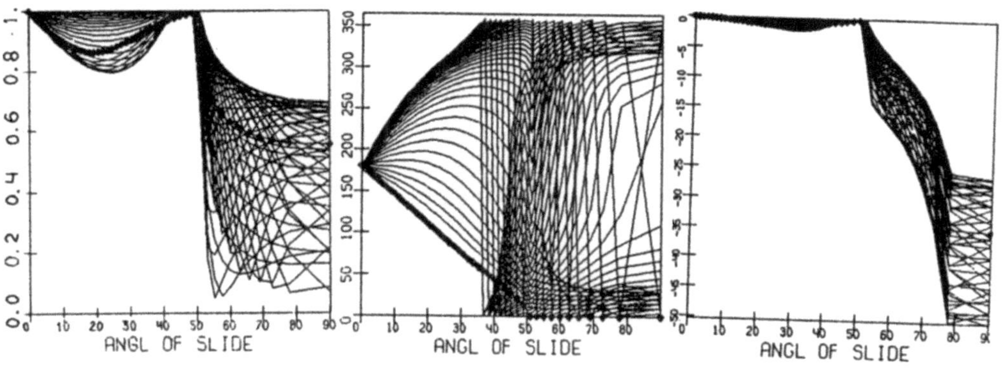

a) b) c)

FIGURE 1. Changes of $V(\theta)$ when the thickness of the sedimentary layer are varied from 0 to 100 meters; a) the modulus of the reflection coefficient at the layered bottom ($|V(\theta)|$) as a function of the sliding angle; b) the phase of the reflection coefficient at the layered bottom ($argV(\theta)$) as a function of the sliding angle; c) losses in dB per 1 km as a function of the sliding angle.

in dB per 1 km $\beta = (1 - |V(\theta)|)^2/D(x_L)$, where $D(x_L)$ is the length of the ray circuit, as functions of the sliding angle θ near the bottom are given for a frequency of 400 Hz, the thickness of the first and the second sedimentary layer being $d_2 = 2d_1$. The curves corresponding to the signal reflection at an elastic half-space $d_1 = 0$ m are indicated by markers. It is evident from Fig. 1 that the introduction of the sedimentary layer and the growth of its thickness up to $d \approx 70$ m results in an increase of losses during the propagation for all angles (except for $\theta_c \approx 50^o$). When the sedimentary layer reaches the thickness of 70 m, the character of $V(\theta)$ dependence changes and the modulus $|V(\theta)|$ begins to increase. It is evidence that an optimal propagation frequency can appear in the case of a sedimentary layer.

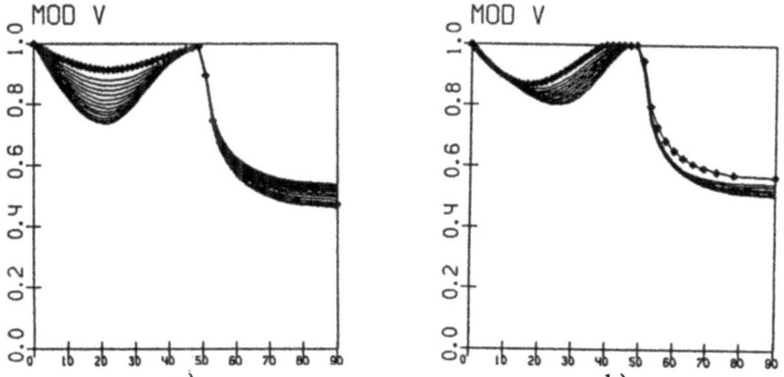

a) b)

FIGURE 2. Dependencies of $V(\theta)$ on a) the longitudinal sound speed in the sedimentary layer: $c_{\ell 1} = 1000 \,\mathrm{div}\, 2200$ m·s^{-1}; b) the sedimentary layer density: $\rho_1 = 1 \,\mathrm{div}\, 3$ g·cm^{-3}.

Similar dependencies $|V(\theta)|$ can be plotted when varying other parameters of the layered bottom such as the longitudinal sound speed (c_ℓ) in the sedimentary layer and the sedimentary layer density (see Fig. 2a,b). The dependence $V(\theta)$ does not essentially change in this case. An increase of $c_{\ell 1}$ and ρ_1 in the layer leads to a decrease of the reflection coefficient at small sliding angles. The minimum of the reflection coefficient at all c_ℓ is observed at the sliding angles $\theta \approx 20^o$. If we increase the sedimentary layer density, the minimum of the reflection coefficient is displaced in the angle. Influence of the parameters of the elastic half-space $(c_{\ell 2}, c_{t2})$ on $|V(\theta)|$ is more essential. If we take into account transverse waves in the elastic half-space, then the total inner reflection disappears (Fig. 3a, the curve at c_{t2}=0 is marked) and $|V(\theta)|$ decreases at small sliding angles. If c_{t2} increases so that it becomes greater than $c_{\ell 1}$ in the layer, there appears an interval of the angles where $|V(\theta)| = 1$. In shallow sea, the smaller angles, the lower numbers of modes, i.e., an increase of c_{t2} results in that of the sound damping at low frequencies, which (alongside high-frequency damping due to volume absorption) creates prerequisites for the appearance of optimal propagation frequency.

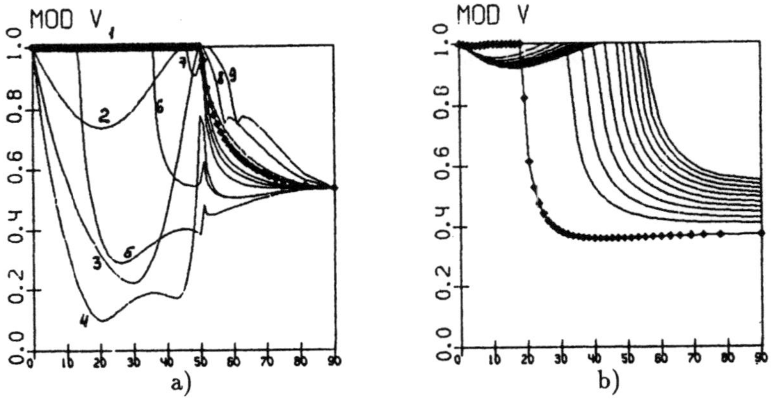

FIGURE 3. Dependencies of $V(\theta)$ on a) transverse waves in the elastic half-space: $c_{t2} = 0$ div 3000 m·s^{-1}; b) longitudinal waves in the elastic half-space: $c_{\ell 2}$ = 1540 div 3000 m·s^{-1}($c_{\ell 1}$ = 1540 m·s^{-1}).

Fig. 3b shows the influence of $c_{\ell 2}$ on $|V(\theta)|$ in the elastic half-space. If $c_{\ell 2}$ in the half-space equals $c_{\ell 1}$ in the layer (the curve is marked), the transverse waves in the half-space affect $|V(\theta)|$ only slightly.

To determine how the sound speed profile in the waveguide affects the acoustic field level, we calculated the acoustic field for various seasonal conditions of propagation. The thickness of sedimentary layers affects both the modulus and the phase of the reflection coefficient and that's why we give the calculations of the acoustic field taking into account the coherent ray interference for "summer" and "winter" sound speed profiles. Figures 4a,b give the acoustic field intensity for "summer" and "winter" profiles of the sound speed. The depth of the waveguide is $H = 200$ m, the depth of the source is $z_s = 50$ m, the depth of the receiver is $z_r = 100$ m, the radiation frequency is $F = 400$ Hz. Each plot gives five curves corresponding to different thicknesses of sediments: $d_1 = 0, 1, 2, 3, 4, 5$ m, $d_2 = 2d_1$. The upper curve corresponds to a sediment-free layer ($d_1 = 0$ m). The analysis of the calculations supports the conclusion that in summer the sound propagation strongly depends on the sea bottom characteristics due to negative rays refraction (downwards) (Fig. 4a). In winter the influence of the sea bottom decreases due to positive rays refraction (upwards) (Fig. 4b).

a)

b)

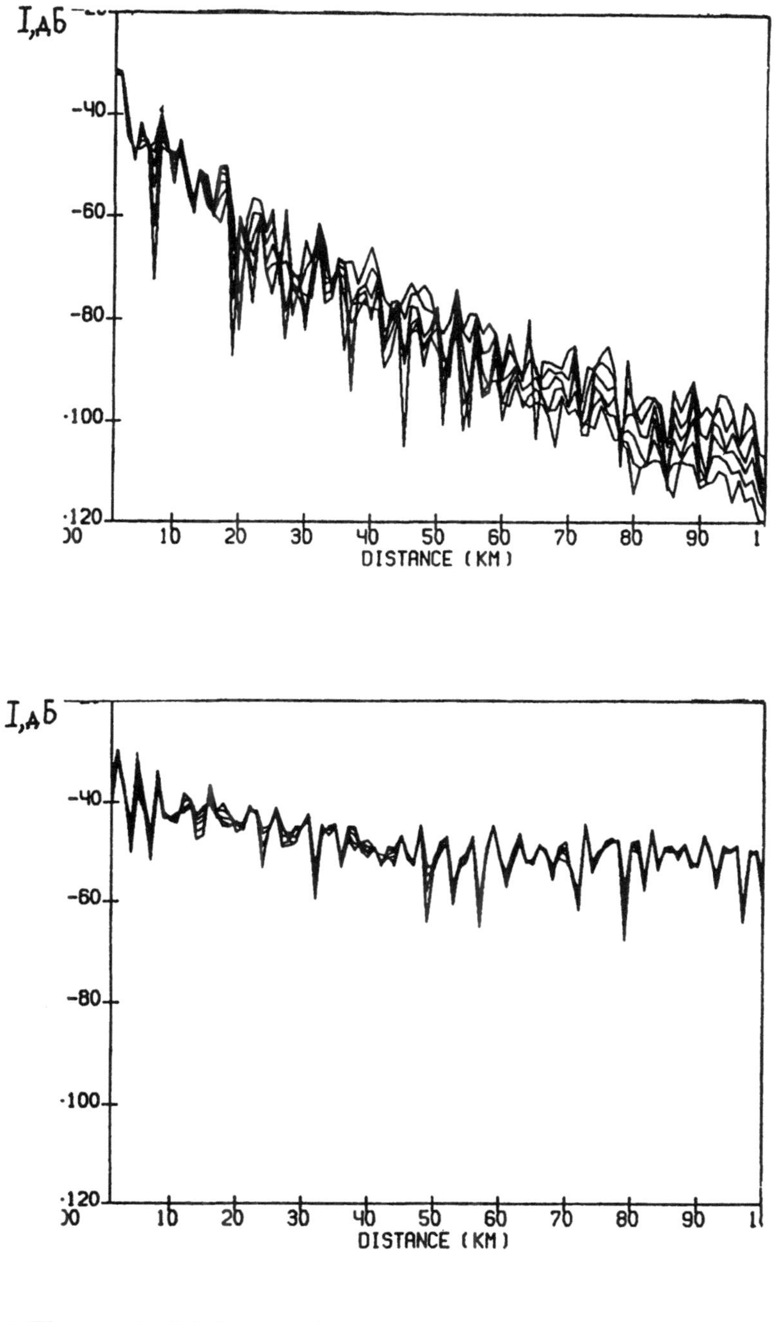

Figure 4. The acoustic field intensity for various seasonal conditions of propagation: a) for the "summer" sound speed profile; b) for the "winter" sound speed profile.

a)

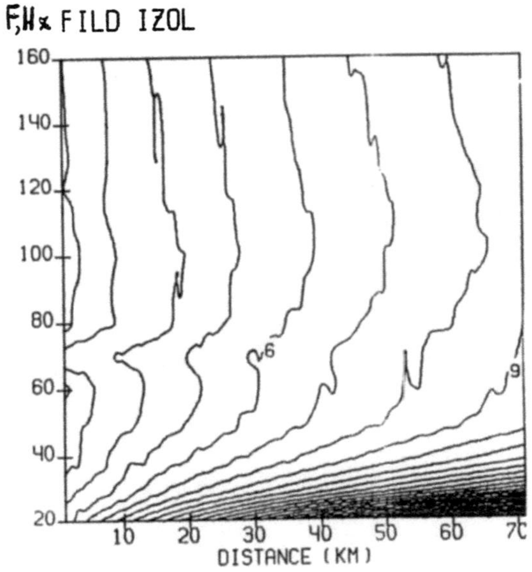

b)

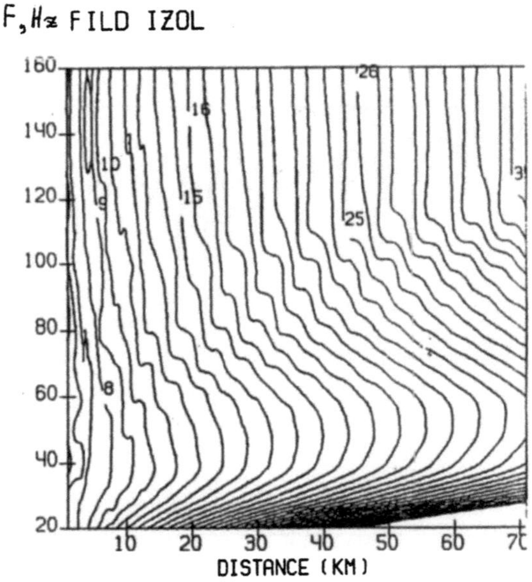

FIGURE 5. Isolines of the acoustic field intensity on the plane: distance-frequency for two types of sediments: a) layer of sandy silt; b) layer of silty clay; – lying on an elastic half-space filled with silty sand.

Consider the influence of different types of sediment material on the acoustic field level. Figures 5a,b give the results of calculations (made by a mode program) of the acoustic field in a wide frequency range for two types of sediments:

1) a layer of sandy silt (Fig. 5a);

2) a layer of silty clay (Fig. 5b);

lying on an elastic half-space filled with silty sand. The profile of sound speed corresponded to the summer period. The correlation of both the acoustic field level and the optimal signal propagation frequency with the sediment properties is evident.

Apart from the direct problems allowing us to predict the regularities of acoustic fields, we are faced with the necessity to solve such problems of ocean acoustic tomography that deal with the determination of physical and acoustic sea bottom characteristics.

In this paper we demonstrate the possibility to determine bottom characteristics from direct acoustic field measurements and an analysis of the above regularities of the angular dependence $V(\theta)$. Here we used the experimental data obtained during the expedition in the Atlantic Ocean.

This region is characterized by rather even bottom and has the depth of the order of 4.5 km. The depth of the receiver and the ship-towed emitter was 50 m.

Preliminary estimations of the number of layers and their parameters were made by analyzing the experimental propagation losses (PL) for three frequencies (of 133, 237 and 533 Hz) (see Fig. 6a,b,c).

The plots representing the losses show that at sliding angles $\theta_c = 25^o - 30^o$ PL decreases at all the frequencies, which can be explained by total inner reflection. The sound speed in the layer corresponding to this critical angle is $c_2 = c_b / \cos\theta_{c2} = 1770$ m·s^{-1}, where c_b is the sound speed in the water near the bottom. At low frequencies of 133 and 237 Hz the observed loss jump at sliding angles $\theta_{c1} = 60^o - 75^o$ is, probably, associated with the interaction of the low-frequency sound with deeper layers with the speed $c_{\ell 1} = c_b / \cos\theta_{c1} = 3000 - 3500$ m·s^{-1}. Besides, we see PL increase at small sliding angles $\theta < \theta_{c2}$. The value of this increase is dependent on frequency: PL level decreases with a frequency decrease and is practically absent at $F = 133$ Hz. It can be connected with the presence of a thin near-surface liquid layer where the sound speed is smaller than that in water and whose substratum is more rigid. The maximum effect of this layer being observed at a high frequency, we shall consider the behavior of PL at $F = 533$ Hz when determining the layer parameters. The analysis showed that we can suggest the next 2-layer model of sediments placed on an elastic half-space:

the 1-st liquid layer: $c_{\ell 1} < c_b = 1539$ m·s^{-1},

the 2-nd liquid layer: $c_{\ell 2} = 1770$ m·s^{-1},

the elastic half-space: $c_\ell = 3500$ m·s^{-1}.

To make "fine" correction of the bottom parameters, the results of numerical simulation carried out according to the program presented in Fokin and Fokina (1991) were "fitted" to the experimental data. As a result, the next geoacoustic bottom model was constructed:

the 1-st layer of sediments: $c_\ell = 1520 + i \cdot 1$ m·s^{-1}, $\rho = 1.5$ g·cm^{-3}, $d = 1.8$ m;

the 2-nd layer of sediments: $c_\ell = 1800 + i \cdot 15$ m·s^{-1}, $\rho = 1.7$ g·cm^{-3}, $d = 15$ m;

the elastic half-space: $c_\ell = 3500 + i \cdot 15$ m·s^{-1}, $c_t = 800 + i \cdot 0$ m·s^{-1}, $\rho = 2.2$ g·cm^{-3}.

Figures 7a,b,c give calculation results for PL at three frequencies. Numerical simulation is in good agreement with experimental measurements of the acoustic field along the whole trace.

a) b) c)

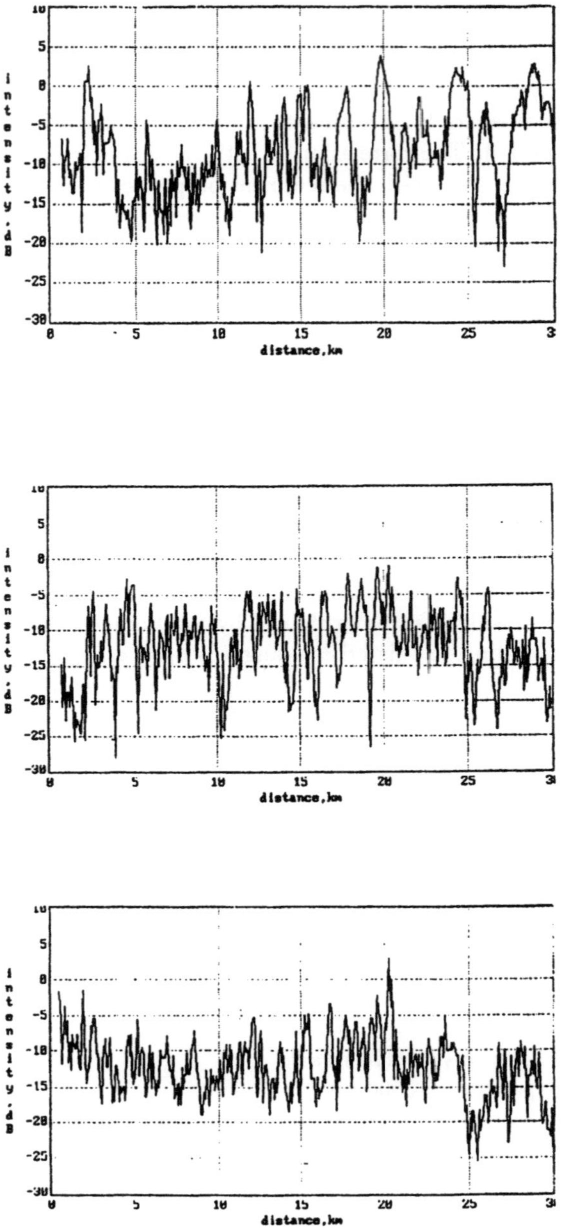

Figure 6. Experimental propagation losses (PL) for three frequencies: a) F = 133 Hz; b) F = 237 Hz; c) F = 533 Hz.

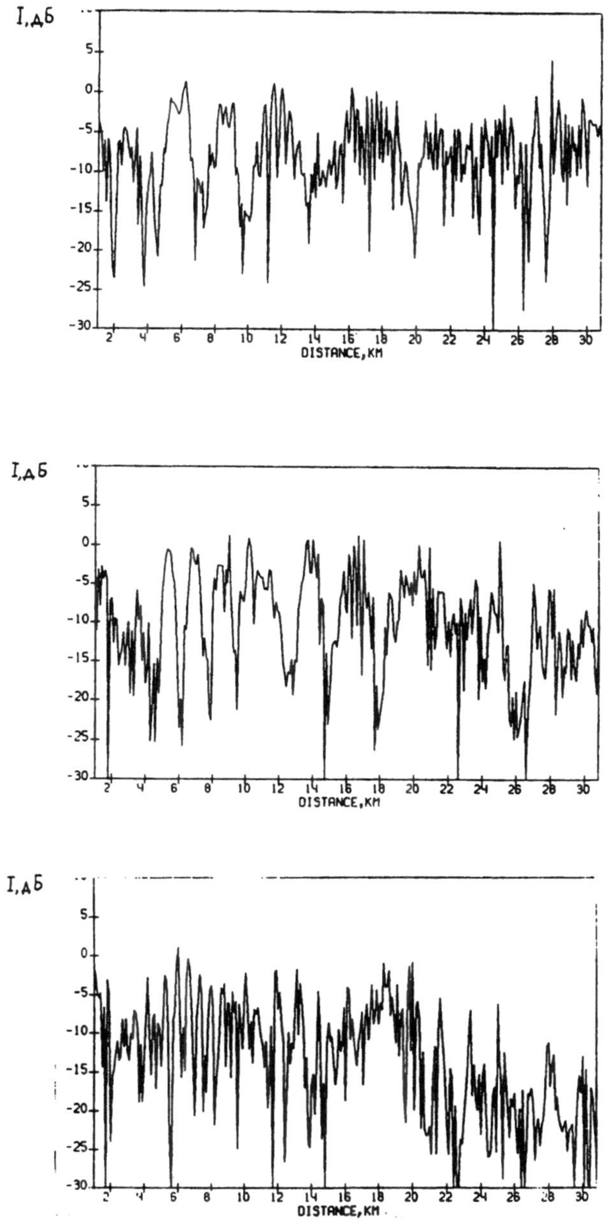

FIGURE 7. Calculation propagation losses (PL) for three frequencies: a) $F = 133$ Hz; b) $F = 237$ Hz; c) $F = 533$ Hz.

REFERENCES

Brekhovskikh L. M. and Godin O. A. (1989), *Acoustics of Stratified Media.* Nauka, Moscow. (in Russian)

Fokin V. N. and Fokina M. S. (1991), Computation of the acoustic field in a non-homogeneous waveguide using the ray method, *Akust. Jurn.*, vol. 37, no. 4, 782–788. (in Russian)

Haskell N. A. (1953), The dispersion of surface waves in multilayered media, *Bull. Seismol. Soc. Amer.*, vol. 43, no. 2, 17–34.

Thomson W. T. (1950), Transmission of elastic waves through a stratified solid material, *J. Appl. Phys.*, vol. 21, no. 2, 89–93.

An Algorithm for Solving a Problem in Elasto–Electrodynamics

O. A. KLIMENKO

Institute of Mathematics, Siberian Branch of Russian Academy of Sciences
Universitetskij prospect 4, Novosibirsk, Russia

ABSTRACT

In the paper we consider the following problem of elasto–electrodynamics. Let an elastic body be located in the half–space and let a deformation of the body generate a current. The problem is to find the current intensity provided the elastic and electromagnetic parameters are known. We propose a finite–difference scheme for solving the problem.

We consider the electromagnetic field with an elastic body in a half–space. We suppose that a deformation of this body generates the current. The aim of the present work is to obtain the quantity which characterizes the current intensity. This problem was originally considered in the paper by Lorenzi and Romanov (1993)

The coefficients æ, ρ, ε, σ characterizing elastic and electromagnetic properties are taken to be constant and known. It allows us to reduce this problem to a problem of determining $\chi(y)$ from the system

$$\frac{\partial^2 w}{\partial t^2} - \frac{\partial^2 w}{\partial y^2} + \frac{\sigma}{\varepsilon}\frac{\partial w}{\partial t} + \frac{\chi(y)}{2\varepsilon æ}\delta'(t - \sqrt{\frac{\rho}{\varepsilon æ}}) = 0, \quad y \in \mathbf{R}, \tag{1}$$

$$w = 0, \quad t < 0, \tag{2}$$

$$[w](0,t) = 0, \tag{3}$$

$$\left[\frac{\partial w}{\partial y}\right](0,t) = 0, \tag{4}$$

$$\frac{\partial w}{\partial t}\,|_{y=0} = g(t), \tag{5}$$

where $g(t)$ is a given function. We suppose that $\alpha = \sqrt{\rho/\varepsilon æ} > 1$ and denote by $[\cdot]$ the jump across the line $y = 0$.

The Direct Problem

Let us find the solution of (1)-(3) in the form

$$w = c_1(y)\theta(t-y)\theta(y) + c_2(y)\theta(t+y)\theta(-y) + c_3(y)\theta(t-\alpha y)\theta(y) + \tilde{w}, \tag{$*$}$$

where $c_1(y)$, $c_2(y)$, $c_3(y)$ are smooth functions, and $\tilde{w}$ is a continuous function. Substitute $(*)$ in (1) and equate to zero coefficients at the singularities $\delta'(t-\alpha y)$, $\delta'(t-y)$, and

$\delta'(t+y)$. Using conditions (3),(4), we can find the functions $c_1(y)$, $c_2(y)$, $c_3(y)$. We shall assume $\sigma = 0$ for $y < 0$ in future. In this case $c_2(y) = c_2(0) = 0$. Thus, we obtain the following problem:

$$\frac{\partial^2 \tilde{w}}{\partial t^2} - \frac{\partial^2 \tilde{w}}{\partial y^2} + \frac{\sigma}{\varepsilon}\theta(y)\frac{\partial \tilde{w}}{\partial t} - c_1''(y)\theta(y)\theta(t-y) - c_3''(y)\theta(y)\theta(t-\alpha y)$$

$$+ \left(2c_3'(y)\alpha + \frac{\sigma}{\varepsilon}c_3(y)\right)\theta(y)\delta(t-\alpha y) = 0, \tag{6}$$

$$\tilde{w} = 0, \quad t < 0,$$

$$[\tilde{w}]\,|_{y=0} = 0,$$

$$\left[\frac{\partial \tilde{w}}{\partial t}\right]\,|_{y=0} = \left(-c_1'(0) - c_3'(0)\right)\theta(t).$$

It is obvious that $\tilde{w} = 0$ if $t < |y|$. We shall seek for the function $\tilde{w}$ in both of the domains $t/\alpha < y < t$ and $0 < y < t/\alpha$ independently taking into account that the functions $\partial\tilde{w}/\partial y$, $\partial\tilde{w}/\partial t$ have jumps across the lines $t = |y|$, $t = \alpha y$. To find the jumps we use the fact that $\tilde{w}$ to be representable as

$$\tilde{w} = \hat{c}_3\theta(y)\theta_1(t-\alpha y) + \hat{c}_1(y)\theta(y)\theta_1(t-y) + \hat{c}_2(y)\theta(-y)\theta_1(t+y) + \hat{w},$$

where $\hat{w}$ is continuous and differentiable. In this case, for $t/\alpha < y < t$ equation (6) has the form

$$\frac{\partial^2 \tilde{w}}{\partial t^2} - \frac{\partial^2 \tilde{w}}{\partial y^2} + \frac{\sigma}{\varepsilon}\frac{\partial \tilde{w}}{\partial t} - c_1''(y) = 0, \tag{7}$$

while in the domain $0 < y < t/\alpha$ it becomes

$$\frac{\partial^2 \tilde{w}}{\partial t^2} - \frac{\partial^2 \tilde{w}}{\partial y^2} + \frac{\sigma}{\varepsilon}\frac{\partial \tilde{w}}{\partial t} - c_1''(y) - c_3''(y) = 0, \tag{8}$$

The conditions can be alternatively written as

$$\left[\frac{\partial \tilde{w}}{\partial y}\right]|_{t=\alpha y} = -\alpha\hat{c}_3(y), \quad \left[\frac{\partial \tilde{w}}{\partial t}\right]|_{t=\alpha y} = \hat{c}_3(y), \quad \left[\frac{\partial \tilde{w}}{\partial y}\right]|_{t=y} = -\hat{c}_1(y),$$

$$\left[\frac{\partial \tilde{w}}{\partial t}\right]|_{t=y} = \hat{c}_1(y), \quad \left[\frac{\partial \tilde{w}}{\partial y}\right]|_{t=-y} = \hat{c}_2(0), \quad \left[\frac{\partial \tilde{w}}{\partial t}\right]|_{t=-y} = \hat{c}_2(0). \tag{9}$$

Description of a Numerical Algorithm

Let us introduce new functions

$$u_1 = \frac{\partial \tilde{w}}{\partial t} + \frac{\partial \tilde{w}}{\partial y}, \quad u_2 = \frac{\partial \tilde{w}}{\partial t} - \frac{\partial \tilde{w}}{\partial y}.$$

We rewrite equations (7), (8) in the form

$$\left(\frac{\partial}{\partial t} - \frac{\partial}{\partial y}\right)u_1 + \frac{\sigma}{2\varepsilon}(u_1 + u_2) = b(y,t),$$

$$\left(\frac{\partial}{\partial t}+\frac{\partial}{\partial y}\right)u_2+\frac{\sigma}{2\varepsilon}(u_1+u_2)=b(y,t); \tag{10}$$

$$b(y,t)=\begin{cases} c_1''(y), & t/\alpha<y<t, \\ c_1''(y)+c_3''(y), & 0<y<t/\alpha. \end{cases}$$

From conditions (9) we obtain

$$[u_1]\,|_{y=0}=-c_1'(0)-c_3'(0), \quad [u_1]\,|_{t=\alpha y}=\hat{c}_3(y)(1-\alpha), \quad [u_1]\,|_{t=y}=0;$$

$$[u_2]\,|_{y=0}=c_1'(0)+c_3'(0), \quad [u_2]\,|_{t=\alpha y}=\hat{c}_3(y)(1+\alpha). \tag{11}$$

It is easy to obtain that u_2=0 for $y<0$ and

$$u_1(y,t)\,|_{y<0}=u_1^-(0,t+y)=u_1^+(0,t+y)-[u_1](0,t+y)=u_1^+(0,t+y)+c_1'(0)+c_3'(0).$$

It means that if we know the solution on the line $y=0$ we can find it in the domain $y<0$. The idea of the numerical algorithm consists in finding the functions u_1, u_2 layer by layer (see Figure 1) . On the first layer which coincides with the line $\tau=-\xi$ the values of u_1, u_2 are known. Then we find u_1, u_2 on the second layer which is at the distance h from the first layer. Then we determine u_1, u_2 on the third layer and so on. We take a point (y,t) on the layer where u_1, u_2 are to be found and assume the point $(\hat{y},\hat{t})$ to belong to the preceding layer where u_1, u_2 have already been determined.

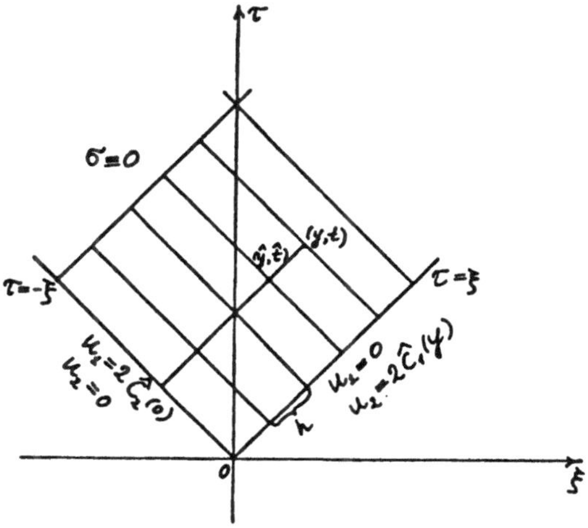

FIGURE 1.

Let us integrate the first equation of system (10) along the characteristics $\tau-t=y-\xi$ from the point $((t+y)/2,(t+y)/2)$ where the values of u_1, u_2 are known to the point (y,t). Then we integrate the second equation of system (10) along the characteristics $\tau-t=\xi-y$ from the point $(\hat{y},\hat{t})$ to the point (y,t). If the characteristics intersects the line across which the desired functions have jumps the integration is performed separately along those parts of characteristics where u_1, u_2 are continuous. At the same time, the jumps of u_i are to be taken into account. It is easy to find these jumps using formulae (11). If a point (y,t) lies on the line $\tau=\alpha\xi$, the functions u_1, u_2 have two limit values.

Let $k_1 = -\frac{\sigma}{2e}$. We obtain the following Volterra integral system :

$$u_1(y,t) = k_1 \int_{\frac{t+y}{2}}^{t} [u_1(y+t-\tau,\tau) + u_2(y+t-\tau,\tau)]\, d\tau + G_{m1},$$

$$u_2(y,t) - u_2(\hat{y},\hat{t}) = k_1 \int_{\hat{t}}^{t} [u_1(y-t+\tau,\tau) + u_2(y-t+\tau,\tau)]\, d\tau + G_{m2}, \qquad (12)$$

The values of the jumps of u_1, u_2 and integrals of the right-hand side of (10) are included in the expressions for G_{m1}, G_{m2}. We shall find functions u_1, u_2 at the nodes of a uniform mesh on the characteristics (see Figure 2).

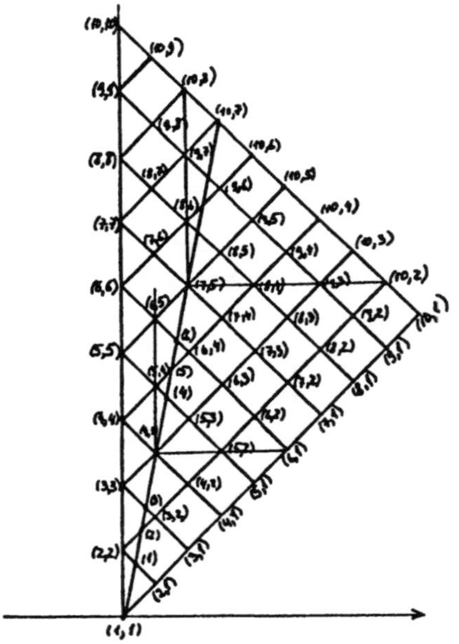

FIGURE 2.

We replace the integrals in (12) by the sums using the trapezoid formula. The presented difference scheme has the accuracy $O(h^2)$. We suppose that solution u_1 and u_2 of the problem (12) belong to a class C^2 everywhere excepting the points of the line $t = \alpha y$. The solution of the discretized problem satisfies the following estimate

$$\varepsilon \leq \mathbf{C}h^2,$$

where

$$\varepsilon = \max_{1\leq i,j\leq N} \left(|u_1(y_i,t_j) - u_1(i,j)| + |u_2(y_i,t_j) - u_2(i,j)|\right),$$

and $\mathbf{C}$ is a constant, $u_1(i,j)$, $u_2(i,j)$ are the numerical values of $u_1(y_i,t_j)$, $u_2(y_i,t_j)$ at the point (i,j). The computational experiments show that the algorithm is workable enough.

The Inverse Problem

To find the unknown coefficient $\chi(y)$ we need the following information:

$$\frac{\partial w}{\partial t}\Big|_{y=0}= g(t).$$

It is clear from the representation $(*)$ that the function $g(t)$ has the form

$$g(t) = a\delta(t) + g_0(t)\theta(t),$$

where a, $g_0(t)$ are known. Consequently, it is possible to find $\chi(0)$, $\chi'(0)$ analytically. In the numerical algorithm we seek for χ' and represent χ using the trapezoid formula:

$$\chi(a_l) = \int_{a_{l-1}}^{a_l} \chi'(æ)dæ + \chi(a_{l-1}) \approx \frac{a_l - a_{l-1}}{2}[\chi'(a_l) - \chi'(a_{l-1})] + \chi(a_{l-1}),$$

where a_l are the points of intersection of the characteristics and the line $t = \alpha y$.

χ' can be reconstructed as follows (see Figure 2):

1. $u_1(i,i)$, $u_2(i,i)$, $\chi'(0)$, $u_1(i,1)$, $u_2(i,1)$, $i = 1,\dots,N$ are known.
2. To determine χ' at the point (1) we use the information $u_1(2,2)$, $u_2(2,2)$,
3. To determine χ' at the point (2) we apply linear extrapolation using $\chi'(0)$ and χ' at the point (1).
4. We solve the direct problem and find u_1, u_2 at the points (3,2), (4,2), (5,2).
5. We find χ' at the point (3) from $u_1(3,3)$, $u_2(3,3)$.
6. We solve the direct problem and find u_1, u_2 at the points (4,3), (5,4), (6,5), (7,6), (8,7), (9,8), (10,9).
7. For the points (6,5), (5,4), (4,3), (5,2), (6,1) we have the above situation, so we can find χ', u_1, u_2 at the points (10,8), (9,7), (8,6), (7,5), (8,4), (9,3), (10,2) analogously.

From steps (1)-(7) it is clear that we can find χ' and u_1, u_2 at all the points (i,j). If α is a rational number, such a point as (4,3) necessarily exists otherwise we use instead of α its rational approximation. The obtained difference scheme has accuracy $O(h^2)$. We suppose that χ belong to the class C^3. The solution of discretized inverse problem satisfies the following estimate

$$E_\chi \leq C_1 h^2,$$

where

$$E_\chi = \max|\chi'(a_l) - \chi'_l|,$$

and C_1 is a constant, χ'_l is the numerical value of χ' at the point a_l. The computational results confirm algorithm's acceptability.

REFERENCES

Lorenzi. A, Romanov V. G. (1993), Identification of an electromagnetic coefficient connected with deformation currents. *Inverse Problems,* vol. 9, no. 2, 301–320.

Three–Dimensional Structure of Lateral Heterogeneities in P-Velocities in the Upper Mantle of the Southern Margin of Siberia from Teleseismic Data

I. Yu. KULAKOV and S. I. KESELMAN

United Institute of Geology, Geophysics and Mineralogy,
Siberian Branch of Russian Academy of Sciences
University av. 3, Novosibirsk, 630090, Russia

ABSTRACT

A three–dimensional model of lateral variations of P-wave velocities in the upper mantle of the Southern margin of Siberia has been obtained using more than 6000 teleseismic arrival times registered at stations of Siberia and Kazakhstan. The method was used which is based on the triangulate block parametrization and permits us to accord the densities of output and input information. The inversion stability has been improved significantly as compared to the traditional approaches. Special attention was paid to the problem of testing the obtained results. A study of the method stability (i.e. sensitivity of results to noise) shows the reconstruction to be stable, even though the noise level exceeds the relevant signal. Another test consisted in a reconstruction of velocity anomalies using two independent sets of events. For these two sets the determined velocity structures were found to be in satisfactory accordance. The preliminary interpretation of the results permits us to suggest some mechanisms of development of the studied region.

1. Introduction

The study of the mantle structure is of great importance for investigating the nature of processes in the Earth, modelling possible ways of its formation and development and for giving an insight into the structure and dynamics of other cosmic bodies.

The Earth can be represented with a sufficient accuracy, as a spherical symmetric body, whose properties change with a depth. The distribution of the main parameters, such as seismovelocities, density, elastic modules and so on, along the radius can be obtained by interpretation of seismological data and lie at the basis for the study of the entrails to a great depth . Though the lateral perturbations of these parameters from the average model are relatively small, they have the great influence on most geodynamical processes and determine, in particular, the large scale surface dynamics: formations of oceans and mountains, movements of continents, etc. Seismic tomography is the most powerful method to study the horizontal deep-earth heterogeneities.

The methods used for examining the structure of lateral heterogeneities in the Earth with seismic rays differ significantly from those of computed tomography applied, for example, in medicine. The effective algorithms based on integral transformations, which permit the problem to be solved quickly enough and with high resolution, have been worked out for medical tomography, where the shape of the used rays is usually straight (Natterer, 1986). The application of such approaches is problematic in seismic tomography, where the shape of the seismic rays depend on the sought–for velocity distribution. As far as we known, only a few of theoretical investigations concern the application of analytical methods to tomographic reconstruction (Alekseev et al., 1969), but we have failed to find their successful practical realization.

The second important peculiarity of seismic tomography is that an explored geophysical object can be examined by seismic rays on one side only. From this point of view the best application of tomography in geology is realized in scanning between wells. In this case rather perfect algorithms whose resolution approaches to that of medical tomography have been developed (Nolet, 1987).

The application of tomographic methods at great depths exceeding 100 km involves additional difficulties. First, only natural sources, that is, earthquakes, can give information on such depths. They are distributed on the surface of the globe in extremely irregular manner. Signals from these sources are registered at the stations, whose number is relatively small.

Therefore, in this case the studied area is covered by the rays very irregularly, which is a serious obstacle to the research.

Second, the accuracy of the arrival times data, is rather low, especially in Siberia where the analog equipment is mainly out of date.

Third, neither coordinates of the natural sources nor their times are known. Their determination is an additional problem which needs a lot of supplementary data and calculations.

So, the interior structure is reconstructed in seismology using a very scanty set of data. In fact, only two types of data are given: the coordinates of the receivers and the arrival times of the rays. Therefore, tomographic approaches used in seismology are less well studied theoretically than, for example, in the case of well–logging.

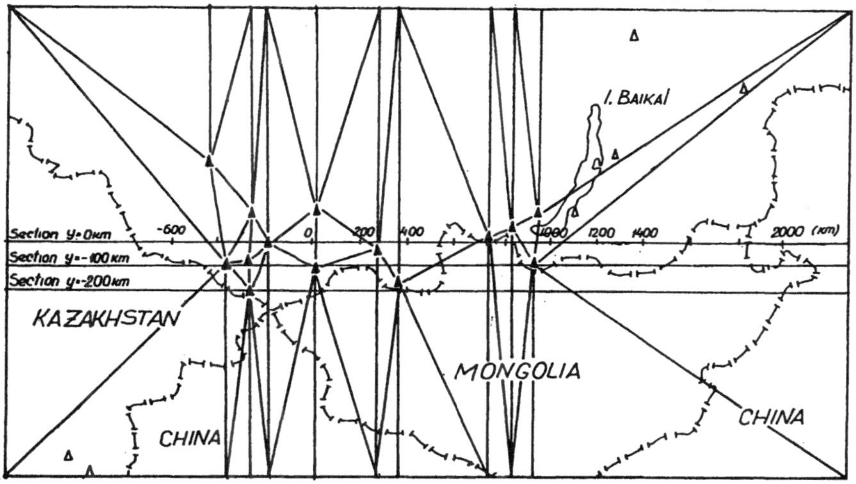

FIGURE 1. The studied territory and the used stations. Black triangles are the stations whose information were employed in inversion. The others have been used only for a statistical analysis. The triangulate network for calculating the station corrections and lines of vertical sections (Fig.6) is shown.

The region taken as an object for this study (Fig.1) is very interesting from a geodynamic point of view. The study of the deep–earth sources of the tectonic activity of Altai and Sayan fold areas and the Baikal rift zone is a very important and complicated problem, which has been the subject of much investigation for the last three decades. One of the first tomographic analysis of the mantle was made in 1969 (Alekseev et al. 1971) just for this region. A lot of researches on the mantle heterogeneities beneath Altai and Sayan (Gobarenko and Yanovskaya, 1983, Mordvinova, 1988 etc.) and Baikal (Krylov et al., 1971, Rogozhina and Kozhevnikov, 1979, Pusyrev, 1981, Zorin et al. 1986) made a great contribution to the study of the mechanism for the recent activity of these regions. But the resolution of the obtained models does not permit one to explain the autonomic development of these three areas. The main aim of this research is to determine more detailed structure of the lateral variations in the upper mantle beneath the southern margin of Siberia at a depth from 100 to 700 km in order to describe the dynamic state and development of this territory. The obtained results will be probably useful in seismology for localizing earthquakes and for examining the seismic activity of the considered region.

2. Formulation of the Problem

The teleseismic technique based on the utilization of the rays from far sources is used for solving the problem. The main criterion for selection of the rays is their intersection with the lower boundary of the studied area (Fig. 2).

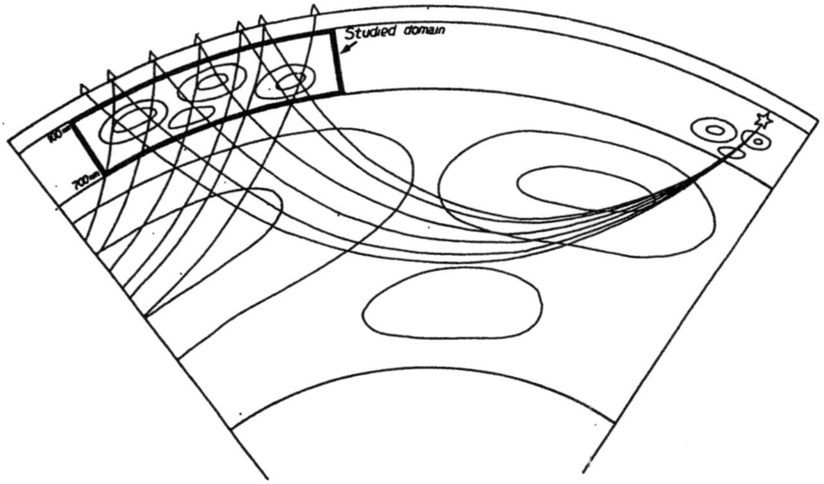

FIGURE 2. An illustration to justification of the teleseismic assumptions.

2.1. The available data.

1. The accurate coordinates of receivers.

2. The absolute arrival times of the rays registered at the stations. Their accuracy was shown, to be low. Only one advantage of this type of data is their large body. More than 6000 rays have been used in this work.

3. Coordinates and origin times of the sources. These data result from the estimations and are not accurate. But the teleseismic approach, as will be shown below, does

not require the accuracy in information on the sources. The information about more than 600 earthquakes has been used.

4. The reference velocity model. The spherically symmetric model PEM-A (Dziewonski et al., 1975) has been chosen. It is important that the teleseismic approach permits researchers to determine the relative velocity anomalies and, in contrast to the case of using local events, does not depend on the choice of a reference model.

2.2. Additional remarks. Important information concerns the character of anomalies inside and outside of the studied area. Assuming that the sought-for mantle anomalies result from the convection and their characteristic horizontal size is directly connected with the thickness of the layer where this convection proceeds, one may conclude that the average size of heterogeneities in the lower mantle is considerably larger than that in the upper one. The results of tomographic study of the lower mantle (Dziewonski, 1984) confirm this supposition.

2.3. Assumptions. Based on available information, an important assumption should be made to solve the problem of determining the velocity structure in the upper mantle beneath the studied territory. Let the residual times δt_{ij} along the ray from i-th source to j-th receiver, caused by anomalies in the studied area be written as

$$\delta t_{ij} = T_{ij}^{pem} + \tau_i^s + \tau_j^r, \qquad (1)$$

(Dziewonski and Anderson, 1983), where T_{ij} is the real travel time along the ray, T_{ij}^{pem} is the travel time along the ray according to the standard model PEM–A, τ_i^s is the source correction, τ_j^r is the station correction. The source correction includes all factors affecting upon the travel time of the ray before its entrance into the studied area: inaccuracy of coordinate and origin time of the source, heterogeneities in the lower mantle and near the source, errors connected with the chosen standard model, and so on. All these factors have the uniform influence on all rays from one source to the stations. The lateral variations of the times inside one ray tube are neglected in view of the assumption about rather large horizontal size of anomalies in the lower mantle.

As for the station correction, it includes the regular features connected with the station: the relief, anomalies in the crust and the lithosphere beneath the station, several regular instrument factors, and others. The scheme of their calculation in this work is somewhat different from (1) and permits us to take into account not only the vertical features but also the lateral ones.

The important assumption, typical for such a sort of problems, consists in linear approximation of the inverse kinematic ray problem. Taking into account that the sought-for anomalies, as a rule, do not exceed 5%, this assumption can be accepted with a sufficient accuracy.

The three-dimensional structure of the velocity variations should be reconstructed from the available data and the accepted assumptions.

3. The Residual Times

An error of determining the arrival times of the rays is several times greater then the level of the relevant signal because of low-quality registration and interpretation of the initial information. To improve the quality of the data, the methods based on the averaging of the times along the rays from the nearest sources are used in this work. For this purpose, the surface of the Earth was divided into the segments of 12° × 12° (Fig. 3), and all earthquakes inside each segment were replaced by one source, whose reliability was reflected in the corresponding weight at the inversion, connected with the number of events in every cell. A special attention here was paid to the problem of rejecting defect information.

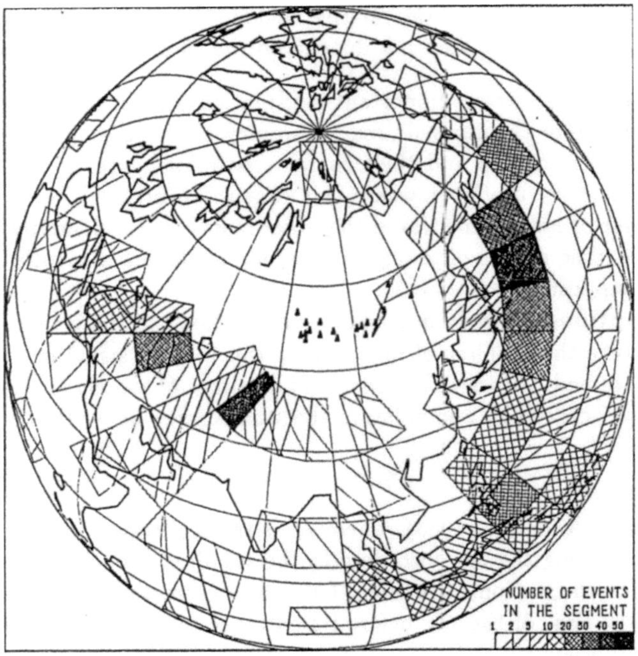

FIGURE 3. Distribution of the used earthquakes on the earth's surface divided into segments of 12° × 12° with the corresponding number of events inside. The triangles label the stations used in this work.

The calculation of the source corrections was based on the assumptions about the smooth character of the sought-for heterogeneities in the studied area and about a random distribution of other factors in travel times.

As a result of the mentioned procedures, the number of data decreased from 6000 to 900 and their quality was improved significantly. However, the level of the errors in times had the same order as the relevant signal. Therefore, special attention was paid to the problem of the algorithm stability to a random noise in data.

4. Algorithm

In studying the mantle by tomographic methods, the problem of reconstruction of the velocity anomalies is reduced, as a rule, to the system of linear equations:

$$A_{ij}\theta_j = x_i, \quad i = 1, \cdots, I, \quad j = 1, \cdots, J, \tag{2}$$

where θ is a vector of unknown parameters describing the velocity distribution in the studied area, x is a data vector, $A_{ij} = \partial x_i / \partial \theta_j$ is the first derivative matrix. The representation of the problem in such a linear form is possible if the values of the components of θ are small enough to neglect the derivatives of higher orders. Taking into account the assumption on the character of the anomalies to be sought, it is quite acceptable for our case. Usually the quasilinearized inversion is used which consists in iterative linearized inversion by scheme (2) till the obtained solution will be sufficiently accurate. The methods based on scheme (2) differ from one another either by the type of parametrization or by the sort of initial data. The residual times caused by heterogeneities in the studied area are taken in this work for the data set x.

Choosing a sort of parametrization we intended to overcome serious shortcomings of the methods used widely in practice. Our parametrization is based on the subdivision of the studied area into tetrahedral blocks, each having a constant gradient of squared slowness inside. Thus, the continuous velocity approximation in the studied area is realized, and one can avoid the problem of velocity discontinuity which is characteristic for one of the most widespread method based on rectangular block parametrization. Analytic formulae of ray tracing are known for a medium described in such a way (Cerveny, 1984).

The key aspect in this method of parametrization is the possibility to accord the densities of output and input information. It is well known that in seismology the density of rays providing information on the deep–earth structure is extremely uneven. Nevertheless, in most of the methods applied in practice the density of output information is constant. For example, in the case of rectangle block parametrization this amounts to the uniform discretization of the considered area, and in the case of parametrization by harmonic series this is the uniform spectrum in any area's part. This fact leads to the situation when relevant information in well sampled regions is partially lost or passes to poorly sampled regions where the false anomalies can appear. The accordance between input and output information in our method is realized as follows: the vertices of the triangulate network are located so as to average the number of the rays investigating the environs of any vertex (i.e., tetrahedrons adjoining one another by this vertex) (Fig. 4).

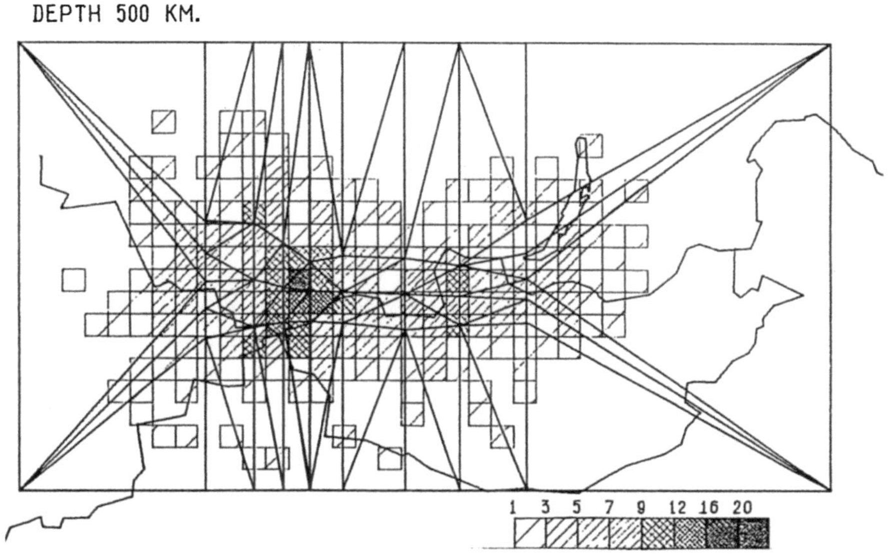

Figure 4. An example of subdivision of the studied area by the tetrahedral network according with the rays density. Squares show the number of the rays passing through the level. The scale gives the number of rays in a square of 100 × 100 km.

In this case the resolution of the obtained velocity model distinctly reflects the quantity of input information: the better is a priori information on a part of the area, the more detailed is the conclusion on its structure. Moreover, such an approach is very favorable in inversion. The accordance between input and output information leads to the almost uniform filling of columns of the matrix A by nonzero elements, and this fact considerably improves some its qualitative characteristics. The results of numerical testing showed that in the case of a uniform network, the value of the condition number of the matrix (the ratio of the maximum eigenvalue to the minimum one) had the order of $10^7 - 10^8$, whereas in the case of accordance between output and input of information it was less than 100. Naturally, the stability of such as approach to a noise in the input data is higher.

When reconstructing the velocity structure in the explored area, a special attention was paid to the problem of determining the station corrections. According to our method, these corrections are calculated simultaneously with the velocity variations as some fictitious velocity anomalies at the points coinciding with the stations and entering to the common network (Fig.1). In this case one can take into account not only vertical factors but also lateral variations in the crust and the lithosphere.

The Method of Singular Value Decompositions has been applied for the inversion of the matrix A. The Quality of Singular Numbers (QSN) played the role of the regularization parameter. The important problem is to find QSN optimal for the best relationship between stability and accuracy of the inversion.

This algorithm has been realized and tested both for two-dimensional (Kulakov, 1992) and three-dimensional cases. The comparison of the obtained results shows that though the number of unknowns in the 2D case was twice less than that in the 3D case, the condition number of the matrix A was of 3-4 orders higher. This proves the fundamental advantage of the three-dimensional algorithm as compared to the two-dimensional one.

5. Results

Using the above algorithm, we obtained the three-dimensional distribution of the P-wave velocities in the upper mantle of the southern margin of Siberia at a depth from 100 to 700 km. The main information about heterogeneities is contained within a significant zone whose shape is determined from the initial data. The obtained result is presented both in the horizontal (Fig.5) and in the vertical sections (Fig.6), whose projections are given in Fig.1. The obtained images resulted from the reconstruction of 210 parameters: 14 station corrections and 196 velocity variations at the vertices network. Analysis of the results shows that the horizontal section at a depth of 100 km does not seem to reflect the real velocity structure as the rays here do not form a sufficient system of intersections. The obtained values are most likely to reflect the lateral variations in the crust and the lithosphere and to be connected with the station corrections.

Special attention was paid to testing the results. To this end, two types of tests have been proposed. The first consists in the testing of the algorithm stability to the random noise in the data vector. The image of heterogeneities obtained from the inversion of real data (Fig. 7a) has been used as an initial model. The residual times were calculated as follows:

$$\delta t_i = A_{ij}\delta v_j + \varepsilon_i, \tag{3}$$

where δv are velocity parameters of the initial model, ε is a random, gaussian error. The average relevant signal and the level of the noise were calculated as

$$\vec{\delta t} = \sum \frac{|A_{ij}\delta v_j|}{I}, \tag{4}$$

$$\vec{\varepsilon} = \sum \frac{|\varepsilon_i|}{I}. \tag{5}$$

The results of reconstruction (Fig. 7b,c) show that the reconstruction is realized stably enough even though $\vec{\varepsilon}$ is twice greater than $\vec{\delta t}$.

The second test consists in reconstruction of the sought-for heterogeneities using two sets of independent data. For this purpose, all events used in the experiment were divided into two groups by an arbitrary criterion (for example, with odd and even numbers) and employed independently for calculating the velocity structure. A comparison of the reconstructed images (Fig.8) shows the satisfactory correlation between the obtained results. So, the results of both of the tests prove the reliability of the obtained results and permit us to suggest that the revealed heterogeneities do reflect the real structure of the mantle.

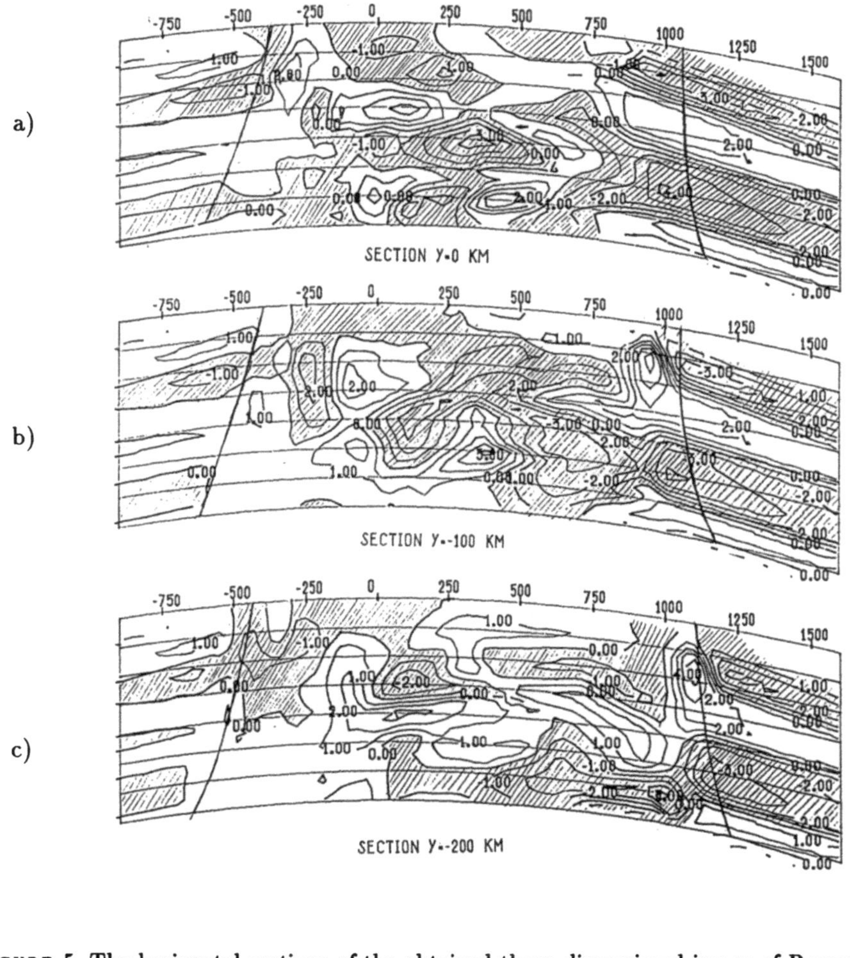

FIGURE 5. The horizontal sections of the obtained three–dimensional image of P-wave velocity variations. The significant zone is shown. The areas with negative perturbations are shaded. The values of contour lines are given as a percentage of the average value of velocity.

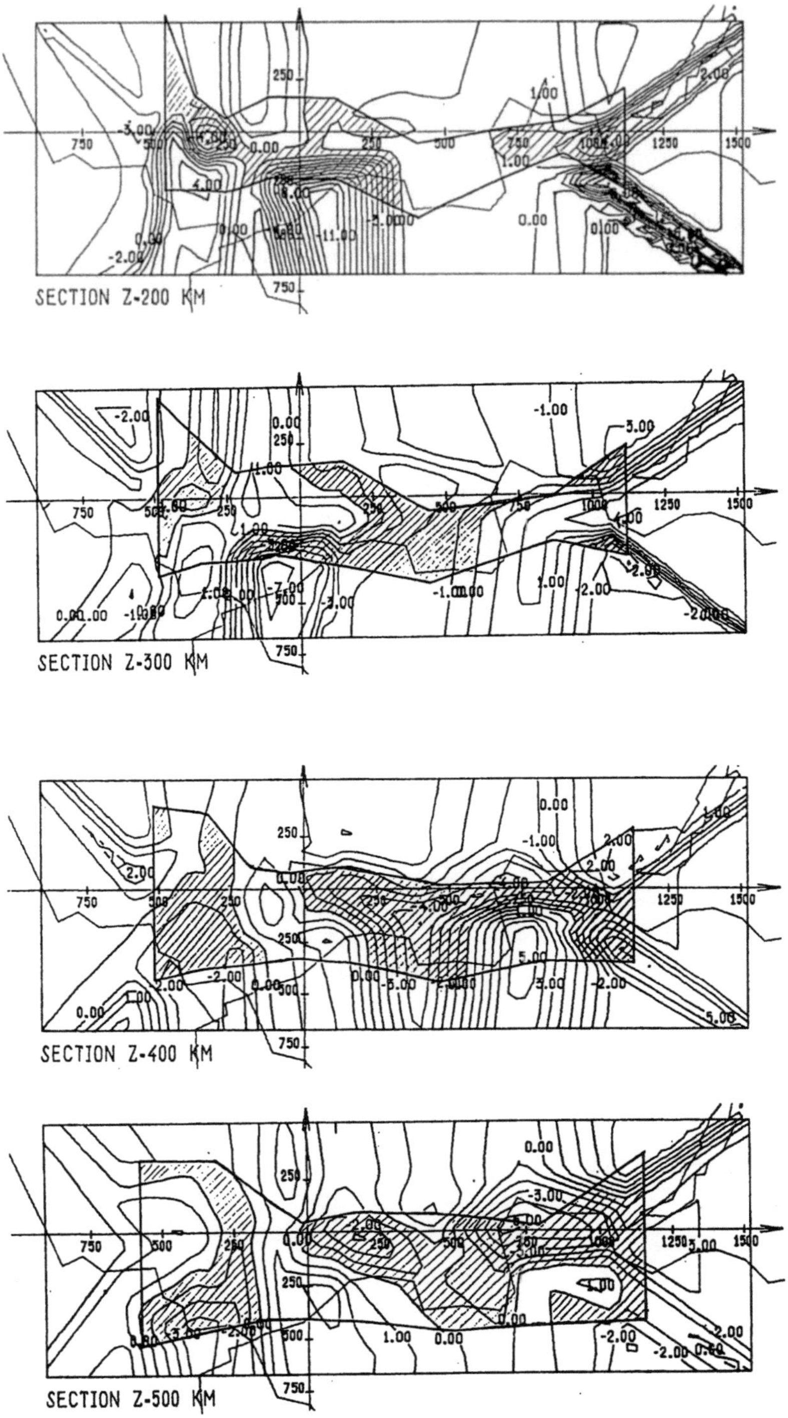

Figure 6 a), b), c), d).

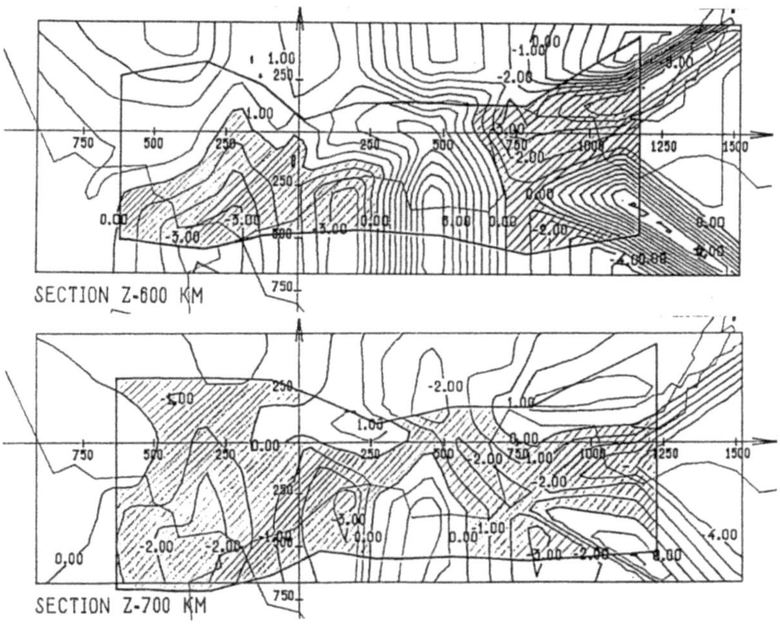

FIGURE 6 a)-f). The vertical sections of the obtained three-dimensional image of P-wave velocity variations. The significant zone is shown. The areas with negative perturbations are shaded. The values of contour lines are given as a percentage of the average value of velocity.

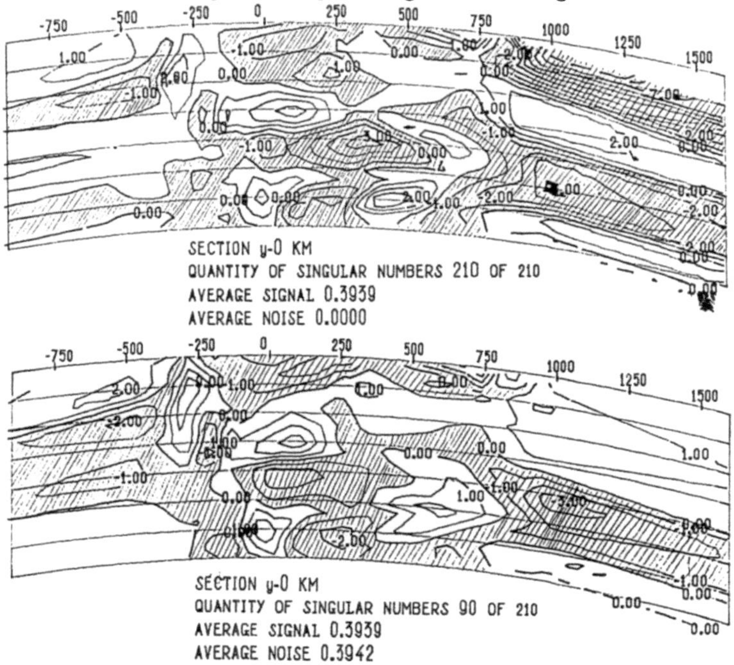

FIGURE 7 a), b).

c)

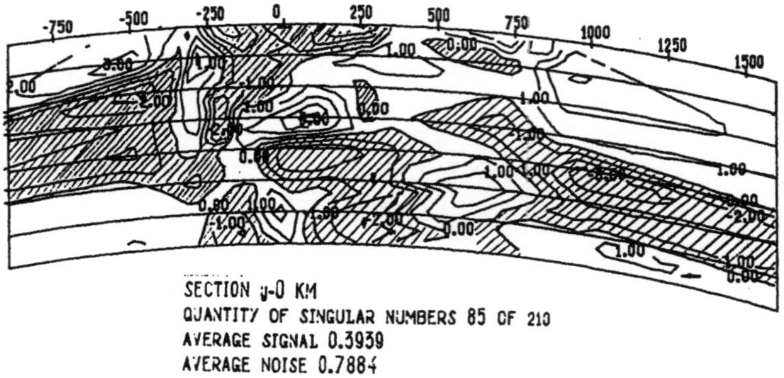

FIGURE 7. The results of the stability testing. a) The initial model. b) Reconstruction from the data with a noise equal to the relevant signal. c) The case when noise is twice higher than the relevant signal.

a)

b)

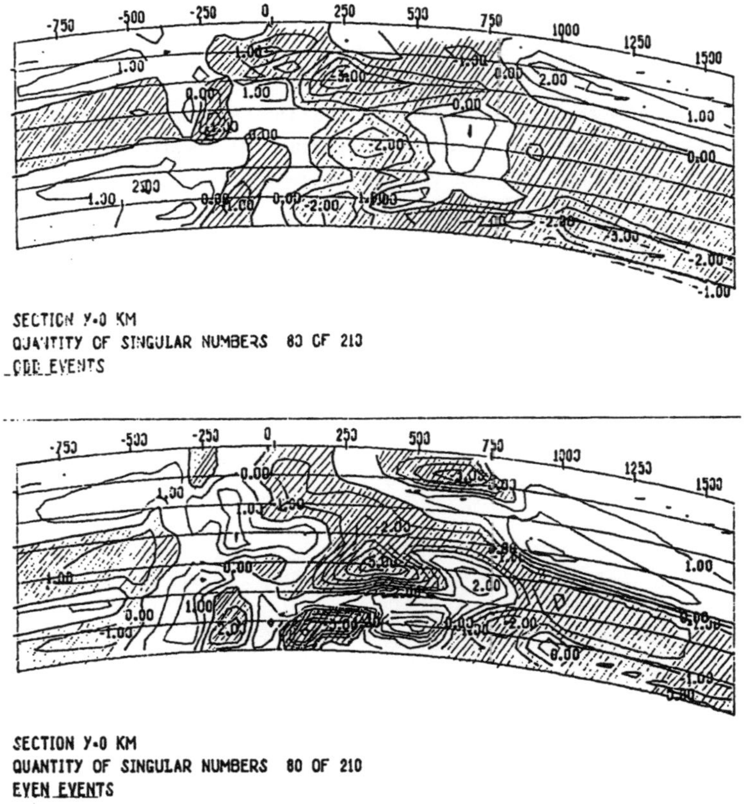

FIGURE 8. The results of reconstruction of real heterogeneities from two independent sets of data (with odd and with even numbers).

6. Discussion

We attempted to correlate the reconstructed structure of heterogeneities in the upper mantle with the main geological estimate of dynamics of the crust and the lithosphere in the studied territory and equally with possible hypotheses of their development. When interpreting geodynamically the tomographic images, the linear relationship between the main parameters, such as seismovelocity, density, temperature, is usually assumed (e.g., Hager and Richards, 1989). In this case the heightened velocity values correspond to the heightened density and the lowered temperature. Under the assumption that the substance viscosity in the mantle has a finite value, these anomalous zones correspond to descended flows, and, on the contrary the parts with the lessened velocity correspond to ascended flows. On this supposition, the image of the flows in the studied area can be reconstructed and calculated correlated with surface processes. An attempt of qualitative geodynamic interpretation is shown in Fig.9.

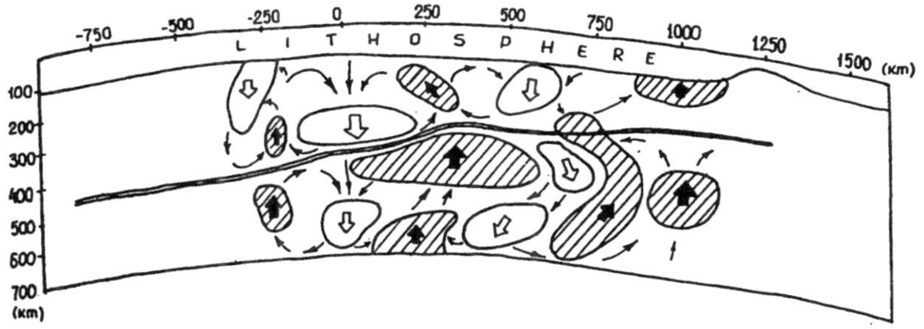

FIGURE 9. Hypothetical directions of the flows following from the calculated velocity structure.

According to the data on recent activity, the studied territory contains three autonomic active regions: the Altai and the Sayan fold area and the Baikal rift zone. The activity of the first two reveals itself in the mountain formation without noticeable manifestation of geothermal processes (Duchkov and Sokolova, 1986). The Baikal rift zone is characterized by the active spreading of the crust with its intensive heating. Judging from the obtained velocity images the activity of these three areas seems to be caused by the existence of three flows at a depth from 700 to 400 km beneath the corresponding regions. The structure of the reconstructed anomalies confirms the possibility of the development of two–layered convection in the upper mantle. One of the variants of such a process is shown in the Fig.10. These ideas well fit in the modern conception of dynamics and development of the studied region.

7. Conclusion

The method reconstructing effectively velocity heterogeneities from the data with an extremely high level of noise has been worked out and realized successfully in practice. The idea of coordination of input and output information was a key factor of optimal utilization of initial information, which permitted us to improve significantly the solution stability. Both the stability of this algorithm to the high level of noise and the reliability of the obtained results have been tested, and shown as quite satisfactory.

The presented algorithm can be applied for processing of seismological information in other regions as well, especially in those where the quality level of information is low. This algorithm does not require much computer time. Using more powerful hardware, one can increase the number of input data and improve the resolution of the obtained images.

This work will continue. We intend to realize the method based on the utilization of local earthquakes in order to study more precisely the upper mantle to a depth of hundreds kilometers. We also plan to widen the studied territory including the data from the stations of Mongolia and Baikal region.

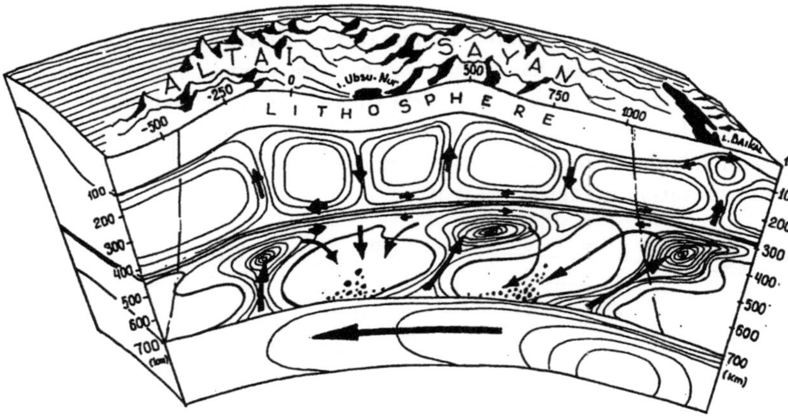

FIGURE 10. Schematic image of a variant of the convective structure in the upper mantle. This model corresponds to the two layered convection with lowered viscosity in the upper layer.

8. Acknowledgments

The authors are very grateful to the staff of Altai–Sayan Seismological Expedition for information on the stations of this area and for their help in its interpretation. We appreciate the fruitful discussions with S.V.Goldin, V.V.Piankov, A.G.Filina and A.B.Druzhinin as very helpful for realizing and creation of the algorithm. This research has been supported by the System of Scientific Grants of Siberian Branch of the Russian Academy of Sciences.

REFERENCES

Alekseev A. S., Lavrent'ev M. M., Mukhometov R. G., Neresov N. L., Romanov V. G. (1969), A numerical method for determining the earth's mantle structure. *In: Mathematical Problems in Geophysics.* vol. 2, Novosibirsk, 179–201. (in Russian)

Alekseev A. S., Lavrent'ev M. M., Mukhometov R. G., Neresov N. L., Romanov V. G. (1971), Numerical method of determination of the earth's mantle structure. *In: Mathematical Problems in Geophysics,* vol. 2, Novosibirsk, 179–201. (in Russian)

Cerveny V. (1986), Seismic ray theory. *Lecture Notes.* Inter. Center for Theoretical Physics, Miramare, Trieste.

Duchkov A. D. and Sokolova L. S. (1986), Thermal regime of the lithosphere of Siberia. *In: Recent Geodynamics of the Lithosphere of Siberia.* Institute Geologii i Geofiziki AN SSSR, Novosibirsk, 46–62. (in Russian)

Dziewonski A. M. (1984), Mapping the lower mantle: determination of lateral heterogeneity in *P* velocity up to degree and order 6. *J. Geophys. Res.*, vol. 89, no. B7, 4381–4399.

Dziewonski A. M., Hales A.L., Lapwood E.R. (1975), Parametrically simple earth models consistent with geophysical data. *Phys. Planet. Inter.*, vol. 10, no. 1, 12–48.

Dziewonski A. M. and Anderson D. L. (1983), Travel time and station corrections for P waves at teleseismic distances. J.Geophys.Res., vol. 88, 3295–3314.

Gobarenko V. S. and Yanovskaya T. B. (1983), The study of the structure of horizontal heterogeneities in the upper mantle of Altai-Sayan zone. Izvestiya Acad. Nauk SSSR, Fizika Zemli, vol. 4, 21–35. (in Russian)

Hager B. H. and Richards M. A. (1989), Long wavelength variations in Earth's geoid: physical models and dynamical implications. Phil. Trans. R. Soc. Lond., vol. A 328, 309-327.

Krylov S. V., Mishenkin B. P., Petrik G. V. and Seleznyov V. S. (1971), On a seismic model of the upper mantle in Baikal rift zone. Geologiya i Geofisica, vol. 12, 108–112. (in Russian)

Kulakov I. Yu. (1992), Linearized seismic tomography problem in application to regional research. *Russian-Norwegian Oil Exploration Workshop*, vol. 2, Expanded abstracts, Voss, Norway.

Mordvinova V. V. (1988), The lithosphere and the crust in the South of Siberia from Earthquakes data; determination of the thickness of the crust and lithosphere beneath the Altai–Sayan fold area and the Baikal rift. Izvestiya AN SSSR, Fizica Zemli, vol. 5. (in Russian)

Natterer F. (1986), *The mathematics of computerized tomography.* Teubrer and John Wiley & Sons Ltd., Stuttgart, 288.

Nolet G., (Editor) (1987), *Seismic tomography (with applications in global seismology and exploration geophysics)*, Reidel, Dordrecht, 386.

Pusyrev N. N. (Editor) (1981), *Interior of Baikal (from seismic data)*, Nauka, Novosibirsk, 105 p. (in Russian)

Rogozhina V. A. and Kozhevnikov V. M. (1979), *The area of the anomalous mantle beneath the Baikal rift.* Novosibirsk, Nauka. (in Russian)

Zorin A. Yu., Mordvinova V. V., Novoselova M. R. and Turutanov E. Kh. (1986), Mantle density inhomogeneity beneath the Baikal rift. *Izvestiya Acad. Nauk SSSR, Fizika Zemli*, vol. 5, 43–58. (in Russian)

Simultaneous Determination of Elastic and Electromagnetic Medium Parameters [1]

M. M. LAVRENT'EV, Jr. and V. I. PRIIMENKO

Institute of Mathematics,

Universitetskii Pr., 4, 630090, Novosibirsk, Russia

ABSTRACT

An inverse problem of determining both the elastic and the electromagnetic (EM) parameters is studied. A specific case when EM waves are generated by elastic deformations is considered. For a model system of two equations with piecewise constant coefficients an approximate algorithm solving the inverse problem is described. The algorithm uses the Fourier transform of measurements.

Introduction

An inverse problem of simultaneous determination of electromagnetic (EM) and elastic parameters is considered. More precisely, we study the Maxwell and the Lame systems in the case when the EM field is generated by elastic oscillations. At the same time, we neglect the reverse influence of the electromagnetic field on the elastic oscillations.

It is of common knowledge that when elastic waves are propagated through an electroconductive medium within the magnetic field, an interaction between EM and elastic oscillations appears. The resulting waves are called the magneto-elastic (ME) waves. In the case of electroconductive fluid, similar waves arise which have been studied properly enough (see, for example, Landau and Lifschitz, 1959). Elastic waves propagation in the magnetic field of the Earth was investigated in Knopoff (1955) as well as some boundary value problems for a layer and for a half space adjacent to elastic non-conducting media. We refer readers also to Novatskii (1986), Keilis-Borok and Monin (1959), Dunkin and Eringen (1963), Rakhmatulin and Shkenev (1985), Balakirev and Gilinskii (1982) where similar questions are studied.

In all the mentioned papers the influence of the EM field on the deformation field is considered as a result of the Lorentz forces arising in the equations of movements. To reflect an increase of the electric field density, an additional term should be included into the equation describing the Ohm law. This term is dependent on the velocity of particles which move in the magnetic field. In this case, the influence of the magneto-elastic field on the elastic waves propagation can be shown (see Landau and Lifschitz, 1959) to be negligible. At the same time, as follows from experimental researches (see Alers and Fleury, 1963), if the original magnetic field is strong enough, the influence of the ME forces may be significant.

[1]The work was partially supported by the Russian Foundation of Fundamental Researches under grant No. 93-01-01-1739

The ME models are of our interest due to their applications in geophysics and acoustics.

1. Formulation of the Problem

Let us consider one of the inverse problems of simultaneous determination of electromagnetic (EM) and elastic medium parameters (the EM field arises due to the action of the displacement field). We will study this problem in the following form.

Let $\mathbf{R}^3$, $\mathbf{R}^3_\pm$ be a three-dimensional space of $x = (x_1, x_2, x_3)$ and half-spaces $\{x \in \mathbf{R}^3, \pm x > 0\}$, respectively. Suppose that $\mathbf{R}^3_-$ corresponds to the Earth atmosphere with constant electromagnetic parameters $\varepsilon_0 > 0$, $\mu_0 > 0$, $\sigma_0 = 0$. We shall describe the EM waves propagation in $\mathbf{R}^3_-$ by the following system:

$$\begin{aligned} \varepsilon_0 \vec{E}_t &= \vec{\nabla} \times \vec{H}, \\ \mu_0 \vec{H}_t &= -\vec{\nabla} \times \vec{E}, \\ \vec{\nabla} \circ (\mu_0 \vec{H}) &= 0. \end{aligned} \tag{1}$$

Likewise, $\mathbf{R}^3_+$ corresponds to the Earth's crust where the interaction processes of elastic and EM waves can be observed. These processes will be simulated by the following equations:

$$\begin{aligned} \mu \vec{H}_t &= -\vec{\nabla} \times \vec{E}, \\ \sigma \vec{E} &= \vec{\nabla} \times \vec{H} + \sigma\mu \vec{H}^0 \times \vec{u}_t, \\ \vec{\nabla} \circ (\mu_0 \vec{H}) &= 0, \end{aligned} \tag{2}$$

$$\rho u_{itt} = \sum_{j=1}^{3} \frac{\partial T_{ij}}{\partial x_j} T_{ij}(\vec{u}), \quad i = 1,2,3. \tag{3}$$

Here we use the following notations: $\vec{\nabla} = (\frac{\partial}{\partial x_1}, \frac{\partial}{\partial x_2}, \frac{\partial}{\partial x_3})$, $\vec{H}^0 = (H_1^0, H_2^0, H_3^0)$ is a constant vector; $\vec{E} = (E_1, E_2, E_3)$ and $\vec{H} = (H_1, H_2, H_3)$ are tension vectors of electric and magnetic fields; $\vec{u} = (u_1, u_2, u_3)$ is a displacement vector; $T(\vec{u}) = (T_{ij}(\vec{u}))$, $i,j = 1,2,3$, where $T_{ij} = 2æS_{ij}(\vec{u}) + \delta_{ij}\lambda\vec{\nabla} \circ \vec{u}$, is a stress tensor, $S_{ij}(\vec{u}) = \frac{1}{2}\left(\frac{\partial u_j}{\partial x_i} + \frac{\partial u_i}{\partial x_j}\right)$, $i,j = 1,2,3$ is a deformation tensor; $\sigma = \sigma(x_3)$ and $\mu = \mu(x_3)$ describe the electric conductivity and permeability of the Earth's crust, respectively; $\lambda = \lambda(x_3)$, $æ = æ(x_3)$ are the Lame parameters; $\rho = \rho(x_3)$ stands for the medium density. Equations (2), (3) were derived from the general model equations of magneto elasticity Novatskii (1986) by linearizing them in a neighborhood of the constant solution and neglecting free charges. We assume the following interface conditions to hold on the surface $x_3 = 0$:

$$\begin{aligned} \mu H_k \,|_{x_3=+0} - \mu_0 H_k \,|_{x_3=-0} &= 0, \\ E_k \,|_{x_3=+0} - E_k \,|_{x_3=-0} &= 0, \end{aligned} \tag{4}$$

where $k = 1,2$.

The boundary conditions for the elasticity system are as follows:

$$T_{k3}(\vec{u}) \,|_{x_3=0} = \delta_{k3}\delta(t, x_1, x_2), \quad k = 1,2,3, \tag{5}$$

where $\delta(\cdot)$ is the Dirac delta-function concentrated at the origin. As to the initial data, it is natural to assume that both of the EM and the elastic fields are absent up to the moment $t = 0$:

$$(\vec{E}, \vec{H}, \vec{u}) \mid_{t<0} \equiv 0. \tag{6}$$

To provide a unique solution to the problem, we should add the radiation condition at infinity:

$$\vec{E} \mid_{|x| \to \infty} \to 0, \quad \vec{H} \mid_{|x| \to \infty} \to 0. \tag{7}$$

The direct problem is to find the vectors $\vec{E}$, $\vec{H}$ satisfying (1)-(7) if the medium parameters are known. The inverse problem is to determine the elastic and the EM parameters from (1)-(7) using additional information on the behavior of some components of the vectors $\vec{u}$, $\vec{E}$ on the Earth surface $x_3 = 0$.

A particular case of the above problem with specified additional information will be considered below.

2. Formulation of a Simplified Problem

Note that the problem for Lame and Maxwell equations, similar to (1)-(7), was studied in Lorenzi and Romanov (1992). In this paper we focus our attention on one of the simplest versions of (1)-(7) which, at the same time, preserves the basic properties of more adequate models. Consider the following system:

$$\begin{gathered} u_{tt} = V^2(z) u_{zz}, \quad (z,t) \in \mathbf{R}^1_+ \times \mathbf{R}, \\ E_t = c^2(z) E_{zz} + \mu H u_{tt}, \\ u_z \mid_{z=0} = b\delta(t), \\ \left(E_z - c_0 \left[\frac{c_0}{c} \right]^2 E_t \right) \mid_{z=0} = 0, \\ E \mid_{z \to \infty} \to 0, \\ (E, u) \mid_{t<0} = 0. \end{gathered} \tag{8}$$

Here $u = \text{Re } F_{x_1x_2}(u_3) \mid_{\nu_1=\nu_2=0}$, $E = \text{Re } F_{x_1x_2}(E_1) \mid_{\nu_1=\nu_2=0}$, where E_1 is the first component of the tension vector for the electric field; u_3 is the third component of the displacement vector; $F_{x_1x_2}$ is the generalized Fourier transform operator with respect to x_1, x_2, and ν_1, ν_2 are variables dual to x_1, x_2; $V(z) = \left(\frac{\lambda + 2æ}{\rho}\right)^{1/2}$ describes the longitudinal seismic wave velocity; $c(z) = (\sigma\mu)^{-1/2}$ characterizes the velocity of propagation of diffusion processes for the EM waves in the Earth's crust; $c_0 = (\varepsilon_0\mu_0)^{-1/2}$ simulates the velocity of the EM waves propagation in the Earth atmosphere; H is a constant which depends on the Earth EM field; $z = x_3$ stands for a spatial variable which characterizes the depth; $b = [\lambda(0) + 2æ(0)]^{-1}$.

Equations (8) are derived from (1)-(7) using the Fourier transform. Note that, due to the boundary conditions (5), we can isolate equations including E_1 and u_3 only. Moreover, we consider equations (8) for $z > 0$ only because for $z < 0$ the coefficients ε_0, μ_0 are supposed to be known. The domain $z < 0$ is included in (8) via the boundary conditions which use the parameter c_0. For further details we refer readers to Yakhno (1990). It was shown that under some assumptions only one component of the solution to the Lame system is non-trivial.

System (8) describes the EM field in the case when it is activated by elastic oscillations generated by an instantaneous vertical point impulse.

Consider the following class of piecewise constant functions.

DEFINITION. The functions $c(z)$, $V(z)$ are said to belong to the class $\mathcal{M}$ if

i). there exist sets c_m, V_m, z_n, z'_n ($m = 1, \dots, k+1$, $n = 1, \dots, k$) of positive numbers such that

$$c(z) = \begin{cases} c_m, \text{for } z \in (z_{m-1}, z_m), & m = 1, \dots, k, \\ c_{k+1}, \text{for } z > z_k, \end{cases}$$

$$V(z) = \begin{cases} V_m, \text{for } z \in (z'_{m-1}, z'_m), & m = 1, \dots, k, \\ V_{k+1}, \text{for } z > z'_k, \end{cases}$$

where $z_0 = z'_0 = 0$;

ii). there exist positive constants τ_V, τ_c such that

$$\tau_V = \frac{z'_1 - z'_0}{V_1} = \frac{z'_2 - z'_1}{V_2} = \dots = \frac{z'_k - z'_{k-1}}{V_1},$$

$$\tau_c = \frac{z_1 - z_0}{c_1} = \frac{z_2 - z_1}{c_2} = \dots = \frac{z_k - z_{k-1}}{c_1}, \tag{9}$$

As we consider the piecewise constant coefficients c, V, we should add to (8) the interface conditions at the discontinuity points, namely:

$$[u] \,|_{z=z'_m} = [V^2 u_z] \,|_{z=z'_m} = 0,$$

$$[E] \,|_{z=z_m} = [c^2 E_z] \,|_{z=z_m} = 0. \tag{10}$$

Now we can proceed to the formulation of the problem.

Inverse problem 1 (IP1). Given functions $u_0(t)$, $E_0(t)$ and constants τ_V, τ_c, find such coefficients $c(z)$, $V(z) \in \mathcal{M}$ (i.e. number sets $\{V_k, c_k, z_k, z'_k\}$) that

$$u \,|_{z=0} = u_0(t), \quad E \,|_{z=0} = E_0(t), \tag{11}$$

where u, E solves equations (8).

3. Inverse Problem for a System of Ordinary Differential Equations

Before proceeding to a description of an approximate algorithm for IP1, let us consider a more simple problem. We modify system (8) replacing E_t by E_{tt}, i.e., we will consider in (8) the following equation for E :

$$E_{tt} = c^2(z) E_{zz} + \mu H u_{tt}.$$

To solve the modified inverse problem, we use the formal Fourier transform with respect to t. In the sequel we will use the same Fourier images u, E of the original functions as above. For convenience, let us decompose our problem into two ones:

$$u_{zz} + \omega^2 V^{-2} u = 0,$$

$$u_z(0,\omega) = h_u(\omega), \quad u_z - \mathrm{i}V^{-1}\omega u \to 0, \; z \to \infty, \tag{12}$$
$$[u]\,|_{z=z'_m} = [V^2 u_z]\,|_{z=\_m'} = 0, \quad m = 1,\dots,k;$$
$$E_{zz} + \omega^2 c^{-2} E = \omega^2 c^{-2} \mu H u,$$
$$E_z(0,\omega) = h_E(\omega), \quad E_z - \mathrm{i}c^{-1}\omega E \to 0, \; z \to \infty, \tag{13}$$
$$[E]\,|_{z=z_m} = [c^2 E_z]\,|_{z=z_m} = 0, \quad m = 1,\dots,k.$$

(here $h_u(\omega)$ is the Fourier image of the corresponding boundary condition on u in (8), and the function $h_E(\omega)$ will be defined below).

Problems (12), (13) can be solved consequently. Consider the following problem.

Inverse problem 2 (IP2). Given a function $\Phi_u(\omega)$, find such a coefficient $V(z) \in \mathcal{M}$ that

$$u(o,\omega) = \Phi_u(\omega), \tag{14}$$

where $u(z,\omega)$ solves equations (12).

Upon solving IP2, we can determine the coefficient $c(z)$ by taking into consideration the fact that the right-hand side of equation (13) is known.

Thus, we have the following problem.

Inverse problem 3 (IP3). Given functions $\Phi_E(\omega)$ and $u(z,\omega)$, find such a coefficient $c(z) \in \mathcal{M}$ that

$$E(0,\omega) = \Phi_E(\omega), \tag{15}$$

where $E(z,\omega)$ solves equations (13).

REMARK. As $E(0,\omega)$ is supposed to be known, $E_t(0,t)$ could be considered as a given function. So, under the assumption that we know the constant c_1, the function $h_E(\omega)$ in (13) is the Fourier image of the quantity $c_0(c_0/c_1)^2 E_t(0,t)$ which can be calculated.

In fact, we have decomposed our original IP1 into IP2 and IP3.

As was shown in Lavrent'ev (1991)., one can solve IP2 recursively, namely:

1. determine the number sets

$$\varphi_m = \int\limits_{\omega_0}^{\omega_0+\pi/\tau_V} h_u^{-1}(\omega)\Phi_u(\omega)\exp\{-2\mathrm{i}m\omega\tau_V\}\mathrm{d}\omega, \tag{16}$$

$$h_n^m = h_{m-1}^m h_{m-n-1}^{m-1} + h_n^{m-1}, \quad (n = 0,\dots,m-2, \; m = 2,\dots,k) \tag{17}$$
$$h_{m-1}^m = -\gamma_{m-1}, \quad h_0^1 = h_m^m = 0,$$
$$\gamma_1 = \varphi_1/(2\varphi_0),$$
$$\gamma_m = (2\varphi_0)^{-1}\left(\varphi_m + \sum_{n=1}^{m-1} h_n^m \varphi_{m-n}\right) / \prod_{n=1}^{m-1}(1-\gamma_n^2), \tag{18}$$

where ω_0 is an arbitrary positive number;

2. find the following numbers

$$V_1 = \frac{\pi}{\tau_V \varphi_0},$$
$$V_{m+1} = \frac{1+\gamma_m}{1-\gamma_m} V_m. \tag{19}$$

Due to equations (9), one can calculate the numbers z'_m from the V_m computed, which completes the solution of IP2.

Let us consider an intersection $(z_{m-1}, z_m) \bigcap (z'_{n-1}, z'_n)$. As the right–hand side of equation (13) is of the form

$$u_n(z,\omega) = u_n^1 + u_n^2 = B_1^n \exp\{i\omega z/V_n\} + B_2^n \exp\{-i\omega z/V_n\},$$

we can use the following representation for the solution of (13):

$$E_m(z,\omega) = E_m^0 + E_m^1 + E_m^2. \tag{20}$$

Here

$$E_m^0 = A_1^m \exp\{i\omega z/c_m\} + A_2^m \exp\{-i\omega z/c_m\} \tag{21}$$

is the general solution of the homogeneous equation (13) and E_m^j are its partial solutions when u in the right–hand side of equation (13) is replaced by u_m^j, respectively.

These partial solutions are

$$E_m^j = r_m^j \exp\{(-1)^{j+1} i\omega z/V_n\}, \tag{22}$$

where (see equation (13))

$$r_m^j = -\frac{\omega^2 V_n^2 H \mu B_j^m}{\omega^2 (V_n^2 - c_m^2)} = -\mu H B_j^m \left(1 - (\frac{c_m}{V_n})^2\right)^{-1}. \tag{23}$$

It is of common knowledge that the ratio c_m/V_n is comparable with 10^{-1} for the majority of materials. So we neglect the term $(c_m/V_n)^2$ in (23) assuming

$$r_m^j = -\mu H B_j^m. \tag{24}$$

Thus, taking into consideration that the interface conditions (13) can be represented in the form of recursive relations on B_j^n (see Lavrent'ev, 1991), we will derive the same for the functions A_j^m from (13), (20)-(22), (24). So, the above algorithm solving IP2 is workable for IP3 also. The only difference is the way of using additional information (15). We have

$$E_1(0,\omega) = E_1^0(0,\omega) + E_1^1(0,\omega) + E_1^2(0,\omega) = A_1^1 + A_1^2 - \mu H(B_1^1 + B_2^1) = \Phi_E(\omega),$$

$$E_z(0,\omega) = i\omega c_1^{-1}(A_1^1 - A_2^1) - i\omega V_1^{-1}\mu H(B_1^1 - B_2^1) = h_E(\omega).$$

As

$$B_1^1 + B_2^1 = \Phi_u(\omega),$$

$$i\omega V_1^{-1}(B_1^2 - B_2^1) = h_u(\omega)$$

(see Lavrent'ev, 1991), the algorithm for IP3 is as follows:

1. determine the numbers

$$\varphi'_m = \int_{\omega_1}^{\omega_1+\pi/\tau_c} (h')^{-1}(\omega)\Phi'(\omega)\exp\{-2im\omega\tau_c\}d\omega, \tag{25}$$

   where ω_1 is a positive number and

$$h'(\omega) = h_E(\omega) + \mu H h_u(\omega), \tag{26}$$

$$\Phi'(\omega) = \Phi_E(\omega) + \mu H \Phi_u(\omega); \tag{27}$$

2. find

$$c_1 = \frac{\pi}{\tau_c \varphi_0'},$$

$$c_{m+1} = \frac{1+\gamma_m}{1-\gamma_m} c_m, \tag{28}$$

where γ_m, h_m^n are given by formulae (17), (18) with φ_m replaced by φ_m'. As mentioned above, one can calculate the quantities z_m (see (9)) from the c_m obtained.

Finally, IP1 can be solved using (16)–(19) and (25)–(28).

REFERENCES

Alers G. A. and Fleury P. A. (1963), Modification of the velocity of sound in metals by magnetic field. *Phys. Rev.*, vol. 129, no. 6, 2435.

Balakirev M. K. and Gilinskii I. A. (1982), *Waves in the "Piezo" Crystals.* "Nauka", Novosibirsk. (in Russian)

Dunkin J. W. and Eringen A. C. (1963), On the propagation of waves in an electromagnetic elastic solid. *J.Engn. Sci.*, vol. 1, no. 4, 461.

Keilis–Borok V. I. and Monin A. S. (1959), Magnetoelastic waves on the boundary of the Earth core. *Izvestiya Akademii Nauk SSSR, ser. geofizika*, no. 11, 1529–1541. (in Russian)

Knopoff L. (1955), The interaction between elastic wave motions and a magnetic field in electrical conductors. *J. Geophys. Res.*, vol. 60, no. 4, 441.

Landau L. D. and Lifschitz E. M. (1959), *Electrodynamics of Continuous Media.* "Fizmatgiz", Moscow. (in Russian)

Lavrent'ev M. M., Jr. (1991), An inverse problem for a wave equation. *Calcolo*, vol. 28, no. 3-4, 249–265.

Lorenzi A. and Romanov V. G. (1992), *Identification of an Electromagnetic Coefficient Connected with Deformation Currents.* Universita degli Studi di Milano Quaderno, no. 16.

Novatskii V. (1986), *Electromagnetic Effects in the Solid Bodies.* "Mir", Moscow.

Rakhmatulin Kh. A. and Shkenev Yu. S. (1985),*The Interaction Between Media and Waves.* "FAN", Tashkent. (in Russian)

Yakhno V. G. (1990), *Inverse Problems for the Differential Equations of Elasticity.* "Nauka", Novosibirsk. (in Russian)

A Modification of ART Method for Cone-Beam Tomography of High Space Resolution

A.V. LIKHACHOV and V.V. PICKALOV

Institute of Theoretical and Applied Mechanics,
Siberian Branch of Russian Academy of Sciences
Novosibirsk 630090, Russia

ABSTRACT

The possibilities of solving a cone-beam tomography problem by algebraic methods are considered. A modified ART is suggested for tomography of high space resolution.

1. Introduction

Three-dimensional (3-D) tomographic algorithms seem to be more preferable for an investigation of inner structure of a 3-D object than the use of the set of independent two-dimensional reconstructions of some its layers. Moreover, the latter way imposes so severe restrictions on the data collection equipment that it fails in some explorations. Thus, the further development of 3-D reconstruction techniques is needed. Nowadays there is a large number of papers devoted to 3-D tomography problems. Some of them have exact inversion formulas as a subject (Tuy, 1983; Grangeat, 1990; Smith, 1990). Others deal with the finite rows expansion methods leading, as a rule, to solving large linear systems (Altschuler and Herman, 1977; Kowalski, 1977; Herman, 1980). For solving the systems, special iterative methods were developed. The algebraic reconstruction technique (ART1) which is in the scope of this paper belongs to these methods (Censor, 1983). Its applications to tomography problems one can find in the papers by Herman et al. (1973), Schlindwein (1978), and Herman (1980).

It is clear that 3-D tomography computer algorithms demand big arrays for the projection data and the function to be reconstructed. These arrays become really terrible if one wants to increase the space resolution of the tomogram. For example, if the area of reconstruction is divided into 512x512x512 voxels, then only the 3-D solution needs 128 Mbite of computer memory (if the solution is presented by integer*1). It is obvious that under such conditions some part of the data has to be placed in the external memory. As will be seen bellow, ART1 is inconvenient when using the external memory. We suggest some modification of ART1 which helps to avoid this drawback.

2. ART1 and its Modification

Consider a system of linear algebraic equations:

$$Ag = f, \tag{1}$$

where A is an $I \times J$ matrix, $g \in R^J$, $f \in R^I$. In computer tomography g and f are the vectors of the function to be reconstructed and of the projection data, respectively. The matrix A is a discretized approximation of the projecting operator (Herman, 1980). Let us solve system (1) by the ART1 method. Following Herman (1980), the $k+1$–th iterative solution in ART1 is

$$g^{(k+1)} = g^{(k)} + \lambda^{(k)} \frac{f_{i(k)} - (a^{i(k)}, g^{(k)})}{\|a^{i(k)}\|^2} a^{i(k)}, \quad \|a^{i(k)}\| \neq 0. \tag{2}$$

Here $a^{i(k)}$ is the $i(k)$–th row of the matrix A, $\lambda^{(k)}$ is a relaxation parameter, $i(k) = [k(mod\ I) + 1]$. The scalar product and Euclidean norm in R^J are defined in a usual way:

$$(a^{i(k)}, g^{(k)}) = \sum_{l=1}^{J} a_l^{i(k)} \, g_l^{(k)}, \quad \|a^{i(k)}\|^2 = (a^{i(k)}, a^{i(k)}). \tag{3}$$

As was shown in Herman (1980), the iterative process (2) converges if $0 < \lambda \leq 2$ for an arbitrary first guess $g^{(0)} \in R^J$. If, in addition, $g^{(0)}$ belongs to the linear span of the vectors $a^{(i)}, i = 1, ..., I$, then this process converges to the solution of system (1) with the least norm if the system is consistent and to the general solution in the case of its inconsistency.

As is seen from (2), all the components of g are to be handled at each iteration. Suppose now that the vector g is so large that it exceeds the computer memory. Evidently, it is necessary to store g in an external file. Using formula (2) in such a case leads to a great number of external exchanges, which in its turn causes the computation time to increase sharply. ART1 is suggested to be modified for only one section of the matrix $g(x, y, z)$ to be stored in computer memory, say, the section $x = const$.

Let us mention that the iterations for which $(a^i, a^j) = 0$ are independent $(nI + 1 \leq i, j \leq (n+1)I, n = 0, 1, \ldots)$. Indeed, at the i–th iteration there change only those elements g_l of g for which $a_l^i \neq 0$. The condition $(a^i, a^j) = 0$ means that there is no l for which $a_l^i \neq 0$ and $a_l^j \neq 0$ simultaneously. Therefore, different elements of the vector g are in processing at the i–th and the j–th iteration. This means that such iterations are independent one from another.

We modified the ART1 algorithm having in mind the high space resolution problem which demands I and J to be large. We suggest to handle simultaneously the iterations $i_1, i_2, ... i_M$ which satisfy the following conditions:
(a) all $i_1, i_2, ... i_M$ belong to one of the intervals $[nI + 1, (n+1)I]$, $n = 0, 1....$,
(b) the scalar products (a^p, a^q) for $\forall p, q \in \{i_1, i_2, ... i_M\}$ are small enough.
The former condition is obvious. The latter is based on the mentioned above. In the next section we illustrate an application of the modified ART to a cone-beam tomography problem where the above sets of iterations appear.

3. An Application of the Modified ART to a Cone-Beam Tomography Problem

The cone-beam problem was taken as in Grangeat (1990). The function $g(x, y, z)$ is reconstructed inside a finite domain $\Omega \subset R^3$ from a finite set of its conic projections $f_m(u, v), m = 1, ..., M$. To explain the definition of a conic projection, let us turn to Fig. 1 where a usual scheme of data collection for transmission tomography is presented. In this case $g(x, y, z)$ describes a density distribution of the object to be tested. The sources of penetrating radiation are placed at M of points S_m with spherical coordinates (L_m, θ_m, ϕ_m)

($M = 1$ in Fig. 1). A planar detector corresponds to each source. The detector front surface is in the plane Pd_m defined by the equation $x\alpha_m = d_m, \alpha_m, x \in R^3$, the spherical coordinates of α_m are $(1, \theta_m + \pi/2, \pi - \phi_m)$. The function $f_m(u,v)$ is connected with the total absorption and can be transformed easily to integrals of the density distribution (Herman, 1980). A coordinate system (O_1, u, v) on the plane Pd_m is determined so that its origin is an orthogonal projection of the point S_m on this plane.

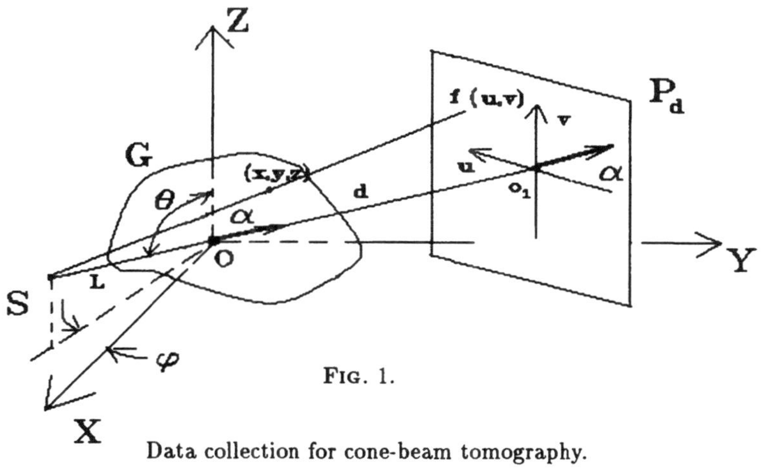

FIG. 1.

Data collection for cone-beam tomography.

A centered at the origin cube of side length 2 is considered as a reconstructed region. It is divided into cubic voxels. The density in each voxel is proposed to be constant. The detector supposed as a square of side length 2 is digitized to square pixels too. The penetrating radiation is represented by the rays starting from a source point and passing through the pixels' tops on the corresponding detector plane. An element a_l^p of the matrix A in (1) is defined as the path length of the i-th ray inside the j-th voxel, $i = 1, ..., I, j = 1, ..., J$. Then

$$(a^p, a^q) = \sum_{l,m}^{J} a_l^p \, a_m^q \, \delta_{lm} \quad , \tag{4}$$

where $\delta_{lm} = 1$ for $l = m$ and $\delta_{lm} = 0$ for $l \neq m$.

As is obvious from simple geometry (see Fig. 1), the scalar product (4) is small for the rays starting from the same source. So, the iterations related to such rays form the set defined in the previous section. Consequently, one can handle all the data from each projection simultaneously.

4. Numerical Simulations

To study the performance of the modified ART (named as mART bellow), some numerical simulations were carried out. The cone-beam tomography problem described in the previous section was solved by mART and by "classical" algorithms of matrix inversion used in computerized tomography: ART1 presented by formula (2), MART and SIRT. The noise influence on performance of the algorithms was also investigated. In this connection, some procedures of regularization were involved and the dependence of reconstruction quality on the relaxation parameter λ was considered.

Two different types of density distribution have been reconstructed. The first consists of the functions of $C^\infty(\mathrm{R}^3)$ which approximate diffusive objects such as plasma. The second is presented by discontinuous functions with compact support corresponding to the hard body model. Two schemes of projecting geometry have been used. For simple phantoms the sources were placed equidistantly on a circle lying in the (x, y)−plane and centered at the origin. The radius of the circle was 5, the number of sources was 10 (Scheme 1). For more complicated models the same sources distribution in the (x, z)−plane was added to the previous scheme (Scheme 2). The object to be reconstructed was supposed to be inside the unit sphere. The line integrals of the model function $g(x, y, z)$ were calculated analytically to obtain the projections $f_m(u, v)$. The reconstructed domain was sampled into 44x44x44 voxels. For mART the 70x70x70 sampling was also used. The grids on the detectors corresponded to those on the reconstructed area, i.e., the detectors were divided into 44x44 and 70x70 pixels. The mean square error Δ was used to evaluate the quality of reconstruction:

$$\Delta = \left(\frac{\sum_{j=1}^{J} (g_j^M - g_j^{(k)})^2}{\sum_{j=1}^{J} (g_j^M)^2} \right)^{\frac{1}{2}} \times 100\% \quad , \tag{5}$$

where g^M and $g^{(k)}$ are the densities of the j−th voxel of the model and of the reconstructed function, respectively. To make the following analysis more transparent, we change the definition of the iteration used in Section 2. Now we name as an iteration the cycle where all projection data are processed at once. Thus, such a "big" iteration corresponds to the I of the former ones.

4.1 ART1

Here the results of the performance of ART as applied to a cone-beam tomography problem are presented. The special attention is paid to the reconstruction from noised projection data and, hence, to the use of regularizing procedures. As expected, for the simulations carried out, the best results have been obtained for the reconstruction of spherically symmetrical Gaussian function of half-width 0.3 (Model 20) from ideal (non-noised) projections. In Fig. 7 the results of reconstruction of Model 20 by both of the algorithms ART and mART for a 44x44x44 sampling and projection Scheme 1 are given. Curves 1-2 relate to ART1 with $\lambda = 1., 0.5$, respectively.

As an example of the reconstruction of more complex phantom, let us consider the reconstruction of Model 25 for a 44x44x44 sampling and Scheme 2. The model consists of three Gaussians, its parameters are given in the following table.

N	x_0	y_0	z_0	α	β	γ	ampl
1	0	0.4	0	0.15	0.15	0.4	1.0
2	0	-0.3	0.3	0.5	0.2	0.2	1.5
3	0	-0.3	-0.5	0.1	0.1	0.1	0.5

The cross-section of Model 25 by the plane $x = 0$ is presented in Fig. 2. In Fig. 3 the same section of the image reconstructed from the ideal projections by ART1 with $\lambda = 0.1$ after 10-th iteration is given (the number of projections is 20). As is seen from Fig. 3, the smoothing is very essential to improve the reconstruction quality. In Fig. 4 a smoothed reconstruction is represented. Here the reconstructed image is averaged in the 3x3x3 moving window after each iteration. The mean square errors for the tomograms presented in Fig. 3 and 4 are 29.24% and 14.65%, respectively.

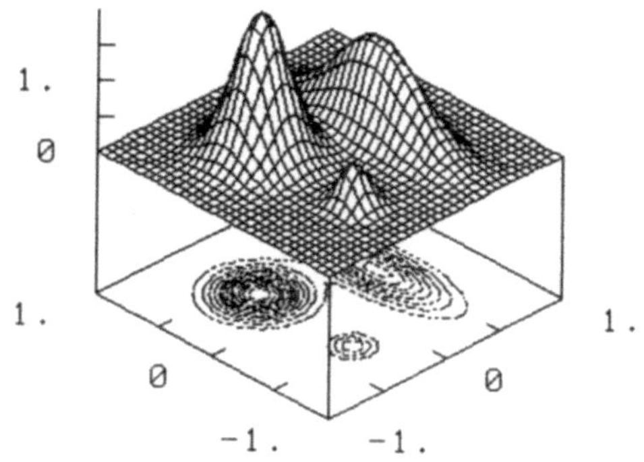

FIG.2.

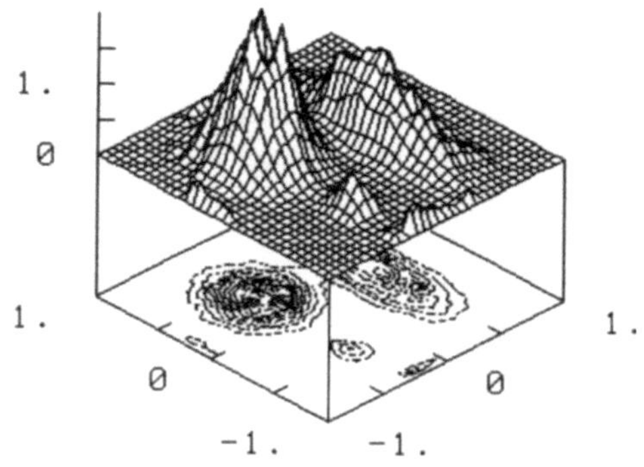

FIG.3.

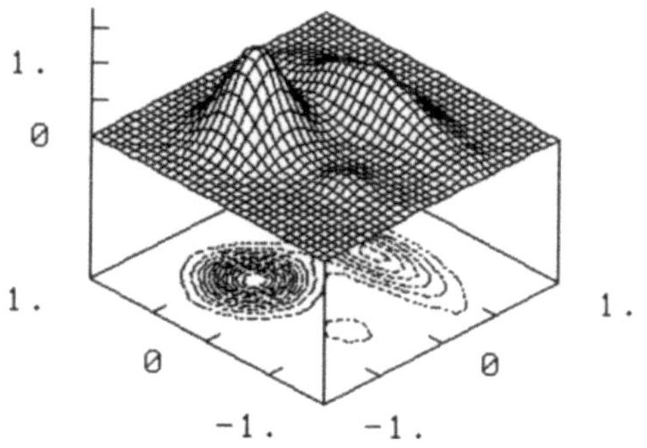

FIG.4.

Cross-sections of 3D-images by the plane $x = 0$.

FIG.2. Model 25.

FIG.3. A reconstruction of Model 25 by ART1.

FIG.4. A smoothed reconstruction of Model 25 by ART1.

To check the resistance of the algorithm to input data errors, two models of accidental noises in projection data were used. The first is the normal distributed noise with zero mean and varying disperse achieving 5% of the signal value at a considered point (type 1). The second is the same, except that the disperse is constant and equal to 5% of the maximal signal value on a projection (type 2). In Fig. 5 the reconstruction results of Model 25 from noised data by ART1 are given. Curves 1-3 present the results for the reconstruction from the ideal projections and from the projections distorted by noises of types 1 and 2, respectively. Curves 4-6 illustrate the reconstruction from the same data, but the smoothing described above is used. One can observe an unexpected result, namely, the smoothed reconstruction by ART1 depends on noise very weakly, if at all (at least for the chosen phantom).

Some interesting results are obtained by varying the set of iterations where the smoothing is used. Fig. 6 demonstrates the dependence of Δ on the iteration number under different smoothing conditions. In Fig.6 the results of reconstruction of Model 25 by ART1 from the projection data distorted by noises of type 2 are represented. Curve 1 corresponds to the averaging after each iteration. Curve 2 illustrates the results obtained in averaging after 1-st, 11-th and every even iteration. Curve 3 deals with the averaging after 1-st, 5-th, 10-th and so on. A more complicate sequence of "smoothing" iterations is presented by Curve 4. They are 1-st, 5-th, 10-th, 12-th, 14-th, 16-th and all the following. As is seen from Fig. 6, it is possible to improve the reconstruction quality by varying a sequence of "smoothing" iterations.

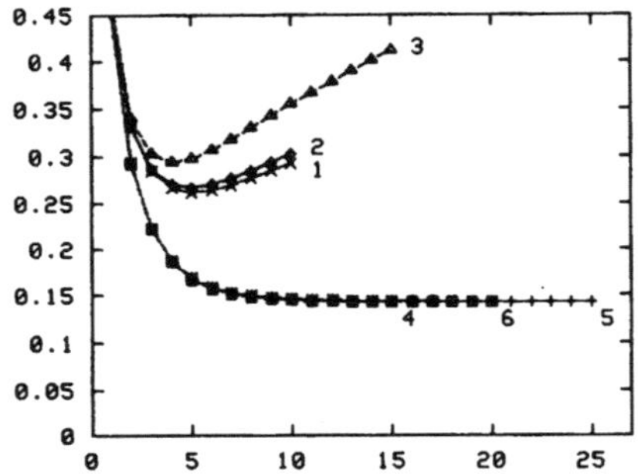

Fig.5

The performance of ART1. Noised projection data, smoothing.

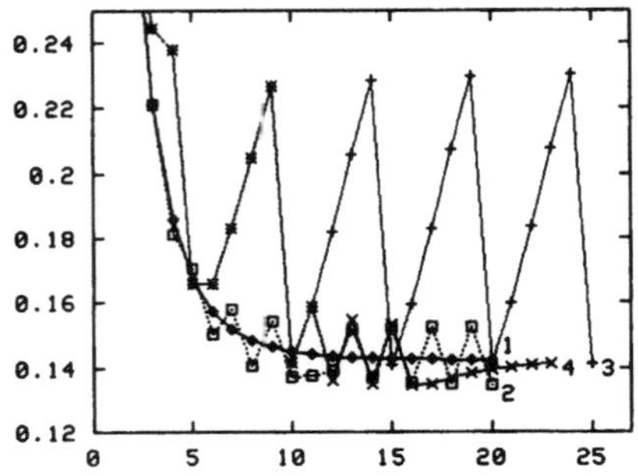

Fig.6

The performance of ART1 under different smoothing conditions.

4.2 mART

As far as mART is a modification of ART1, their behavior has to be similar. The multiple numerical simulations support this supposition. In particular, both of the algorithms have the same response to the smoothing described in the previous section. But their behavior differs when varying the relaxation parameter λ. The illustration is given in Fig. 7. In Fig. 7 the results of reconstruction of Model 20 (a spherical Gaussian function of half-width 0.3) from 10 ideal projections (Scheme 1) by ART and mART for a 44x44x44 sampling are given. Curves 1-2 relate to ART1 with $\lambda = 1., 0.5$, respectively. Curves 3-5 illustrate the reconstruction of Model 20 by mART for $\lambda = 1., 0.5, 0.25$, respectively.

In Fig. 8 the results of performance of mART for a 70x70x70 sampling are presented. Here Model 20 is reconstructed for Scheme 1. Curves 1-3 correspond to $\lambda = 1., 0.5, 0.25$, respectively.

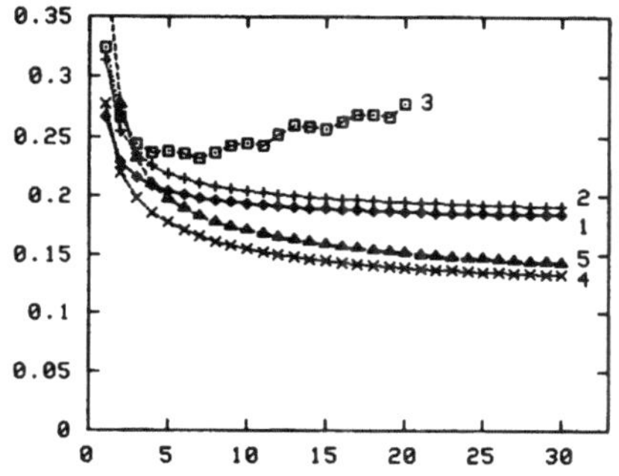

FIG.7.

A comparison of mART and ART1.
A 44x44x44 sampling.

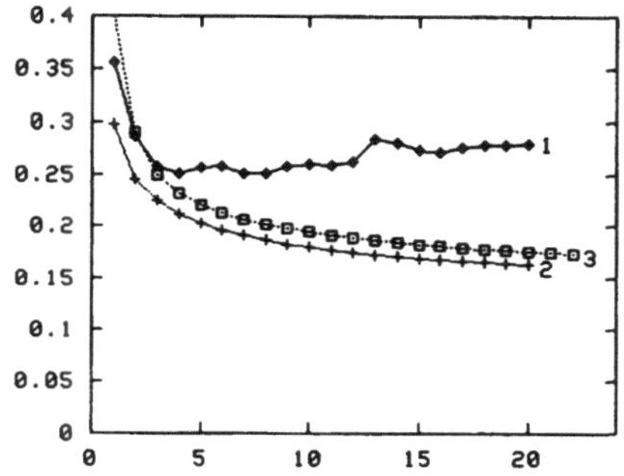

FIG.8.

The performance of mART.
A 70x70x70 sampling.

4.3 A comparison of ART1, MART and SIRT

It is interesting to compare the performance of ART1 in reconstructing an object from its conic projections with that of the algorithms MART and SIRT which are widely used in computerized tomography.

Following Censor (1983), the $k+1-$th iterative solution in MART is

$$g_j^{(k+1)} = \left(\frac{f_{i(k)}}{(a^{i(k)}, g^{(k)})} \right)^{\lambda^{(k)} a_j^{i(k)}} g_j^{(k)} . \tag{6}$$

Here the first guess is chosen as $g^{(0)} = e^{-1}\vec{1}$ where e is the base of the natural logarithm, $\vec{1}$ is the unit vector.

The algorithm SIRT is representable in the form

$$g^{(k+1)} = (E - \lambda A^* A) g^{(k)} + \lambda A^* f , \tag{7}$$

where E is the unit operator, A^* is an operator conjugate to the operator A.

Using Fig. 9-10, one can compare the performance of the algorithms ART1, MART and SIRT. Fig. 9 concerns the reconstruction from ideal projections. Curves 1-3 illustrate the reconstruction of Model 25 by ART1, MART and SIRT, respectively. Curves 4-6 present the same reconstruction but the smoothing is involved after each iteration. The results of reconstruction from the projections distorted by noise of type 2 are presented in Fig. 10. Here, Curves 1-3 are the results obtained by ART1, MART and SIRT with the smoothing after each iteration. Curves 4-6 are the performance of ART1, MART and SIRT with the smoothing after 1-st, 11-th and each even iteration.

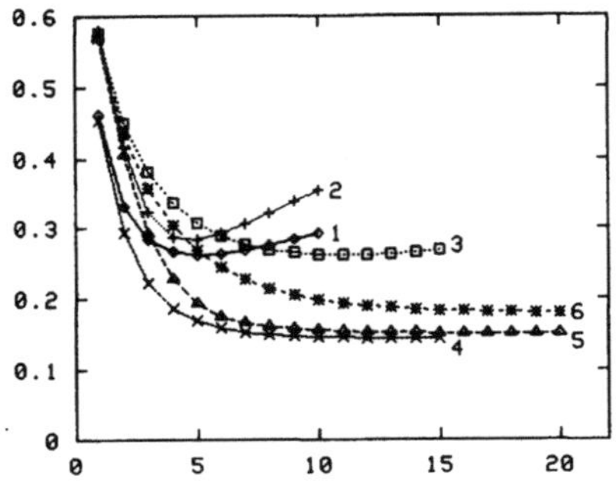

FIG.9.

A comparison of ART1, MART, and SIRT. Ideal projection data.

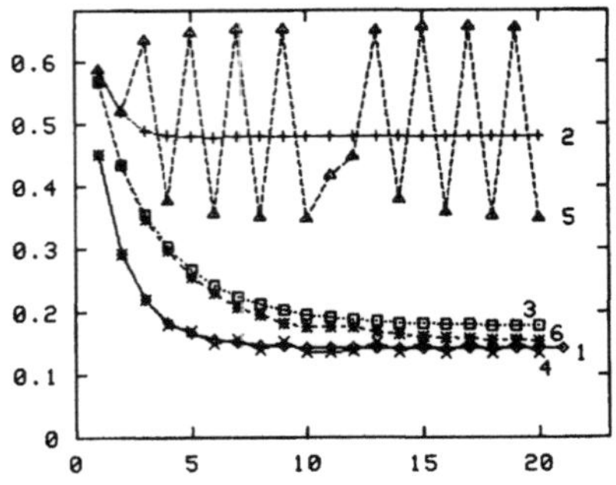

FIG.10.

A comparison of ART1, MART, and SIRT. Noised projection data.

From Fig. 9 one can conclude that all the algorithms are comparable for ideal data. For the considered example ART1 provides the best results (in sense of the minimum of Δ). If we turn to the distorted data (Fig. 10), we find the performance of MART to be significantly worse than that of the two other algorithms. The main conclusion that we got by comparing the algorithms is that MART diverges fastest (in sense of increasing Δ) and is non-resistant to noises; on the contrary, SIRT diverges most slowly.

Finally, we comment on the computer simulation process. The computer codes written in Fortran-77 run on the IBM PC/AT486 under MS DOS.

5. Conclusions

The possibilities of solving a cone-beam tomography problem by algebraic methods are considered. In particular, a modification of ART algorithm is suggested. It allows us to process efficiently large data arrays which exceed the computer memory. The performance of ART1, MART, SIRT and the modified ART was tested in numerical simulations. The resistance of the algorithms to noises and the influence of smoothing on the reconstruction process were investigated. All the algorithms can be used in cone-beam tomography, but some additional regularizing procedures are needed for the reconstruction from real (noised) data. In the simulations carried out, ART1 and the modified ART produced the best results. Thus, a means to increase the space resolution in 3D-tomography was developed.

Acknowledgments

The work was supported in part by the Scientific Research Organization of the Netherlands, NWO, grant No. 713-097.

REFERENCES

Altschuler M.D. and Herman G.T. (1977), Fully three-dimension image reconstruction using series expansion methods. In: *A Review of Information Procession in Medical Imaging*. Oak Ridge National Lab., Oak Ridge, Tenn., 124–142.

Censor Y. (1983), Finite series-expansion reconstruction methods. *Proceedings of IEEE*. 71, 409–419.

Grangeat P. (1990), Mathematical framework of cone-beam 3D reconstruction via the first derivative of the Radon transform. In: *Mathematical Methods in Tomography*. Proceedings of conference, Oberwolfach, Germany, 5-11 June, 1990. Herman G.T., Louis A.K. and Natterer F. (Eds.). Springer-Verlag, 66–97.

Herman G.T. (1980), *Image Reconstruction from Projections: The Fundamentals of Computerized Tomography*. Academic Press, New-York.

Herman G.T., Lent A. and Rowland S.W. (1973), ART: Mathematics and Applications. *J. Teor. Biol.* 42, 1–32.

Kowalski G. (1977), Fast 3-D scanning systems using a limited titling angle. *Appl. Optics*. 16, 1686–1690.

Schlindwein M. (1978), Iterative three-dimensional reconstruction from twin-cone beam projections. *IEEE Trans. Nucl. Sci.* NS-25, 1135–1143.

Smith B.D. (1990), Cone-beam tomography: recent advances and tutorial review. *Opt. Eng.* 29(5), 524–534.

Tuy H.K. (1983), An inversion formula for cone-beam reconstruction. *SIAM J. Appl. Math.* 43(3), 546–552.

A New Approach to the Problem of 3D Interpolation from an Arbitrary Set of Points

A.V. LIKHACHOV and V.V. PICKALOV

Institute of Theoretical and Applied Mechanics,
Siberian Branch of Russian Academy of Sciences
Novosibirsk 630090, Russia

ABSTRACT

The problem of 3-D interpolation from an arbitrary set of points is considered. A new iterative regularizing algorithm for the interpolation is suggested.

1. Introduction

Many applications in physics require an interpolation of experimental data to more convenient regular grids. For example, the interpolation is very essential for performance of computer tomography algorithms which demand projection data to be defined on the regular sets of points (Herman, 1980). One of the main difficulties encountered in processing experimental data is a noise influence. Therefore, the interpolation of data requires some special procedures for noise suppress. Available techniques such as the smoothing-spline interpolation (Marchuk, 1977) solve the problem successfully for one- or two-dimensional cases. The further increase of the problem dimension leads to significant complication of formulae, which makes them useless for numerical computations. Moreover, the above techniques require experimental data to be defined on regular grids. The problem of interpolation from irregular grids is known to be far from its resolution.

In this paper we discuss a new approach to the interpolation problem which has an advantage over the traditional schemes, especially in multidimensional case and when interpolating from irregular grids. The method consists in using some regularizing algorithm for inversion of an interpolation operator. A procedure for constructing the interpolation operators is described in Section 2. The results of numerical simulations are presented in Section 3.

2. A New Approach to the Interpolation Problem

Let M be a finite set of points in a bounded domain $\Omega \subset R^3$ at which values of a limited function $f(x, y, z)$ of three variables are determined. The problem is to evaluate $f(x, y, z)$ at the nodes of a regular grid defined on Ω. Note that unless M is a set of nodes of any other regular grid, the traditional interpolation methods fail. Indeed, most interpolation algorithms need a regular structure of the data to be interpolated. These schemes interpolate functions defined on a regular grid to an arbitrary set of points but fail in the inverse order.

Suppose that μ is the number of elements of M and ν is the number of grid nodes. As was mentioned, one can easily construct an interpolation operator $A : R^\nu \to R^\mu$. Suppose A to be linear and limited. These restrictions aren't so severe, as might appear at first sight. Indeed, the latter is obvious. The former is satisfied for a wide class of interpolation schemes. For example, the interpolation by polinomials of power P determines some linear operator A. To construct A, consider the following scheme:

$$f_m = \sum_{j,\,k,\,l} \alpha_{jkl}\, x_m^j y_m^k z_m^l, \quad 0 \le j,k,l \le P, \quad j+l+k \le P. \tag{1}$$

Here f_m is the value of f at the m-th point of M, and x_m, y_m, z_m are its coordinates. Let i denote an ordered triple of integers j,k,l. The coefficients α_i are found from the values f_n at N nodes of a regular grid: $\alpha_i = \sum b_{in} f_n$. Substituting α_i into (1) and rearranging, we get $f_m = \sum a_{mn} f_n$. The a_{mn} are the elements of the matrix operator A.

We give a more detailed description of the matrix a_{mn} for the three-linear interpolation from a cubic grid. The three-linear interpolation is a development of the be-linear scheme. Suppose that the $m-$th point C of M is situated inside a cubic cell with the tops having the normalized coordinates $(q,r,s),...,(q+1,r+1,s+1)$ (Fig.1). The value f_m of the function f at this point is determined in the following way. The values f_m^1 and f_m^2 at the points C_1 and C_2 which are orthogonal projections of the point C on the planes $z = z_s$ and $z = z_{s+1}$, respectively, are calculated by the be-linear interpolation from the tops belonging to the corresponding planes. Then the usual linear interpolation from the points C_1 and C_2 is used to obtain f_m.

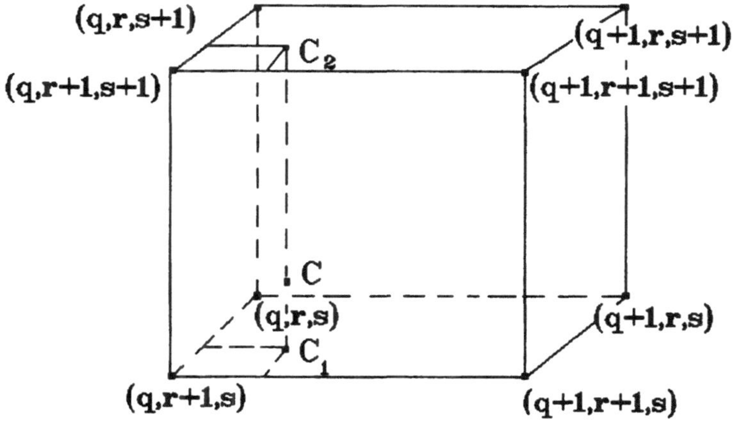

Fig.1

Three-linear interpolation scheme.

In such a case, the number of non-zero elements in each row of the matrix A does not exceed 8. If one chooses the numeration of grid nodes as $n = (s-1)RQ+(r-1)Q+q, 1 < q < Q, 1 < r < R, 1 < s < S$, the significant elements of the $m-$th row of the matrix A

are determined by the formulae:

$$\begin{aligned} a_{m,(s-1)RQ+(r-1)Q+q} &= (x_{q+1}-x_m)(y_{r+1}-y_m)(z_{s+1}-z_m) \\ a_{m,(s-1)RQ+(r-1)Q+q+1} &= (x_m-x_q)(y_{r+1}-y_m)(z_{s+1}-z_m) \\ & \dots \\ & \dots \\ a_{m,sRQ+rQ+q+1} &= (x_m-x_q)(y_m-y_r)(z_m-z_s) \ . \end{aligned} \tag{2}$$

Here the non-normalized coordinates are used. These formulae can be derived from (1) by the universal procedure described above, but it is more easy to obtain them using the three-linear interpolation.

So, we have the linear system

$$f_M = Af \ , \tag{3}$$

where f_M is a vector of values of the function defined at the points of the set M, f is a vector of values of the same function on the regular grid. A linear interpolation operator is defined as an $m \times n$ matrix A. One can solve equation (3) as a usual linear system. Let us define an operator A^{-1} as some regularizing algorithm for inverting the operator A. The operator A^{-1} can be regarded as the sought-for interpolation process.

3. Numerical Simulations

3.1 *A comparison with traditional interpolation methods*

The new interpolation algorithm was compared with traditional ones to assess its relative accuracy and performance. The zeroth-order and the three-linear interpolation algorithms were chosen for comparison because they are mostly used in 3-D case. The three-linear interpolation was described in detail in the previous section. The zeroth-order scheme is the simplest method for estimating the function values at the points where it is unknown. Namely, the value is assumed to be the same as at the nearest point where the function is determined.

The matrix A of the linear system (3) was constructed using formulae (2) for the three-linear interpolation. It is clear that the performance of the interpolation technique depends on the algorithm used for solving system (3). For this purpose the so-called ART method was chosen as quite simple and widely used in applications (Herman et al., 1973).

The spherical gaussian distribution g^0 of half-width 0.3 was taken to test the algorithms. We considered the interpolation of g^0 from a spherical 44x44x44 grid dividing the unit ball to a cubic 44x44x44 grid defined on the cube of side length 2 with the center at the origin. Two series of simulations were carried out. Initially the distribution g^0 was centered at the origin. Then it was shifted by 0.3 along the $z-$axis.

To evaluate the interpolation accuracy, we used the mean square error:

$$\Delta = \left(\frac{\sum_{i=1}^{\nu} (g_i^0 - g_i)^2}{\sum_{i=1}^{\nu} (g_i^0)^2} \right)^{\frac{1}{2}} \times 100\% \ , \tag{4}$$

where g_i^0 is the analytically calculated value of the model function g^0 at the $i-$th node of the cubic grid, g_i is the result of an interpolation into the same node.

In Table 1, the performance of the algorithms is given. The error Δ may have a minimum when using ART (Herman et al., 1973). Therefore, we present the minimum value of Δ for ART in Table 1 and below. The corresponding iteration number is given in brackets. The results of the ART-interpolation are tabulated for three different relaxation parameters. Here we give the CPU time per iteration for ART. All computations were performed on IBM PC/AT 486/66.

model		zeroth-order int.	three-linear int.	ART-interpolation		
				$\lambda = 0.5$	$\lambda = 0.1$	$\lambda = 0.05$
centered	Δ,	11.86	11.43	5.29(2)	4.38(5)	4.38(9)
shifted	%	13.03	12.67	9.50(3)	5.24(7)	5.11(11)
CPU time,s.		7.90	31.58	18.78		

Table 1:

Based on the three-linear interpolation, the suggested algorithm is seen from Table 1 to provide the mean square error almost three times less than it is in the three-linear interpolation. In addition, it is possible to decrease the CPU time for the ART-interpolation by choosing a suitable relaxation parameter λ.

3.2 *An application to a cone-beam tomography problem*

Here we consider an application of the new interpolation algorithm to a cone-beam tomography problem. When the cone-beam projection data are inverted by the technique suggested by Grangeat (1990), the interpolation from a special set of points to a regular grid is involved. The points of the set have the spherical coordinates $(S_m\mathbf{n}_k, \mathbf{n}_k)$ $m = 1, ..., M$, $k = 1, ..., K$, where $\mathbf{n}_k \in S^2$, and $S_m \in R^3$ is the point at which the m–th radiation source is located. These points do not determine any regular grid and, as was mentioned above, the traditional interpolation methods fail in this case.

To test the performance of the interpolation algorithms, the model functions with different properties have been used. There were smooth functions, or discontinuous ones, or functions of various symmetry classes. Both the three-linear interpolation and the zeroth-order scheme failed. Some results of the performance of the new algorithm are given in Table 2, where the value of Δ (in per cent) and the corresponding iteration number are represented for different relaxation parameters and model's types.

λ	0.50	0.25	0.10	0.05	0.01
model 1	34.19(2)	33.24(2)	32.74(3)	32.99(3)	36.18(5)
model 2	12.65(2)	10.31(2)	8.13(2)	7.60(2)	7.50(6)
model 3	11.28(10)	9.03(11)	7.67(14)	8.55(19)	17.70(40)

Table 2:

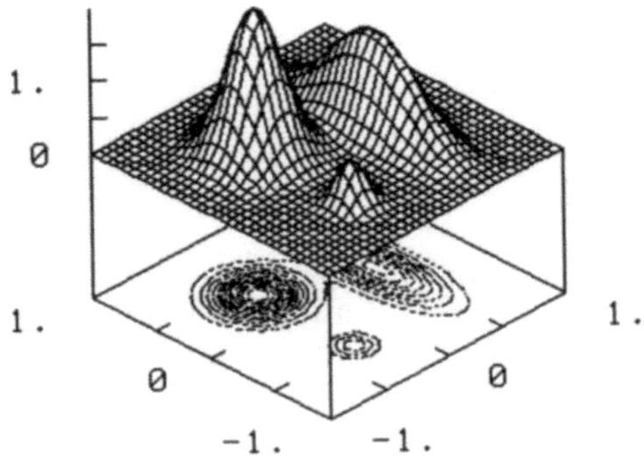

Fig.2

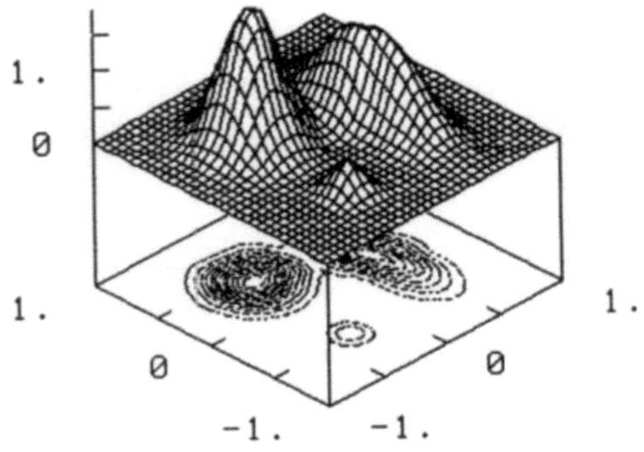

Fig.3

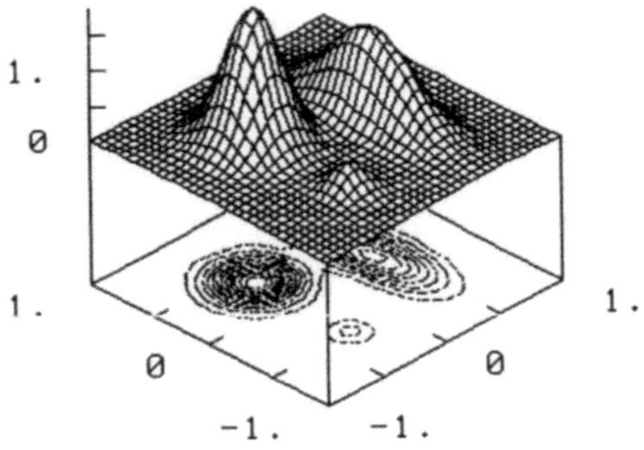

Fig.4

The cross-section of 3D-images by the plane $x = 0$.

Fig.2. Model 3.

Fig.3. The interpolation with $\lambda = 0.5$.

Fig.4. The interpolation with $\lambda = 0.05$.

Here Model 1 is determined by the characteristic function of a ball:

$$f(x,y,z) = 1 \quad , \quad |r| \leq 0.3 \ ,$$
$$f(x,y,z) = 0 \quad , \quad |r| > 0.3 \ ,$$

where $|r| = \sqrt{(x^2+y^2+z^2)}$. Model 2 is the spherical gaussian distribution of half-width 0.3 centered at the origin. Model 3 is determined by

$$f(x,y,z) = \sum_{i=1}^{3} a_i \exp\left(-\frac{\ln 2}{\alpha_i^2}(x-x_{0i})^2 - \frac{\ln 2}{\beta_i^2}(y-y_{0i})^2 - \frac{\ln 2}{\gamma_i^2}(z-z_{0i})^2\right) \ ,$$

where the parameters of the gaussian functions are given in the following table.

N	x_0	y_0	z_0	α	β	γ	a
1	0	0.4	0	0.15	0.15	0.4	1.0
2	0	-0.3	0.3	0.5	0.2	0.2	1.5
3	0	-0.3	-0.5	0.1	0.1	0.1	0.5

The following parameters were used in calculating the data for Table 2. The sources number M was 20. The sources were placed equidistantly on two perpendicular circles (each having ten sources) lying in the (x, y)- and (x, z)- plane and centered at the origin. The radii of the circles are equal to 5. For each source, the vector **n** runs $\mathbf{n} = (\sin\theta_{k1}\cos\phi_{k2}, \sin\theta_{k1}\cos\phi_{k2}, \cos\theta_{k1})$, $k1 = 1, \ldots, 256$, $k2 = 1, \ldots, 256$.

The data in Table 2 confirm the obvious supposition that discontinuous functions are not interpolated so well as smooth ones. One may get more interesting results by comparing the interpolation of symmetrical and non-symmetrical functions (Models 2 and 3). For spherically symmetrical functions the value of Δ which characterizes the reconstruction quality decreases monotonically with decreasing λ. For non-symmetrical models, the error Δ has a minimum in λ. It results from some properties of ART and, may be, from the symmetrical geometry of sources that has been chosen in modeling.

Fig. 2-4 demonstrate the reconstruction quality. The cross-section of Model 3 by the plane $x = 0$ is presented in Fig. 2. In Fig. 3-4 we give the same section of the ART-interpolated function from the above set of points for $\lambda = 0.5$ and $\lambda = 0.05$, respectively.

4. Conclusions

A new approach to the interpolation problem is suggested. It allows us to evaluate a function at nodes of a regular grid when the function is known on an arbitrary set of points. In particular, the new method can be useful for solving a cone-beam tomography problem by the Grangeat's technique. A numerical algorithm based upon ART and the three-linear interpolation formulae was constructed. It has been found experimentally that the suggested algorithm is essentially more accurate than the traditional zeroth- and first-order methods.

Acknowledgments

The work was supported by the Scientific Research Organization of the Netherlands, NWO, grant No. 713-097.

REFERENCES

Grangeat P. (1990), Mathematical framework of cone-beam 3D reconstruction via the first derivative of the Radon transform. In: *Mathematical Methods in Tomography.* Proceedings of conference, Oberwolfach, Germany, 5-11 June, 1990. Herman G.T., Louis A.K. and Natterer F. (Eds.). Springer-Verlag, 66–97.

Herman G.T. (1980), *Image Reconstruction from Projections: The Fundamentals of Computerized Tomography.* Academic Press, New-York.

Herman G.T., Lent A. and Rowland S.W. (1973), ART: Mathematics and Applications. *J. Teor. Biol.* 42, 1–32.

Marchuk G.I. (1977), *Methods of Numerical Mathematics.* Nauka, Moscow. (in Russian)

A Space-time Tomography Algorithm for the Five-Camera Soft X-Ray Diagnostic at the RTP Tokamak

E.S. LYADINA[1], C.P. TANZI, D.F. da CRUZ, L.C. INGESSON, A.J.H. DONNÉ

FOM-Instituut voor Plasmafysica "Rijnhuizen"

P.O. Box 1207, 3430 BE Nieuwegein, The Netherlands

[1]On leave from Russian Research Center 'I.V. Kurchatov Institute', Moscow, Russia

Present address: JET Joint Undertaking - Abingdon OXON OX14 3EA - UK

ABSTRACT

A five-camera soft x-ray tomography system is one of the main tools of plasma diagnosis of RTP, a medium-sized tokamak. The relatively small number of cameras is caused by the limited access to the plasma. Systematic errors and noise distort mainly the highest angular harmonics. Elimination of noise can be excellently performed by means of space-time tomography processing. An additional optimization program for the determination of the camera positions allows to decrease systematic errors in the inputs of the space-time tomography procedure. A system of control detectors can be recommended for the elimination of uncertainties in the camera positions in future multi-camera tomography systems.

Introduction

The study of transport mechanisms and magnetohydrodynamic (MHD) perturbations in tokamaks is an important area of work in controlled thermonuclear fusion research. The program of RTP, a medium-sized tokamak, is dedicated to these key issues.

For plasmas with electron temperatures of the order of 1 keV, as in RTP, soft x-rays are produced during collisions between electrons and ions through the mechanisms of bremsstrahlung, radiative recombination, and line radiation (Hutchinson I.H, 1987).

A soft x-ray diagnostic, installed on a cross-section of the RTP tokamak, detects soft x-ray radiation in the 1 to 10 keV range by a system of five cameras. Each camera consists of an array of sixteen Si photodiodes which view the plasma through a pin-hole (da Cruz D.F. and Donné A.J.H., 1990). A schematic view of the five-camera soft x-ray diagnostic is given in Fig. 1. The time resolution of the system is 5 μs. It is therefore well suited for studies of MHD plasma perturbations, which in RTP have frequencies in the range of tens of μs to a few ms. Soft x-rays are mainly emitted by the central part of the plasma, while visible light emission originates mainly from the plasma edge. The soft x-ray tomography system is complementary to the 80-channel visible light system of RTP described by V.V. Pickalov.

The widths of the viewing chords of each detector are in general much narrower than the spatial variation of the emissivity, and the plasma is optically thin in the x-ray spectral region. It follows that the x-ray signals are approximated by line integrals, so that the local soft x-ray emissivity as a function of time can be reconstructed by means of tomographic inversion methods. The diagnosis of tokamak plasmas is in general complex, and is further hindered by limited access to the plasma. This is the reason for the relatively limited number of independent views, while the limited number of channels per camera has been determined both by size and cost.

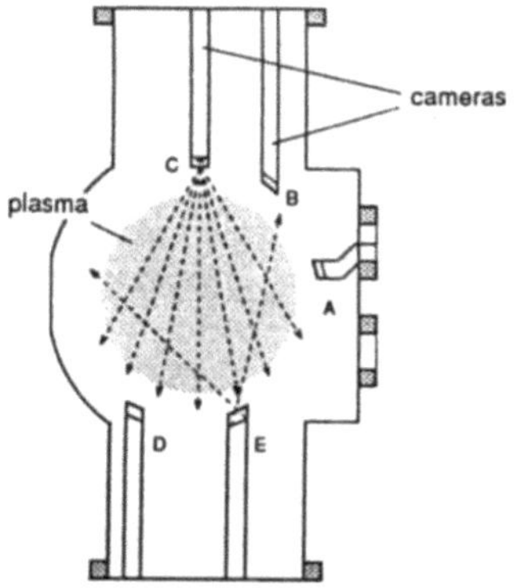

FIGURE 1. Schematic view of the five-camera soft x-ray diagnostic. The shaded region represents the plasma.

With this type of experimental set-ups, which view a circular cross-section, the tomographic inversion method of Cormack is generally preferred (Granetz R.S. and Smeulders P., 1988). Moreover, the representation of Cormack is quite useful in reconstructing the emission from a tokamak plasma because MHD theory predicts the presence of structures with angular periodicity in the plasma, and that only the lowest angular harmonics are involved in the equilibrium and principal instabilities.

Numerical tomographic reconstruction algorithms are constrained by the characteristics of the system. Especially the highest angular harmonics can be strongly distorted by stochastic and systematic errors. Furthermore, small uncertainties in the relative position of the cameras in a multi-camera system can lead to strong artefacts in the reconstruction. In the next sections, the steps applied to reduce the influence of these effects are detailed.

Noise Filtration by Space-Time Processing

Elimination of noise can be excellently performed by means of space-time tomography processing (Dnestrovskij Yu.N. et al., 1990). This method is based on a statistical analysis of the time traces of the spatial expansion coefficients used in the Cormack reconstruction algorithm. The time information can be used because the radiation emitted by the plasma has an a *priori* continuous time evolution. Numerical simulations have been performed for the RTP five-camera set-up using complex spatial-temporal phantoms. Figure 2 presents the result of a simulation illustrating the advantages of the space-time tomography algorithm above the normally used Cormack algorithm, where the reconstruction is performed on separate time slices. Space-time tomography processing uses an adaptive filter, which effectively suppresses noise effects in the reconstructed image (Dnestrovskij Yu.N. et al., 1990).

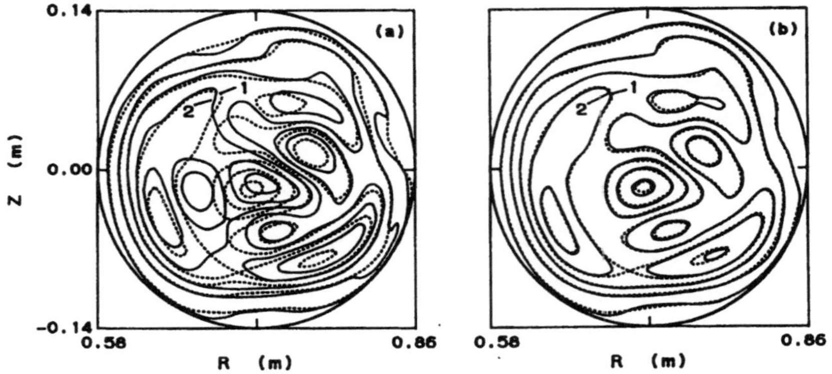

FIGURE 2. Contour plots of phantom (1, dotted line) and reconstructed soft x-ray emissivity (2, continuous line) given by the standard Cormack method (a) and by the space-time tomography method (b), using poloidal modes up to nine Fourier components with 1% added noise.

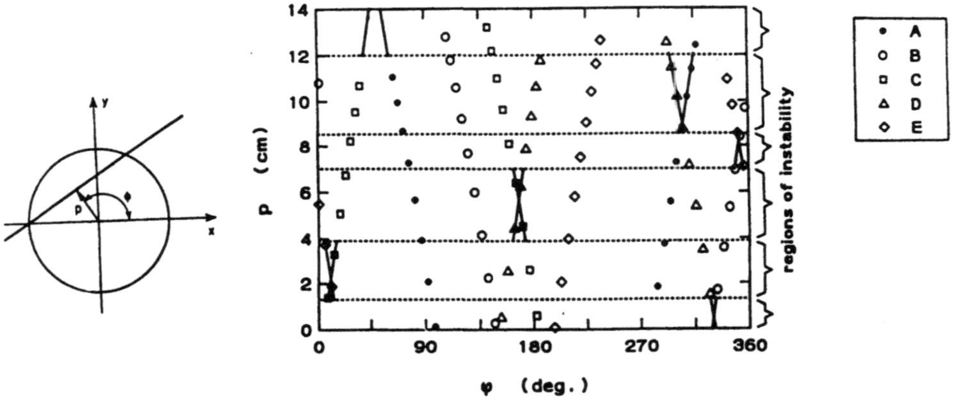

FIGURE 3. Positions in the p-ϕ space of the viewing chords – marked by different symbols to represent the five cameras shown in Fig. 1 – of the soft x-ray detectors at RTP. The definition of the impact parameter p and of the angle ϕ is also shown. The sets of detectors from each of the five cameras form lines of which the intersections are marked by crosses. The crosses indicate coinciding viewing chords, which, due to noise, cause instability. For each value of p, our method removes the worst instability.

Elimination of Singularities

Each viewing chord is represented by a point in the p-ϕ diagram (Fig. 3), and each camera by a line. Because the viewing angle of each camera is large, detectors of other cameras fall within it. This results in coincident lines of sight (i.e. points of intersection in the p-ϕ diagram). Each coincidence leads to a singularity in the reconstruction algorithm and implies that only nine (instead of ten) Fourier components are available for calculation at the values of p corresponding to these points. The presence of noise and systematic errors (e.g. uncertainties in the alignment of the cameras) gives rise to areas of instability that have a finite spread around the points of intersection. We have adopted a method which is based on finding, for each value of p, the two lines which are closest along ϕ

(Fig. 3). The data from one of the two lines are omitted and nine Fourier components are calculated for this value of p using data from all other cameras. The same is done again, this time omitting the data of the second line. The average of the two results is used in the reconstruction.

Reducing Systematic Errors due to Uncertainties in Camera Positions

The presence of detector elements from different cameras which are looking face-to-face along the same chord can have certain advantages. The line-integrated emissivity should, in first order, be the same for these coincident chords. However, if the design positions of the cameras are used, the calculation of the signals in the points of intersection (by means of interpolation of the signals from neighboring detectors) shows a significant difference (Fig. 4a).

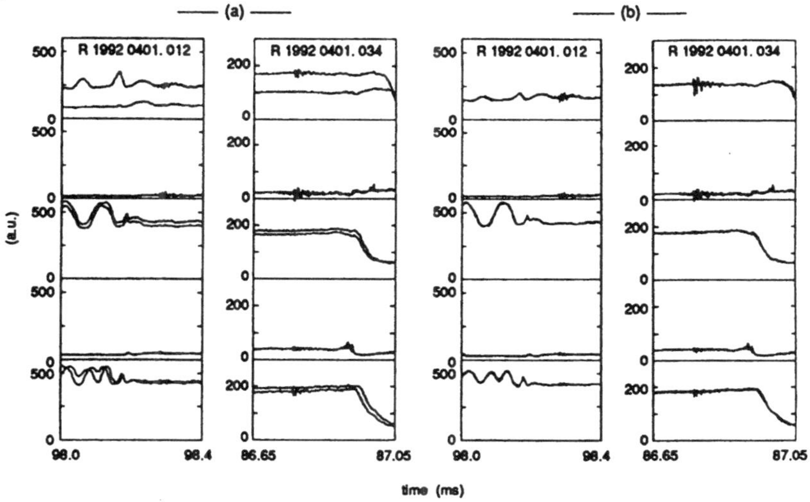

Figure 4. Discrepancy between the signals of two different RTP discharges in the five points of intersection before (a) and after (b) the determination of the camera positions.

This difference is mainly caused by uncertainties in both the spatial arrangement and the calibration of the cameras. Because each camera has a point of intersection with one or more other cameras, the reduction of the discrepancy between these signals could yield a direct criterion for reducing the uncertainty both on the camera positions and the relative calibration coefficients between cameras. Note that, due to the fan-like arrangement of chords, these coefficients may not be determined by means of normalization of the integral projections to the same value. An optimization program has been developed to deduce the positions of the cameras and calibration coefficients that yield a minimal discrepancy between the signals in the points of intersection. It is assumed that the main source of uncertainties on the positions of the cameras is a combination of rigid displacement and rotation of the whole camera, i.e. of the system consisting of the array of detectors and the pin-hole through which the plasma is viewed.

The program includes some regularizing terms, which do not allow camera positions that are unrealistically far from design values. Furthermore, it is possible to incorporate additional information about the alignment of all integral projections for discharges with an almost axially symmetric distribution. The program uses experimental data from several time windows from different discharges, chosen to include some characteristic variations (oscillations, internal and major disruptions, pellet injection). The total amount of experimental data used by the optimization program reaches 24000 time values which give a sufficiently large statistical base for making a reliable decision. Figure 4b presents the signals along the coincident chords after refining the camera positions and calibration factors for typical examples of discharges used by the program. The optimization has resulted in a significant decrease of systematic errors on the inputs of the space-time tomography procedure. An additional system of control detectors providing identical signals can be recommended for the elimination of uncertainties in the camera positions in future multi-camera tomography systems.

Results

The first results of space-time tomography applied to major disruptions on RTP have been presented in Rommers J.H. et al. (1993). In Fig. 5 the time signals from the central channels of the side camera during a sawtooth crash in an RTP plasma heated by means of Electron Cyclotron Resonance are shown. Some reconstructed contours of equal emissivity are given in Fig. 6. Figure 7 shows the time signals from the central channels of the side camera during a sawtooth crash in an ohmic plasma. Figure 8 shows equal emissivity contours during the precursor oscillation (rotating structure), Fig. 9 shows the subsequent evolution. The presence of singularities with almost the same p value decreases the effective number of harmonics that can be reconstructed to seven Fourier components. Figures 6, 8 and 9 show that the method described here is effective in reducing stochastic and systematic errors which may have caused artificial structures in the reconstructions with the same number of Fourier components in other multi-camera systems (Janicki C. et al., 1989, 1990).

Acknowledgments

This work was performed under the Euratom-FOM association agreement with financial support from NWO and EURATOM .

REFERENCES

da Cruz D.F. and Donné A.J.H. (1990), *Rev. Sci. Instrum.* **61** 3067.

Dnestrovskij Yu.N., Lyadina E.S., Savrukhin P.V. (1990), *Proc. IAEA Techn. Comm.*, Meeting on Time-resolved Plasma Diagnostics (Nagoya, 1990), pp. 232-292.

Granetz R.S. and Smeulders P. (1988), *Nucl. Fusion*, **28**, 457.

Hutchinson I.H (1987), *"Principles of Plasma Diagnostics"*, Cambridge University Press.

Janicki C., Décoste R. and Simm C. (1989), *Phys. Rev. Lett.*, **62**, 3038.

Janicki C. et al. (1990), *Nucl. Fusion*, **30**, 950.

Pickalov V.V., Ingesson L.C., Donné A.J.H. (1993), this conference.

Rommers J.H. et al. (1993), *Proc. of the 20th EPS Conference on Controlled Fusion and Plasma Physics*, Lisboa, Portugal, 26-30 July 1993, I-275.

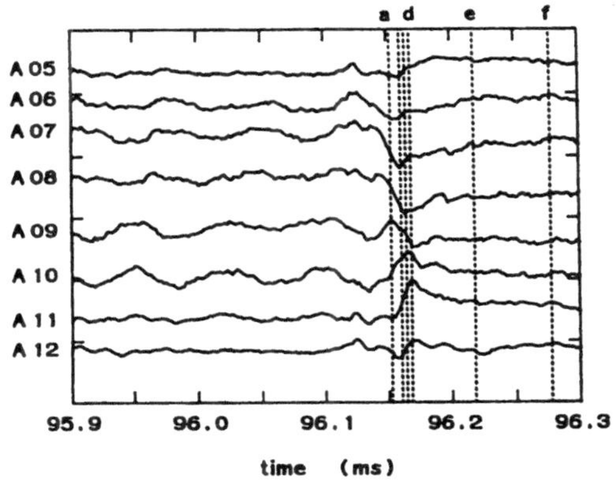

FIGURE 5. Time signals from the central channels of the side cameras during a sawtooth crash in an RTP plasma heated by means of Electron Cyclotron Resonance.

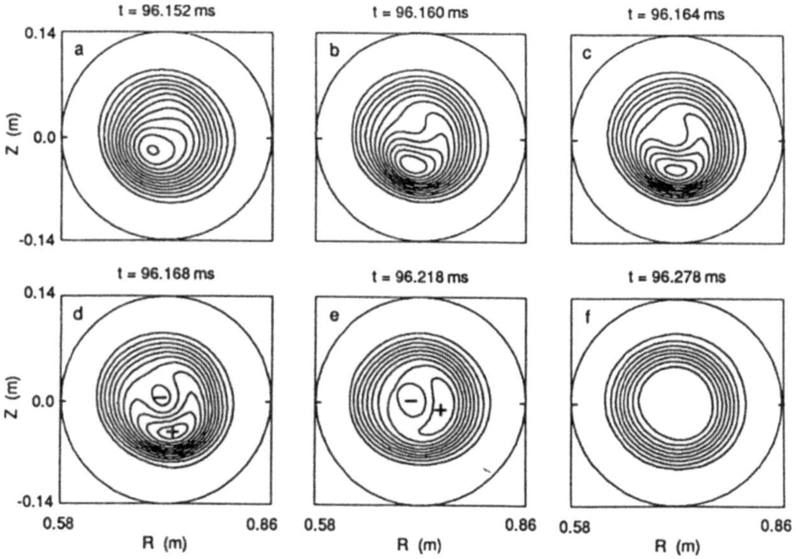

FIGURE 6. Contours of equal emissivity for several time frames from the time evolution of the RTP plasma during a sawtooth crash in a plasma heated by means of Electron Cyclotron Resonance. The reconstruction is made with seven Fourier components.

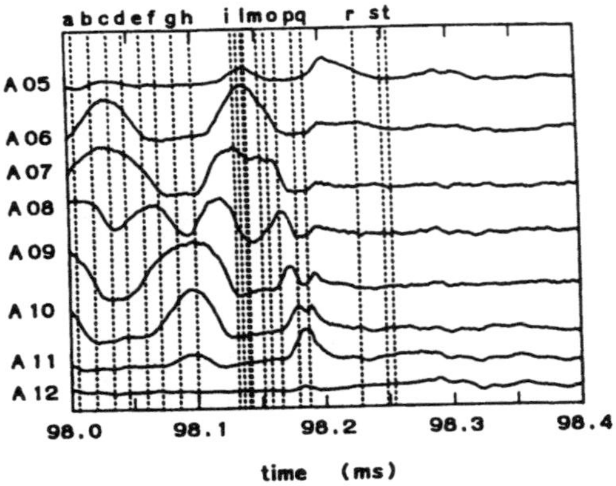

FIGURE 7. Time signals from the central channels of the side cameras during a sawtooth crash in an ohmic RTP plasma.

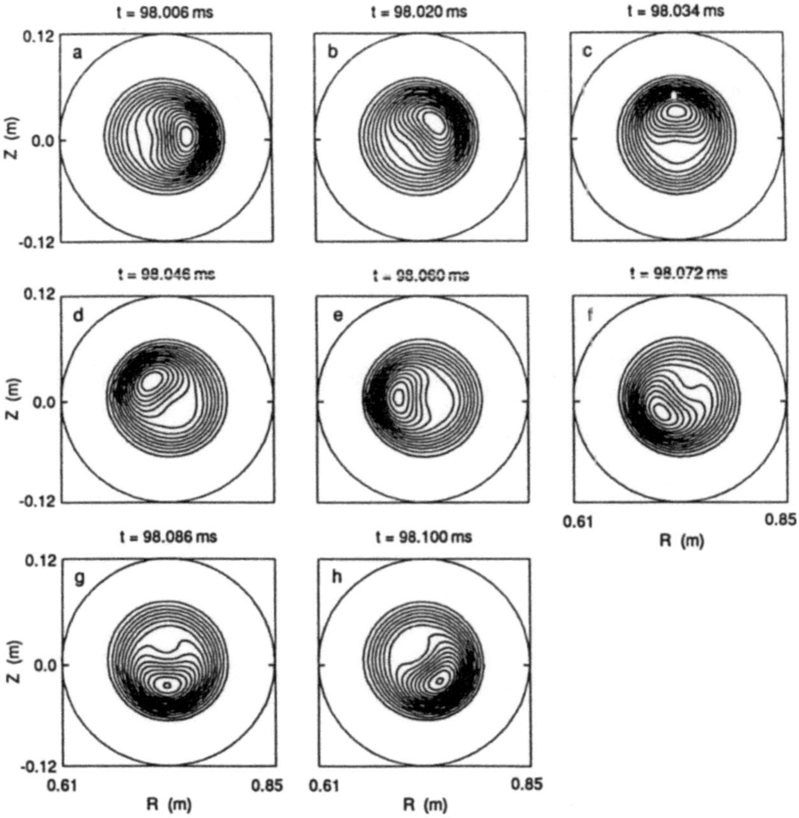

FIGURE 8. Contours of equal emissivity during the phase of the sawtooth precursor oscillation (rotating structure) in an ohmic RTP plasma. The reconstruction is made with seven Fourier components.

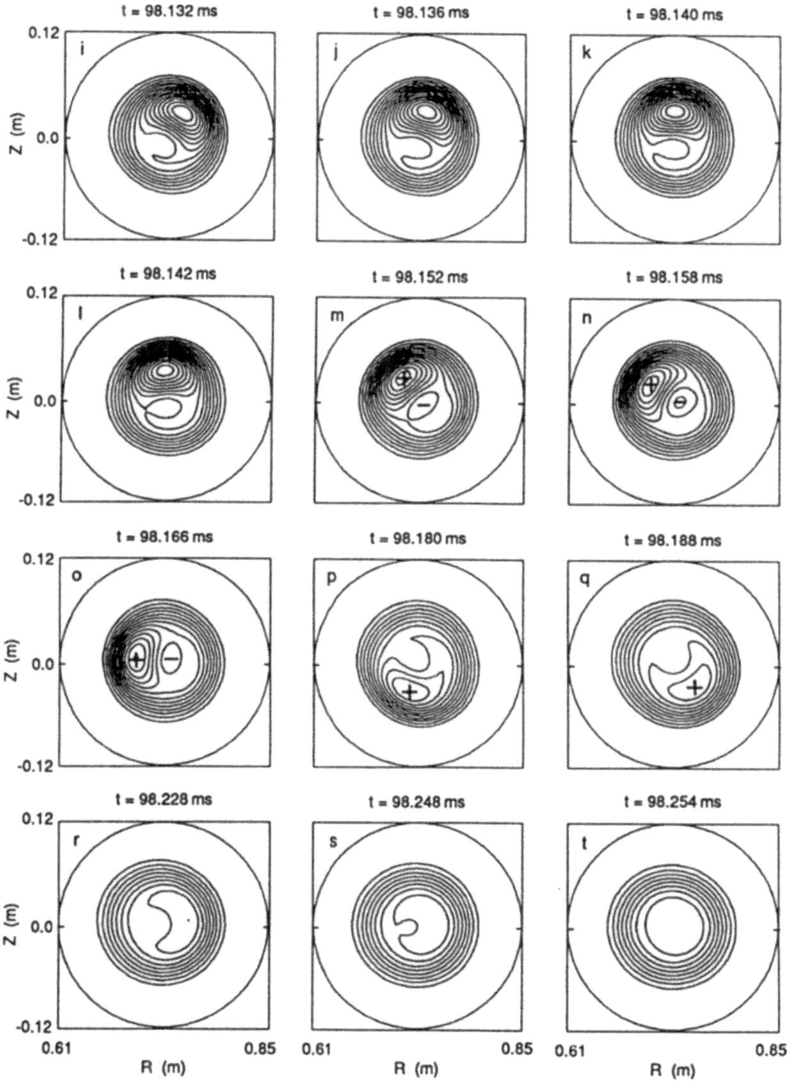

FIGURE 9. Evolution of the sawtooth crash following the precursor oscillation of Fig. 8. The reconstruction is made with seven Fourier components.

Direct and Inverse Problems for System of Electromagneto–Elasticity Equations

I. Z. MERAZHOV and V. G. YAKHNO

Institute of Mathematics, Novosibirsk University,

Universitetskii Pr., 4, 630090, Novosibirsk, Russia

ABSTRACT

A system of electromagneto-elasticity is shown to reduce to a system of symmetric t–hyperbolic Fridrix equation system of the first order. As a result, we obtain the existence and uniqueness theorem for direct problems of electromagneto-elasticity such as Cauchy problem or initial boundary problems.

We consider also a number of inverse problems of determining some combinations of density, magnetic permeability, elasticity modules, piezoelectric and dielectric modules as the functions depending on x_3 only.

We consider a system of electromagneto - elasticity differential equations. Let $x = (x_1, x_2, x_3) \in \mathbf{R}^3$, $t \in \mathbf{R}_+$, $x_3 > 0$,

$$\rho \frac{\partial^2 u_i}{\partial t^2} = \sum_{j=1}^{3} \frac{\partial Tij}{\partial x_j}, \quad i = 1, 2, 3; \tag{1}$$

$$\text{curl } H = \frac{\partial D}{\partial t}, \quad \text{curl } E = -\mu \frac{\partial H}{\partial t},$$

$$\text{div } D = 0, \quad \text{div}\,(\mu H) = 0. \tag{2}$$

Here $u = (u_1, u_2, u_3)$ is the displacement vector function, $u_i = u_i(x,t)$, $i = 1, 2, 3$;

$$T_{ij} = \sum_{k,l=1}^{3} c_{ijkl} \frac{\partial u_k}{\partial x_l} - \sum_{k=1}^{3} e_{kij} E_k, \quad i = 1, 2, 3, \quad j = 1, 2, 3,$$

are the mechanic stresses; $\rho = \rho(x)$ is the density, $\rho(x) > 0$; $c_{ijkl} = c_{ijkl}(x)$ are the elastic modules; $e_{kij} = e_{kij}(x)$ are the piezoelectric modules; $E = (E_1, E_2, E_3)$ and $H = (H_1, H_2, H_3)$ are the electric and magnetic intensity vector functions, $E_i = E_i(x,t)$, $H_i = H_i(x,t)$, $i = 1, 2, 3$;

$$D_j = \sum_{k=1}^{3} \varepsilon_{ij} E_k + \sum_{k=1}^{3} e_{jkl} \frac{\partial u_k}{\partial x_l}, \quad j = 1, 2, 3,$$

are the components of the electric induction vector functions, $D = (D_1, D_2, D_3)$, $D_i = D_i(x,t)$, $i = 1, 2, 3$; $\varepsilon_{ij} = \varepsilon_{ij}(x)$ are the dielectric modules; and $\mu = \mu(x)$ stands for the magnetic permeability.

Let us suppose $c_{ijkl} = c_{jikl} = c_{ijlk} = c_{klij}$, $e_{kij} = e_{kji}$; $\varepsilon_{ij} = \varepsilon_{ji}$.

System (1), (2) describes the bound electromagnetic and elasticity waves. The connection between elasticity and electromagnetic processes is defined by the piezoelectric modules of media.

This system of electromagneto - elasticity will be shown to be equivalent to a system of symmetric t-hyperbolic Fridrix equations of the first order. For this purpose change the indexes $\alpha = (ij)$, $\beta = (kl)$ in the following way: $(11) \to 1$, $(22) \to 2$, $(33) \to 3$, $(23) = (32) \to 4$, $(13) = (31) \to 5$, $(12) = (21) \to 6$ and introduce the functions

$$\tau_\alpha = \tau_{ij} = \sum_{k,l=1}^{3} c_{ijkl} \frac{\partial u_l}{\partial x_k}, \quad i = 1,2,3; \quad j = 1,2,3; \tag{3}$$

$$U_i = \frac{\partial u_i}{\partial t}, \quad i = 1,2,3.$$

Differentiating (3) with respect to t, we get six equations

$$\frac{\partial \tau_\alpha}{\partial t} = \sum_{k,l=1}^{3} c_{ijkl} \frac{\partial^2 U_l}{\partial x_k}.$$

System (1), (2) when rewritten in terms of the functions τ_α, U_i becomes

$$A \frac{\partial U}{\partial t} + \sum_{i=1}^{3} A_i \frac{\partial U}{\partial x_i} = G. \tag{4}$$

where $U = [u_1, u_2, u_3, \tau_1, \tau_2, \tau_3, \tau_4, \tau_5, \tau_6, E_1, E_2, E_3, H_1, H_2, H_3]^*$ (the symbol $(\star)$ denotes transposition), A is a block-diagonal matrix of the form

$$A = \operatorname{diag}(\rho I_3, C^{-1}, \varepsilon, \mu I_3),$$

I_3 is the identity 3×3 matrix, C^{-1} is a 6×6 matrix inverse to $C = [c_{\alpha\beta}]$,

$$\varepsilon = \begin{pmatrix} \varepsilon_{11} & \varepsilon_{12} & \varepsilon_{13} \\ \varepsilon_{12} & \varepsilon_{22} & \varepsilon_{23} \\ \varepsilon_{13} & \varepsilon_{23} & \varepsilon_{33} \end{pmatrix}, \quad A_i = \begin{pmatrix} 0_{3,3} & A_i^1 & A_i^2 & 0_{3,3} \\ A_i^{1\star} & 0_{6,6} & 0_{6,3} & 0_{6,3} \\ A_i^{2\star} & 0_{3,6} & 0_{3,3} & A_i^3 \\ 0_{3,3} & 0_{3,6} & A_i^{3\star} & 0_{3,3} \end{pmatrix}, \quad i = 1,2,3;$$

$$A_1^1 = \begin{pmatrix} -1 & 0 & 0 & 0 & 0 & 0 \\ 0 & 0 & 0 & 0 & 0 & -1 \\ 0 & 0 & 0 & 0 & -1 & 0 \end{pmatrix}, \quad A_2^1 = \begin{pmatrix} 0 & 0 & 0 & 0 & 0 & -1 \\ 0 & -1 & 0 & 0 & 0 & 0 \\ 0 & 0 & 0 & -1 & 0 & 0 \end{pmatrix},$$

$$A_3^1 = \begin{pmatrix} 0 & 0 & 0 & 0 & -1 & 0 \\ 0 & 0 & 0 & -1 & 0 & 0 \\ 0 & 0 & -1 & 0 & 0 & 0 \end{pmatrix},$$

$$A_1^2 = \begin{pmatrix} e_{11} & e_{21} & e_{31} \\ e_{16} & e_{26} & e_{36} \\ e_{15} & e_{25} & e_{35} \end{pmatrix}, \quad A_2^2 = \begin{pmatrix} e_{16} & e_{26} & e_{36} \\ e_{12} & e_{22} & e_{32} \\ e_{14} & e_{24} & e_{34} \end{pmatrix}, \quad A_3^2 = \begin{pmatrix} e_{15} & e_{25} & e_{35} \\ e_{14} & e_{24} & e_{34} \\ e_{13} & e_{23} & e_{33} \end{pmatrix},$$

$$A_1^3 = \begin{pmatrix} 0 & 0 & 0 \\ 0 & 0 & -1 \\ 0 & 1 & 0 \end{pmatrix}, \quad A_2^3 = \begin{pmatrix} 0 & 0 & 1 \\ 0 & 0 & 0 \\ -1 & 0 & 0 \end{pmatrix}, \quad A_3^3 = \begin{pmatrix} 0 & 1 & 0 \\ -1 & 0 & -1 \\ 0 & 0 & 0 \end{pmatrix};$$

$0_{m,n}$ is zero $m \times n$ matrix, where m and n are the numbers of rows and columns, respectively.

The existence and uniqueness theorems for Cauchy problem (Mizohata 1973) and initial boundary problems (Godunov 1978) follow from the general theory of differential equations.

REMARK. It was shown in (Godunov 1978) that there exist a unique solution in $C^1(\mathbf{R}^3 \times \mathbf{R}_+)$ of the direct problem of determining a vector-function Y satisfying (3) and the following conditions

$$Y(x,0) = \varphi(x); \tag{5}$$

$$Y(0,t) = \sum_{j=n_0+1}^{n_1} \alpha_{ij} Y_j(0,t), \quad i = 1,2,\cdots,n_0,$$

$$Y(H,t) = \sum_{j=n_0+1}^{n_1} \beta_{ij} Y_j(H,t), \quad i = n_0+1,\cdots,n_1; \tag{6}$$

where $\varphi(x)$, $\alpha_{ij}(t)$, $\beta_{ij}(t)$ are smooth functions, n_0 and $n_1 - n_0$ are the numbers of the outgoing and incoming characteristics, respectively.

Hence, we can apply the Fourier transform F_{x_1,x_2} to the solution.

The operator F_{x_1,x_2}, with zero parameters ν_1 and ν_2, when applied to the right- and left-hand sides of (4)-(6) gives

$$A\frac{\partial u}{\partial t} + A_3 \frac{\partial u}{\partial x_3} = g, \tag{7}$$

where $u(x_3,t) = F_{x_1,x_2}[U]\,|_{\nu_1=\nu_2=0}$, $\quad g(x_3,t) = F_{x_1,x_2}[f]\,|_{\nu_1=\nu_2=0}$ . .

It is known from linear algebra that, if the matrices A and A_3 are symmetric and A is positive definite, then there exists a singular matrix T such that $T^{-1}AT = I$, $T^{-1}A_3T = D$, where I is the identity matrix and D is diagonal. The columns of T obey $(k_jA - A_3)t = 0$, where k_j are roots of the equation

$$\det(k_jA - A_3) = 0.$$

Setting $u = TW$ in (7) and multiplying (7) from the left by T^{-1} we get

$$I\frac{\partial W}{\partial t} + D\frac{\partial W}{\partial x_3} + QW = R, \tag{8}$$

where

$$Q = T^{-1}A_3\frac{\partial T}{\partial x_3}T, \ R = T^{-1}g.$$

For media of the cubical structure of crystals, that is, in the case when $c_{11} = c_{22} = c_{33}$, $c_{12} = c_{13} = c_{23}$, $c_{44} = c_{55} = c_{66}$, $e_{14} = e_{25} = e_{36}$, $\varepsilon_{11} = \varepsilon_{22} = \varepsilon_{33}$, $\mu_{11} = \mu_{22} = \mu_{33}$, and all the others c_{ij}, e_{ij}, ε_{ij}, μ_{ij} are equal to zero, the matrix D has the form

$$D = \operatorname{diag}\left(\sqrt{\frac{c_{44}}{\rho}}, \lambda_1, 0, 0, -2\sqrt{\frac{c_{11}}{\rho}}, 2\sqrt{\frac{c_{11}}{\rho}}, \lambda_2, \sqrt{\frac{c_{44}}{\rho}}, 0, -\lambda_1, \frac{-1}{\sqrt{\varepsilon\mu}}, 0, \frac{1}{\sqrt{\varepsilon\mu}}, \lambda_2, 0\right),$$

where

$$\lambda_j = \frac{1}{2}\left\{\frac{1}{\varepsilon\mu} + \frac{c}{\rho} + \frac{\varepsilon_{14}^2}{\rho\varepsilon} + (-1)^{j+1}\left[\left(\frac{1}{\varepsilon\rho} + \frac{c}{\rho} + \frac{e_{14}^2}{\rho\varepsilon}\right)^2 - 4\frac{c}{\rho}\frac{1}{\varepsilon\mu}\right]^{\frac{1}{2}}\right\}, \ j = 1,2.$$

We consider system (8) in the domain $\Omega = \{(x,t): \; 0 \le x \le H, \; t > 0\}$ under the following conditions

$$W(x,0) = \psi(x) = T^{-1}\varphi(x); \tag{9}$$

$$W_i(0,t) = p_i(t), \; t \ge 0, \; i = 2,6,7,8,13; \tag{10}$$

$$W_i(H,t) = p_i(t), \; t \ge 0, \; i = 1,5,10,11,14;$$

where $p_i(t)$ is known functions.

As in Romanov (1984), we consider the following inverse problem.

Inverse problem. Let us consider fifteen problems of type (8)-(10), each having its own set of the functions ψ, p but the same matrixes D and Q. All the functions involved in the problem numbered l, $1 \ge l \ge n$, will be marked with superscript l. Therefore,

$$\left(E\frac{\partial}{\partial t} + D\frac{\partial}{\partial x_3} + Q\right) W^l = R^l;$$

$$W^l(x,0) = \psi^l(x); \tag{11}$$

$$W_i^l(0,t) = p_i^l(t), \; i = 2,6,7,8,13;$$

$$W_i^l(H,t) = p_i^l(t), \; i = 1,5,10,11,14; \; l = 15.$$

The problem is to find the matrix $Q = Q(x_3)$, $x \in [0,H]$, provided that the following information concerning the solution of problem (11) is known:

$$W_i^l(H,t) = h_i^l(t), \; i = 2,6,7,8,13; \tag{12}$$

$$W_i^l(0,t) = h_i^l(t), \; i = 1,5,10,11,14; \; l = 15.$$

We suggest that the agreement conditions on initial, boundary and additional data hold:

$$\psi_i^l(0) = p_i^l(0) = h_i^l(0), \; i = 2,6,7,8,13; \tag{13}$$

$$\psi_i^l(H) = p_i^l(H) = h(H), \; i = 1,5,10,11,14; \; l = 15.$$

Then the below theorem follows from the results by Romanov (1984).

Theorem. *Assume that* $G^l \in C^1(\Omega)$, $\psi \in C^1[0,H]$, $p_i^l \in C^1(0,\infty)$, *the necessary conditions (13) on the inverse problem data are fulfilled and, besides,*

$$\det \psi(x_3) \ne 0, \; x \in [0,H].$$

Then there exists such an $H^* > 0$ *that for* $H \in (0,H^*)$ *the inverse problem has a unique solution determined from* $h_i^l(t) \in C^1[0, H/\,|d_i|]$ *where* d_i *are elements of the matrix* D.

REFERENCES

Dieulesaint E.and Royer D. (1974), Ondes Elastiques dans les Solides. *Application au Tratement du Signal.* Masson et Cie, (in Franch)

Godunov S. K. (1978), *Ecuacion de la Fisica Matematica.* Mir publishers, Moscow, (in Spanish)

Mizohata S. (1973), *The Theory of Partial Differential Equations.* Cambridge University Press, New York.

Romanov V. G. (1987), *Inverse Problems of Mathematical Physics.* VNU Science Press BV. Utrecht.

A Method of the Numerical Ray Tracing for a 3D-Inhomogeneous Isotropic Medium

A. K. MOROZOV, A. A. SNEGOVOY[1], S. A. TISHCHENKO[2]

Institute of Oceanology RAS, VNIIGeosystem, Scientific- Production Center "System", (NPO "Almaz", Inc.)

Krasikova street 23, Moscow, 117851, Russia

[1] Varshavskoe shosse 8, Moscow, 113105, Russia

[2] Leningradsky prospect 80, Moscow, 125178, Russia

ABSTRACT

We consider an initial value problem for differential ray equations in a 3D-inhomogeneous medium with a smooth refractive index $n(x, y, z)$ given on a uniform grid in the rectangular system of coordinates. The system of the second-order nonlinear ray equations of geometrical optics, when integrated, becomes a system of quasilinear equations, whose solution is written in such a form that an effective numerical scheme can be constructed for the ray tracing in neighborhoods of turning points. The results of numerical testing illustrate the performance of the proposed method.

1. Introduction

While solving numerically the inverse problems of geometrical optics by applying the variational principle of choice of the solution with minimal discrepancy, the multiple solution of the direct problem is necessary to obtain a large set of ray trajectories. Meanwhile, the time needed for the ray tracing calculation is the major part of the total computer CPU time necessary for solving the inverse problem. That's why the development of the effective numerical procedures for the ray tracing in a 3D-inhomogeneous isotropic medium is of vital importance.

Nowadays, the method of ray tracing is widely used which is based on the triangulation procedure of the refractive index $n(x, y, z)$ defined on a rectangular grid and uses the explicit analytic form of the solution in elementary cells of triangulation.

The Runge-Kutta's type methods with the adaptive choice of step size have also found wide use in solving the ray equation.

We remark that the high-performance triangulation method is often employed in solving the direct problem. However, a failing of this method is that false caustics may appear and rays split at the discontinuity points of the gradient of $n(x, y, z)$. It should be also noted that the inverse problems of wave propagation are ill-posed according to the Hadamard's classification. So, the common practice in solving such inverse problems is the use of the Tikhonov's regularization method based on the stabilizing functional technique, which requires the desired solution to be smooth enough. This makes the use of the triangulation method difficult (in the case of discontinuity of the gradient of $n(x, y, z)$) for solving the inverse problems.

The Runge-Kutta's type methods need much computational time since the step size has to be reduced essentially in neighborhoods of the ray's reflection (turning) points.

In this paper we propose a new numerical method for solving the ray equations in the rectangular system of coordinates for a 3D-inhomogeneous isotropic medium whose refractive index $n(x,y,z)$ defined on a uniform grid is approximated by interpolation polynomials of the 5-th order.

The proposed method uses the predictor-corrector scheme. When finding the solution in neighborhoods of turning points, the approximate quadrature of the exact solution is used to avoid reducing the step size to excess.

The numerical experiments had shown that the method proposed is of high performance not only in accuracy, which is typical for the Runge-Kutta's type numerical schemes with an adaptive choice of the step size, but also in computational time, due to triangulation of the refractive index procedure.

2. The Base Analytical Background

We consider the main analytic equations for the rays propagating in an isotropic medium with a refractive index $n(x,y,z)$. The function $n(x,y,z)$ is assumed to be bounded, single-valued and smooth enough to guarantee the uniqueness of the solution, say, $n(x,y,z) \in C^2$. The ray trajectory connecting two given points A and B in the medium with the refraction index $n(x,y,z)$ is defined according to the Fermat's stationary time principle, i.e., the ray is the extremal curve for the Fermat's functional

$$\delta J = 0, \quad J = \int_A^B n \mathrm{d}s, \tag{1}$$

where $\mathrm{d}s$ is an element of the path length along the ray. If (x,y,z) is the rectangular system of coordinates in the 3D medium and the ray trajectory parametrization is made in x, then equation (1) can be rewritten in the form

$$\delta \int_A^B n\sigma_1 \mathrm{d}x = 0, \tag{2}$$

where σ_1 is the path length along the ray

$$\sigma_1^2 = 1 + {y'_x}^2 + {z'_x}^2. \tag{3}$$

Using the Euler equations (Kravtsov and Orlov, 1983), we can obtain a system of ordinary differential equations for determining the ray trajectories in the rectangular system of coordinates (x,y,z):

$$\begin{cases} \frac{\mathrm{d}}{\mathrm{d}x}\left(\frac{\partial(n\sigma_1)}{\partial y_x'}\right) - \frac{\partial(n\sigma_1)}{\partial y} = 0 \\ \frac{\mathrm{d}}{\mathrm{d}x}\left(\frac{\partial(n\sigma_1)}{\partial z_x'}\right) - \frac{\partial(n\sigma_1)}{\partial z} = 0 \end{cases} \tag{4}$$

which after some transformations will be as

$$\begin{cases} n{y_{xx}}'' = (1 + {y_x}'^2 + {z_x}'^2)(\frac{\partial n}{\partial y} - {y_x}'\frac{\partial n}{\partial x}) \\ n{z_{xx}}'' = (1 + {y_x}'^2 + {z_x}'^2)(\frac{\partial n}{\partial z} - {z_x}'\frac{\partial n}{\partial x}) \end{cases} \tag{5}$$

This system when partly integrated takes the form

$$\frac{n^2}{(1+y_x'^2+z_x'^2)} = \frac{n_0^2}{(1+y_{0x}'^2+z_{0x}'^2)} + \int\limits_{x_0}^{x} \frac{\partial n^2}{\partial x} \mathrm{d}x. \tag{6}$$

This equation can be obtained by straightforward calculation or by using the optico-mechanical analogy (Born and Wolf, 1964; Kravtsov and Orlov, 1983). Similar equations can also be derived for the spherical and the polar system of coordinates, but in most cases they are much more complicated and cannot be separated into independent variables as in the case of rectangular system of coordinates.

Transforming (6), we have

$$1+y_x'^2+z_x'^2 = \frac{n^2}{[\frac{n_0^2}{(1+y_{0x}'^2+z_{0x}'^2)} + \int\limits_{x_0}^{x} \frac{\partial n^2}{\partial x} \mathrm{d}x]}. \tag{7}$$

Substituting (7) in (5) yields

$$y_{xx}'' + y_x' \frac{1}{2} \frac{\partial n^2}{\partial x} \varphi(x,y,z) - \frac{1}{2} \frac{\partial n^2}{\partial y} \varphi(x,y,z) = 0, \tag{8}$$

$$z_{xx}'' + z_x' \frac{1}{2} \frac{\partial n^2}{\partial x} \varphi(x,y,z) - \frac{1}{2} \frac{\partial n^2}{\partial z} \varphi(x,y,z) = 0, \tag{9}$$

with

$$\varphi(x,y,z) = \frac{n^2}{[\frac{n_0^2}{(1+y_{0x}'^2+z_{0x}'^2)} + \int\limits_{x_0}^{x} \frac{\partial n^2}{\partial x} \mathrm{d}x]}. \tag{10}$$

So, in the 3D case the variables can be separated, and equations (8),(9) are quasilinear in $\{y_{xx}'', y_x'\}$ and $\{z_{xx}'', z_x'\}$, respectively. Denoting

$$\psi_x'(x,y,z) = \frac{\partial n^2}{\partial y}, \quad \chi_x'(x,y,z) = \frac{\partial n^2}{\partial z}, \tag{11}$$

we have

$$y_{xx}'' + y_x' \frac{\varphi'}{2\varphi} = \frac{\psi_x'}{2\varphi}, \quad y(0) = y_0, \quad y_x'(0) = y_{0x}', \tag{12}$$

$$z_{xx}'' + z_x' \frac{\varphi'}{2\varphi} = \frac{\chi_x'}{2\varphi}, \quad z(0) = z_0, \quad z_x'(0) = z_{0x}'. \tag{13}$$

Then

$$y_x' = \frac{1}{\varphi^{\frac{1}{2}}} (\frac{1}{2} \int\limits_0^x \frac{\psi_x'}{\varphi^{\frac{1}{2}}} \mathrm{d}x + y_{0x}' C^{\frac{1}{2}}), \tag{14}$$

$$z_x' = \frac{1}{\varphi^{\frac{1}{2}}} (\frac{1}{2} \int\limits_0^x \frac{\chi_x'}{\varphi^{\frac{1}{2}}} \mathrm{d}x + z_{0x}' C^{\frac{1}{2}}), \tag{15}$$

where

$$C = \frac{n_0^2}{(1+y_{0x}'^2+z_{0x}'^2)} = const. \tag{16}$$

Integrating (14)-(15) with respect to x, we obtain

$$y = y_0 + \int_0^x \frac{1}{\varphi^{\frac{1}{2}}}\left(\frac{1}{2}\int_0^x \frac{\psi_x{}'}{\varphi^{\frac{1}{2}}}\mathrm{d}x + y_{0x}{}'C^{\frac{1}{2}}\right)\mathrm{d}x, \tag{17}$$

$$z = z_0 + \int_0^x \frac{1}{\varphi^{\frac{1}{2}}}\left(\frac{1}{2}\int_0^x \frac{\chi_x{}'}{\varphi^{\frac{1}{2}}}\mathrm{d}x + z_{0x}{}'C^{\frac{1}{2}}\right)\mathrm{d}x. \tag{18}$$

Relations (17)–(18) give implicitly the analytic solution of the problem of finding the ray trajectory $(y(x), z(x))$ parametrized in x. They are used as the base relations in the numerical algorithm based on the proposed high-precision method.

3. The Numerical Algorithm

For simplicity, we shall consider the 2D case (in the 3D case everything is done similarly). To improve the algorithm accuracy, the parts of the ray trajectory, which contain the turning points or the points where the value of $y_x{}'$ in modulus is small, are calculated by the improved algorithms of numerical integration (the higher accuracy is achieved by additional computation of the values of $y_{xx}{}''$). To calculate the other parts of the ray trajectory (up-going and down-going) we use more simple (in computational aspect) algorithms of numerical integration.

The common predictor (used for all the parts of the trajectory) is

$${}^{0}y_{2x}{}' = \frac{12(y_0 - y_1)}{h} + 5y_{0x}{}' + 8y_{1x}{}'; \tag{19}$$

$${}^{0}y_2 = y_0 + \frac{h(y_{0x}{}' + 4y_{1x}{}' + y_{2x}{}')}{3}; \tag{20}$$

where h is the integration step size, the right subscript 2 denotes the number of a current point, the left superscript stands for the iteration number. Further we shall use the following auxiliary values

$$C_0 = \frac{{n_0}^2}{(1 + {y_{0x}{}'}^2)}; \tag{21}$$

$$V_0 = {C_0}^{\frac{-1}{2}}; \tag{22}$$

$$W_0 = V_0 \frac{\partial {n_0}^2}{\partial x}. \tag{23}$$

Consider now the parts of the ray trajectory containing no turning points (where the sign of $y_x{}'(x)$ doesn't change). In this case, the iterations are performed using the relation

$$y_x{}' = \left(\frac{n^2}{(C_0 + \int_0^x \frac{\partial n^2}{\partial x}\mathrm{d}x)} - 1\right)^{\frac{1}{2}}. \tag{24}$$

Relation (24) follows from (7) taking in 2D case the form

$$1 + {y_x{}'}^2 = \frac{n^2}{\left(\frac{{n_0}^2}{(1+{y_{0x}{}'}^2)} + \int_0^x \frac{\partial n^2}{\partial x}\mathrm{d}x\right)}.$$

After determining ${}^{j-1}y_2$, we find $n_2{}^2$, $\frac{\partial n_2{}^2}{\partial x}$ and the numerical analogue of the subradical expression in (24) (the corrector):

$$ {}^j y_{2x}{}'^2 = \frac{{}^{j-1}n_2{}^2}{(C_0 + \frac{h(\frac{\partial n_0{}^2}{\partial x} + 4\frac{\partial n_1{}^2}{\partial x} + \frac{\partial^{j-1} n_2{}^2}{\partial x})}{3})} - 1; \tag{25} $$

$$ {}^j y_2 = y_0 + \frac{h(y_{0x}{}' + 4y_{1x}{}' + {}^j y_{2x}{}')}{3}. \tag{26} $$

The iterations are repeated until the required accuracy of ${}^j y_2$ is achieved.

On the ray trajectory part containing a turning point we use the relation

$$ {}^j y_{2x}{}' = \frac{[\frac{h}{6}(\frac{\partial n_0{}^2}{\partial y}C_0{}^{\frac{-1}{2}} + 4\frac{\partial n_1{}^2}{\partial y}C_1{}^{\frac{-1}{2}} + \frac{\partial^{j-1} n_2{}^2}{\partial y}{}^{j-1}C_2{}^{\frac{-1}{2}}) + y_{0x}{}'C_0{}^{\frac{1}{2}}]}{{}^{j-1}C_2{}^{\frac{1}{2}}}, \tag{27} $$

where

$$ C_1 = \frac{n_1{}^2}{(1 + y_{1x}{}'^2)}, \tag{28} $$

$$ {}^{j-1}C_2 = \frac{{}^{j-1}n_2{}^2}{(1 + {}^{j-1}y_{2x}{}'^2)}. \tag{29} $$

From the 2D analogue of equation (5) we can find

$$ y_{0xx}{}'' = (1 + y_{0x}{}'^2)\frac{(\frac{\partial n_0{}^2}{\partial y} - y_{0x}{}'\frac{\partial n_0{}^2}{\partial x})}{2n_0{}^2}, \tag{30} $$

$$ {}^j y_{2xx}{}'' = (1 + {}^j y_{2x}{}'^2)\frac{(\frac{\partial^{j-1} n_2{}^2}{\partial y} - {}^j y_{2x}{}'\frac{\partial^{j-1} n_2{}^2}{\partial x})}{2{}^{j-1}n_2{}^2}, \tag{31} $$

$$ {}^j y_2 = y_0 + \frac{h}{15}(7(y_{0x}{}' + {}^{j-1}y_{2x}{}') + 16y_{1x}{}' - h({}^{j-1}y_{2xx}{}'' - y_{0xx}{}'')). \tag{32} $$

The values of ${}^j n_2{}^2$, $\frac{\partial^j n_2{}^2}{\partial y}$, $\frac{\partial^j n_2{}^2}{\partial x}$ are determined from ${}^j y_2$.

Relation (27) is the corrector in the case under consideration.

We think of the trajectory as passing through a turning point at a current step if the value of $y_x{}'$ computed by (19), (24) and (27) changes sign at this step.

Unless the computed value of the ${}^j y_2$ has the sufficient accuracy, the iterations have to be repeated.

4. The Results of Numerical Testing

The numerical experiment was performed for the case when the medium refraction index modelled as $n^2(x,y) = a + bx + cy$ was defined on a uniform grid with the steps h_x and h_y in x and in y, respectively. The model parameters were $a = 1, b = -0.001, c = -0.01, h_y = 0.1$. The value of h_x varied from 2 to $0.5 (h_x = 2, 1, 0.5)$. The ray trajectory had the initial values $x = 0, y(0) = 0, y_x{}'(0) = 0.05$ (see also Fig.1). The discretization points of $n(x,y)$ were located at the grid points. In Fig.1 the case when $h_x = 2$ is shown. The numerical solution was compared with the theoretical one given by the relation

$$ y(x) = x\frac{c}{b} - \frac{2}{b}(\frac{a}{(1 + y_{x0}{}'^2)})^{\frac{1}{2}}(\frac{c}{b} - y_{x0}{}')[(\frac{a}{(1 + y_{x0}{}'^2)} - bx)^{\frac{1}{2}} - (\frac{a}{(1 + y_{x0}{}'^2)})^{\frac{1}{2}}]. $$

Figure 2 gives the values of the computational errors along the trajectory for different values of h_x. The numerical experiment proved the reasonableness of using the improved algorithm of numerical integration in neighborhoods of turning points where the absolute value of y_x' is small. Nevertheless, at the turning points, the computation error increases step-wise and thus influences on the total error for the remaining part of the ray trajectory. The error at a turning point is determined by the "phase" of the ray trajectory relative to the grid points position. If the turning point is passed successfully (when it is situated at a very short distance from a grid point), the error can have even less value (at $h_x = 2$) than in the case of more dense grid ($h_x = 1, 0.5$).

If the ray passes the turning point between the grid points, then there appear no computation problems. One can check the value of

$$q = \frac{n^2}{\left(\frac{n_0{}^2}{(1+y_{0x}{}'^2)} + \int\limits_0^x \frac{\partial n^2}{\partial x} dx\right)} - 1$$

and in the case when its modulus is small we do not compute the trajectory and proceed to the next step. But it should be pointed out that the proposed method looking very attractive due to its simplicity loses its efficiency when the value of the horizontal component $\frac{\partial n}{\partial x}$ of the refractive index is very large.

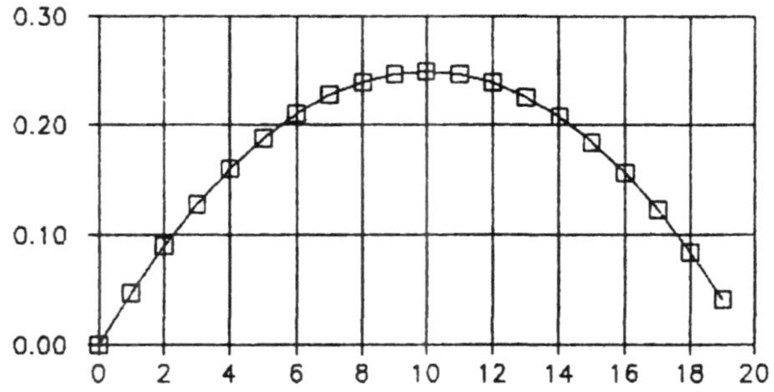

FIGURE 1. Ray trajectory.

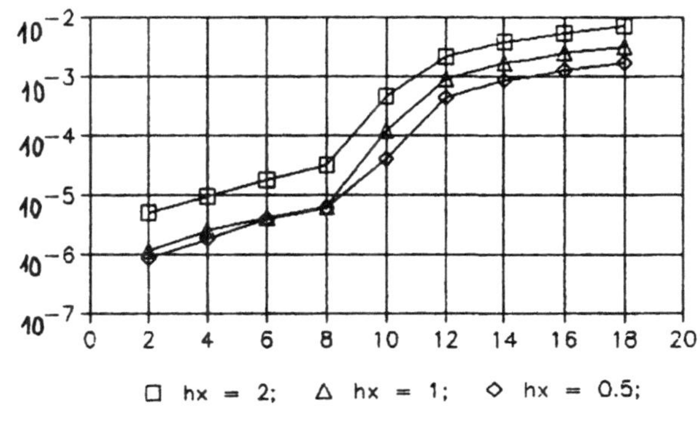

FIGURE 2. Calculation errors.

REFERENCES

Born M., Wolf E. (1964), *Principles of Optics.* Pergamon Press, Oxford, London.

Kravtsov Yu. A., Orlov Yu. I. (1983), *The Geometrical Optics of the Inhomogeneous Media.* Moscow: Nauka. (in Russian)

Morse P. M., Feshbach H. (1953), *Methods of Theoretical Physics.* McGrow-Hill Book Company Inc., New York, Toronto, London.

Seismic Tomography With Applications in Global
Seismology and Exploration Geophysics. Edited by Guust Nolet, 1987, by D.Reidel Publishing Company, Dordrecht, Holland.

Sampling and Resolution in CT

F. NATTERER

Westfälische Wilhelms-Universität Münster,
Institut für Numerische und instrumentelle Mathematik
Einsteinstrasse 62, D-48149 Münster
AMS (MOS) subject classification: 92C55, 44A12, 65R10, 65R20, 45L10

0. Introduction

Questions of sampling and resolution play an important role in tomography. Thus it is not surprising that many authors dealt with this problem. We mention only Logan (1975), Louis (1981, 1982, 1984), Joseph and Schulz (1980), Rattey and Lindgren (1981), Tretiak (1975). Even the early paper of Bracewell and Riddle (1956) contains a result on resolution.

We measure resolution in terms of the band-limited theory of communication theory. Let f be a function supported in the unit ball of $I\!R^n$, and define its Fourier transform by

$$\hat{f}(\xi) = (2\pi)^{-n/2} \int_{\mathbb{R}^n} f(x) e^{-ix\cdot\xi} dx.$$

Assume f to be essentially Ω-band-limited, i.e., $\hat{f}(\xi) \sim 0$ for $|\xi| > \Omega$ in an appropriate sense, to wit

$$\varepsilon_0(f,\Omega) = \int_{|\xi|>\Omega} |\hat{f}(\xi)| d\xi \leq \varepsilon \int_{\mathbb{R}^n} |\hat{f}| d\xi$$

with $\varepsilon << 1$. A somewhat loose application of the Shannon-Kotelnikov-Whittaker sampling theorem (see e.g. Jerry, 1977) tells us that the smallest detail in f is of size $2\pi/\Omega$, and that a sampling distance of π/Ω suffices to reconstruct f reliably from its samples. In the spirit of this theorem our question is how tomographic data of f have to sampled in order to reconstruct f reliably.

A typical result is as follows. Let R be the Radon transform, i.e.,

$$(Rf)(\theta, s) = \int_{x\cdot\theta=s} f(x) dx, \quad \theta \in S^{n-1}, \quad s \in I\!R^1.$$

Assume $(Rf)(\theta_j, \cdot) = 0$ for p directions $\theta_1, \ldots, \theta_p$ which are sufficiently general. Then, for $0 < \vartheta < 1$

$$\|f\|_{L_\infty(|x|<1)} \leq \frac{(2\pi)^{-n/2}}{1-\eta(\vartheta,\Omega)} \varepsilon_0(f, \vartheta\Omega) \tag{0.1}$$

provided that $p \geq \binom{\Omega+n-1}{n-1}$. Here, $\eta(\vartheta, \Omega)$ decays exponentially as $\Omega \to \infty$. For details see Louis (1981), Natterer (1986, Theorem 2.4).

Since

$$\binom{\Omega+n-1}{n-1} = \frac{1}{(n-1)!}\Omega^{n-1}(1+O(\Omega^{-1}))$$

we may interpret (0.1) as follows: If the Radon transform of an essentially Ω-band-limited function is known for $p \geq \frac{1}{(n-1)!}\Omega^{n-1}$ sufficiently general directions, then f is determined reliably.

For the X-ray-transform

$$Pf(\theta,x) = \int_{\mathbb{R}^1} f(x+t\theta)\mathrm{d}t, \quad \theta \in S^{n-1}, \quad x \in \theta^\perp$$

Maaß (1987) showed that f can be recovered reliably from p complete projections $Pf(\theta_j,\cdot)$, $j=1,\ldots,p$ provided that $p \geq \Omega$. Finally, for the divergent beam transform

$$Df(a,\theta) = \int_0^\infty f(a+t\theta)\mathrm{d}t, \quad \theta \in S^{n-1}, \quad |a| > 1$$

we showed in Natterer (1993) for $n=2$ that f can be recovered reliably from p complete projections $Df(a_j,\cdot)$, $j=1,\ldots,p$, the a_j uniformly distributed on $|a|=r>1$, provided that $p\frac{2r}{r+1}\Omega$.

In this paper we study fully discrete $2D$ sampling schemes. We present the results of Natterer (1993) in a simplified form. We give a reconstruction algorithm of the circular harmonic type which can be proved to achieve the claimed resolution.

1. Two-dimensional Sampling Schemes

Depending on the specific application the lines we integrate over are usually arranged in a certain pattern which we refer to as scanning geometry. In parallel beam tomography the lines come in parallel bundles. This situation is conveniently modelled by the Radon transform

$$Rf(\varphi,s) = \int_{x\cdot\theta=s} f(x)\mathrm{d}x, \quad \theta = \begin{pmatrix}\cos\varphi\\ \sin\varphi\end{pmatrix} \tag{1.1}$$

of a function f in $\mathbf{R}^2$. Here, the lines are parametrized by the angle φ and the real number s. In fan beam tomography the lines are arranged into fans emanating from points ("sources") distributed over a circle of radius $r>0$ around the origin. This gives rise to a reparametrization of the Radon transform, called the fan beam (or divergent beam, Smith et al., 1977) transform D. Let $b = r\begin{pmatrix}\cos\beta\\ \sin\beta\end{pmatrix}$ be a source and let $L(\beta,\alpha)$ be the ray through b making an angle α with the line (the "centre line") joining b with the origin. α is taken to be positive if $L(\beta,\alpha)$, viewed from b, is left to the centre line. Then,

$$Df(\beta,\alpha) = \int_{L(\beta,\alpha)} f(x)\mathrm{d}x. \tag{1.2}$$

D and R are related by

$$Df(\beta,\alpha) = Rf(\beta+\alpha-\pi/2, r\sin\alpha). \tag{1.3}$$

Now we are able to describe some scanning geometries used in tomography. In the standard parallel geometry we select p directions $\varphi_j = j\Delta\varphi$, $\Delta\varphi = \pi/p$, and lateral sampling points $s_\ell = \ell\Delta s$, $\Delta s = 1/q$. The measured line integrals are

$$g_{j\ell} = Rf(\varphi_j, s_\ell) \quad j = 0,\dots,p-1, \quad \ell = -q,\dots,q. \tag{1.4}$$

In the standard fan beam geometry we have p sources $b_j = r\begin{pmatrix}\cos\beta_j\\ \sin\beta_j\end{pmatrix}$, $\beta_j = j\Delta\beta$, $\Delta\beta = 2\pi/p$ and detector angles $\alpha_\ell = \ell\Delta\alpha + \delta$, $\Delta\alpha = \overline{\alpha}/q$, $\overline{\alpha} = \arcsin(1/r)$. δ is the detector set-off. It is usually chosen either as 0 or as $\Delta\alpha/4$. The measured line integrals are

$$g_{j\ell} = Df(\beta_j, \alpha_\ell), \quad j = 0,\dots,p-1, \quad \ell = -q,\dots,q. \tag{1.5}$$

Special cases are the PET geometry (which comes up naturally in positron emission tomography (PET)) which is obtained for $r = 1$, $\Delta\alpha = \Delta\beta/2$ and $\delta = 0$ (p even), $\delta = \Delta\alpha/2$ (p odd). It is obtained by placing p sources equidistantly on the circle and joining each pair of sources by straight lines.

It is well known that for (1.4) to have resolution $2\pi/\Omega$ it is necessary and sufficient that

$$\Delta\varphi \le \frac{\pi}{\Omega}, \qquad \Delta s \le \frac{\pi}{\Omega}. \tag{1.6}$$

This has been shown by Bracewell and Riddle (1956) simply by proving that the filtered backprojection algorithm, which is the most widely used reconstruction algorithm, recovers accurately an Ω–band-limited function if (1.7) is satisfied. It follows that the number of data we need for resolution $2\pi/\Omega$ is

$$\frac{\pi}{\Delta\varphi}\frac{2}{\Delta s} = \frac{2}{\pi}\Omega^2. \tag{1.7}$$

It has been observed by Rattey and Lindgren (1981) that this is not best possible. They showed by a 2D Fourier analysis of the Radon transform that the same resolution can be obtained with only one half of the data by using what came to be known as the parallel interlaced geometry. For p even, this geometry is obtained simply by dropping those lines in (1.4) for which $j + \ell$ odd, leaving us with only

$$\frac{1}{\pi}\Omega^2 \tag{1.8}$$

data. This number is best possible. Kruse (1989) and Faridani (1991) found versions of the filtered backprojection algorithm which actually achieve the correct resolution for the interlaced geometry, and Klaverkamp (1991) recently suggested a clever implementation of the direct algebraic algorithm (see Natterer, 1986, Chapt. 5) with the same property.

The purpose of the present paper is to extend these results to the fan beam geometry. We shall show that for (1.5) to have resolution $2\pi/\Omega$ it is necessary and sufficient that

$$\Delta\beta \le (r+1)\frac{\pi}{r\Omega}, \quad \Delta\alpha \le \frac{\pi}{r\Omega}. \tag{1.9}$$

Thus the minimal number of data for resolution $2\pi/\Omega$ is

$$\frac{2\pi}{\Delta\beta}\frac{2\arcsin(1/r)}{\Delta\alpha} = \frac{4}{\pi}\frac{r}{r+1}\frac{\arcsin(1/r)}{1/r}\Omega^2. \tag{1.10}$$

Many suboptimal results have been obtained in the literature, see Natterer (1986), p. 84. The question arises wether there exists a fan beam analogue of the parallel interlaced geometry which needs only one half of this data.

We shall show that this is in fact the case. There are efficient fan beam geometries which need only

$$\frac{1}{\pi}\frac{\arcsin(1/r)}{1/r}\Omega^2 \tag{1.11}$$

data. All of these efficient geometries are of the form

$$\begin{aligned} \beta_j &= j\Delta\beta, \\ \alpha_{j\ell} &= \ell\Delta\alpha + (j \bmod 2m)\gamma + \delta \end{aligned} \tag{1.12}$$

with certain numbers $\Delta\beta$, $\Delta\alpha$, γ, δ, m depending on $1/r$ and Ω. We call (1.12) a fan beam geometry with dynamic detector shift. Such a geometry has equidistant source positions and a fixed detector spacing. When the source moves from one position to the next one, the detector array undergoes a shift. It returns to its original position after $2m-1$ shifts. Thus the realization of (1.12) necessitates a non-rigid source-detector system. Practically this can be done by the flying spot technique of CT scanners such as SIEMENS SOMATOM PLUS (Alexander and Krumme, 1988). It turns out that (1.12) can be rearranged into a parallel interlaced geometry with sinusoidal (as opposed to equidistant as in (1.4)) lateral sampling. Also, (1.12) is a subset of a standard fan beam geometry and also of the PET geometry for the disk $|x| < r$ as reconstruction region.

2. Sampling of Periodic Functions

Df is a periodic function in each of its arguments. We shall give a version of the Petersen–Middleton sampling theorem (Petersen and Middleton, 1962) for such functions.

More generally we consider functions g in $\mathbf{R}^n$ having periods $p_1, \ldots, p_n$ with $P = (p_1, \ldots, p_n)$ non singular. We call such functions P-periodic. We do not aim at the greatest possible generality and assume all functions to be infinitely differentiable.

Let $L_W = W\mathbf{Z}^n$ be the lattice generated by the real non singular (n,n)-matrix W. Let $L_W^\perp = L_{2\pi W^{-T}}$ be the reciprocal lattice of L_W. We remark that W is not determined uniquely by L_W, but $L_W^\perp$ and $|\det(W)|$ are.

We shall sample P-periodic functions on L_W. For this to make sense we must assume $L_P \subseteq L_W$. This is the case iff $P = WM$ with an integer matrix M. It follows that $L_W^\perp \subseteq L_P^\perp$.

The Fourier transform of a P-periodic function g is a function on $L_P^\perp$, namely

$$\hat{g}(\xi) = |\det(P)|^{-1} \int_{P[0,1)^n} g(x)\,\mathrm{e}^{-\mathrm{i}x\cdot\xi}\mathrm{d}x, \quad \xi \in L_P^\perp. \tag{2.1}$$

g can be recovered from its Fourier transform by the Fourier series

$$g(x) = \sum_{\xi \in L_P^\perp} \hat{g}(\xi)\,\mathrm{e}^{\mathrm{i}x\cdot\xi}. \tag{2.2}$$

Our treatment relies entirely on Poisson's summation formula for the group $\mathbf{R}^n/L_P$ and its subgroup L_W/L_P:

$$\sum_{y\in L_W/L_P} g(y) = \left|\frac{\det(P)}{\det(W)}\right| \sum_{\xi\in L_W^\perp} \hat{g}(\xi). \tag{2.3}$$

This is a well known result in commutative harmonic analysis on locally compact groups, see e.g. Terras (1985). For a direct proof see Kruse (1989), Natterer (1993).

Theorem 2.1. *Let K be a finite set in $L_P^\perp$ such that $K+\xi \cap K+\xi' = \phi$ for $\xi, \xi' \in L_W^\perp$ and $\xi \neq \xi'$.*

(i) If g is a P-periodic function such that $\hat{g}(\xi) = 0$ for $\xi \notin K$, then g is uniquely determined by its values on L_W/L_P.

(ii) If g_1, g_2 meet the assumption on g in (i), then

$$\int_{P[0,1)^n} g_1(x)g_2(x)dx = |\det(W)| \sum_{y\in L_W/L_P} g_1(y)g_2(y).$$

This is an immediate consequence of (2.3), see Natterer (1993).

We shall apply Theorem 2.1 to functions whose Fourier transform is not strictly zero outside K. Then, the conclusions of the theorem hold only approximately. The error caused in this way is called aliasing in communication theory. From the proof of the theorem we see that aliasing is small if the replica $\eta \to \hat{g}(\xi+\eta)$ of $\hat{g}$ do not interfere significantly with $\hat{g}$ in K for $\xi \in L_W^\perp$, $\xi \neq 0$.

3. Asymptotic Estimates

In this section we consider functions in $\mathbf{R}^2$ which are P-periodic with $P = 2\pi \begin{pmatrix} 1 & 0 \\ 0 & 1/2 \end{pmatrix}$. The first one of these functions is $g = Df$ where f is a function supported in $|x| \le \rho$ and essentially Ω-band-limited. The second one is the function

$$h(\beta,\alpha) = v(x\cdot\theta - r\sin\alpha)\cos\alpha, \qquad \theta = \begin{pmatrix} \cos(\beta+\alpha-\pi/2) \\ \sin(\beta+\alpha-\pi/2) \end{pmatrix} \tag{3.1}$$

where x is fixed and

$$\hat{v}(\sigma) = |\sigma|\hat{\phi}\left(|\frac{\sigma}{\Omega}|\right) \tag{3.2}$$

with a filter function $\hat{\phi}$ vanishing outside $[0,1]$. h is given by (3.1) for $|\alpha| \le \pi/2$ only and is extended as a π-periodic function of α beyond this interval.

The Fourier transforms of g, h are defined on $L_P^\perp = \mathbf{Z}\times 2\mathbf{Z}$. Thus we can write

$$\hat{g}(k,m) = \frac{1}{2\pi^2}\int_{-\pi}^{\pi}\int_{-\pi/2}^{\pi/2} g(\beta,\alpha)\,\mathrm{e}^{-ik\beta-im\alpha}\mathrm{d}\alpha\mathrm{d}\beta$$

where k, m are integers, m even, and correspondingly for h. Let K be the set

$$|k-m| < \Omega r, \quad |k|r < |k-m|. \tag{3.3}$$

It has been shown in Natterer (1993) that for $1 < r$, $\hat{g}$, $\hat{h}$ decay fast away from K. Hence these functions can be considered as being small in a region slightly larger than K.

4. Efficient Sampling Geometries and Reconstruction Algorithms

We now combine the results of the preceding sections to find efficient sampling lattices for $g = Df$ and appropriate reconstruction algorithms. According to Theorem 2.1 (i), a lattice L_W with $L_W \supseteq L_P$ is a good sampling lattice for g if the translates of the set K of (3.3) with respect to $L_W^\perp$ are mutually disjoint. We see that this is the case if we choose as generator of $L_W^\perp$ the matrix

$$2\pi W^{-T} = \Omega \begin{pmatrix} 1 & 0 \\ 1-r & 2r \end{pmatrix}. \tag{4.1}$$

From (4.1) we obtain immediately

$$W = \frac{\pi}{\Omega} \begin{pmatrix} 2 & 1-\vartheta \\ 0 & \vartheta \end{pmatrix}, \qquad \vartheta = 1/r. \tag{4.2}$$

Let P be as in section 3. For $L_P \subseteq L_W$ to hold it is necessary and sufficient that

$$\begin{pmatrix} 2\pi \\ 0 \end{pmatrix}, \begin{pmatrix} 0 \\ \pi \end{pmatrix} \in W\mathbf{Z}^2.$$

This is the case iff

$$\Omega, r\Omega \quad \text{integer of the same parity.} \tag{4.3}$$

This is practically not a serious restriction since it can be fulfilled by slight changes in Ω, r. Also, a closer inspection of the proof reveals that the second of these conditions, $r\Omega$ an integer of the same parity as Ω, can be disposed of. The lattice points can be written as (β_k, α_k), $k = \begin{pmatrix} k_1 \\ k_2 \end{pmatrix}$, k_i integer, with

$$\beta_k = \frac{\pi}{\Omega}(2k_1 + (1-\vartheta)k_2), \quad \alpha_k = \frac{\pi}{\Omega r}k_2. \tag{4.4}$$

In §5 we shall study these lattices from a geometric point of view.

In order to derive a reconstruction algorithm for the scanning geometry (4.4) we start out from a regularized version of Radon's inversion formula. In parallel geometry it reads

$$V * f(x) = \int_0^{2\pi} \int_{-\rho}^{\rho} v(x \cdot \theta - s) Rf(\varphi, s) ds d\varphi \tag{4.5}$$

where θ is as in (1.1) and

$$\begin{aligned} \hat{v}(\sigma) &= \frac{1}{2}(2\pi)^{-3/2}|\sigma|\hat{\phi}\left(\frac{|\sigma|}{\Omega}\right) \\ \hat{V}(\xi) &= \frac{1}{2\pi}\hat{\phi}\left(\frac{|\xi|}{\Omega}\right), \end{aligned}$$

see e.g. Natterer (1986), Chapt. V.1. Here, $\hat{\phi}$ is a low pass filter vanishing outside $[0,1]$ and V is an approximate δ-function.

The parallel coordinates φ, s are related to the fan beam coordinates β, α by

$$\varphi = \beta + \alpha - \pi/2, \qquad s = r \sin\alpha. \tag{4.6}$$

Note that the two coordinate systems are related in a one-to-one fashion. But in each system, each straight line is represented exactly twice.

Substituting (4.6) into (4.5) we get

$$V * f(x) = r \int_0^{2\pi} \int_{-\pi/2}^{\pi/2} v(x \cdot \theta - r \sin \alpha) Df(\beta, \alpha) \cos \alpha d\alpha d\beta,$$

$$\theta = \begin{pmatrix} \cos(\beta + \alpha - \pi/2) \\ \sin(\beta + \alpha - \pi/2) \end{pmatrix}.$$

On the right hand side we have an inner product to which we can apply Theorem 2.1 (ii) with $g_1 = h$ and $g_2 = Df$. According to §3 the support of the Fourier transform of both factors is essentially K. Thus the inner product can be evaluated on L_W/L_P, yielding apart from aliasing

$$V * f(x) = \det(W) \sum_k v(x \cdot \theta_k - r \sin \alpha_k) Df(\beta_k, \alpha_k) \cos \alpha_k, \tag{4.7}$$

$$\theta_k = \begin{pmatrix} \cos(\beta_k + \alpha_k - \pi/2) \\ \sin(\beta_k + \alpha_k - \pi/2) \end{pmatrix}.$$

Here the summation is restricted to those $k \in \mathbf{Z}^2$ for which $(\beta_k, \alpha_k) \in L_{W/P}$. In order to prevent aliasing the remarks following (4.1) should be observed.

We conclude this section by deriving the conditions (1.10) for the standard fan beam scanning geometry (1.5). We see that

$$2\pi W^{-T} = 2\Omega r \begin{pmatrix} r+1 & 0 \\ 0 & 1 \end{pmatrix}, \quad \text{i.e.,} \quad W = \frac{\pi}{\Omega r} \begin{pmatrix} r+1 & 0 \\ 0 & 1 \end{pmatrix}$$

defines a good sampling lattice for g. This leads to a standard fan beam geometry with

$$\Delta\beta = (r+1)\frac{\pi}{\Omega r}, \quad \Delta\alpha = \frac{\pi}{\Omega r}, \tag{4.8}$$

proving (1.9). The translates of K do not cover all of the Fourier plane, showing that standard fan beam scanning is not best possible.

Apart from aliasing we have in this case

$$V * f(x) = r \Delta\alpha \Delta\beta \sum_{j=0}^{p-1} \sum_{\ell=-q}^{q} v(x \cdot \theta_{j\ell} - r \sin \alpha_\ell) Df(\beta_j, \alpha_\ell) \cos \alpha_\ell \tag{4.9}$$

$$\begin{aligned} \beta_j &= j\Delta\beta &&, \ j = 0, \ldots, p-1, \\ \alpha_\ell &= \ell\Delta\alpha &&, \ \ell = -q, \ldots, q, \\ \theta_{j\ell} &= \begin{pmatrix} \cos(\beta_j + \alpha_\ell - \pi/2) \\ \sin(\beta_j + \alpha_\ell - \pi/2) \end{pmatrix} && . \end{aligned}$$

A naive evaluation of (4.9) requires $2pq = O(\Omega^2)$ operations for each reconstruction point x. Since $V * f$ has to be evaluated on a grid with step size π/Ω this amounts to $O(\Omega^4)$ operations. This is much more than the $O(\Omega^3)$ operations of the usual filtered backprojection algorithm (Herman, 1980) for fan beam data. However, exploiting the rotational symmetry of (4.9) we obtain an algorithm of complexity $O(\Omega^3 \log \Omega)$ which is competitive in speed with the filtered backprojection algorithm. This algorithm evaluates $V * f$ at the points $x_{ik} = t_i \begin{pmatrix} \cos\beta_k \\ \sin\beta_k \end{pmatrix}$, i.e.,

$$V * f(x_{ik}) = r\Delta\alpha\Delta\beta \sum_{e=-q}^{q} \sum_{j=0}^{p-1} v\left(t_i \cos(\beta_k - \beta_j - \alpha_\ell + \tfrac{\pi}{2})\right) \cdot g(\beta_j, \alpha_\ell) \cos\alpha_\ell.$$

The j-sum is a cyclic convolution which can be done by FFT in $O(p \log p)$ operations. This has to be done for each i, ℓ, leading to the complexity $O(q^2 p \log p) = O(\Omega^3 \log \Omega)$. The resulting algorithm is of the circular harmonic type, see e.g. Hansen (1981). In numerous numerical experiments this algorithm turned out to be much faster (in spite of the additional $\log \Omega$) than the filtered backprojection algorithm and also much more accurate.

5. Geometrical Properties of the Efficient Scanning Geometry

The lattice L_W from (4.2) enjoys some nice geometrical properties some of which may be useful in practice. We give the results for a reconstruction region $|x| < \rho$ where $\rho \le r$.

Theorem 5.1. *L_W is identical to the lattice*

$$\begin{aligned} \varphi_j &= \frac{\pi}{\Omega\rho} j \quad , \quad j = 0, \dots, 2\Omega\rho - 1, \\ s_{j\ell} &= r \sin \frac{\pi}{\Omega r} \ell, \quad \ell + j \text{ even} \end{aligned} \tag{5.1}$$

in parallel coordinates.

PROOF. Using the relation (4.6) between the parallel coordinates φ, s and the fan beam coordinates β, α we find from (4.4) that L_W can be represented in parallel coordinates by

$$\varphi_k = \frac{\pi}{\Omega\rho}(2k_1 + k_2) - \frac{\pi}{2}, \quad s_k = r \sin \frac{\pi}{\Omega r} k_2.$$

Putting $j = 2k_1 + k_2$, $\ell = j - 2k_1$ we get

$$\varphi_k = \frac{\pi}{\Omega\rho} j - \frac{\pi}{2} \quad , \quad s_k = r \sin \frac{\pi}{\Omega r} \ell$$

and $j + \ell$ is even. Vice versa, for j, ℓ with $j + \ell$ even, $k_2 = \ell$ and $k_1 = (j - \ell)/2$ are integers, showing that each pair $\varphi = \frac{\pi}{\Omega\rho} j - \frac{\pi}{2}$, $s = r \sin \frac{\pi}{\Omega r} \ell$ with $j + \ell$ even belongs to L_W. By (4.3), $\Omega\rho$ is an integer, hence we can restrict j to $0, \dots, 2\Omega\rho - 1$.

We call (5.1) a parallel interlaced geometry with sinusoidal lateral sampling. Since f is supported in $|x| < \rho$, ℓ can be restricted to

$$|\ell| \le \frac{\Omega r}{\pi} \overline{\alpha}, \quad \overline{\alpha} = \arcsin(\rho/r). \tag{5.2}$$

Theorem 5.1 together with (4.1) leads directly to a reconstruction algorithm of the filtered backprojection type.

The following theorem shows that L_W is a fan beam geometry with dynamic detector shift. For its formulation we assume $\vartheta = N/M$ with relatively prime integers $N \le M$. It is easy to see that

$$\gcd(2M, M-N) = \begin{cases} 1 & , \quad M-N \text{ odd}, \\ 2 & , \quad M-N \text{ even}. \end{cases}$$

We now define

$$n = \frac{N}{\gcd(2M, M-N)} = \begin{cases} N & , \quad M-N \text{ odd}, \\ N/2 & , \quad M-N \text{ even} \end{cases} \tag{5.3}$$

In other words we have $\vartheta = n/m$ with integers or half integers $n \le m$. We take integers if this representation of ϑ is possible with n, m relatively prime and of different parity. Otherwise we take half integers with numerators relatively prime. A few examples are

$$\begin{aligned} \vartheta &= 1 &&: \ n = m = 1/2, \\ \vartheta &= 1/2 &&: \ n = 1, \quad m = 2, \\ \vartheta &= 1/3 &&: \ n = 1/2, \quad m = 3/2. \end{aligned}$$

Obviously, $m - n$ and $2m$ are always integers and $\gcd(m-n, 2n) = 1$. Thus there is exactly one positive integer $k_0 < 2m$ such that

$$(m-n)k_0 \equiv j \mod 2m \tag{5.4}$$

except in the case $m = n (= \frac{1}{2})$ for which we put $k_0 = 0$.

Theorem 5.2. *Let n, m be as in (5.3) and k_0 as in (5.4). Then, L_W is identical to the lattice*

$$\beta_j = \frac{\pi}{\Omega \rho m} j \quad , \quad \alpha_{j\ell} = \frac{2\pi m}{\Omega r}\ell + \frac{\pi k_0}{\Omega r}(j \bmod 2m), \qquad j = 0, \ldots, 2\Omega\rho m - 1. \tag{5.5}$$

PROOF. Using (5.3) we can represent the points of L_W from (4.4) in fan beam coordinates as

$$\beta_k = \frac{\pi}{\Omega \rho m}(2mk_1 + (m-n)k_2), \qquad \alpha_k = \frac{\pi}{\Omega r} k_2.$$

As k_1, k_2 run through the integers, $j = 2mk_1 + (m-n)k_2$ assumes al integer values since $2m, m-n$ are relatively prime. Thus, $(\beta, \alpha) \in L_W$ iff for some integer j

$$\beta = \frac{\pi}{\Omega \rho m} j, \quad \alpha = \frac{\pi}{\Omega r} k_2. \tag{5.6}$$

where k_2 is any solution of $(m-n)k_2 \equiv j \mod (2m)$. The general solution of this congruence is

$$k_2 = (j \bmod 2m)k_0 + 2m\ell; \tag{5.7}$$

this holds also in the case $m = n$ $(= \frac{1}{2})$ since $k_0 = 0$ in that case. Plugging (5.7) into (5.6) and changing the notation yields (5.5).

Now we turn to the question of redundancy. In the coordinate systems we used every line is represented exactly twice. Thus it may well happen that two different lattice points represent the same line. We call a lattice redundant if it contains at least one such pair.

Theorem 5.3. *L_W is redundant iff $\Omega\rho$ even. In that case, every line is represented by exactly two points of L_W.*

PROOF. Two lattice points (β_k, a_k), $(\beta_{k'}, \alpha_{k'})$ represent the same line iff

$$\beta_k + \alpha_k = \beta_{k'} + \alpha_{k'} \pm \pi, \quad a_k = -\alpha_{k'}.$$

Inserting (4.4) and dividing by $\pi/(\Omega\rho)$ we get

$$2k_1 + k_2 = 2k_1' + k_2' \pm \Omega\rho, \quad k_2 = -k_2'$$

or, equivalently,

$$\pm\,\Omega\rho = 2(k_1 - k_1') + 2k_2, \quad k_2 = -k_2' \tag{5.8}$$

Thus, if L_W is redundant, then $\Omega\rho$ in necessarily even. Vice versa, if $\Omega\rho$ is even, then (5.8) has a unique solution $k' \neq k$ for such k.

L_W contains essentially

$$2\Omega\rho\frac{\Omega r}{\pi}\overline{\alpha} = \frac{2\arcsin\vartheta}{\pi\,\vartheta}\rho^2\Omega^2$$

points, as can be seen from (5.1), (5.2). If $\Omega\rho$ is odd, this also the number of lines generated by L_W. If $\Omega\rho$ is even, then the number of lines is only one half of this number, leaving us with the optimal number (1.12) of data. Thus L_W should only be used with $\Omega\rho$ even. In that case one can exploit redundancy in the following way. In the parallel arrangement (5.1) of L_W, one simply restricts j to $0, \ldots, \Omega\rho-1$. Then, $\varphi_j, s_{j\ell}$ and $\varphi_{j+\Omega\rho}, s_{j,-\ell}$ represent the same line. In the fan beam arrangement (5.5), one restricts j such that $\beta_j \in [0, \pi+2\overline{\alpha})$ with α from (5.2). Then, for each lattice point $(\beta_{j'}, \alpha_{j'\ell'}$ with $\beta_{j'}$ in the complementary interval $[\pi+2\overline{\alpha}, 2\pi)$, $(\beta_j - \alpha_{j'\ell'})$ with $\beta_j = \beta_{j'} + 2\alpha_{j'\ell'} - \pi$ is a lattice point in the restricted set which represents the same line.

In order to have redundancy also in the case of $\Omega\rho$ odd we introduce the lattice

$$\overline{L}_W = L_W + \begin{pmatrix} 0 \\ \delta \end{pmatrix}, \quad \delta = \frac{\pi}{2\Omega r}.$$

$\overline{L}_W$ is a lattice with a detector set-off δ.

Theorem 5.4. *$\overline{L}_W$ is redundant iff $\Omega\rho$ odd. In that case, every line is represented by exactly two points of $\overline{L}_W$.*

PROOF. Exactly as in the preceding proof we see that $(\beta_k, \alpha_k + \delta)$, $(\beta_{k'}, \alpha_{k'} + \delta)$ represent the same line iff

$$2k_1 + k_2 = 2k_1' + k_2' = \pm\Omega\rho, \qquad k_2 + \frac{1}{2} = -(k_2' + \frac{1}{2}), \tag{5.9}$$

or, equivalently,

$$2(k_1 - k_1') + 2k_2 + 1 = \pm\Omega\rho, \quad k_2 + k_2' = -1. \tag{5.10}$$

Thus, if $\overline{L}_W$ is redundant, then $\Omega\rho$ is necessarily odd. Vice versa, if $\Omega\rho$ is odd, then (5.10) has a unique solution $k' \neq k$ for each k.

The nice thing with $\overline{L}_W$ for $\Omega\rho$ odd is that it contains non-redundant sublattices representing the same scanning geometry while containing only one half of the points of $\overline{L}_W$. One of those sublattices is

$$\overline{L}_W = \left\{ Wk + \begin{pmatrix} 0 \\ \delta \end{pmatrix} : 2k_1 + k_2 \text{ even} \right\}. \tag{5.11}$$

In fact we see from (5.9) that for each lattice point $(\beta_{k'}, \alpha_{k'} + \delta)$ in the complementary set $L_W - \tilde{L}_W$ there is a lattice point in $\tilde{L}_W$ which represents the same line. Note that (5.11) has the optimal number (1.12) of points. A look at the proof of Theorem 5.1 shows that $\tilde{L}_W$ may be represented in parallel coordinates by

$$\begin{aligned} \varphi_j &= \frac{\pi}{\Omega\rho} j + \delta - \frac{\pi}{2}, \quad j = 0, 2, \ldots, 2\Omega\rho - 2, \\ s_\ell &= r \sin \frac{\pi}{\Omega r}\left(\ell + \frac{1}{2}\right), \quad \ell \text{ even}. \end{aligned} \tag{5.12}$$

In fan beam coordinates, the points of $\overline{L}_W$ can be represented either by $(\beta_k, \alpha_k + \delta)$ or by $(\beta_j, \alpha_{j\ell} + \delta)$. These points are identical iff

$$j = 2mk_1 + (m - n)k_2, \quad k_2 = (j \bmod 2mk_0) + 2m\ell,$$

as can be seen from the proof of Theorem 5.2 it follows that

$$j = m(2k_1 + k_2) - n((j \bmod 2m)k_0 + (2m\ell).$$

Thus, if $(\beta_k, \alpha_k + \delta) \in \tilde{L}_W$, i.e., $2k_1 + k_2$ even, then

$$j + nk_0(j \bmod 2m) + 2nm\ell \equiv 0(2m),$$

or, since $nk_0 \equiv mk_0 - 1\ (2m)$,

$$mk_0(j \bmod 2m) + 2nm\ell \equiv 0(2m).$$

Since $n - m$ is an integer this is equivalent to

$$k_0(j \bmod 2m) + 2m\ell \text{ even}. \tag{5.13}$$

Because of $\gcd(2m, k_0) = 1$ this is the case precisely for one half of the lattice points $(\beta_j, \alpha_{j\ell} + \delta)$. We conclude that $\tilde{L}_W$ is given by

$$\begin{aligned} \beta_j = \frac{\pi}{\Omega\rho m} j, \quad \alpha_{j\ell} &= \frac{2\pi m}{\Omega r}\ell + \frac{\pi k_0}{\Omega r}(j \bmod 2m) + \frac{\pi}{2\Omega r} \\ j = 0, \ldots, 2\Omega\rho m - 1 \quad &, \quad k_0(j \bmod 2m) + 2m\ell \text{ even}. \end{aligned} \tag{5.14}$$

The latter condition is easily seem to be equivalent to

$$\begin{array}{rlll} j & \text{even} & if\ 2m & \text{even}, \\ \ell & \text{even} & if\ 2m & \text{odd and } k_0 \text{ even}, \\ \ell + j & \text{even} & if\ 2m & \text{odd and } k_0 \text{ odd}. \end{array} \tag{5.15}$$

We remark that for $\rho = r$ (i.e., $n = m = \frac{1}{2}$) and $\delta = 0(\Omega\rho$ even), $\delta = \frac{\pi}{2\Omega r}$ ($\Omega\rho$ odd) the efficient schemes are just the well-known PET scheme mentioned in the introduction. We also see that a fan beam scheme with dynamic detector shift is a sublattice of a PET scheme iff $2\Delta\alpha \,/\, \Delta\beta$, $2\gamma \,/\, \Delta\beta, 4\delta \,/\, \Delta\beta$ are integers, the last one with the same parity as $2\pi \,/\, \Delta\beta$. A quick perusal shows that this in fact the case for all efficient schemes. Thus the PET scheme is not only the efficient scheme for $\rho = r$, it also contains all the other efficient schemes for subsets.

REFERENCES

Alexander J., Krumme H. J. (1988), Einführung in die Technik von SOMATOM·PLUS. *Electromedica* **56**, 50–56.

Bracewell R. N., Riddle A. C. (1956), Strip integration in radio astronomy. *Aus. J. Phys.* **9**, 198–217.

Faridani A. (1991), Reconstructing from efficiently sampled data in parallel-beam computed tomography. In: G.F. Roach (ed.): *Inverse Problems and Imaging*, Pitman Research Notes in Mathematics Series, vol. 245, Longman 1991, pp. 68–102.

Hansen E. W. (1981), Circular harmonic image reconstruction, *Applied Optics* **20**, 2266–2274.

Herman G. T. (1980), Image Reconstruction from Projections. *The Fundamentals of Computerized Tomography*. Academic Press.

Jerry A. J. (1977), The Shannon sampling theorem - its various extensions and applications: a tutorial review. *Proc. IEEE* **65**, 1565–1596.

Joseph P. M., Schulz R. A. (1980), View sampling requirements in fan beam computed tomography. *Med. Phys.* **7**, 692–702.

Klaverkamp W. (1991), Tomographische Bildrekonstruktion mit direkten algebraischen Verfahren. *Dissertation*, Fachbereich Mathematik der Universität Münster, Einsteinstrasse 62, 4400 Münster, Germany.

Kruse H. (1989), Resolution of reconstruction methods in computerized tomography. *SIAM J. Sci. Stat. Comp.* **10**, 447–474.

Logan B. F. (1975), The uncertainty principle in reconstructing functions from projections. *Duke Math. J.* **42**, 661–706.

Louis A. K. (1981), Ghosts in tomography - The null space of the Radon transform. *Math. Meth. in the Appl. Sci.* **3**, 1–10.

Louis A. K. (1982), Optimal sampling in nuclear magnetic resonance (NMR) tomography. *J. of Computer Assisted Tomography* **6**, 334–340.

Louis A. K. (1984), Orthogonal function series expansions and the null space of the Radon transform. *SIAM J. Math. Anal.* **15**, 621–633.

Maaß P. (1987), The X-ray transform: singular value decomposition and resolution. *Inverse Problems* **3**, 729–741.

Natterer F. (1986), *The Mathematics of Computerized Tomography*. Wiley-Teubner.

Natterer F. (1993), Sampling in fan beam tomography. *SIAM J. Appl. Math.* **53**, 358–380.

Petersen P. P., Middleton D. (1962), Sampling and reconstruction of wave-number-limited functions in N-dimensional euclidean space. *Inf. Control* **5**, 279–323.

Rattey P. A., Lindgren A. G. (1981), Sampling the $2-D$ Radon transform. *IEEE Trans. Acoust. Speech Signal Procesing*, **ASSP-29**, 994–1002.

Scudder H. J. (1978), Introduction to computer aided tomography. *Proceedings of the IEEE* **66**, 628–637.

Smith K. T., Solmon D. C., Wagner S. L. (1977), Practical and mathematical aspects of the problem of reconstructing a function from radiographs. *Bull. AMS*, **83**, 1227–1270.

Terras A. (1985), *Harmonic analysis on symmetric spaces and applications I.* Springer.

Tretiak O. J. (1975), The point speed function for the convolution algorithm. In: Gordon, R. (ed.): *Image processing for 2-D and 3-D reconstruction from Projections.* Stanford.

Some Problems of Computerized Tomography in Static Potential Fields [1]

A. A. SCHWAB

Institute of Hydrodynamics

Pr. Lavrent'ev, Novosibirsk, 630090, Russia

ABSTRACT

The paper is concerned with a problem of restoration of a stress tensor. A straightforward method is considered as applied to this problem.

Formulation of the Inverse Problem

The paper deals with a nonclassical problem of restoration of a total stress tensor and with a problem of discerning defects in a body when loads and displacements are applied simultaneously to a part of the body surface. Such problems arise in diagnostics of the strain-stress state of objects, holographic defectoscopy, geomechanics, and seismology when estimating tectonic stresses in the Earth's crust. In geophysics such problems are known as the continuation problems of magnetostatic, electrostatic, gravitation and other fields.

At present these problems are solved by a semi-inverse method, which is rather efficient especially for the problems with a lack of experimental data.

1. Let us consider an elastic body Ω with a surface $\partial\Omega$. Let $\vec{P}$ be a load vector, and $\vec{U}$ be displacement vectors which are assumed to be given on a part $S_{u\sigma}$ of the surface $\partial\Omega$:

$$\vec{U} = \{u_1^V, u_2^V, u_3^V\}, \quad \vec{P} = \{p_1^V, p_2^V, p_3^V\}. \tag{1}$$

DEFINITION. The problem of determining a tension and a strain field in a body under conditions (1) will be called the Problem $(\vec{U}, \vec{P})$ (see Schwab, 1989).

The Problem $(\vec{U}, \vec{P})$ can be reduced to the Cauchy problem for the Laplace equation (see Schwab, 1989). The main attention is paid to the solution of two-dimensional problem by solving the Cauchy problem. The solution of such problems is obtained by the regularization method due to their numerical instability (see Strakhov, 1990). The idea to use the representation of the general solution of the elasticity problem by a holomographic vector was developed in Schwab (1988, 1989). The main definitions are given below.

Let $\vec{u}$ be a vector of elastic displacement, i.e.,

$$L\vec{u} \equiv (\lambda + 2\mu)\nabla(\nabla \cdot \vec{u}) - \mu\nabla \times (\nabla \times \vec{u}) = 0. \tag{2}$$

[1]The work was supported by the Russian Foundation of Fundamental Research under the grant No. 94-01-00-896.

Let us introduce a function $\vec{f}(\vec{r}) = -\mu\nabla\times\vec{u}$, $f_0(\vec{r}) = (\lambda+2\mu)\nabla\cdot\vec{u}$. Then from (2) it follows that

$$\nabla\cdot\vec{f}(\vec{r}) = 0, \quad \nabla f_0(\vec{r}) + \nabla\times\vec{f}(\vec{r}) = 0. \tag{3}$$

System (3) is the Moisil-Theodoresco system (see Moisil and Theodoresco, 1931). The quaternion function $f(\vec{r}) = f_0(\vec{r}) + \vec{f}(\vec{r})$ is a left-regular function of incomplete quaternion variable (Fueter, 1932). The four-component vector $f(r) = \{f_0, f_1, f_2, f_3\}$ satisfying (3) is called the holomorphic vector (see Bitsadze, 1984). In this case $f_0 = (\lambda+2\mu)\nabla\times\vec{u}$, $f_i = -\mu(\nabla\times\vec{u})_i$. We recall one of the basic theorem of the theory of complex variable functions (Bitsadze, 1984).

Theorem 1. *If a vector $f(\vec{r}) \in C^1$ within a region $\Omega^+ \subset E_3$ with the Lyapunov piecewise smooth boundary solves MTS in Ω, then*

$$\Delta_{00}(\vec{r}) = \begin{cases} f(\vec{r}), & \vec{r}\in\Omega^+, \\ 0, & \vec{r}\in\Omega^-, \end{cases} \tag{4}$$

where

$$\Delta_{00} = \frac{1}{4\pi}\iint\limits_{\partial\Omega} M(\vec{r},\vec{r_0}) f(\vec{r_0})\mathrm{d}S_{\vec{r_0}},$$

$M(\vec{r},\vec{r_0})$ is a differential matrix operator

$$M(\vec{r},\vec{r_0}) = D^*\left(\frac{\partial}{\partial y_1}, \frac{\partial}{\partial y_2}, \frac{\partial}{\partial y_3}\right)\frac{1}{|\vec{r}-\vec{r_0}|}D(n_1, n_2, n_3).$$

As follows from Schwab (1988), the problem $(\vec{U}, \vec{P})$ can be formulated as a problem of reconstructing the holomorphic vector in Ω from its values on $S_{u\sigma}$. This problem can be reduced by means of the integral Cauchy formula to integral equation

$$-\frac{1}{2}q(\vec{r}) + \frac{1}{4\pi}\iint\limits_{\partial\Omega\setminus S_{u\sigma}} M(\vec{r},\vec{r_0})q(\vec{r_0})\mathrm{d}S_{\vec{r_0}} = \begin{cases} F^*(\vec{r}), & \vec{r}\in\Omega^+\setminus S_{u\sigma}, \quad (5) \\ F^*(\vec{r}), & \vec{r}\in S_{u\sigma}, \quad (6) \end{cases}$$

$$F^*(\vec{r}) = \iint\limits_{S_{u\sigma}} M(\vec{r},\vec{r_0})q^*(\vec{r_0})\mathrm{d}S_{\vec{r_0}}.$$

The value $q^*(\vec{r})$ can be calculated using vectors $\vec{U}$, $\vec{P}$ on $S_{u\sigma}$. Problem (1), (5), (6) was solved numerically. Figures 1, 2 present the results of numerical calculations the case when $\vec{u}$ is given on a face of the unit cube: $\vec{u} = \operatorname{grad}\psi(x,y,z)$, where

$$\psi(x,y,z) = \left((x-x_0)^2 + (y-y_0)^2 + (z-z_0)^2\right)^{-\frac{1}{2}}, \quad (x_0,y_0,z_0) = (0.5, 0.5, 0.5).$$

The error was about 5–10 %. The solution was constructed without regularization.

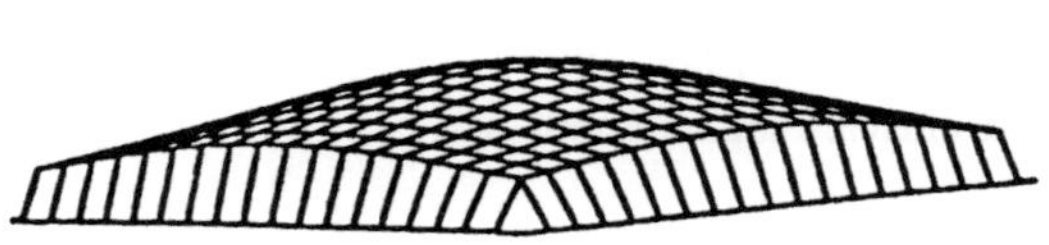

FIGURE 1. The surface of the initial data $u = u(x,y,0)$.

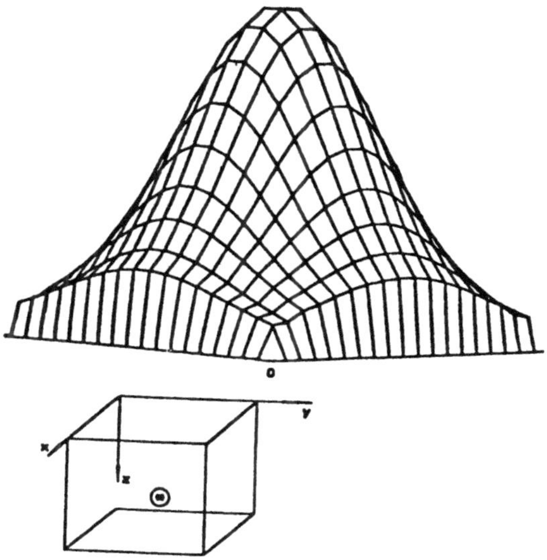

FIGURE 2. The solution surface $u = u(x, y, 0.375)$.

2. Let us consider the problem $(\vec{U}, \vec{P})$ for an elastic inhomogeneous medium. The inhomogeneity was shown (Schwab, 1989) to bring anomalies into the tension field. The numerical discernment of the anomalies is questionable due to conditional well-posedness of the problem.

Theorem 2. *(On the boundary values of the holomorphic vector). For the value of the four-component vector $q(\vec{r})$ given on a closed Lyapunov smooth surface $\partial\Omega$ to be the boundary value of a holomorphic vector in Ω^+, it is necessary and sufficient that the condition*

$$\iint\limits_{\partial\Omega} M(\vec{r}, \vec{r}_0) q(\vec{r}_0) \mathrm{d}S_{\vec{r}_0} = 0, \quad \vec{r} \in \Omega^-, \tag{7}$$

holds.

PROOF. The necessity follows immediately from (4). To prove the sufficiency, define the vector $Q(r)$ on $\Omega^+ \cup \Omega^-$ as

$$Q(r) = \frac{1}{4\pi} \iint\limits_{\partial\Omega} M(\vec{r}, \vec{r}_0) q(\vec{r}_0) \mathrm{d}S_{\vec{r}_0}.$$

Then from an analogy of the Sokhotsky-Plemeli formula we have $Q^+(\vec{r}_i) - Q^-(\vec{r}_e) = q(\vec{r}_0)$, $\vec{r}_0 \in \partial\Omega$, where $\vec{r}_e$, $\vec{r}_i$ are external and internal limiting points to $\vec{r}_0$. From (7) it follows that $Q^-(\vec{r}_e) = 0$, hence, $Q^+(\vec{r}_i) = q(\vec{r}_0)$, $\vec{r}_0 \in \partial\Omega$. The theorem is proved.

A class of four-component vectors satisfying (7) is called the class $B(\partial\Omega)$. If the boundary value belongs to $B(\partial\Omega)$, the vector can be continuously extended into Ω. If defects are present in the region, they bring anomalies into the solution (Schwab, 1990), i.e., the vector can not be continuously extended into Ω^+ or $q(\vec{r}) \notin B(\partial\Omega)$. Thus by verification of the criterion given in Theorem 2, we can answer the question about the existence of defects in Ω^+.

A numerical analysis was performed to estimate the criterion efficiency. The values of $q(\vec{r}) \in B(\partial\Omega)$ and $q_1(\vec{r}) \notin B(\partial\Omega)$ were prescribed on the surface of a unit cube. The four-component vector $q_1(\vec{r})$ described the field in the case when two pressure centers were located in the cube. Figures 3, 4 show the surfaces $|\delta(\vec{r})|$ of deviation from the criterion (7) when the point moves above the cube surface. The peaks correspond to the strongest influence of the pressure centers. The coordinates of the pressure centers can be indicated approximately by analyzing the peaks on the whole surface.

The analysis performed has shown that the criterion (7) is essentially used in solving the problem of defectoscopy by the straightforward method.

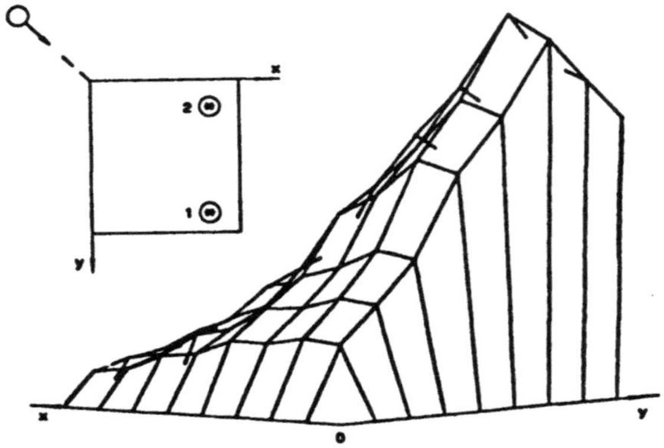

Figure 3. The surface $|\delta(\vec{r})|$ for $x = 1.1$.

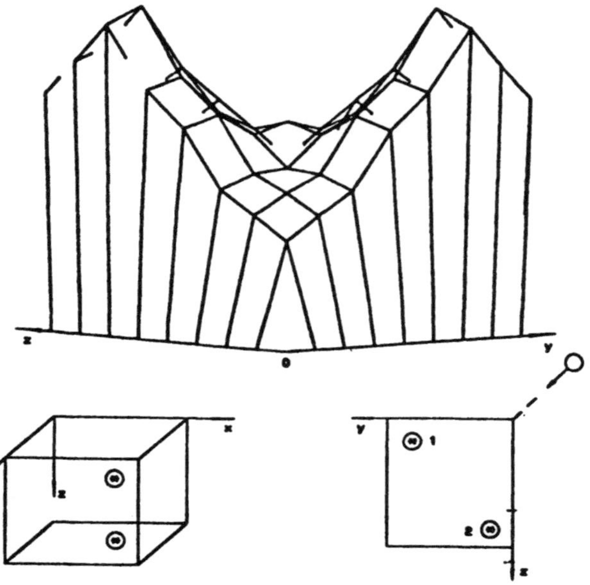

Figure 4. The surface $|\delta(\vec{r})|$ for $z = 1.1$.

Conclusion

An efficient straightforward method was proposed for solving a class of inverse problems.

The combination of straightforward and semi-inverse method was shown to allow us to get rid of the limitations of each method and to extend essentially a class of problems that can be solved numerically.

REFERENCES

Bitsadze A. A. (1984), *Foundation of the theory of analytical functions of complex variable.* Nauka, Moscow. (in Russian)

Fueter R. (1932), Uber die analitishen Darstellungender regularen Functionen einer Quaternionen variable. *Comm. Math. Helv.*, **4**, 9–20.

Mousil Gr. C. and Theodoresco, N. (1931), Fonetions holomorphes dans l'espase. *Mathematicka*, 141–153.

Schwab A. A. (1988), Nonclassical problems of deformed solid body. *Symposium "Verformung und Bruch".* Magdeburg, Germany.

Schwab A. A. (1989), On a problem of local continuation. *Dynamics of a continuous medium.* Inst. Hydrodynamics, Novosibirsk, 92. (in Russian)

Schwab A. A. (1989), Noncurrent static problems in the elasticity theory. *Izv. AN SSSR, Mekhanika Tverdogo Tela*, no. 6, 140-146, (in Russian) (in Russian)

Schwab A. A. (1989), A problem of computerized tomography based on the holographic interferometry for defectoscopy of elastic bodies. *Studies of ill-posed problems in mathematics and physics.* Inst. Mathemat., Novosibirsk. (in Russian)

Shiro Kubo (1988), Inverse problems related to the mechanics and fracture of solids and structures. *ASME Int. J., Ser. 1*, **31**, no. 2, 157–175.

Strakhov V. N. (1990), On stable methods for solving linear problems in geophysics. *Izv. AN SSSR, Fizika Zemli*, no. 8, 123-131, (in Russian)

Tanaka M. and Masuda Y. (1986), Boundary element method applied to some inverse problems. *Int. J. Eng. Anal.*, **3**, 143–148.

Luminescent Tomography of Semiconductor Materials

Sergey V. SELISHCHEV, Sergey A. TERESHCHENKO, Yuri P. MASLOBOEV

Moscow Institute of Electronic Engineering, Moscow, Russia

ABSTRACT

We propose a new approach to the reconstruction of the $A^{\mathrm{III}}B^{\mathrm{V}}$ – type semiconductor depth structure using photoluminescence spectra. In particular it is possible to find the dependence of the band gap on the distance of the absorbing point from the wafer surface, after measuring the luminescent radiation spectral distribution and solving a nonlinear integral equation under the condition that the functional dependence of the absorption coefficient on the incident radiation energy and on the band gap at the absorbing point is known. Using information on the spectrum of outgoing luminescent radiation one can reconstruct the 3-D distribution of such a characteristic of the semiconductor material as the band gap.

It is practically impossible to use the traditional geometries of the computed tomography for researching 3-D characteristics distributions of a semiconductor wafers material because of the strong asymmetry between the semiconductor wafers transverse sizes and the active layer thickness ($\sim 10^4$). The transformation of the geometrical quantities into physical ones with an acceptable range of varying values is a way to overcome this difficulty. For example, it is possible to measure the current induced by the light beam (LBIC) or the electron beam (EBIC). We propose a new approach to the reconstruction of the $A^{\mathrm{III}}B^{\mathrm{V}}$ – type semiconductor depth structure using photoluminescence spectra.

We shall consider the following simplified model. Let the laser beam with the quantum energy E_i and the intensity $I_0(E_i)$ incidents perpendicularly to the surface of the semiconductor heterostructure with an unknown distribution of the band gap $E_g(x)$ that depends monotonously on the depth x (Figure 1). Laser radiation is absorbed in the semiconductor thus producing electron-hole pairs, which recombine partially with the emitting of the luminescent radiation that returns to the semiconductor plate surface. The luminescent radiation is also absorbed on its way to the surface.

We shall take into account only radiation emitted almost normally to the surface of the semiconductor heterostructure. The absorption coefficient of the light with the energy E at the depth x is marked as $\alpha(E_g(x), E)$. Then the intensity $I_f(E_i, E_f)$ of the luminescent radiation with the energy E_f is expressed through

$$I_f(E_i, E_f) = I_0(E_i) \int_0^{+\infty} \left\{ e^{-\int_0^x \alpha(E_g(x_1), E_i)\mathrm{d}x_1} \alpha\left(E_g(x), E_i\right) \times \varphi\left(E_g(x), E_f\right) e^{-\int_0^x \alpha(E_g(x_2), E_i)\mathrm{d}x_2} \right\} \mathrm{d}x, \qquad (1)$$

where $\varphi(\cdot,\cdot)$ is the spectrum of the luminescent radiation emitted from the depth x.

Upon measuring spectral distribution of the luminescent radiation, one can find the dependence of the band gap $E_g(x)$ on the depth x solving equation (1) if the functional dependence of the absorption coefficient on the energy of incident radiation and the band gap at the absorption point are known. It is possible to define the 3-D band gap distribution function throughout the semiconductor after repeating analogous measurements at the other points of the semiconductor plate surface.

Thus, using information on the spectrum of outgoing luminescent radiation, it is possible to reconstruct the 3-D distribution of such characteristics of the semiconductor material as the band gap, which allows one to find other characteristics, for example, the Al concentration in $Al_xGa_{1-x}As$.

However, the laser radiation is absorbed in the narrow layer near the point at which $E_g = E_i$. Therefore, the reconstructed distribution will be related to this narrow layer only. To get over this difficulty we propose to carry out measurements with different values of the exciting energy E_i. Varying the exciting energy E_i enables one to scan the band gap range and, consequently, the band gap depth distribution.

Let us consider a set of the spectra $I_f^{(j)}\left(E_i^{(j)}, E_f, E_g(x)\right)$ for different values of the exciting energy E_i ($j = 1, 2, \ldots, n$). In this expression we pointed out the evident dependence of the photoluminescence spectra on the band gap depth distribution $E_g(x)$. If only the distribution $E_g(x)$ is unknown, then the model spectra $\mathrm{J}_f^{(j)}\left(E_i^{(j)}, E_f, \mathbf{E}_g(x)\right)$ can be computationally determined for an arbitrary band gap depth distribution $\mathbf{E}_g(x)$ from (1), and thereafter the deviations F_j of the model spectra $\mathrm{J}_f^{(j)}\left(E_i^{(j)}, E_f, \mathbf{E}_g(x)\right)$ from the experimental spectra $I_f^{(j)}\left(E_i^{(j)}, E_f, E_g(x)\right)$ can be found by

$$F_j = \int \left[I_f^{(j)}\left(E_i^{(j)}, E_f, E_g(x)\right) - \mathrm{J}_f^{(j)}\left(E_i^{(j)}, E_f, \mathbf{E}_g(x)\right)\right]^2 \mathrm{d}E_f. \tag{2}$$

It is obvious that the functionals F_j have global minimums if $E_g \equiv \mathbf{E}_g(x)$. However, because of the nonlinearity of equation (1), these functionals may have response surfaces of very complicated shapes, for example, additional local minimums. Therefore, it is difficult to use the standard minimization techniques. Computer simulation program was elaborated for searching for the depth distribution $E_g(x)$. This program constructs the photoluminescence spectra for a given distribution $E_g(x)$ and a given exiting energy $E_i^{(j)}$. Then the estimation $\mathbf{E}_g(x)$ of the initial distribution $E_g(x)$ can be found using the above functionals and the constructed spectra as experimental data.

The described method was used for the reconstruction of the quasi-continuous band gap distribution in the semiconductor depth. The continuity was simulated by a big number (~ 70) of small steps (see Figure 2, the curve E_g). At the same time, the reconstruction was performed in a lesser number (7) of the steps. The reconstructed distribution (see Figure 2, the curve $\mathbf{E}_g$) approximates the given distribution E_g well enough. It proves the efficiency of the method proposed.

This work was supported by the Russian Foundation of Fundamental Research under grant No. 93-02-2215.

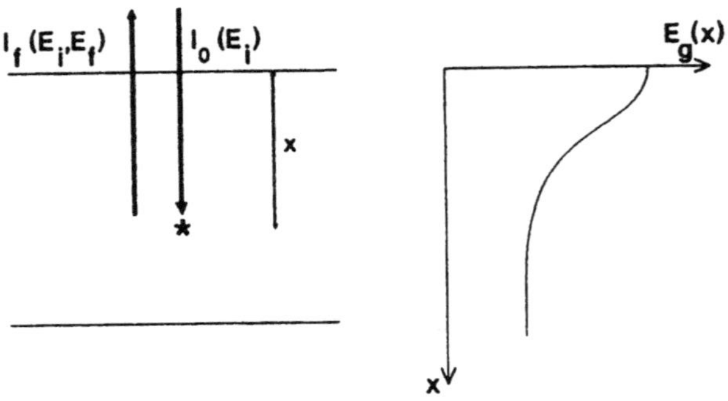

Figure 1. Luminescent tomography measurement scheme.

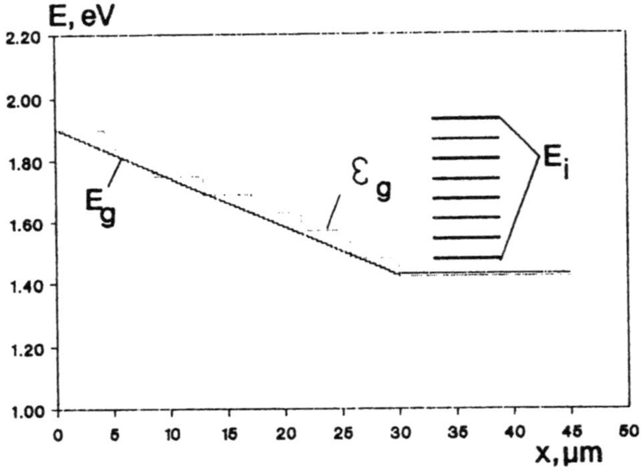

Figure 2. Given (E_g) and reconstructed ($\mathbf{E}_g$) band gap semiconductor depth distribution.

REFERENCES

Bajaj J., Tennant W. E. (1990), Remote contact LBIC imaging of defects in semiconductors. *J.of Crystal Growth*, no. 103, 170–178.

Bondarenko I. E., Likharev S. K., Rau E. I., Yakimov E. B. (1990), Microtomography of semiconductor crystals in the EBIC mode. *J. of Crystal Growth*, no. 103, 197–199.

On a Reconstruction Formula for Ray Transform after Discrete Incomplete Data Given on a Line

Susumu TANABE

Department of Mathematical Sciences, University of Tokyo, Tokyo, Japan

Institute of Control Sciences, Russian Academy of Sciences, Moscow, Russia

ABSTRACT

We consider a problem of reconstructing a function with a compact support provided that the values of its ray transform are given at discrete points located on a line apart from the function support. We show that two assumptions have to be imposed on the location of an infinite number of point detectors in order to obtain an exact reconstruction formula. The first is a uniform distribution of the detectors at infinity. The second requires the detectors location to be symmetric with respect to some point on the line. In deriving the reconstruction formula, we used mainly the Lagrangian integration of the theory of entire functions.

0. Introduction

Let us think of the following problem: an unknown function $f \in H^{\sigma+\frac{n-1}{2}}(\mathbf{R}^n)$ (for some $\sigma > 1/2$) with compact support is given on $\mathbf{R}^n$ space. Suppose we take a line l apart from the supp$\{f\}$, and the values of integrals taken along the lines issued from certain discrete set of points situated on the line l are known. That is to say, we know

$$(I_f)(s_j,\theta) = \int_{\mathbf{R}} f(s_j+\theta_1 t, \theta_2 t, ..., \theta_{n-1}t, t)\mathrm{d}t, \qquad \text{for all} \quad \theta \in \mathbf{R}^{n-1},$$

where $\{s_j\}$ is a discrete set of points on l.

The question is whether it is possible to recover the unknown function f from the data $\{(I_f)(s_j,\theta)\}$ for $(s_j,\theta) \in l \times S^{n-1}$. It is a classical fact that there exists a reconstruction formula if we know $I_f(y,\theta)$ for all $y \in \mathbf{R}^n$ and $\theta \in S^{n-1}$ (Gel'fand et al., 1962). Therefore we call the above problem of reconstruction "problem of incomplete data".

The background of this setting of problem is as follows. Let C be a curve with a parameter expression $\gamma(\lambda)$, $\lambda \in \mathbf{R}$, in $\mathbf{R}^n$ that is distant from supp$\{f\}$. The reconstruction of f is possible, for example, if the following so-called Kirillov-Tuy condition is satisfied: for any plane $H_{\omega,p} = \{x \in \mathbf{R}^n; <\omega, x>= p\}$ which has a non-empty intersection with supp$\{f\}$, there is a value of parameter $\lambda = \lambda(\omega,p)$ such that the curve C intersects with $H_{\omega,p}$ at the point $\gamma(\lambda(\omega,p))$. (Kirillov ; Natterer, 1986; and Denisjuk, 1991). If all the data on the curve

$$g(\lambda,\xi) = \int_{-\infty}^{\infty} f(\gamma(\lambda)+t\xi)\mathrm{d}t, \quad \lambda \in \mathbf{R}, \quad \xi \in S^{n-1}$$

are known, one can use the effective reconstruction formulae given in the following theorem by Denisjuk (1991).

Theorem 0.1. *Let K be a complex of lines in $\mathbf{R}^n$ intersecting with the given curve C and the Kirillov-Tuy condition be satisfied for C with respect to the compact support of $f(x)$. Assume we know the integrals of f along all the lines of complex K. Then we have the following reconstruction formulae:*

a) if $n = 2k$,

$$f(x) = c_n\sqrt{-\Delta}\int_{S^{n-1}} \Omega_{n-1}(\omega)\int_{S^{n-2}(\omega)} \Omega_{n-2}(\xi)L(\omega, D)g(\lambda(\omega, <\omega, x>), \xi),$$

$$L(\omega, D) = \frac{1}{\sum\omega_i^{n-2}}\sum_{i=1}^{n}\left(\frac{\partial}{\partial\xi_i}\right)^{n-2},$$

b) if $n = 2k+1$,

$$f(x) = c_n\int_{S^{n-1}} \Omega_{n-1}(\omega)\int_{S^{n-2}(\omega)} \Omega_{n-2}(\xi)L(\omega, D)g(\lambda(\omega, <\omega, x>), \xi),$$

$$L(\omega, D) = \frac{1}{\sum\omega_i^{n-1}}\sum_{i=1}^{n}\omega_i(\frac{\partial}{\partial\xi_i})^{n-2};$$

$S^{n-2}(\omega) = \{\xi \in S^{n-1}; <\omega, \xi>= 0\}$, $c_n = \frac{(-1)^{(n-1)/2}}{2^n\pi^{n-1}}$,
$\Omega_d :=$ standard volume form on a unit sphere S^d,
$G(p; \mathcal{A})$ means the so-called canonical function for the set of nodes $\mathcal{A} = \{s_j \in \mathbf{R} : j \in \mathbf{Z}\}$ to be defined in Section 2.

On the other hand, in Lavrent'ev et al. (1981) the following uniqueness theorem is given.

Theorem 0.2. *Let $f \in C_0^\infty(\mathbf{R}^n)$ and K be a compact set in $\mathbf{R}^n$ such that $\operatorname{supp}\{f\} \subset K$. Suppose that there exists an infinite sequence β_k $(k = 1, 2, \ldots)$ that converges to a point $\beta_0 \in \mathbf{R}^n \cup \{\infty\} \setminus K$. If $Rf(\beta_k, \theta) = 0$ for all k and all $\theta \in S^{n-1}$, then $f(x) \equiv 0$.*

Under these circumstances, the reconstruction formulae a) and b) can be specified if two conditions are imposed on the location of the points s_j.

Main Theorem. *In the above situation, we suppose that the points $\{s_j\}$ are situated on the line ℓ in the following way:*

(S.1) $$s_j \approx Cj + C_1, \quad \text{as } j \to \infty \qquad (0.1)$$

(S.2) $$s_j = -s_{-j}$$

where $C > 0$ and C_1 are real constants.

Then

a) if $n = 2k$, $f(x) = c_n\sqrt{-\Delta}$

$$\times \int_{S^{n-1}-\{\omega_1=0\}} \Omega_{n-1}(\omega)\int_{S^{n-2}(\omega)} \Omega_{n-2}(\xi)L(\omega, D)I_f\left(\frac{<\omega, x>}{\omega_1}, \xi_1/\xi_n, \ldots, \xi_{n-1}/\xi_n\right), \qquad (0.2\text{even})$$

$$L(\omega, D) = \frac{1}{\sum\omega_i^{n-2}}\sum_{i=1}^{n}\left(\frac{\partial}{\partial\xi_i}\right)^{n-2},$$

b) if $n = 2k+1$,

$$f(x) = c_n \int\limits_{S^{n-1}-\{\omega_1=0\}} \Omega_{n-1}(\omega) \int\limits_{S^{n-2}(\omega)} \Omega_{n-2}(\xi) L(\omega, D) I_f\left(\frac{<\omega, x>}{\omega_1}, \xi_1/\xi_n, ..., \xi_{n-1}/\xi_n\right), \tag{0.2$_{odd}$}$$

$$L(\omega, D) = \frac{1}{\sum \omega_i^{n-1}} \sum_{i=1}^{n} \omega_i \left(\frac{\partial}{\partial \xi_i}\right)^{n-2};$$

where

$$I_f(p, \omega) = (2\pi)^{\frac{-(n-1)}{2}} \int\limits_{\mathbf{R}^{n-1}} e^{iz\omega} \sum_{-\infty}^{\infty} \frac{F(s_j, z) G(p)}{G(s_j)(p - s_j)} \mathrm{d}z \tag{0.3}$$

$$F(s_j, z) = \int e^{-iz\theta} I_f(s_j, \theta) \mathrm{d}\theta.$$

Other notations are the same as in Theorem 0.1.

We would like to remark that the problem on the possibility of reconstruction after discrete data has come into our view thanks to Prof. V. G. Romanov during his stay in Tokyo on February 1992 and that the outline of the proof of Theorem 1.1 was given by Prof. A. Kaneko. We wish to express our gratefulness to Profs. A. S. Denisjuk, V. P. Palamodov, V. A. Sharafutdinov for their useful discussions.

1. Fourier Transform

First of all let us specify the disposition of the $\mathrm{supp}\{f\}$ and the line l. We will use the coordinate system so that the line l coincides with the x_1 axis. The lines issued from a point $(p, 0, ..., 0)$ on the x_1 axis can be expressed as

$$(x_1, x_2, ..., x_n) = (p + \theta_1 t, \theta_2 t, ..., \theta_{n-2} t, t)$$

where $(\theta_1, ..., \theta_{n-1}) \in \mathbf{R}^{n-1}$. By this choice we can express all the lines issued from the points on the x_1-axis and passing through $\mathrm{supp}\{f\}$, with finite values of θ. Thus we define the ray transform

$$I_f(p, \theta_1, ..., \theta_{n-1}) = \int_{\infty}^{\infty} f(p + \theta_1 t, \theta_2 t, ..., \theta_{n-1} t, t) \mathrm{d}t. \tag{1.1}$$

It shall be remarked that $I_f(p, \theta_1, ..., \theta_{n-1})$ is a function with a compact support of θ -variables for each fixed p. As we aim to use the argument of analytic continuation, we will consider the Fourier transform of the data $I_f(p, \theta)$ with respect to the variables θ:

$$F(p, z_1, ..., z_{n-1}) = \int\limits_{\mathbf{R}_\theta^{n-1}} \int\limits_{\mathbf{R}_t} e^{iz\theta} f(p + \theta_1 t, \theta_2 t, ..., \theta_{n-1} t, t) \mathrm{d}t \tag{1.2}$$

$$= \int\limits_{\mathbf{R}^{n-2}} d\theta' \int\limits_{\mathbf{R}} \mathrm{d}x_1 \int\limits_{\mathbf{R}} e^{i(z_1(x_1-p)/t + z'\cdot\theta')} f(x_1, \theta_2 t, ..., \theta_{n-1} t, t) \mathrm{d}t/t,$$

where $z' = (z_2, ..., z_{n-1})$ and $\theta' = (\theta_2, ..., \theta_{n-1})$. Here the factor $1/t$ in the integrand does not bother the convergence of the integration as $f(x_1, \theta_2 t, ..., \theta_{n-1} t, t) \equiv 0$ for $|t| < L$, where L denotes the distance between the x_1-axis and $\mathrm{supp}\{f\}$.

The following proposition is easily seen from (1.2) and the remark on the compactness of the support of $I_f(p,\theta)$ as a function of θ -variables.

Proposition 1.1. *The function $F(p,z)$ is an entire function of (p,z). Moreover, it is entire of order 1 with respect to the $p-$variable for fixed z-variables. In other words, we have an estimate*

$$|(d/dp)^m F(0,p)| < C_f(z')(|z_1|/L)^m \tag{1.3}$$

where $C_f(z')$ denotes a positive constant depending on f and z' only.

By using the Fourier image $F(p,z)$, the injectivity of the map

$$\mathcal{E}'(\mathbf{R}^n) \longrightarrow \mathcal{D}'(\mathbf{R}_p \times \mathbf{R}_\theta^{n-1}),$$

$$f \longmapsto I_f,$$

can be shown. Though this statement is a special version of Theorem 0.2, we give it here because of the simplicity of the proof.

Theorem 1.1. *If $I_f(s_j,\theta) \equiv 0$ for all $\theta \in \mathbf{R}^{n-1}$ and $j \in \mathbf{Z}$, then $f(x)$ is identically zero.*

PROOF.

Let us express the Fourier transform (1.2) by a change of variables in the integral, i.e.

$$F(p,z) = \int e^{i(rz_1+sz_1p+z'\cdot\theta')} f(r/s,\theta_2/s,\dots,\theta_{n-1}/s,1/s)1/s^2 dsdrd\theta'.$$

Here we set $s = 1/t$ and $r = x_1/t$. Let

$$g(s,r,\theta') = 1/s^2 f(r/s,\theta_2/s,\dots,\theta_{n-1}/s,1/s),$$

then $g(s,r,\theta')$ has the compact support as a distribution on the space $\mathbf{R}_r^1 \times \mathbf{R}_s^1 \times \mathbf{R}^{n-2}$. Let us denote the Fourier transform of $g(s,r,\theta')$ by $\hat{g}\ (z_1,w,z')$. Suppose that $I_f(s_j,\theta) \equiv 0$ for all $j \in \mathbf{Z}$ and $\theta \in \mathbf{R}^{n-1}$, since $F(p,z) = \hat{g}\ (z_1,pz_1,z')$, we have

$$\hat{g}(z_1,s_jz_1,z') = 0 \tag{1.4}$$

for all $j \in \mathbf{Z}$ and $z \in \mathbf{C}^{n-1}$. Because of the compactness of supp$\{g\}$, $\hat{g}\ (z_1,w,z')$ is an entire function of exponential type on $\mathbf{C}_z^n \times \mathbf{C}_w^1$. From (1.4) it follows that $\hat{g}\ (z_1,w,z')$ vanishes on an infinite number of affine surfaces $\{w = s_jz\}$ of codimension one that approach $\{z_1 = 0\}$ as $j \to \infty$. An entire function $\hat{g}$ which is not identically zero can not possess this property. Thus $\hat{g}\ (z_1,w,z') \equiv 0$, hence f is identically zero.

2. Lagrangian Interpolation

In view of Theorem 0.1 and Theorem 1.1, it is enough to recover $F(p,z)$ from $F(s_j,z)$ to prove our main theorem. Namely, the main task of this section is to prove

Proposition 2.1. *The Fourier transform $F(p,z)$ of $I_f(p,\omega)$ with respect to the ω variable (see (1.2)) coincides with*

$$\Phi(p,z) = \sum_{j=1}^{\infty} \frac{F(s_j,z)G(p;\mathcal{A})}{G'(s_j;\mathcal{A})(p-s_j)}.$$

For this purpose we employ some theorems in the theory of entire functions. From now on, all the notations and notions concerning the theory of entire functions are those taken from Levin (1964).

Let $f(z)$ be an entire function which has simple zeros at all the nodes $\{s_j\}(|s_j| \to \infty)$ and let $\{b_j\}$ be values assigned at the nodes. If the series

$$\sum_{j=1}^{\infty} \frac{b_j f(z)}{f'(s_j)(z-s_j)} \tag{2.1}$$

converges uniformly in every bounded region, it represents an entire function which solves the interpolation problem. Series (2.1) is called a Lagrangian interpolation series. It is convenient to use the canonical function $G(z;\mathcal{A})$ with the nodes $\mathcal{A} = \{s_j; j \in \mathbf{Z}\}$ as the function $f(z)$ in (2.1). The definition of the canonical function gives following expression (Chapter II, Section 1) when the index $\rho = \inf\{\lambda \in \mathbf{R}^1 : \sum_j (1/s_j)^\lambda < \infty\}$ appears to be a positive integer:

$$G(z;\mathcal{A}) = \exp(C_\rho z) \prod_j G(z/s_j,\rho).$$

Here $G(u;\rho) := (1-u)\exp(u + u^2/2 + \dots + u^\rho/\rho)$ and C_ρ is a constant such that there exists

$$\delta_G = \lim_{r\to\infty} 1/L(r)\{C_\rho + \sum_{|s_j|\le r} 1/s_j\},$$

where $L(r) = r^{\rho(r)-\rho}$ is defined by the indices $\rho(r)$ and $\rho = \lim_{r\to\infty} \rho(r)$ that are introduced in Chapter I of Levin (1964). From assumptions (S.1) and (S.2) imposed on the set $\mathcal{A} = \{s_j\}$, we get the following properties of the canonical function $G(p,\mathcal{A})$.

Lemma 2.2. *It holds that*

1) $\rho = 1$.

2) $h_{G(p;\mathcal{A})}(\omega) = \begin{cases} \pi\Delta(0)\sin\omega, & 0 \le \omega \le \pi, \\ -\pi\Delta(0)\sin\omega, & \pi \le \omega \le 2\pi, \end{cases}$

where $h_{G(p;\mathcal{A})}$ *shall be defined below in (2.2),* $\omega = \arg p$.

Particularly, $h_{G(p;\mathcal{A})}(0) = h_{G(p;\mathcal{A})}(\pi) = 0$.

PROOF.

The assertion 1) follows from (S.1). The following theorem of Levin (1964) yields 2).

Theorem 2 of Chapter II. *Let* $\rho(r)$ *be an order of growth, let* $\rho = \lim_{r\to\infty}\rho(r)$ *be an integer, and let us assume that the set* $\{a_k\}$ *of the zeros of the entire function*

$$f(z) = z^m e^{P(z)} \prod_k G(z/a_k;\rho)$$

has an angular density with the index $\rho(r)$. *Then for all* z *not belonging to a certain* C^0*-set, the following asymptotic equation will hold:*

$$\log|f(z)| \approx h_f(\omega) r^{\rho(r)}, \tag{2.2}$$

where

$$h_f(\omega) = \int_{\omega-2\pi}^{\omega} (\psi-\omega)\sin\rho(\psi-\omega)\,d\Delta(\psi) + \tau_f \cos\rho(\omega-\omega_f),$$

$$\delta_f = \tau_f e^{i\omega_f} = \lim_{r\to\infty} 1/L(r)\left[C_\rho + 1/\rho \sum_{|a_n|\le r} a_n^{-\rho}\right],$$

and $L(r) = r^{\rho(r)-\rho}$.

Applying this theorem to our situation, we see that δ_G and C_ρ must be zero because of condition (S.2) that the nodes s_j are located on the real axis symmetrically with respect to the origin.

In Theorem 12 of Chapter II (Levin, 1964) a sufficient condition for the Lagrangian interpolation to give a reconstruction formula was formulated. This theorem, however, can not be applied to our situation directly. Therefore, we will show Proposition 2.1 by other means.

Firstly we shall check the convergence of the expression

$$\sum_{j\in\mathbf{Z}} \frac{F(s_j,z)G(p)}{G'(s_j;\mathcal{A})(p-s_j)}. \tag{2.3}$$

To do this, we study the behavior of the canonical function $G(p;\mathcal{A})$ as $|p|\to\infty$ in an angular domain containing the positive real axis. From Lemma 2.2, on the boundary of C_R-circles with the center s_n and the radius R the following asymptotic behavior holds:

$$\ln\frac{|G(p;\mathcal{A})|}{|p-s_j|} \approx H_{G(z;\mathcal{A})}(0)\cdot r^{\rho(r)} \approx \pi\Delta(0)\sin(R/s_j)r^{\rho(r)}, \quad |p| = r, \quad \operatorname{Re} p > 0$$

By the principle of maximum for subharmonic functions,

$$\ln|G(s_j;\mathcal{A})| \approx R\Delta(0)|s_j|r^{\rho(r)-1}.$$

For a fixed p^0, there exists $N(p^0) > N$ such that

$$|p^0 - s_j| > |s_j| - |p^0| > \frac{1}{2}|s_j|.$$

Thus, outside the C_R-circles on the p-plane, we have

$$\left|\frac{F(s_j,p^0)}{G(s_j;\mathcal{A})(p-s_j)}\right| \le \frac{|F(s_j,p^0)|}{(R/2)\Delta(0)|s_j|^{\rho(|s_j|)}}, \quad n > N(p^0). \tag{2.4}$$

Evidently, a similar estimate holds for $s_j < 0$.

Now we see the decay of $F(s_j,p^0)$ as $j\to\infty$. From the trace theorem for Sobolev spaces, if $f(x)\in H^{\frac{n-1}{2}+\sigma}(\mathbf{R}^n)$ and $\sigma\ge 1/2$, then

$$f(x_1,\theta_2 t,\dots,\theta_{n-1}t,t)\in H^\sigma(\mathbf{R}^1_t) \quad \text{for fixed} \quad (x_1,\theta_2,\dots,\theta_{n-1}).$$

In other words,

$$\int |\hat{F}(x_1,\theta',\tau)|^2(1+|\tau|^2)^\sigma d\tau < \infty$$

for every $(x_1,\theta_2,\dots,\theta_{n-1})$. Here

$$\hat{F}(x_1,\theta',\tau) = \int e^{i(z_1x_1/t+t\tau)} f(x_1,\theta_2 t,\dots,\theta_{n-1}t,t)dt.$$

In view of the compactness of supp$\{f\}$, we see

$$\int \left|\hat{F}(x_1,\theta',\tau)\right|^2(1+|\tau|^2)^\sigma \mathrm{d}\tau \mathrm{d}x_1 \mathrm{d}\theta' < \infty.$$

According to the fact that $L^1(\Omega) \subset L^2(\Omega)$ for a compact Ω,

$$\int \left|\int \hat{F}(x_1,\theta',\tau)\mathrm{d}x_1\mathrm{d}\theta'\right|^2 (1+|\tau|^2)^\sigma \mathrm{d}\tau$$

$$\leq |\Omega|^{1/2}\left(\int \left|\hat{F}(x_1,\theta',\tau)\right|^2(1+|\tau|^2)^\sigma \mathrm{d}\tau \mathrm{d}x_1 \mathrm{d}\theta'\right)^{1/2},$$

where $|\Omega|$ denotes the volume of supp$\{f\}$ measured by the natural measure of the variables (x_1, θ'). Thus the assertion

$$\int e^{\mathrm{i}(z_1^0 x_1/t + z'^0\cdot\theta'^0)} f(x_1, \theta_2 t, \ldots, \theta_{n-1}t, t)\mathrm{d}x_1\mathrm{d}\theta \in H^\sigma(\mathbf{R}^1)\cap \mathcal{E}'(\mathbf{R}^1_t)$$

can be seen if we take into account that the integrand vanishes in the neighborhood of $t = 0$. In other words, for $F(p, z)$ defined in (1.2),

$$\int |F(s,z^0)|^2(1+s^2)^\sigma \mathrm{d}s < \infty,$$

which yields

$$|F(s,z^0)| \leq C(z^0)(1+s^2)^{-\sigma-1/2} \quad \text{for large } s. \tag{2.5}$$

This estimate and (2.4) yield that the series

$$\sum_{j\in\mathbf{Z}} \frac{F(s_j,z^0)}{G'(s_j,\mathcal{A})(p-s_j)} \tag{2.6}$$

converges outside the C_R-circles to some function $\phi(p, z^0)$ on the p-complex plane. By applying the maximum principle to the series (2.6) on each C_R-circle, we see that in every bounded domain the series (2.3) converges to an entire function

$$\Phi(p,z^0) = \phi(p,z^0)G(p;\mathcal{A}).$$

The convergence with respect to the p-variable implies that the order of integration and summation can be exchanged in the reconstruction formula, so

$$\Phi(p,z) = \int e^{\mathrm{i}z\cdot\theta} \sum_{j\in\mathbf{Z}} \frac{f(p+t\theta_1, \theta_2 t, \ldots, \theta_{n-1}t, t)G(p;\mathcal{A})}{G'(s_j,\mathcal{A})(p-s_j)} \mathrm{d}t\mathrm{d}\theta$$

This expression implies the entireness of $\Phi(p, z)$ with respect to the z-variables.

Furthermore we shall check that $\Phi(p, z)$ coincides with $F(p, z)$. For this purpose, we shall argue as in Palamodov (1991a). Let us think of the meromorphic form

$$\Theta_p(\zeta) = \frac{F(\zeta,z)}{G(\zeta;\mathcal{A})(\zeta-p)}\mathrm{d}\zeta.$$

We see that $\Theta_p(\zeta)$ has poles of order one at the points $\zeta = \pm s_1, \pm s_2 \ldots$. Let us integrate $\Theta_p(\zeta)$ along the boundary of a domain $\{\zeta : |\operatorname{Im}\zeta| \leq B, |\operatorname{Re}\zeta| \leq A\}$ so that no poles should be located on the boundary. From the Paley-Wiener theorem and (2.5) we see that

$$|F(\zeta, z^0)| \leq C_1(1+|\zeta|^2)^{-\sigma-1/2} \exp(L \operatorname{Im}(z_1^0\zeta)).$$

On the other hand, by Lemma 2.2, 2),

$$|G(\zeta; \mathcal{A})| \geq \exp(\pi\Delta(0)|\operatorname{Im}\zeta|).$$

Thus we have an estimate

$$\frac{|F(\zeta, z)|}{|G(\zeta; \mathcal{A})(\zeta - p)|} \leq \begin{cases} C_1(1+|A|^2)^{-\sigma-1/2} \exp(L \operatorname{Im}(z_1\zeta)) & \text{on } |\operatorname{Re}\zeta| = A, \\ C_2 A^{1-2\sigma} B^{-2\sigma} \exp(L|z_1| - \pi\Delta(0))B & \text{on } |\operatorname{Im}\zeta| = B. \end{cases}$$

That is to say,

$$\left| \int_{|\operatorname{Re}\zeta|=A} \Theta_p \right| \leq \frac{4C_1}{L|z_1|} A^{-1/2-\sigma} \exp(L|z_1|)B,$$

$$\left| \int_{|\operatorname{Im}\zeta|=B} \Theta_p \right| \leq C_2 A^{1-2\sigma} B^{-2\sigma} \exp(L|z_1| - \pi\Delta(0))B.$$

We can choose constants A and B so that

$$A^{-1/2-\sigma} \exp(L|z_1|) \to 0,$$

$$A^{1-2\sigma} B^{-2\sigma} \exp(L|z_1| - \pi\Delta(0))B \to 0,$$

as $A, B \to \infty$. Namely, we shall take A so that the following two inequalities hold:

$$A >> \exp\left(\frac{L|z_1|}{\sigma + 1/2} B\right),$$

$$A >> B^{\frac{2\sigma}{1-2\sigma}} \exp\left(\frac{L|z_1| - \pi\Delta(0)}{2\sigma - 1}\right) B.$$

In calculating the residuals at all the poles of $\Theta_p(\zeta)$, we get the equality

$$\Phi(p, z) \equiv F(p, z).$$

Thus Proposition 2.1 has been shown.

To prove Main theorem, we have only to apply Theorem 0.1 to the case where the complex of lines K is made of all the lines that intersect with a fixed line l. At the same time, we shall remark that the violation of the Kirillov-Tuy condition causes insufficiency in the data corresponding to $\{\omega_1 = 0\}$. Taking into account that the set $\{\omega_1 = 0\}$ has measure zero in $S^{n-1} = \{|\omega| = 1\}$ and that the integrand

$$\int_{S^{n-2}(\omega)} \Omega_{n-2}(\xi) L(\omega, D) I_f\left(\frac{<\omega, x>}{\omega_1}, \xi_1/\xi_n, \ldots, \xi_{n-1}/\xi_n\right)$$

is regular, we see that this insufficiency can be neglected in the reconstruction formulae.

3. Some Remarks

REMARK 1. The argument of analytic continuation to write down reconstruction formulae for Radon transforms after incomplete data has already been adopted by Denisjuk and Palamodov (1988). In Palamodov (1991a) a special version of (2.3) named the Kotel'nikov–Shannon sampling theorem has been used. Namely, if we know the data of Radon transforms, we have

$$(R_{n-1}f)(p,\omega) = \int_{<x,\omega>=p} f(x)\mathrm{d}x$$

only for $|\omega_1| \geq \mathrm{tg}\ t\lambda \cdot |\omega'|$, where λ is a sharp angle.

But to our knowledge, no attempt to use the general version of Lagrangian interpolation (2.3) has been made for the reconstruction after incomplete data.

REMARK 2. Here we remark that the investigated problem can be reduced to another kind of problem after incomplete data, as suggested by V. P. Palamodov.

In Palamodov (1991b) it is mentioned that the bundle of lines $\lambda(x,\omega), \cup_{\omega\in S^{n-1}}\{x + t\omega; t \in \mathbf{R}\}$ which pass through a fixed point $x \in l$ can be transformed into a family of parallel lines by means of a projective transform L_P that throw the line l to the infinity.

In general, the projective transform L_P on $\mathbf{RP^n}$ is induced by a nonsingular linear transform L acting on $\mathbf{R}^{n+1}$:

$$L : \mathbf{R}^{n+1} \longrightarrow \mathbf{R}^{n+1},$$

$$\xi = (\xi_0, \xi_1, ..., \xi_n) \longmapsto y = (y_0, y_1, ..., y_n) = (< \mathbf{u_0}, \xi >, ..., < \mathbf{u_n}, \xi >),$$

$$L_P : \mathbf{RP}^n \longrightarrow \mathbf{RP}^n,$$

$$(x_1, ..., x_n) = (\xi_1/\xi_0, ..., \xi_n/\xi_0) \longmapsto (y_1/y_0, ..., y_n/y_0),$$

where $\{\mathbf{u}_j; j = 0, 1, \ldots, n\}$ denotes a set of (n+1)-vectors that are linearly independent of one another. Further, we will use the notation $\mathbf{u}_j = (u_j^0, ..., u_j^n)$.

In this situation we see that L_P transforms the line

$$\lambda(\omega, p) = \{(p + t\omega_1, t\omega_2, ..., t\omega_{n-1}, t); t \in \mathbf{R}\}$$

into

$$L_P(\lambda(\omega,p)) = \left(\frac{(u_1^0 + u_1^1 p)s}{< \mathbf{u_0'}, \omega >} + \frac{< \mathbf{u_1'}, \omega >}{< \mathbf{u_0'}, \omega >}, \frac{(u_n^0 + u_n^1 p)s}{< \mathbf{u_0'}, \omega >} + \frac{< \mathbf{u_n'}, \omega >}{< \mathbf{u_0'}, \omega >}\right).$$

Here we have used the notation $\mathbf{u}_j' = (u_j^1, ..., u_j^n)$, $s = 1/t$, $< \mathbf{u_j'}, \omega >= u_j^1\omega_1 + ... + u_j^{n-1}\omega_{n-1} + u_j^n, (\omega_n \equiv 0)$ and remarked that $u_0^0 = u_0^1 = 0$ under the condition that the line l shall be thrown out to the infinity. As we know how the parameter intervenes into the description of the line $L_P(\lambda(\omega,p))$, it can be seen that the following relationship holds:

$$f(x(t))\frac{\mathrm{d}\tau}{|\omega|} = -f(x(y(s)))\frac{< \mathbf{u_0'}, \omega > \mathrm{d}\sigma \mathbf{x_n}(\mathbf{y}(\mathbf{s}))^2}{||u^0 + u^1 p||}.$$

Here $\mathrm{d}\tau$(resp. $\mathrm{d}\sigma$) denotes the Lebesgue measures on the line $\lambda(\omega,p)$(resp. $L_P(\lambda(\omega,p))$) naturally induced from that of $\mathbf{PR^n}$, $|\omega| = (\omega_1^2 + ... + \omega_{n-1}^2 + 1)^{1/2}$, $||u^0 + u^1 p|| = \left(\sum_{j=1}^{n}(u_j^0 + u_j^1 p)^2\right)^{1/2}$. This formula is nothing but a special case of Proposition 3, (6) in Palamodov (1991b).

Let us denote by $I_f(L_P\lambda(\omega,p))$ the integral $\int_{L_P(\lambda(\omega,p))} f(x(y(s)))d\sigma$. Then we have

$$\int_{\lambda(\omega,p)} f(x(t))x_n(t)^{-2}\frac{d\tau}{|\omega|} = \frac{<\mathbf{u}_0',\omega>}{||u^0+u^1p||}I_f(L_P\lambda(\omega,p)).$$

Therefore, if we recover the values of $I_f(L_P\lambda(\omega,p))$ for all $(\omega,p) \in S^{n-1}\times\mathbf{R}$, we can reconstruct $f(x)x_n^2$ and hence $f(x)$ itself. Thus the problem is reduced to finding $J_g(\theta,x')$ for $\forall(\theta,x') \in S^{n-1}\times\mathbf{R}^{n-1}$, $x'=(x_2,...,x_n)$, from $J_g(\theta^{(j)},x') \in S^{n-1}\times\mathbf{R}^{n-1}$, where

$$\theta^{(j)} = \left(\frac{u_1^0+u_1^1p}{<\mathbf{u}_0',\omega>}s_j,...,\frac{u_n^0+u_n^1p}{<\mathbf{u}_0',\omega>}s_j\right),$$

$$J_g(\theta,x') = \int g(t\theta_1,x_2+t\theta_2,...,x_n+t\theta_n)dt.$$

As the Fourier transform of $J_g(\theta,x')$ with respect to the x'-variables turns out to be $\hat{g}(-\theta'\cdot\xi',\xi_2,...,\xi_n)$, $\theta'=(\theta_2,...,\theta_n)$, $\xi'=(\xi_2,...,\xi_n)$, i.e.,

$$\int e^{-ix'\xi'}J_g(\theta,x')dx' = \hat{g}(-\theta'\cdot\xi',\xi_2,...,\xi_n),$$

we can use an argument of the Lagrangian interpolation to reconstruct an entire function $\hat{g}(\xi)$ departing from the data $\{\hat{g}(\theta'^{(j)}\cdot\xi',\xi_2,...,\xi_n);\ j\in\mathbf{Z}\}$.

It will be an interesting problem to compare the stability (i.e., how much the error in the given discrete data may cause the error in the recovered function) of the reconstruction formula (0.2) and the formula that shall be obtained by using the projective transform.

REFERENCES

Denisjuk A. S. (1991), *Integral Geometry on Real Manifolds*, Moscow, Ph.D.thesis. (in Russian)

Denisjuk A. S., Palamodov V. P. (1988), *Inversion de la Transformation de Radon d'après des Données Incomplètes*, C. R. Acad. Sci., Paris, vol. 307, Série I, 181–183. (in French)

Gel'fand I. M., Graev M. I., Vilenkin N. Ya. (1962), *Generalized Functions*, vol. 5, Moscow, Fizmatgiz, (English translation, Academic Press, NY, 1973).

Kirillov A. A., On a problem of I. M. Gel'fand. *Soviet Math. Dokl.* **2**, 268–269

Lavrent'ev M. M., Derevtsov E. V., Sharafutdinov V. A. (1981), Determination of isotropic source in a homogeneous medium by means of its imaging, *Dokl. Akad. Nauk SSSR*, **260**, no. 1, 799–803. (in Russian)

Levin B. Ya. (1956), *Distribution of Zeros of Entire Functions*, Moscow, Gosudarstvennoje Izdatelśtvo Tekhniko-Teoreticheskoj Literatury (English translation, AMS, Providence, 1964).

Natterer F. (1986), *The Mathematics of Computerized Tomography*, Wiley-Teubner.

Palamodov V. P. (1991a), *Some Singular Problems in Tomography*, AMS Transactions of Mathematical Monographs, vol. 81, 123–140.

Palamodov V. P. (1991b), *Inversion Formulas for Three Dimensional Ray Transform*. Lecture Notes in Mathematics, no. 1497, Springer Verlag.

Construction of a Velocity Model from Seismological Data

T. A. TUSHKO

Computer Center, Krasnoyarsk-36, Russia

ABSTRACT

A search for efficient approaches to reconstruction of a velocity function in a medium is of great interest in seismotomography. Here an iterative approximate scheme of the data inversion is suggested which based on the known general solution. As an example, it is applied to a three-dimensional velocity problem using arrival times of seismic waves as data. The paper presents some practical results concerning the Kaspian sea region.

Method

We present a velocity function by a finite number of unknown parameters describing a medium.

Let $P = \{p_i\}$, $i = 1, \ldots, M$, be a vector of model parameters.

Let the observed values of the wave travel times $t = \{t_j\}$, $j = 1, \ldots, N$, are fixed. The measurements may contain some errors which are supposed to be additive and Gaussian. They are defined by a covariance matrix of measurement $C_{t_0 t_0} = \text{diag}\,\{\sigma_{0j}^2\}$.

A priori information is given in the form of an initial approach P_0 to the solution and its covariance matrix $C_{p_0 p_0} = \text{diag}\,\{\sigma_{p_i}^2\}$.

The relation between the measurements and the model parameters is known as $g(P) = t$.

The problem is to estimate the vector P from the observations $\{t_j\}$ and a priori information.

The stochastic approach allows us to linearize this inverse problem in a neighborhood of the initial approximation and to use the classical solution [1]:

$$\hat{P} = P_0 + \left(G^T C_{t_0 t_0}^{-1} G + C_{P_0 P_0}^{-1}\right)^{-1} G^T C_{t_0 t_0} \Delta t,$$
$$C_{\hat{P}\hat{P}} = \left(G^T C_{t_0 t_0}^{-1} G + C_{P_0 P_0}^{-1}\right)^{-1}, \qquad (1)$$

where G is the matrix of the first derivatives of $g(P)$ and Δt is the travel time residual.

This approach is widely used in inverse problems because of good quality of the velocity function restoration. But if the number of model parameters and measurements is large, this approach needs rather large memory capacity and computational time. So, there appears the question as how to reduce the problem dimensionality. To this end, recurrence formulas for the solution (1) were obtained in case of independent measurements:

$$\Delta p_i^l = \frac{\sum_j c_{ij} q_j}{\sigma_{0l}^2 + \sum_i \left[\sum_j q_j c_{ij}\right] \cdot q_i} \cdot \Delta t_l', \quad l = 1, \ldots, N,$$

$$c_{i,K}^{l} = c_{i,K}^{l-1} - \frac{\sum_j c_{ij}q_j \cdot \sum_j q_j c_{jK}}{\sigma_{0l}^2 + \sum_i \left[\sum_j q_j c_{ij}\right] \cdot q_i}, \quad i = 1,\ldots,M, \quad j = 1,\ldots,M. \tag{2}$$

$$\hat{P} = P_0 + \Delta P^n, \quad C_{\hat{P}\hat{P}} = C_{PP}^n,$$

where l is the equation number, $\Delta t_l' = t_l - G^l P_0 - G^l \Delta P^{l-1}$, q is an element of the matrix G.

Such an approach is known in filtration theory as the Kalman–Bucci filter. It saves the computer memory, but it is numerically unstable and needs much computational time.

Thus, using the diagonal form of the matrix C_{PP}, we suggest an iterative algorithm such that only diagonal elements $\{s_i^2\}$ of this matrix are taken into account:

$$\Delta p_i^l = \left[\frac{s_i^2 q_i}{\sigma_{0l}^2 + \sum_j s_j^2 q_j^2} \cdot \Delta t_l\right], \quad l = 1,\ldots,N,$$

$$\left(\Delta s_i^2\right)^l = \frac{s_i^4 q_i^2}{\sigma_{0l}^2 + \sum_j s_j^2 q_j^2}, \quad i = 1,\ldots,M,$$

$$\overline{\Delta p_i} = \frac{\sum_i \Delta p_i^l}{N}; \quad \overline{\Delta s_i^2} = \frac{\sum_i \left(\overline{\Delta s_i^2}\right)^l}{N}, \quad j = 1,\ldots,M,$$

$$\hat{p}_i = p_i^0 + \overline{\Delta p_i}, \quad \hat{s}_i^2 = s_i^2 - \overline{\Delta s_i^2}. \tag{3}$$

This approach allows us to find the solution (1) approximately. Moreover, it has an advantage in computational time and stability over the recurrent algorithm (2).

This algorithm has been tested extensively on the basis of real problems. For example, the Caucasus model parameters were improved by solving an overdetermined equation system of dimensions 8585 * 2116. The computational experience confirms the reliability of the algorithm.

The main point of the algorithm is the choice of its parameters $\{s_i^2\}$ in the case when the errors of a priori values of the model parameters are unknown. We construct a "basic" set $\{s_i^2\}^*$ as follows:

$$(s_i^2)^* \simeq \frac{\sigma_0^2}{(\partial g(P)/\partial p_i)^2}; \quad s_i^2 = n(s_i^2)^*; \quad \sum s_i^2 q_i^2 \sim \sigma_0^2,$$

where n is an integer depending on the accuracy of a priori value of the i-th parameter.

Results

Using the algorithm (3), a new block-layered crust model for the Kaspian sea region was constructed.

The first arrival times of the P-waves generated by the Kaspian earthquake of March 6, 1986, and its aftershock sequence were taken for input data. The hypocenter parameters were taken from the catalogue. The previous DSS exploration data allowed us to parametrize the medium and to find the first approximation of the velocity function parameters.

Parametrization of the model.

As follows from geophysical experience, the Kaspian sea region has a block-layered structure of the crust. So, it is natural to consider a three-dimensional block-layered model of the medium.

The region was divided into blocks, each having its velocity function. The crust in a block was assumed to consist of three horizontal layers. In the upper crust and the mantle the velocity is taken to be constant, while in the remainder it increases linearly in depth. Thus, each block is determined by 6 parameters. Such a model was originally suggested by Slavina and Pivovarova (1985) for describing the Caucasus earth's crust.

The Kaspian sea region was divided into three blocks whose velocity functions was calculated according to the Deep Seismic Sounding data smoothed so that the earth's crust in each block was described by six parameters. Block sizes were chosen on the basis of the data given in Krasnopevtzeva (1984) and with regard to the capabilities of the observational network.

The obtained velocity parameters were specified by solving an inverse kinematic problem whose data are presented by the traveltime residuals and hypocenters parameters. The results are shown in Figure 1.

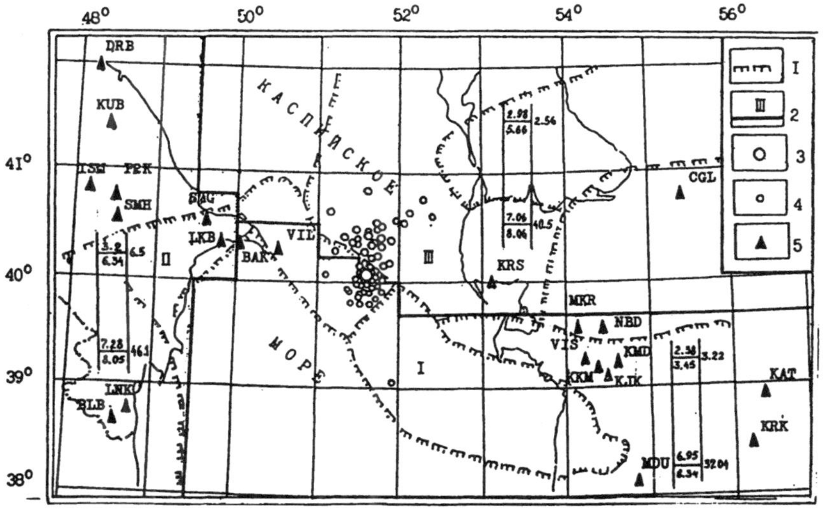

Figure 1. The map-scheme of subdividing the Kaspian sea region into velocity blocks. 1 – boundary of the blocks according to Krasnopevtzeva (1984), 2 – boundary and numbers of blocks in present work, 3 – main earthquake, 4 – aftershocks, 5 – seismostations.

Evaluation of adequacy and accuracy of the velocity model.

To test the adequacy of a new model for the actual structure, the traveltime residuals and the station corrections were analyzed. So, the histogram of the residuals distribution shows their non-shift and a decrease of their mean square value by 10%. The station corrections in the block-layered model are less than those in the Jeffreys-Bullen model for all stations except for may be one (KRK).

The accuracy of the velocity model was estimated in the L_2-norm as compared to an arbitrary experimental velocity curve of DSS. As a result, we obtained $\sigma_v^2 = 0.24$.

Application of the block-layered model.
The hypocenters ot the Kaspian earthquake sequence were relocated in the new crust model (Kondovskaya and Tushko, 1993). As a result, many hypocenters were displaced in depth while the epicentral parameters remained virtually unchanged. The mean square value of the time residuals was diminished by 40%.

Conclusions

1. The proposed iterative algorithm is efficient for solving large-dimensional systems and can be employed for seismological data inversion.
2. The block-layered model is adequate for the medium beneath the Kaspian sea and can be used in seismological practice.

Acknowledgments

We are grateful to Dr. Privalova N. B. for useful suggestion and discussions. We want to thank also Prof. Kondovskaya N. V. and Krasnopevtzeva G. V. for providing the experimental data and discussion on the results.

REFERENCES

Kondovskaya N. V. and Tushko T. A. (1993), *A block-layered model of a medium and determination of hypocenters of the Kaspian earthquake.* Physics of the Earth, no. 7. (in Russian)

Krasnopevtzeva G. V. (1984), *The in-depth structure of the Caucasus seismoactive region.* Moscow, Nauka, 110 p. (in Russian)

Pivovarova N. B. and Slavina L. B. (1985), *A numerical algorithm for finding an earthquake center and computational results.* Moscow, Nauka, 53–75. (in Russian)

Tarantola A. and Valetter B. (1982), *Generalized nonlinear inverse problems solved using the least squares criterion.* Rew. of Geoph. and Space Ph., vol. 20, no. 2, 219–232.

Solution of Convolution Integral Equations of the First Kind on a Segment

A. F. VORONIN

Institute of Mathematics
Novosibirsk, Russia

ABSTRACT

In the paper the first-kind integral equation

$$\int_0^b k(|x-y|)u(y)\mathrm{d}y = f(x), \quad x \in [0,b],$$

is reduced to the integral Wiener-Hopf equation

$$\gamma u_2(x) + \int_0^\infty k_2(|x-y|)u_2(y)\mathrm{d}y = f(x), \quad x \in [0,\infty],$$

where

$$k_2(x) = \frac{\partial^2}{\partial x^2}k(x), \quad \gamma = \mathrm{const},$$

and

$$u(x) = \frac{\partial^2}{\partial x^2}u_2(x), \quad x \in [0,b].$$

Introduction

Let us consider an integral Wiener-Hopf equation

$$\gamma u_2(x) + \int_0^\infty k_2(|x-y|)u_2(y)\mathrm{d}y = f(x), \quad x \in \mathbf{R}^+ \equiv [0,\infty], \tag{0.1}$$

where

$$k_2(x) = \frac{\mathrm{d}^2}{\mathrm{d}x^2}k(x), \quad f(x), k(x) \in W_2^2(\mathbf{R}^+), \quad \gamma = \mathrm{const} \neq 0, \tag{0.2}$$

and a convolution integral equation of the first kind on a segment $[0,b]$:

$$\int_0^b k(|x-y|)u(y)\mathrm{d}y = f(x), \quad x \in [0,b]. \tag{0.3}$$

Equation (0.1), unlike equation (0.3), can be solved exactly (in a closed form) (Gakhov and Chersky, 1978).

In the present paper equation (0.3) is reduced to equation (0.1) under some additional restrictions on the kernel $k(x)$ behavior outside the segment $[0,b]$. This allows the Wiener-Hopf equation theory (Gakhov and Chersky, 1978) to be used for investigating and solving equations of type (0.3).

1. Notation and Formulation of the Main Result

Let us introduce a class of functions $L_1(\mathbf{R};\lambda)$, to which the kernel $k(x)$ belongs:

$$L_1(\mathbf{R};\lambda) = \Big\{k(x): \quad k(x) = e^{-\lambda x}\int_{-\infty}^{\infty} e^{-iyx} dg(y), \quad x\in\mathbf{R}\Big\}, \tag{1.1}$$

where λ is a parameter, Re $\lambda = \lambda_0 > 0$; $g(y)$ is the Lebesgue-Stieltijes measure on $\mathbf{R}$, which has an absolutely continuous component $g_c(y)$ such that

$$y^j g_c'(y) \in L_1(\mathbf{R}), \quad j = 0,1,2, \tag{1.2}$$

and a discrete component $g_d(y)$:

$$dg_d(y) = \sum_{i=-\infty}^{\infty} \alpha_i \delta(y-y_i)dy, \quad y_i\in\mathbf{R}, \tag{1.3}$$

where

$$\sum_{i=-\infty}^{\infty} | \alpha_i y_i^n | < \infty, \quad n = 0,1,2, \quad \alpha_i = \text{const}_i,$$

$$[\ dg(y) = dg_c(y) + dg_d(y)\ ].$$

There exists $\varepsilon > 0$ such that in the region $\varepsilon > \text{Im } z > 0$ the function $g_c'(z)$ is regular and

$$g_c'(z) < C(1 + | z |)^{-l}, \quad l > 2, \quad 0 < C = \text{const} . \tag{1.4}$$

Let us give an example of functions in $L_1(\mathbf{R};\lambda)$:

$$k(x) = e^{-\lambda x}\sum_{i=1}^{\infty} a_i \sin(ix), \quad x\in\mathbf{R}, \quad \text{where} \quad \sum_{i=1}^{\infty} i^2 | a_i | < \infty,$$

whence

$$W_2^2[0,b] \subset L_1(\mathbf{R};\lambda).$$

Set

$$k_{2h}(x) = \frac{d^2}{dx^2}(k(x) - k(-x)),$$

$$\mathcal{L}k_{2h}(p) = \int_0^{\infty} e^{-px} k_{2h}(x)dx, \quad \text{Re } p > \lambda, \tag{1.5}$$

where $\mathcal{L}$ denotes Laplace transform. Let

$$\sigma(p) = 1 + \frac{1}{\gamma}\mathcal{L}k_{2h}(p). \tag{1.6}$$

Assertion 1. *The function $\mathcal{L}k_{2h}(p)$ can be analytically continued to a half-plane* Re $p > \lambda_0 - \varepsilon$ *with punctured points $p = \lambda + iy_j$, $j = 0, \pm1, \pm2, \ldots$, and in this half-plane the following equalities*

$$\mathcal{L}k_{2h}(p) = \int_{-\infty}^{\infty} e^{-px} k_2(| x |)dx = \int_{-\infty}^{\infty} \Big(\frac{(\lambda+iy)^2}{p-\lambda-iy} - \frac{(\lambda+iy)^2}{p+\lambda+iy}\Big)dg(y) \tag{1.7}$$

hold.

Denote the index of equation (0.1) by a , i.e.,

$$a = \frac{1}{2\pi i}\int_{-\infty}^{\infty} d\ln(D(x)), \quad D(x) = \frac{1}{\sigma(-ix+s)}, \quad x \in \mathbf{R}, \tag{1.8}$$

and assume that the following condition is fulfilled

$$\exists\, s \in (0, \lambda_0) : \sigma(p) \neq 0 \ \forall \ \mathrm{Re}\, p \geq s. \tag{1.9}$$

Denote by $L_2(\mathbf{R}^+, s)$ the space of functions $v(x)$ equipped with the norm $\|v\|^2 = \int_0^\infty |\, v\,|^2 (x)\, e^{-2sx} dx$.

Theorem. *Let conditions (0.2) be fulfilled for*

$$k(x) \in L_1(\mathbf{R}; \lambda), \quad \gamma = 2k'(0) \neq 0, \tag{1.10}$$

and condition (1.9) hold.

Then for every solution $u(x) \in L_2[0,b]$ of equation (0.3) there exists a solution $u_2(x)$ of equation (0.1) in $L_2(\mathbf{R}^+, s)$ such that

$$u(x) = \frac{d^2}{dx^2} u_2(x), \quad x \in [0, b]. \tag{1.11}$$

If equation (0.3) admits a solution in $L_2[0,b]$, and $a \leq 0$ (see (1.8)), then the solutions of equations (0.3) and (0.1) are unique in the corresponding spaces; if $a > 0$, then the solution of equation (0.1) depends on a of arbitrary constants, and the solution of equation (0.3) depends on a_0 of arbitrary constants, $0 \leq a_0 \leq a$.

If equation (0.1) has no solution in $L_2(\mathbf{R}^+, s)$, then equation (0.3) has no solution in $L_2[0,b]$ as well.

REMARK. The theorem is true also in the case when the solution of equation (0.3) belongs to the generalized function space.

2. Proof of the Theorem

To begin with we prove Assertion 1. From (1.1) and (1.5) it follows that

$$\mathcal{L}k_{2h}(p) = \int_0^\infty e^{-px} \int_{-\infty}^{\infty} (\lambda + iy)^2 \left[e^{-\lambda x} e^{-iyx} - e^{\lambda x} e^{iyx}\right] dg(y) dx.$$

By the Fubini's theorem, in view of properties (1.2)–(1.3), we change the integration order in the right-hand side of the previous equality. As a result, we obtain:

$$\mathcal{L}k_{2h}(p) = \int_{-\infty}^{\infty} (\lambda + iy)^2 \left[\frac{1}{p+\lambda+iy} - \frac{1}{p-\lambda-iy}\right] dg(y). \tag{2.1}$$

The right-hand side of equality (2.1) is a regular function in the half-plane Re $p > \lambda_0$. On the other hand, by the Jordan lemma and the Cauchy theorem (Lavrent'ev and Shabat, 1987), from (1.4) it follows that

$$\int_{-\infty}^{\infty} (\lambda + \mathrm{i}y)^2 \frac{1}{p - \lambda - \mathrm{i}y} \mathrm{d}g_c(y)$$

is a regular function in the half-plane Re $p > \lambda_0 - \varepsilon$.
So, the function $\mathcal{L}k_{2\,h}(p)$ is analytic in the half-plane Re $p > \lambda_0 - \varepsilon$ except for the poles at the points $p = \lambda + \mathrm{i}y_j$, $j = 0, \pm1, \pm2, \dots$.

We have

$$\int_{\infty}^{\infty} \mathrm{e}^{-px} k_2(|\, x\,|)\mathrm{d}x = \int_{0}^{\infty} \mathrm{e}^{-px} k_2(x)\mathrm{d}x + \int_{0}^{\infty} \mathrm{e}^{px} k_2(x)\mathrm{d}x,$$

whence due to (0.2) and (1.1), we obtain

$$\int_{\infty}^{\infty} \mathrm{e}^{-px} k_2(|\, x\,|)\mathrm{d}x = \int_{-\infty}^{\infty} (\lambda + \mathrm{i}y)^2 \left[\frac{1}{p + \lambda + \mathrm{i}y} - \frac{1}{p - \lambda - \mathrm{i}y}\right] \mathrm{d}g(y). \tag{2.2}$$

Equalities (2.1) and (2.2) yield (1.7). The statement is proved.

Let equation (0.3) have a solution in $L_2[0, b]$. Then we rewrite equation (0.3) in the form

$$\int_{0}^{x} [k(x - y) - k(-(x - y))]u(y)\mathrm{d}y = -\int_{0}^{b} k(-(x - y))u(y)\mathrm{d}y + f(x). \tag{2.3}$$

We shall consider equation (2.3) for all $x \in \mathbf{R}^+$ since the function $k(x)$ is defined for $x \in \mathbf{R}$. We transform the left-hand side of equality (2.3) using integration by parts twice an denoting

$$M_2(y) = \int_{0}^{y}\int_{0}^{z} u(x)\mathrm{d}x\mathrm{d}z. \tag{2.4}$$

As a result, we obtain

$$\gamma M_2(x) + \int_{0}^{x} k_{2\,h}(x - y)M_2(y)\mathrm{d}y = f(x) - v_2(x), \tag{2.5}$$

where

$$v_2(x) = \int_{0}^{b} k(-(x - y))u(y)\mathrm{d}y. \tag{2.6}$$

Assertion 2. *Equation (2.5) has a unique solution $M_2(x)$ such that*

$$M_2(x)\,\mathrm{e}^{-sx} \in L_2(\mathbf{R}^+), \quad s \in (\lambda_0 - \varepsilon, \lambda_0).$$

PROOF. The Volterra equation (2.5) is equivalent to

$$\mathcal{L}M_2(p) = \frac{\mathcal{L}f(p) + \mathcal{L}v_2(p)}{\gamma + \mathcal{L}k_{2\,h}(p)}, \quad \text{Re}\, p > \lambda_0. \tag{2.7}$$

Let us transform the expression for $\mathcal{L}v_2(p)$ for Re $p > \lambda_0$. Taking into account (2.6), (1.1) and the Fubini's theorem, we have

$$\mathcal{L}v_2(p) = \int_0^\infty e^{-px} \int_0^b \left[e^{-\lambda(y-x)} \int_{-\infty}^\infty e^{-iz(y-x)} dg(z) \right] u(y) dy dx$$

$$= \int_{-\infty}^\infty U(z) \frac{1}{p - \lambda + iz} dg(z),$$

where

$$U(z) = \int_0^b e^{-y(\lambda + iz)} u(y) dy.$$

Then it follows from the proof of Assertion 1 that the function $\mathcal{L}v_2(p)$ is analytic in the half-plane Re $p > \lambda_0 - \varepsilon$. Hence, the right-hand side of equality (2.7) is a holomorphic function in the half-plane Re $p > \lambda_0 - \varepsilon$, and tends to 0 as $|p| \to \infty$. So, according to the Laplace transform properties, Assertion 2 is true.

Let us denote, for $p \in \mathbf{R}$, $s \in (\lambda_0 - \varepsilon, \lambda_0)$,

$$F^+(p) = \mathcal{L}M_2(-ip + s), \quad F^-(p) = \frac{1}{\gamma}\mathcal{L}v_2(-ip + s), \tag{2.8}$$

$$D(p) = \frac{1}{\sigma(-ip + s)}, \quad H(p) = \frac{1}{\gamma} D(p) \mathcal{L}f(-ip + s). \tag{2.9}$$

From (2.8) we have

$$F^+(p) = \int_0^\infty e^{ipx} e^{-sx} M_2(x) dx, \quad F^-(p) = \int_{-\infty}^\infty U(z) \frac{1}{s - \lambda + iz - ip} dg(z). \tag{2.10}$$

Using notation (2.8)-(2.9), we rewrite equation (2.7) as follows:

$$F^+(p) = D(p) F^-(p) + H(p), \quad p \in \mathbf{R}. \tag{2.11}$$

According to (2.10) and Assertion 2, $F^+(p), F^-(p)$ are boundary values of holomorphic functions in the half-planes Im $p > 0$, Im $p < 0$, respectively, and

$$\int_{-\infty}^\infty |F^+(x + iy)|^2 dx \le C^+, \quad \forall y \ge 0,$$

$$\int_{-\infty}^\infty |F^-(x + iy)|^2 dx \le C^-, \quad \forall y \le 0, \tag{2.12}$$

where $C^\pm$ are some positive constants,

$$D(p) - 1, \ H(p) \in L_2(\mathbf{R}); \quad D(p) \ne 0 \quad \forall p \in \mathbf{R}. \tag{2.13}$$

Therefore, boundary condition (2.11) leads to the Riemann's problem on $\mathbf{R}$ whose solution can be constructed (as equation (0.3) has a solution). All the solutions of the Riemann's problem can be written in a closed form and, moreover, unless the index of equation (0.1) is positive, i.e., $a \leq 0$, then the solution of the Riemann's problem is unique, and therefore the solution of equation (0.3) is unique. If $a > 0$, then the Riemann's problem (2.11) depends linearly on a of arbitrary complex constants (Gakhov and Chersky, 1978). Hence, there exists at least one solution of this problem which satisfies equation (2.5).

On the other hand, the Wiener-Hopf equation (0.1) is equivalent to the following Riemann's problem (Gakhov and Chersky, 1978, p. 132–134):

$$F_1^+(p) = D_1(p)F_1^-(p) + H_1(p), \quad p \in \mathbf{R}, \tag{2.14}$$

where

$$F_1^+(p) = \int_0^\infty e^{ipx} e^{-sx} u_2(x)dx,$$

$$D_1(p) = (1 + \frac{1}{\gamma}\int_{-\infty}^{\infty} e^{ipx} e^{-sx} k_2(|x|)dx)^{-1}, \quad H_1(p) = \frac{1}{\gamma} D_1(p)\mathcal{L}f(-ip+s).$$

Note that, due to Assertion 1, $D_1(p) = D(p)$, whence $H_1(p) = H(p)$. Therefore, the Riemann's problems (2.11) and (2.14) coincide, and equality (1.11) holds. The theorem is proved. The validity of Remark follows from (1.11).

3. Example

Consider the equation

$$\int_0^1 e^{-c|x-y|} u(y)dy = \chi(x-x_0)e^{-c(x-x_0)}, \quad x, x_0 \in [0,b], \quad \text{Re } c > 0, \tag{3.1}$$

where

$$\chi(x) = \begin{cases} 1, & \text{for } x > 0, \\ 0, & \text{for } x < 0. \end{cases}$$

It is easy to check that the solution of equation (3.1) has a form

$$u(x) = \frac{1}{2}(\delta(x-x_0) - \delta'(x-x_0)/c). \tag{3.2}$$

Note that in the class of generalized functions this solution is unique. This fact follows from the uniqueness of the Cauchy problem solution for a system of ordinary differential equations of the first order, to which equation (3.1) reduces by differentiating it in x twice. The same result can be obtain if we apply to equation (3.1) the methods used in solving convolution integral equations on a segment with rational symbols (see Zabreiko et al., 1968, Sections 11.2, 11.3).

Let us check that equation (3.3) obeys the theorem conditions (1.9)–(1.10). We have

$$k(x) = e^{-cx} = e^{-\lambda x} \int_{-\infty}^{\infty} e^{ixy} dg_d(x), \tag{3.3}$$

where $\lambda = c$, $dg_d(x) = \delta(x)dx$ ($g_c(x) = 0$). Therefore $k(x) \in L_1(\mathbf{R};\lambda)$, $\gamma = -2c$.

From (3.3), combined with (1.5)–(1.6), it follows that

$$k_2(x) = c^2\,\mathrm{e}^{-cx}, \quad k_{2h}(x) = c^2(\,\mathrm{e}^{-cx} - \mathrm{e}^{cx}\,), \quad \sigma(p) = \frac{p^2}{p^2 - c^2}.$$

Hence, $\sigma(p) \neq 0, \ \forall\ Re\ p > 0$. Therefore, $s \in (0, \lambda_0)$ $(\lambda_0 = \mathrm{Re}\ c)$ and

$$D(x) = \frac{(-\mathrm{i}x + s - c)(-\mathrm{i}x + s + c)}{(-\mathrm{i}x + s)^2}. \tag{3.4}$$

Equation (3.4) implies that the index of equation (0.1) is equal to 1. Therefore, the solution of equation (0.1) depends on an arbitrary constant (Gakhov and Chersky, 1978).

Let us find all the solutions of equation (0.1). To this end, following Gakhov and Chersky (1978), we need to solve the Riemann's problem

$$F^+(x) = F^-(x)D(x) + H(x), \quad \left[F^+(x) = \mathcal{L}u_2(-\mathrm{i}x + s)\right], \tag{3.5}$$

where

$$D(x) = \frac{p_+(x)p_-(x)}{q_-(x)}, \quad H(x) = -\frac{p_+(x)}{2cq_-(x)}\,\mathrm{e}^{-x_0(-\mathrm{i}x+s)},$$

$$p_\pm(x) = -\mathrm{i}x + s \mp c, \quad q_-(x) = (s - \mathrm{i}x)^2.$$

Problem (3.5) has the following solution:

$$F^+(x) = -\frac{p_+(x)}{2cq_-(x)}\,\mathrm{e}^{-x_0(-\mathrm{i}x+s)} + c_1\frac{p_-(x)}{q_-(x)}, \quad F^-(x) = \frac{c_1}{p_+(x)},$$

where c_1 is an arbitrary constant. Therefore,

$$u_2(x) = c_1cx + c_1 + \frac{1}{2}\chi(x - x_0)\left[x - x_0 - 1/c\right].$$

Hence, according to (1.11), one of the solutions of equation (3.1) has the form (3.2). Therefore, equation (3.1) has no solution in $L_2[0, b]$ but admits the unique solution (3.2) in the space of generalized functions.

REFERENCES

Gakhov F. D., Chersky J. I. (1978), *Equations of the Convolution Type.* Moscow, Nauka. 295 p. (in Russian)

Lavrent'ev M. A., Shabat B. V. (1987), *Methods of the Theory of Complex Variable Functions.* Moscow, Nauka. 688 p. (in Russian)

Zabreiko P. P., Koshelev A. I. et al. (1968), *Integral Equations.* Moscow, Nauka. 448 p. (in Russian)

Image Reconstruction for Fan Beam X-Ray Tomography Using a New Integral Transform Pair

Shuang-Ren ZHAO and Horst HALLING

Zentrallabor für Elektronik, Forschungszentrum Jülich GmbH
Postfach 1913, D–52425 Jülich Germany

ABSTRACT

A new integral transform pair has been introduced for the reconstruction of slices from fan beam projection data. This integral transform plays the same role as the Hankel transform (or Fourier Bessel transform) in the parallel beam case and includes the Hankel transform as a special case. The relations between the new integral transform and the Hankel transform pair are derived and the implementation of the algorithm is described. Memory space requirements and speed of the transform are discussed and the advantage of precalculations are mentioned. The transform pair is especially useful in case the two dimensional object slice to be reconstructed has only few low frequency components in the azimuthal direction. Furthermore the method is capable of treating the case where non analytic a priori information is to be taken into account. In this case projections into the measurement directions are performed and the differences between these projection data and the measured data are reconstructed in order to improve the object data. Alternating transformations between projection and object space are efficient and fast. Finally some results of simulations including the influence of noise are shown and possible applications of the method are discussed.

1. Introduction

The problem of reconstructing an object from a set of its projections arises in among other fields, computer-aided tomography CAT, radio astronomy, electron microscopy, and spotlight-mode synthetic aperture radar. Traditional analytic algorithms of reconstruction for parallel or fan beam projections are the convolution backprojection method (Herman ,1980; Lewitt et al., 1978; Mersereau and Oppenheim, 1974; Ramachandran and Lakshmin-arayanan, 1971; Scudder, 1978; Shepp and Logan, 1974) the Fourier method (Peng and Stark, 1987) and the Hankel transform method (Higgins and Munson ,1988). The convolution-backprojection method is fast and widely used. The Fourier method is faster than the convolution-backprojection method, but the reconstruction quality is often not satisfactory because interpolation in the Fourier domain easily introduces artifacts (Natterer, 1986). Recently a Fourier method has been introduced for the fan beam case (Zhao and Halling, 1993) by our group. It has no interpolation in the Fourier domain, so that good reconstruction quality can be obtained. For a special case where the object to be reconstructed has no high frequencies along the angle θ direction, the Hankel transform method can be used and good quality of reconstruction can be achieved by few projections. However, up to now the Hankel method is only used for parallel beam projections.

This paper is extending the Hankel method for the fan beam case. We derived a new integral transform pair which plays exactly the same role in the fan-beam case as the Hankel transform pair in the parallel-beam case.

2. Two Integral Formulas for Fan Beam Tomography

The spirit of the Hankel transform method for parallel beam tomography is first to obtain the integral transform from the object function $f(r, \phi)$ to the projections $P_\theta(u)$ and the inverse transform from the projections $P_\theta(u)$ to the object function $f(r, \phi)$, then expand these into Fourier series. We follow this spirit to obtain the general Hankel Transform pair which can be applied to fan beam geometry.

From the paper by Kak and Slaney, the method of back projection for the case of equally spaced collinear detectors is available

$$f(r, \phi) = \int_0^{2\pi} \frac{1}{U^2} \int_{-\infty}^{\infty} R'_\beta(s) \eta(s' - s) \, \mathrm{d}s \mathrm{d}\beta \tag{1}$$

where

$$h(t) = \frac{1}{4\pi} \int_{-\infty}^{\infty} |\eta| \exp(j\eta t) \, \mathrm{d}\eta \tag{2}$$

$$U = 1 + \rho \sin(\beta - \phi), \quad s' = r \cos(\beta - \phi)/U \tag{3}$$

$$R'_\beta(s) = \frac{D}{\sqrt{D^2 + s^2}} R_\beta(s) \tag{4}$$

and $f(r, \phi)$ is the image function to be reconstructed in polar coordinates (r, ϕ). $R_\beta(s)$ denotes a fan projection as show in Figure 1 where s is the distance along the straight line corresponding to the detector bank and β is the rotation angle, $\rho = \frac{r}{D}$, where D is the distance from the source point F to the origin O.

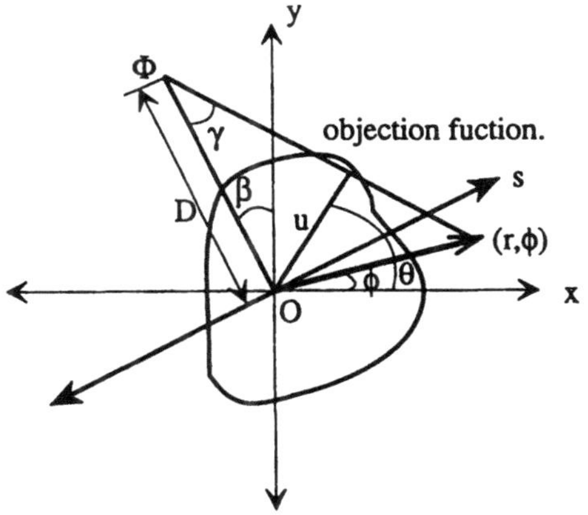

Figure 1. Fan beam geometry

Define the one dimensional Fourier transform of $R'_\beta(s)$ as

$$G'_\beta(\eta) = \mathcal{F}_{\eta\leftarrow s}(R'_\beta(s)) \tag{5}$$

where

$$\mathcal{F}_{\eta\leftarrow s}(.) = \int_{-\infty}^{\infty} (.) \exp(-j\eta s)\, \mathrm{d}s; \quad \mathcal{F}^{-1}_{s\leftarrow\eta}(.) = \frac{1}{2\pi}\int_{-\infty}^{\infty} (.) \exp(j\eta s)\, \mathrm{d}s$$

Combining (1) , (2) and (5) and changing the order of the integral ,we have

$$f(r,\phi) = \frac{1}{4\pi}\int_{-\infty}^{\infty}\int_{0}^{2\pi} G'_\beta(\eta)\frac{1}{U^2}\exp(j\eta s')\, \mathrm{d}\beta\, |\eta|\, \mathrm{d}\eta \tag{6}$$

Equation (4), (5) and (6) gives a integral transform from $R_\beta(s)$ to $f(r,\phi)$, now we derive the other integral transform from $f(r,\phi)$ to $R_\beta(s)$.

The parallel beam projection $P_\theta(u)$ is (Rattey and Lindgren, 1981)

$$P_\theta(u) = \int_{\phi=0}^{2\pi}\int_{r=0}^{R} f(r,\phi)\delta(u - r\cos(\theta-\phi))r\, \mathrm{d}r\mathrm{d}\phi; \tag{7}$$

the fan beam projection $R_\beta(s)$ (Kak and Slaney) is

$$R_\beta(s) = P_\theta(u)|_{u=D\sin(\gamma(s));\quad \theta=\beta+\gamma(s)} \tag{8}$$

where

$$\gamma = \arctan(s/D). \tag{9}$$

Considering

$$u = D\sin(\gamma); \quad \theta = \beta+\gamma, \tag{10}$$

from the paper by Zhao et al. we have

$$\delta(u - r\cos(\theta-\phi)) = \frac{1}{U\cos(\gamma)}\delta(s-s') \tag{11}$$

where U, s' are defined in equation (3) and

$$\cos(\gamma) = \frac{D}{\sqrt{D^2+s^2}} \tag{12}$$

Using equations (8) (9) and (11) ,equation (5) becomes

$$G'_\beta(\eta) = \int_{-\infty}^{\infty}\frac{D}{\sqrt{D^2+s^2}}R_\beta(s)\exp(-j\eta s)\, \mathrm{d}s$$

$$= \int_{-\infty}^{\infty}\cos(\gamma)\int_{\phi=0}^{2\pi}\int_{r=0}^{R} f(r,\phi)\frac{1}{U\cos(\gamma)}\delta(s-s')\, r\, \mathrm{d}r\, \mathrm{d}\phi \exp[-j\eta s]\, \mathrm{d}s$$

$$= \int_{\phi=0}^{2\pi}\int_{r=0}^{R}\frac{1}{U}f(r,\phi)\exp(-j\eta s')\, r\, \mathrm{d}r\, \mathrm{d}\phi \tag{13}$$

where $f(r,\phi) = 0$ if $r > R$. Equation (6) and (13) are two basic integral formulas. (6) can be used for image reconstruction. (13) can be used to calculate the projection data from the image function as required for simulation. These two integral formulas can be expanded using Fourier series,

$$f(r,\phi) = \sum_{m=-\infty}^{\infty} f_m(r)\exp(jm\phi) \tag{14}$$

$$R'_\beta(s) = \sum_{m=-\infty}^{\infty} R'_m(s)\exp(jm\beta) \qquad (a)$$

$$R'_m(s) = \frac{1}{2\pi}\int_0^{2\pi} R'_\beta(s)\exp(-jm\beta)\,\mathrm{d}\beta \quad (b) \tag{15}$$

$$G'_\beta(\eta) = \sum_{m=-\infty}^{\infty} G'_m(\eta)\exp(jm\beta) \qquad (a)$$

combining definition (5) and equations (15) (16) gives,

$$G'_m(\eta) = \int_{-\infty}^{\infty} R'_m(s)\exp(-js\eta)\,\mathrm{d}s \qquad (b) \tag{16}$$

Now we obtain the new integral transform pair

$$f_m(r) = \frac{1}{2}\exp(jm\frac{\pi}{2})\int_{\eta=-\infty}^{\infty} G'_m(\eta)H_m(r,\eta)|\eta|\,\mathrm{d}\eta \qquad (a) \tag{17}$$

where

$$H_m(r,\eta) = \frac{1}{2\pi}\int_{\epsilon=0}^{2\pi} \frac{1}{U^2}\exp[j(m\epsilon - \frac{r\sin(\epsilon)}{U}\eta)]\,\mathrm{d}\epsilon \qquad (b)$$

and

$$G'_m(\eta) = 2\pi\exp(-jm\frac{\pi}{2})\int_0^a f_m(r)I_m(r,\eta)r\,\mathrm{d}r \qquad (a) \tag{18}$$

where

$$I_m(r,\eta) = \frac{1}{2\pi}\int_0^{2\pi} \frac{1}{U}\exp[-j(m\epsilon - \frac{r\sin(\epsilon)}{U}\eta)]\,\mathrm{d}\epsilon \qquad (b)$$

In these two formulas the substitution $\beta - \phi = \epsilon + \pi/2$ has been used and $U = 1 + r\cos(\epsilon)$.

Equations

$$(4) \longrightarrow (15-b) \longrightarrow (16-b) \longrightarrow (17) \longrightarrow (14)$$

can be used as an algorithm for the fan beam image reconstruction. The equation (18) can be used deriving the for fan beam sampling theorem as in Rattey and Lindgren (1981).

3. The Connection Between the New Integral Transform and the Hankel Transform

In this section the connection between the new integral transform and the Hankel transform is derived. This connection gives an alternative to obtain the new integral transform pair.

One algorithm for the parallel-beam image reconstruction is the Hankel transform method (Higgins and Munson, 1988):

$$f_m(r) = \exp(jm\frac{\pi}{2}) \int_0^\infty F_m(\eta) J_m(r\eta)\eta \, \mathrm{d}\eta) \tag{19}$$

$$F_m(\eta) = 2\pi \exp(-jm\frac{\pi}{2}) \int_0^a f_m(r) J_m(r\eta)\, r \, \mathrm{d}r \tag{20}$$

where $F_m(\eta)$ is defined by

$$F_\theta(\eta) = \mathcal{F}_{\eta\leftarrow u}\{P_\theta(u)\} \tag{21}$$

$$F_\theta(\eta) = \sum_{m=-\infty}^{\infty} F_m(\eta) \exp(jm\theta) \tag{22}$$

J_m is the first class Bessel function

$$J_m(r\eta) = \frac{1}{2\pi} \int_{\epsilon=0}^{2\pi} \exp[\pm j(m\epsilon - r\eta \sin(\epsilon))] \, \mathrm{d}\epsilon \tag{23}$$

First we note that if $r = \frac{r}{D} \longrightarrow 0$ then

$$H_m(r,\eta) \longrightarrow J_m(r\eta)$$
$$I_m(r,\eta) \longrightarrow J_m(r\eta)$$

This means that the Hankel transform is a special case (where $r \longrightarrow 0$) of the new integral transform.

The rebinning relation between the fan beam projection and the parallel projection is (Kak and Slaney)

$$R_\beta(s) = P_\theta(u)|_{u=sD/\sqrt{D^2+s^2}, \quad \theta=\beta+\arctan(\frac{s}{D})} \tag{24}$$

$$P_\theta(u) = R_\beta(s)|_{s=uD/\sqrt{D^2-U^2}, \quad \beta=\theta-\arcsin(\frac{u}{D})} \tag{25}$$

Using equations (5) and (21), combining equation (24) and (25) we obtain

$$F_\theta(\eta) = \mathcal{F}_{\eta\leftarrow u}\{\frac{\sqrt{D^2+s^2}}{D}\mathcal{F}^{-1}_{s\leftarrow\eta'}\{G'_\beta(\eta')\}|_{s=\frac{Du}{\sqrt{D^2-U^2}},\quad \beta=\theta-\arcsin(\frac{u}{D})}\} \tag{26}$$

$$G'_\beta(\eta) = \mathcal{F}_{\eta\leftarrow s}\{\frac{D}{\sqrt{D^2+s^2}}\mathcal{F}^{-1}_{u\leftarrow\eta'}\{F_\theta(\eta')\}|_{u=\frac{sD}{\sqrt{D^2+s^2}},\quad \theta=\beta+\arctan(\frac{S}{D})}\} \tag{27}$$

Substituting (16-a) into (26) and comparing with (22), we obtain

$$F_m(\eta) = \mathcal{F}_{\eta\leftarrow u}\{\frac{D}{\sqrt{D^2-U^2}}e^{-jm\arcsin(\frac{u}{D})}\mathcal{F}^{-1}_{s\leftarrow\eta'}G'_m(\eta')|_{s=\frac{uD}{\sqrt{D^2-U^2}}}\} \tag{28}$$

Substituting (22) into (27) and comparing with (16-a) , we obtain

$$G'_m(\eta) = \mathcal{F}_{\eta\leftarrow s}\{\frac{D}{\sqrt{D^2+s^2}}e^{jm\arctan(\frac{s}{D})}\mathcal{F}^{-1}_{u\leftarrow\eta'}\{F_m(\eta')\}|_{u=\frac{sD}{\sqrt{D^2+s^2}}}\} \tag{29}$$

Substituting (28) into (19) and comparing with (17-a) we obtain

$$H_m(r,\eta)|\eta| = \mathcal{F}_{\eta\leftarrow s}\{e^{-jm\arctan(\frac{s}{D})}\frac{D^2}{D^2+s^2}\mathcal{F}^{-1}_{u\leftarrow\eta'}J_m(r\eta')|\eta'||_{u=\frac{sD}{\sqrt{D^2+s^2}}}\} \tag{30}$$

Substituting (20) to (29) and comparing with (18-a) we obtain

$$I_m(r,\eta) = \mathcal{F}_{\eta\leftarrow s}\{e^{jm\arctan(\frac{s}{D})}\frac{D}{\sqrt{D^2+s^2}}F^{-1}_{u\leftarrow\eta'}\{J_m(r\eta')\}|_{u=\frac{sD}{\sqrt{D^2+s^2}}}\} \tag{31}$$

The two formulas above are connections between the Bessel functions J_m and the new functions H_m and I_m. We have obtained the two function H_m and I_m from formulas (6) and (13). This can also be done directly from (30) and (31). Practically we first obtained (17-b) this way. See Appendix.

4. Implementation of This Algorithm

1) The implementation of (4) calculates the weighted projections:

$$R'(\beta_i, s_k) = \frac{D_1}{\sqrt{D_1^2+s_k^2}}R(\beta_i, s_k) \tag{32}$$

where $R(\beta_i, s_k)$ –data of fan-beam projections, $\beta_i = i\Delta\beta$, $\Delta\beta = 2\pi/(N_p+1)$, (N_p+1) –number of projections, which is chosen as a power of 2 so that the FFT algorithm can be easily used. $i \in [0, N_p]$, $s_k = k\Delta s$, $\Delta s = (D_1/(D_1+D_2))\Delta t$, D_1–distance from the source F to the center O, D_2– distance from the center to the detector plane, Δt detector distance, $k \in [-N_{s2}, N_{s2}]$, $N_{s2} = N_s$ div 2, div – integer division, N_s – detector number. See Figure 2.

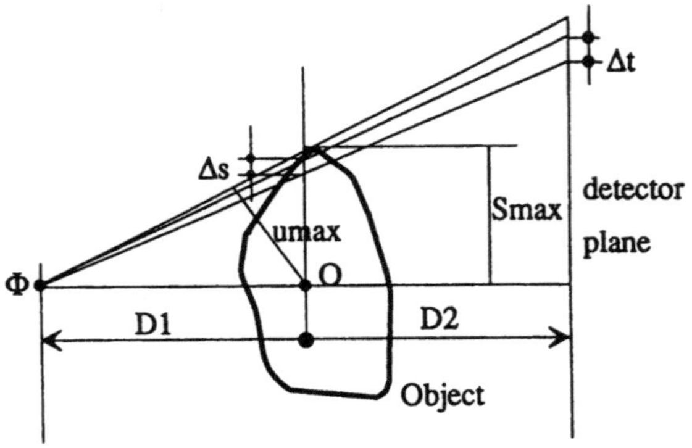

FIGURE 2. Object, projections and detector distance.

2) The implementation of (15-b) calculates the components of the Fourier series. First we define

$$\mathrm{MOD}^N_{m \longrightarrow n} \tag{33}$$
$$\mathrm{IMOD}^N_{m \longrightarrow n} \tag{34}$$

as operators, which change the integer value m to n according to

$$n = m \bmod N \tag{35}$$

and

$$n = \begin{cases} m & : \quad m \in [0, N_2 - 1] \\ m - N & : \quad m \in [N_2, N - 1] \end{cases} \tag{36}$$

respectively, where $N_2 = N$ div 2. Define the FFT and IFFT operators for any series X_m,

$$\begin{aligned} \mathrm{FFT}^N_{m \longrightarrow n}\{X_m\} &= \sum_{m=0}^{N-1} X_m \exp[-jmn\frac{2\pi}{N}] \quad (a) \\ \mathrm{IFFT}_{m \longrightarrow n}\{X_m\} &= \sum_{m=0}^{N-1} X_m \exp[jmn\frac{2\pi}{N}] \quad (b) \end{aligned} \tag{37}$$

since we have

$$R'_m(s_k) = \frac{1}{N_p + 1} \mathrm{IMOD}^{N_p+1}_{n \longrightarrow m} \mathrm{FFT}^{N_p+1}_{i \longrightarrow n}\{R'(\beta_i, s_k)\} \tag{38}$$

where i, $n \in [0, N_p]$, $m \in [-N_{p2}, N_{p2} - 1]$, $N_{p2} = (N_p + 1)$ div 2.

3) The implementation of (17-b) calculates the Fourier components G_m.

$$G_m(\eta_l) = \Delta s \, \mathrm{IMOD}^{N_v+1}_{i \longrightarrow l} \mathrm{FFT}^{N_v+1}_{j \longrightarrow i}\{\mathrm{MOD}^{N_v+1}_{k \longrightarrow j} R''_m(s_k)\} \tag{39}$$

where

$$R''_m(s_k) = \begin{cases} R'_m(s_k) & : \; k \in [-N_{s2}, N_{s2}] \\ 0 & : \; k \in [-N_{v2}, N_{v2}-1] - [N_{s2}, N_{s2}]; \end{cases} \tag{40}$$

and $\eta_l = l\Delta\eta$, $\Delta\eta = 2\pi/s_T$, $s_T = (N_v+1)\Delta s$, l, $k \in [-N_{v2}, N_{v2}-1]$, $N_{v2} = N_v$ div 2, (N_v+1) is size of FFT algorithm, j, $i \in [0, N_v]$, and $N_v > N_s$.

4) The implementation of (17-b) calculates the elements of the Matrix H_m

$$H_m(r_n, \eta_l) = \frac{1}{I}\sum_{i=0}^{I-1}\frac{1}{U^2}\cos[m\epsilon_i - \frac{r_n\sin(\epsilon_i)}{U}\eta_l] \tag{41}$$

where $U = 1 + \rho\cos(\epsilon_i)$, $\rho = r_n/D_1$, $r_n = n\Delta r$, $\Delta r = u_{max}/N_{f_2}$, $N_{f_2} = N_f$ div 2, N_f–this number can be chosen, for example, $N_f = N_s$, $\epsilon_i = i\Delta\epsilon$, $\Delta\epsilon = 2\pi/I$, I–defined by a large integer, $u_{max} = \frac{D_1 s_{max}}{\sqrt{D_1^2+s_{max}^2}}$, $s_{max} = (N_{s2}+0.5)\Delta s$. Note that only $H_m(r_n, \eta_l)$ where $m \in [0, N_{p2}-1]$ needs to be calculated.

5) The implementation of (17-a) calculates $f_m(r_n)$.

$$f_m(r_n) = \frac{1}{2}\exp(jm\frac{\pi}{2})\sum_{l=-(N_{v2}-1)}^{N_{v2}-1} H_m(r_n, \eta_l)\Phi_l G'_m(\eta_l)\Delta\eta \tag{42}$$

where

$$\Phi_l = \begin{cases} \Delta\eta/4 & : \; l = 0 \\ \eta_l & : \; \text{else} \end{cases} \tag{43}$$

in which the dc value of the object function has been taken into account.

6)The implementation of (14) reconstructs the object function.

$$f(x_i, y_i) = f_0(r) + 2\,\mathrm{Re}\{\sum_{m=0}^{N_{p2}-1} f_m(r)\exp[jmf]\} \tag{44}$$

where $f_m(r)$ is a linear interpolation of $f_m(r_n)$.

$$f_m(r) = \frac{\frac{f_m(r_n)}{r_0} + \frac{f_m(r_1)}{r_1}}{\frac{1}{r_0} + \frac{1}{r_1}} = \frac{f_m(r_n)r_1 + f_m(r_{n+1})r_0}{\Delta r} \tag{45}$$

$$\begin{aligned} r_n = \mathrm{trunc}\,(r/\Delta r)\Delta r, \quad & r_{n+1} = r_n + \Delta r \\ r_0 = r - r_n, \quad & r_1 = r_{n+1} - r \end{aligned} \tag{46}$$

trunc takes the integer value and

$$r = \sqrt{x_i^2 + y_i^2}; \quad \phi = \arctan(y_i/x_i). \tag{47}$$

$x_i = i\Delta x$, $y_i = j\Delta y$, $\Delta x = \Delta y = u_{max}/(N_{v2}+0.5)$, $i, j \in [-N_{v2}, N_{v2}-1]$.

It is remarkable that if the matrix $H_m(r_n, \eta_l)$ has been precalculated, most of the computation time is spent in Step 6. In order to reduce the computation time, the implementation of Step 6. can be modified as follows.

6'. Alternative of 6.

$$f(x_i, y_i) = \frac{\frac{f_1}{r_1} + \frac{f_2}{r_2} + \frac{f_3}{r_3} + \frac{f_4}{r_4}}{\frac{1}{r_1} + \frac{1}{r_2} + \frac{1}{r_3} + \frac{1}{r_4}} \tag{48}$$

where

$$f_1 = f(\phi_k, r_n); \quad f_2 = f(\phi_{k+1}, r_n); \quad f_3 = f(\phi_k, r_{n+1}); \quad f_4 = f(\phi_{k+1}, r_{n+1}) \tag{49}$$

$$f(\phi_k, r_n) = \mathrm{IMOD}_{k' \longrightarrow k}^{N_p+1} \mathrm{FFT}_{m' \longrightarrow k}^{N_p+1} \{\mathrm{MOD}_{m \longrightarrow m'}^{N_p+1} f_m(r_n)\} \tag{50}$$

and

$$\begin{aligned} r_1 &= \sqrt{r^2(\phi - \phi_k)^2 + (r - r_n)^2} \\ r_2 &= \sqrt{r^2(\phi_{k+1} - \phi_k)^2 + (r - r_n)^2} \\ r_3 &= \sqrt{r^2(\phi - \phi_k)^2 + (r_{n+1} - r)^2} \\ r_4 &= \sqrt{r^2(\phi_{k+1} - \phi)^2 + (r_{n+1} - r)^2} \end{aligned} \tag{51}$$

and

$$\begin{aligned} r_n &= \mathrm{trunc}\,(r/\Delta r)\Delta r; \quad r_{n+1} = r_n + \Delta r \\ \phi_k &= \mathrm{trunc}\,(\phi/\Delta\phi)\Delta\phi; \quad \phi_{k+1} = \phi_k + \Delta\phi \end{aligned} \tag{52}$$

and r, ϕ is defined in (47).

If only few projections are available, Step 6. is the best choice and the quality of the reconstruction is better than with the convolution backprojection method. If more projections are available, Step 6'. is best choice and the same quality of reconstruction as with the convolution backprojection method is achieved with only a quarter of computation time or even less because of the FFT algorithm (the matrix $H_m(r_n, \eta_l)$ has been precalculated).

5. Conclusion and Outlook

This section examines the performance of the above algorithm. The reconstructions are done with the data produced by a simulation program (Fuhrmann). Figure 1 gives the original object. $u_{max} = 180mm$, the display is 128 * 128. Figure 2 gives the display of the projection data. Projection number $(N_p + 1)$ is 256. The distance from Source F to the rotation center O is $D_1 = 720mm$, the distance form O to the detector plane is $D_2 = 1080mm$, the detector element distance is $3.645161mm$, the detector element number is 127. The FFT size $(N_v + 1)$ is 128. The large number I for the calculation of $H_n(r_n, \eta_l)$ is set to 1000. Figure 3 gives the reconstructions with Step 6 and Step 6' without noise.

FIGURE 3. a) the reconstructions with Step 6, b) with Step 6'.

6. Conclusion and Outlook

We derived an new integral transform pair. We explored an algorithm based on the integral transform which modified the Hankel algorithm used for the reconstruction from the parallel-beam projections to suit the case of the reconstruction from the fan beam projections. The results show that this algorithm can give high quality reconstructed.

The fan beam geometry has many different ray-sampling schemes, equally spaced collinear detectors and equiangular rays (Kak and Slaney), circle camera used for PET and SPECT and so on. In this paper we only consider the case of equally spaced collinear detectors. The case of equiangular rays is consider by another paper.

It is well know that if the projection data are noisy EM methods give a better reconstruction quality than the other methods but it is time consuming. The EM method iterates with a projection operator and a backprojection operator. In a next step we intend to apply the new integral transform as fast projection and backprojection operators to speed up the EM method and study noisy projection data.

In the future we will try to find the corresponding mathematical applications for the new transform pair.

Acknowledgments

Zhao Shuang-Ren wishes to thank the central laboratory for electronics of the research center Juelich (Germany) for the three years financial support as a visiting scientist, Dr G. Fuhrmann for the meaningful discussion, Mr. G. Brandenburg, Mr. R. Moeller and Rapp Jürgen for their help.

Appendix

1. Derive (17–8) from (30) Substitute equation (23) to (30) and change integrals,

$$H_m(r,\eta)|\eta| = \frac{1}{2\pi}\int_{\epsilon=0}^{2\pi}\frac{1}{2\pi}\int_{s=-\infty}^{\infty} e^{j(m\epsilon - m\arctan(\frac{s}{D}+\eta s)}$$

$$\frac{D^2}{D^2+s2}\int_{\eta'=-\infty}^{\infty} e^{-j\eta'(\frac{sD}{\sqrt{D^2+s^2}}+r\sin(\epsilon))}|\eta'|\,d\eta' ds d\epsilon \qquad (1-1)$$

since

$$\int_{-\infty}^{\infty} e^{-j\eta'\omega}|\eta'|\,d\eta' = -\frac{1}{\omega^2} \qquad (1-2)$$

where

$$\omega = \frac{sD}{\sqrt{D^2+s^2}} + r\sin(\epsilon) = \cos(\gamma)U(s-s') \qquad (1-3)$$

and

$$U = 1+\rho\cos(\epsilon') \quad s' = \frac{-r\sin(\epsilon')}{U} \quad \epsilon = \epsilon'+\gamma \quad \gamma = \arctan\left(\frac{s}{D}\right) \quad \cos(\gamma) = \frac{D}{\sqrt{D^2+s^2}} \qquad (1-4)$$

considering the integral transform of $\epsilon = \epsilon' + \gamma$, we have

$$H_m(r,\eta)|\eta| = \frac{1}{2\pi}\int_{\epsilon'=0}^{2\pi}\frac{1}{U^2}e^{jm\epsilon'}\frac{1}{2\pi}\int_{s=-\infty}^{\infty} e^{j\eta s}\frac{1}{(s-s')^2}\,ds d\epsilon' \qquad (1-5)$$

considering the inverse of equation (1–2) we have

$$\frac{1}{2\pi}\int_{s=-\infty}^{\infty} e^{j\eta s}\frac{1}{(s-s')^2}\,ds = e^{j\eta s'}|\eta| \qquad (1-6)$$

$$H_m(r,\eta) = \frac{1}{2\pi}\int_{\epsilon'=0}^{2\pi}\frac{1}{U^2}e^{jm\epsilon'+s'\eta}\,d\epsilon' \qquad (1-7)$$

This is equation (17-b).

2. Derive equation (18-b) from (31)
Substitute (23) to (31) and change integrals

$$I_m(r,\eta) = \frac{1}{2\pi}\int_{\epsilon=0}^{2\pi}\frac{1}{2\pi}\int_{s=-\infty}^{\infty} e^{-j(m\epsilon - m\arctan(\frac{s}{D})+\eta s)}\frac{D}{\sqrt{D^2+s^2}}\cdot$$

$$\int_{\eta'=-\infty}^{\infty} e^{-j\eta'(\frac{sD}{\sqrt{D^2+s^2}}+r\sin(\epsilon))}\,d\eta' ds d\epsilon \qquad (2-1)$$

since

$$\frac{1}{2\pi}\int_{\eta'=-\infty}^{\infty} e^{j\eta'\omega}\,d\eta' = \delta(\omega) \qquad (2-2)$$

ω is defined in (1–3), therefore

$$\delta(\omega) = \delta\left(\frac{Ds}{\sqrt{D^2+s^2}} + r\sin(\epsilon)\right) = \frac{\delta(s-s')}{U\cos(\gamma)} \qquad (2-3)$$

where U, $\cos(\gamma)$, s' are defined in (1-4). Considering the integral transform of $\epsilon = \epsilon' + \gamma$, we obtain

$$I_m(r,\eta) = \frac{1}{2\pi} \int_{\epsilon'=0}^{2\pi} \int_{s=-\infty}^{\infty} e^{-j(m\epsilon'+s\eta)} \cos(\gamma) \frac{\delta(s-s')}{U\cos(\gamma)} \, ds \, d\epsilon'$$

$$= \frac{1}{2\pi} \int_{\epsilon'=0}^{2\pi} \frac{1}{U} e^{-j(m\epsilon'+s'\eta)} \, d\epsilon' \qquad (2-4)$$

This is equation (18-b).

REFERENCES

Fuhrmann G. *Testphantom-Projektionsdatensaetze Fuer die 2D-Computer-Tomographie.* KFA internal publication.

Herman G. T., Ed. (1980), *Image Reconstruction from Projections: The Fundamentals of Computer Tomography.* New York: Academic.

Higgins W. E., Munson D. C. (1988), A Hankel Transform Approach to Tomographic Image Reconstruction. *IEEE Trans. On Medical Imaging*, vol. 7, no. 1.

Kak A. C. and Slaney M., *Computerized Tomographic Imaging.* Printed in the United States of America. IEEE order Number: Pco2071, 90 p.

Lewitt R. M., Bates R. H. T. and Peters T. M. (1978), Image reconstruction from projections, II Modified back-projection methods. *Optik*, vol. 50, 85–109.

Mersereau R. N. and Oppenheim A. V. (1974), Digital reconstruction of multidimensional signals from their projections. *Proc. IEEE*, vol. 62, 1319–1338.

Natterer F. (1986) *The mathematics of computerized Tomography.* Copyright 1986 by John Wiley & Sons Ltd and BG Teubner, Stuttgart, 119 p.

Peng H., Stark H. (1987), Direct Fourier Reconstruction in fan beam tomography. *IEEE trans. on Medical Imaging*, vol. MI-6 no. 3.

Ramachandran G. N. and Lakshmin-arayanan A. V. (1971), Three-dimensional reconstruction from radiographs and electron micrograph: application of convolutions instead of Fourier transform. *Proc. Not Acad. Sci. USA*, vol. 68, 2236–2234.

Rattey P. A. (1984), Tomographic Image Reconstruction from a Limited Number of Projections. *Signal Processing*, **7**, 371–387, North-Holland.

Rattey P. A. and Lindgren A. G. (1981), Sampling the 2-D Random Transform. *IEEE Transactions on Acoustics, Speech, and Signal Processing*, vol. Ascp-29, no. 5.

Scudder H. J. (1978), Introduction to computer aided tomography. *Proc IEEE*, vol. 66, 628–637.

Shepp L. A. and Logan B. F. (1974), The Fourier reconstruction of a head section. *IEEE trans, Nucl, Sci.*, vol. NS-21, 21–43.

Zhao S. R. and Halling H. (1993), *A New Fourier Reconstruction Method for Fan Beam Geometry.* Published in the same meeting.

Zhao S. R. and Halling H., *General Hankel Transform Pair for Fan Beam X-ray Tomography.* (to appear)

Zhao S. R., Halling H., Fuhrmann G., *Sampling the Fan beam Projection Using "fan" function.* KFA internal publication.

Bei Fragen zur Produktsicherheit wenden Sie sich bitte an:
If you have any questions regarding product safety,
please contact:

Walter de Gruyter GmbH
Genthiner Straße 13
10785 Berlin
productsafety@degruyterbrill.com